Introduction to Chemical Reactor Analysis

SECOND EDITION

Introduction to Chemical Reactor Analysis

SECOND EDITION

R.E. Hayes
J.P. Mmbaga

CRC Press
Taylor & Francis Group
Boca Raton   London   New York

CRC Press is an imprint of the
Taylor & Francis Group, an **informa** business

CRC Press
Taylor & Francis Group
6000 Broken Sound Parkway NW, Suite 300
Boca Raton, FL 33487-2742

Printed in the United States of America on acid-free paper
Version Date: 20120831

International Standard Book Number: 978-1-4398-6700-6 (Paperback)

Library of Congress Cataloging-in-Publication Data

Hayes, R. E. (Robert E.)
    Introduction to chemical reactor analysis. -- Second edition / R.E. Hayes, J.P. Mmbaga.
    pages cm
    Includes bibliographical references and index.
    ISBN 978-1-4398-6700-6 (hardcover)
    1. Chemical reactors--Analysis. 2. Chemical reactors--Design. I. Mmbaga, J. P., author. II. Title.

TP157.H35 2013
660'.2832--dc23   2012028122

Visit the Taylor & Francis Web site at
http://www.taylorandfrancis.com

and the CRC Press Web site at
http://www.crcpress.com

Para Goizizar, mi esposa, y mis padres. This book is dedicated to the memory of my parents, and to my wife, who together have shaped my destiny.

R. E. Hayes

I am deeply indebted to all of those who have helped me to become what I am today. My immediate nuclear family as well as my extended family back in Tanzania. I dedicate this book to my wife Edna, and my daughter Joan.

Bwana Apewe Sifa na Utukufu.

J. P. Mmbaga

Contents

About the Book

This book provides an introduction to the basic concepts of chemical reactor analysis and design. It is aimed at both the senior level undergraduate student in Chemical Engineering and the working professional who may require an understanding of the basics of this area.

This book starts with an introduction to the area of chemical reactions, reactor classifications, and transport phenomena, which is followed by a summary of the important concepts in thermodynamics that are used in reactor analysis. This introduction is followed by a detailed development of mole and energy balances in ideal reactors, including multiple reactor systems. A detailed explanation of homogeneous reaction kinetics is then given, including a discussion of experimental techniques in kinetic data analysis. The next chapter describes techniques used in analyzing nonideal reactors, including the residence time distribution and mixing effects. The remainder of the book is devoted to catalytic systems. This latter part of the book contains chapters on the kinetics of catalytic systems, heat and mass transfer effects, and heterogeneous reactor analysis. The final chapter is devoted to experimental methods in catalysis.

There are many worked-out examples and case studies presented in the text. Additional problems are given at the back of each chapter.

At the end of reading this book, and working the problems and examples, the reader should have a good basic knowledge sufficient to perform most of the common reaction engineering calculations that are required for the typical practicing engineer.

Preface

It has been about 10 years since the first edition of this book was published, and it is probably appropriate to begin by offering a justification for writing the book initially, and generating a second edition. As noted in the preface to the first edition, there are many good textbooks on chemical reaction engineering in existence. Many of the existing books on chemical reaction engineering are both excellent and comprehensive. (*Elements of Chemical Reaction Engineering* by Scott Fogler [1999] and *Chemical Reaction Engineering* by Octave Levenspiel [1999] are both considered classics in the field.) However, it can be this very comprehensiveness that may make them confusing to the neophyte. Most books contain material sufficient for several courses on chemical reaction engineering, although in some books the more complex topics are touched on only lightly. Other texts contain a mix of undergraduate and graduate level material, which can also make it difficult for the beginner in this topic to progress easily. This book, therefore, is not meant to be either comprehensive or complete, nor is it intended to offer a guide to reactor appreciation or give detailed historical perspectives. Rather, it is intended to provide an effective introduction to reactor analysis, and contains sufficient material to be covered in two terms of about 35–50 min lectures each on reactor analysis. At the end of reading this book, and working the problems and examples, the reader should have a good basic knowledge sufficient to perform most of the common reaction engineering calculations that are required for the typical practicing engineer.

Chemical kinetics and reactor design probably remain as the engineering specialization that separates the chemical engineer from other types of engineer. Detailed mastery of the subject is not, however, essential for the typical chemical engineering graduate at the Bachelor level, because only a few percent of such graduates become involved in research and design careers that involve complex chemical reaction engineering calculations. On the other hand, a significant number of chemical engineers are employed during some stage in their careers with responsibilities involving the operation of a chemical process plant; as process engineers they are expected to provide objective interpretations regarding reactor performance and means of improving its operation. In this regard some training in the area of chemical reaction engineering is essential to enable the understanding of the factors that affect the performance of a chemical reactor, and thus to effect performance-enhancing measures. This book is thus directed to the majority of students who will be generalists, rather than the small minority who will become specialists in the art of reactor design.

For this, the second edition, the scope of the material has been significantly enhanced. In addition to rearranging the material from the first edition, and the correction of the inevitable errors, five chapters have been added on catalytic reaction engineering. Whereas the first edition was designed for a single course in reaction analysis, and thus focused on homogeneous systems, this edition is designed for two courses in reaction engineering.

This book grew from a set of lecture notes in reactor analysis that was used in the senior year undergraduate courses in reaction engineering. As mentioned, it was designed to be used for two courses in reaction analysis. It should be possible to cover all of the material presented in the text in two courses consisting of around 35 lectures, each of 50 min duration, and 10–12 1 h problem-solving periods.

As noted, this textbook is based on the contents of two undergraduate courses in chemical reaction engineering taught in the Department of Chemical and Materials Engineering at the University of Alberta. The students taking these courses have previously taken courses in thermodynamics, numerical methods, heat transfer, and fluid flow. Some knowledge in all of these areas is assumed, and these basics are not discussed in detail, if at all. The exception is thermodynamics of chemical reactions. Because of the importance of this topic to chemical reaction engineering, more than a passing mention is made to this area. It should be noted, however, that the material on chemical reaction thermodynamics is brief, and is intended only to provide a refresher to the student whose knowledge is a bit rusty. Some prior knowledge of kinetics is useful, for understanding the material in this book.

The order in which the chapters are presented reflects the sequence of the material as presented in the author's courses in chemical reaction engineering. Some readers may be surprised to see that basic reactor mole balances are discussed prior to the detailed development of kinetics and rate expressions. This choice is deliberate, and reflects the authors' experience. If simple reactors, especially flow reactors, are introduced early, it tends to eliminate the impression that a reaction rate is defined as the rate of change of concentration with time. This appears to be an artifact from kinetics as taught in many chemistry classes, where only batch reactors are considered. However, Chapter 5 on kinetics does stand on its own, and could probably be taught prior to Chapter 3 if the instructor feels strongly about it. If this book is used for two courses, then it is suggested that the first course covers Chapters 1 through 4, and either Chapter 5 or 6. The second course would then include Chapter 5 or 6 and Chapters 7 through 11.

Numerical methods are presented in an Appendix. They are essential to the solution of most realistic problems in chemical reaction engineering. There is much good software available on the market today, which makes the solution of the nonlinear problems encountered in CRE relatively straightforward. The reader is expected to have sufficient understanding of numerical methods to be able to solve systems of linear and nonlinear algebraic equations, as well as systems of first-order nonlinear ordinary differential equations.

There are many worked out examples presented in the text, as experience has shown that these are one of the most effective learning tools. Additional problems are given at the back of each chapter.

As this is by design a simple text, and in no way intended to be a comprehensive reference text, suggestions for further reading have been provided. There are many books available, so the student should be able to find one that suits his or her style.

MATLAB® is a registered trademark of The MathWorks, Inc. For product information, please contact:

The MathWorks, Inc.
3 Apple Hill Drive
Natick, MA 01760-2098 USA
Tel: 508 647 7000
Fax: 508-647-7001
E-mail: info@mathworks.com
Web: www.mathworks.com

Acknowledgments

Few books, or indeed any major work, are developed in a vacuum, and this book is no exception. It is necessary to acknowledge some of the people and other works that have influenced the development of this text. In a general as well as a specific sense, a number of the classical textbooks in chemical reaction engineering and related areas have influenced our learning of this interesting field, and these books have had a major influence on the work that follows. These books (in no particular order) are: *Chemical Reaction Engineering* by Octave Levenspiel (1999), *Chemical Engineering Kinetics* by J. M. Smith (1981), *Catalytic Reaction Engineering* by J. J. Carberry, *Elements of Chemical Reaction Engineering* by Scott Fogler (1999), and *Chemical Reactor Analysis and Design* by G. F. Froment and K. B. Bischoff (2011).

The books listed above have had a very general influence on the book as a whole, and in particular Chapters 3 and 4. In addition to these books, it is appropriate to acknowledge books that have had special influence on specific chapters. The book *Chemical Kinetics* by K. Laidler (1987) was drawn on extensively in the development of Chapter 5, especially for the development of free radical mechanisms. Chapter 6 was developed using two main sources; *Mixing in Continuous Flow Systems* by E. B. Nauman and B. A. Buffham (1983) and *Chemical Reaction Engineering* by Octave Levenspiel (1999). Chapter 7 takes material from *Introduction to the Principles of Heterogeneous Catalysis* by J. M. Thomas and W. J. Thomas (1967), and *Heterogeneous Catalysis in Practice* by G. N. Satterfield (1980). Chapter 8 was strongly influenced by *Introduction to the Principles of Heterogeneous Catalysis* by J. M. Thomas and W. J. Thomas (1967). The complete bibliographic details of these and other books are given in the following Recommended Reading section.

The first edition of this book was developed shortly after one of us (REH) had finished co-writing the book *Introduction to Catalytic Combustion* with Professor S. T. Kolaczkowski (Hayes and Kolaczkowski, 1997) of the University of Bath, UK. Some of the material developed for the catalytic combustion book has been incorporated into this book, with some modification.

It is also appropriate to repeat the acknowledgments for the first edition. Professor Ivo Dalla Lana of the Department of Chemical and Materials Engineering, University of Alberta who has many years of experience in teaching undergraduate courses in reactor analysis, and gracefully consented to read the first draft of the first edition of this book. He made many very useful comments and suggestions, which were incorporated into the final version. Bob Barton also proof read the manuscript and made many helpful suggestions. Diane Reckow typed the first draft of the manuscript into the computer and prepared most of the figures. She also prepared many of the figures that have been used in the new material added for the second edition.

A number of problems and examples are given in the book, all of which were taken from our collections that has been used over several years of lecturing. These were originally sourced from a variety of books, usually with some modification, and the origins of some have been forgotten. Nonetheless, many of the problems are based on those presented in the four books mentioned above (Levenspiel, 1999; Fogler, 1999; Froment et al., 2011; Smith, 1981), whilst others were based on problems from the files of Professor Wanke of the University of Alberta. The textbooks mentioned in the list of reading all provide many additional practice problems.

Notwithstanding any efforts made by others, all errors and omissions in this text remain the responsibility of the authors.

Recommended Reading

The reader who requires a more comprehensive treatment of reactor analysis and design will want to consult other textbooks for a complete understanding of this area. The following list of books provides a source of reference for a more detailed study of various areas related to reaction engineering, as well as the background material required both for reactor analysis as well as for system design. The books are grouped by subject area, although there may be some overlap between books in some cases. This list is by no means inclusive, and a trip to the library is always rewarded. The reader should also note that we have included in the list some books that are out of print, but may be available in the library or published on the web.

Reaction Engineering, General, and Basic Books

The following textbooks cover most aspects of chemical reaction engineering to a greater or lesser extent at a predominately undergraduate level. Some of these books also cover more advanced concepts not typically encountered in a single semester course in chemical reaction engineering.

Aris, R., 1965, *Introduction to the Analysis of Chemical Reactors*, Prentice-Hall, Englewood Cliffs.

Aris, R., 1969, *Elementary Chemical Reactor Analysis*, Prentice-Hall, Englewood Cliffs.

Butt, J.B., 2000, *Reaction Kinetics and Reactor Design*, 2nd Ed., CRC Press, New York.

Davis, M.E. and R.J. Davis, 2002, *Fundamentals of Chemical Reaction Engineering*, McGraw-Hill, New York.

Fogler, H.S., 1999, *Elements of Chemical Reaction Engineering*, 3rd Ed., Prentice-Hall, Englewood Cliffs.

Froment, G.B., K.B. Bischoff, and J. De Wilde, 2011, *Chemical Reactor Analysis and Design*, 3rd. Ed., Wiley, New York.

Harriot, P., 2002, *Chemical Reactor Design*, CRC Press, New York.

Hill, C.G. Jr., 1977, *An Introduction to Chemical Engineering Kinetics and Reactor Design*, John Wiley & Sons, New York.

Laidler, K.J., 1987, *Chemical Kinetics*, 3rd Ed., Benjamin-Cummings Publishing Co.

Levenspiel, O., 1999, *Chemical Reaction Engineering*, 3rd Ed., John Wiley & Sons, New York.

Metcalfe, I.S., 1997, *Chemical Reaction Engineering: A First Course*, Oxford University Press, New York.

Missen, R.W., C.A. Mims, and B.A. Saville, 1999, *Chemical Reaction Engineering and Kinetics*, Wiley, Toronto.

Nauman, E.B., 1987, *Chemical Reactor Design*, Wiley, Toronto.

Salmi, T.O., J. Mikkola, and J.P. Warna, 2009, *Chemical Reaction Engineering and Reactor Technology*, CRC Press, Boca Raton, FL.

Schmidt, L.D., 1998, *The Engineering of Chemical Reactions*, Oxford University Press, New York.

Smith, J.M., 1981, *Chemical Engineering Kinetics*, 3rd Ed., McGraw-Hill, New York.

Walas, S.M., 1995, *Chemical Reaction Engineering Handbook of Solved Problems*, Gordon and Breach, Reading.

Winterbottom, J.M. and M. King, 1999, *Reactor Design for Chemical Engineers*, CRC Press, Boca Raton, FL.

Reaction Engineering and Related Issues, More Advanced Books

The following books cover aspects of chemical reaction engineering to a greater extent than those mentioned above. These books are more advanced, and should probably be approached after the material in this book is mastered. This list also includes references to books that may not have reaction engineering as the primary focus, but contain related material, such as mixing.

Baldyga, J. and J.R. Bourne, 1999, *Turbulent Mixing and Chemical Reactions*, John Wiley & Sons, New York.

Kunii, D., and O. Levenspiel, 1991, *Fluidization Engineering*, 2nd Ed., Butterworth-Heinemann, Newton.

Nauman, E.B. and B.A. Buffham, 1983, *Mixing in Continuous Flow Systems*, Wiley, New York.

Ottino, J.M., 1989, *The Kinematics of Mixing: Stretching, Chaos and Transport*, Cambridge University Press, Cambridge.

Westerterp, K.R., W.P.M. van Swaaij, and A.A.C.M. Beenackers, 1984, *Chemical Reactor Design and Operation*, John Wiley & Sons, New York.

Reaction Engineering, Catalysis, and Catalytic Reactors

The following selection of references have a strong emphasis on the catalytic aspects of chemical reaction engineering, including fundamental catalysis.

Carberry, J.J., 1976, *Chemical and Catalytic Reaction Engineering*, McGraw-Hill, New York.

Bruce C.G., J.R. Katzer, and G.C.A. Schuit, 1979, *Chemistry of Catalytic Processes*, McGraw-Hill, New York.

Gates, C.G., 1992, *Catalytic Chemistry*, John Wiley & Sons, New York.

Hayes, R.E. and S.T. Kolaczkowski, 1997, *Introduction to Catalytic Combustion*, Gordon and Breach, Reading.

Lee, H.H., 1985, *Heterogeneous Reactor Design*, Butterworth, London.

Satterfield, C.N., 1970, *Mass Transfer in Heterogeneous Catalysis*, MIT Press, Cambridge, MA.

Satterfield, C.N., 1980, *Heterogeneous Catalysis in Practice*, McGraw-Hill, New York.

Somorjai, G.A., 1994, *Introduction to Surface Chemistry and Catalysis*, John Wiley & Sons, New York.

Srivastava, R.D., 1988, *Heterogeneous Catalytic Science*, CRC Press, Boca Raton, FL.

Thomas, J.M. and W.J. Thomas, 1967, *Introduction to the Principles of Heterogeneous Catalysis*, Academic Press, London.

Thomas, J.M. and Thomas, W.J., 1996, *Principles and Practice of Heterogeneous Catalysis*, VCH, Weinheim.

Twigg, M.V., 1989, *Catalyst Handbook*, Wolfe Publishing Ltd., London.

Wjingaarden, R.J., A. Kronberg, and K.R. Westerterp, 1998, *Industrial Catalysis: Optimizing Catalysts and Processes*, Wiley-VCH, New York.

Catalyst Preparation

The following books cover some of the aspects of catalyst design and manufacture.

Acres, G.J.K., A.J. Bird, J.W. Jenkins, and F. King, 1981, The design and preparation of supported catalysts, *Catalysis (London)*, **4**, 1–30.

Che, M. and C.O. Bennett, 1989, The influence of particle size on the catalytic properties of supported metals, *Adv. Catal.*, **36**, 55–172.

Komiyama, M., 1985, Design and preparation of impregnated catalysts, *Catal. Rev. Sci. Eng.*, **27**(2), 341–372.

Lee, S.-Y. and R. Aris, 1985, The distribution of active ingredients in supported catalysts prepared by impregnation, *Catal. Rev. Sci. Eng.*, **27**(2), 207–340.

Le Page, J.F, J. Cosyns, P. Courty, E. Freund, J.P. Franck, Y. Jacquin, B. Juguin, et al. 1987, *Applied Heterogeneous Catalysis—Design, Manufacture, Use of Solid Catalysts*, Translated from the French edition and distributed by Gulf Publishing Company, Houston, Texas. Chapter 5: The Preparation of Catalysts, pp. 75–123.

Richardson, J.T., 1989, *Principles of Catalyst Development*, Plenum Press, New York.

Trimm, D.L., 1980, *Design of Industrial Catalysts*, Elsevier, Amsterdam.

Thermodynamics

The following thermodynamics textbooks emphasize chemical engineering thermodynamics, and include detailed discussion of the material covered briefly in Chapter 2 of this text.

Daubert, T.E., 1985, *Chemical Engineering Thermodynamics*, McGraw-Hill, New York.

Elliot, R.J. and C.T. Lira, 1999, *Introductory Chemical Engineering Thermodynamics*, Prentice-Hall, Toronto.

Kyle, B.G., 1999, *Chemical and Process Thermodynamics*, 3rd Ed., Prentice-Hall, Englewood Cliffs.

Sandler, S.I., 1999, *Chemical and Engineering Thermodynamics*, 3rd Ed., John Wiley & Sons, New York.

Smith, J.M., H.C. van Ness, and M.M. Abbot, 1996, *Introduction to Chemical Engineering Thermodynamics*, 5th Ed., McGraw-Hill, New York.

Transport Phenomena, Including Mass, Heat, and Momentum Transfer

The following references cover the transport phenomena, including fluid flow, heat transfer, and mass transfer. Special attention should be paid to the text by Bird, Stewart, and Lightfoot (2002), which is often considered to be the defining text in transport phenomena.

Bennett, C.O. and J.E. Myers, 1962, *Momentum, Heat and Mass Transfer*, McGraw-Hill, New York.

Bird, R.B., W.E. Stewart, and E.N. Lightfoot, 2002, *Transport Phenomena*, 2nd Ed., John Wiley & Sons, New York.

Fahien, R.W., 1983, *Fundamentals of Transport Phenomena*, McGraw-Hill, New York.

Gebhart, B., 1993, *Heat Conduction and Mass Diffusion*, McGraw-Hill, New York.

Holman, J.P., 2010, *Heat Transfer*, 10th Ed., McGraw-Hill, New York.

Incropera, F.P., D.P. DeWitt, T.L. Bergman, and A.S. Lavine, 2006, *Introduction to Heat Transfer*, 6th Ed., John Wiley & Sons, New York.

Middleman, S., 1998, *An Introduction to Fluid Dynamics*, John Wiley & Sons, New York.

Mills, A.F., 1999, *Basic Heat and Mass Transfer*, Prentice-Hall, Upper Saddle River.

Taylor, R. and R. Krishna, 1993, *Multicomponent Mass Transfer*, John Wiley & Sons, New York.

Thomas, L.C., 1980, *Fundamentals of Heat Transfer*, Prentice-Hall, Englewood Cliffs.

Physical and Chemical Properties

Physical and chemical properties of substances are necessary for the analysis and design of any chemical engineering system. Most of the books on thermodynamics, reactor analysis, transport phenomena, and so on contain appendices which give some property data. For more comprehensive coverage, including correlations for many commonly required properties, the following reference sources are recommended.

Perry, R.H. and D. Green, 1984, *Chemical Engineer's Handbook*, 6th Ed., McGraw-Hill, New York.
Reid, R.C., J.M. Prausnitz, and B.E. Poling, 1987, *The Properties of Gases and Liquids*, 4th Ed., McGraw-Hill, New York.

Mathematical Methods

There are many excellent books in existence on numerical and other mathematical methods, ranging in complexity from elementary books for the novice to advanced texts for the more knowledgeable reader. Some examples are:

Allen III, M.B. and E.L. Isaacson, 1998, *Numerical Analysis for Applied Science*, Wiley, New York.
Chapra, S.C. and R.P. Candle, 1998, *Numerical Methods for Engineers*, 3rd Ed., McGraw-Hill, New York.
Conte, S.D. and C. de Boor, 1980, *Elementary Numerical Analysis: An Algorithmic Approach*, 3rd Ed., McGraw-Hill, New York.
Evans, G., 1995, *Practical Numerical Analysis*, Wiley, New York.
Faires, J.D. and R.L. Burden, 1993, *Numerical Methods*, PWS-Kent, Boston.
Finlayson, B.A., 1980, *Non-linear Analysis in Chemical Engineering*, McGraw-Hill, New York.
Jenson, V.G. and G.V. Jeffries, 1977, *Mathematical Methods in Chemical Engineering*, Academic Press, London.
Shampine, L.F., 1994, *Numerical Solution of Ordinary Differential Equations*, Chapman & Hall, New York.
Walas, S.M., 1991, *Modelling with Differential equations in Chemical Engineering*, Butterworth-Heinmann, Boston.

Nomenclature

Some commonly used symbols are given in the following list. The list is not all inclusive, and in most cases symbols are defined when they are introduced. It should be pointed out that various books on chemical reaction engineering may use different symbols or give different meanings to the symbols used here. Also, some symbols may have more than one meaning (sometimes many): the correct one depends on the context. The units given below are for the most part the standard SI units. Other units are used in various books and also in this text. Symbols representing variables are usually written in italics, whilst symbols representing chemical species (a, B, C, etc.) are written in normal script. Separate lists are given for Roman letters and Greek letters; however, where the Greek Δ is used to indicate a change in the value of a property, the corresponding variable appears in the main variable list, according to the alphabetical order determined by the letter following the Δ symbol.

The use of double subscripts has been avoided as much as possible. Rather, when two subscripts are required to identify a quantity, the subscripts are separated by commas when necessary. For example, F_{t0} is used to denote the total molar flow rate at the inlet to a reactor.

$\hat{a}_j$	activity of component j in solution, dimensionless
a_m	wetted surface of packed bed, m^{-1}
a_v	surface area per unit volume, m^{-1}
A	pre-exponential factor, units vary
C	concentration, mol/m^3
C_j	concentration of species j, mol/m^3
C_j0	concentration of species j at time zero or at reactor inlet, mol/m^3
C_{jf}	final concentration of species j at time t (batch reactor), mol/m^3
C_{jE}	concentration of species j at reactor outlet (flow reactor), mol/m^3
C_P	constant pressure heat capacity, J/mol K
C_V	constant volume heat capacity, J/mol K
d_p	pore diameter, m
D	diameter, m
D	diffusion coefficient, m^2/s
D_{eff}	effective diffusivity, m^2/s
D_K	Knudsen diffusion coefficient, m^2/s
D_p	Particle diameter, m
E	total energy, J/mol
E	activation energy, J/mol
f	friction factor, dimensionless
f_j	fugacity of component j, bar
$\hat{f}_j$	fugacity of component j in solution, bar
f_j°	fugacity of component j at the standard state, bar
$f(t)$	residence time density function, dimensionless
F_j	molar flow rate of species j, mol/s
F_t	total molar flow rate, mol/s
$F(t)$	cumulative residence time distribution function, dimensionless
g	acceleration owing to gravity, 9.81 m/s^2

G	superficial mass velocity, $kg/m^2 \, s$
$\bar{G}_j$	molar free energy of component j, J/mol
$\bar{G}_j^\circ$	molar free energy of component j in the standard state, J/mol
ΔG_f°	standard free energy change of formation, J/mol
$\Delta G_{f,298}^\circ$	standard free energy change of formation at 298 K, J/mol
ΔG_R	free energy of reaction for arbitrary process, J/mol
ΔG_R°	standard free energy change of reaction, J/mol
$\Delta G_{R,298}^\circ$	standard free energy change of reaction at 298 K, J/mol
h	heat transfer coefficient, $W/m^2 \, K$
H	enthalpy, J or J/mol
H	heat of adsorption, J or J/mol
ΔH_f°	standard enthalpy change of formation, J/mol
$\Delta H_{f,298}^\circ$	standard enthalpy change of formation at 298 K, J/mol
ΔH_R	enthalpy of reaction for arbitrary process, J/mol
ΔH_R°	standard enthalpy change of reaction, J/mol
$\Delta H_{R,298}^\circ$	standard enthalpy change of reaction at 298 K, J/mol
k	rate constant (various subscripts used), various units
k	thermal conductivity, W/m K
k_a	adsorption rate constant, various units
k_d	desorption rate constant, various units
k_m	mass transfer coefficient, m/s
K	equilibrium constant, dimensionless
K	permeability of porous medium, m^2
K_P	pressure equilibrium constant ideal gas, bar^n
K_y	equilibrium constant ideal gas, dimensionless
L	length, m
m	mass, kg
$\dot{m}$	mass flow rate, kg/s
m_j	mass of species j, kg
M_j	molar mass of species j, kg/kmol
M_m	average molar mass of mixture, kg/kmol
N	number of moles, mol
N_j	number of moles of species j, mol
N_j	molar flux of species j, $mol/m^2 \, s$
N_T	total number of moles, mol
P	pressure bar
q	rate of heat transfer, $W \equiv J/s$
q	incremental volumetric flow rate, m^3/s
Q	volumetric flow rate, m^3/s
r	radial coordinate or radius, m
r_m	rate of mass transfer, $mol/m^2 \, s$
r_h	rate of heat transfer, W/m^2
r_{ads}	rate of adsorption, $mol/m^3 \, s$
r_j	rate of formation of species j, units vary
$\left(-r_j\right)$	rate of disappearance of species j, units vary
R	recycle ratio
R_g	gas constant, $8.314/mol \cdot K$
s	steric factor, dimensionless
S_P	point selectivity, dimensionless

ΔS_R°	standard entropy change of reaction, J/mol · K
t	time, s
t_m	mean residence time, s
T	temperature, °C or K
u	velocity, m/s
u_m	mean velocity, m/s
U	internal energy, J or J/mol
U	overall heat transfer coefficient, W/m² K
v	velocity, m/s
v	volume adsorbed, m³
v_m	volume adsorbed in a monolayer, m³
v_m	average velocity, m/s
v_s	superficial velocity, m/s
V	volume, m³
w_j	mass fraction of species j, dimensionless
W	mass of catalyst, kg
W	work, W $\equiv$ J/s
W_E	expansion (pressure volume) work, W $\equiv$ J/s
W_f	flow work, W $\equiv$ J/s
W_S	shaft work, W $\equiv$ J/s
$W(t)$	washout function, dimensionless
x_j	liquid phase mole fraction of species j, dimensionless
X_j	fractional conversion of species j, dimensionless
y_j	gas phase mole fraction of species j, dimensionless
Y_j	gas phase mole fraction of species j, dimensionless
z	height above a reference point, distance along reactor m
z_j	mole fraction of component j, dimensionless
Z_{AB}	frequency of collisions between A and B, collisions/m³s

Greek Letters

α	stoichiometric coefficient, positive for reactants and products
β	coefficient of thermal expansion, K⁻¹
ε	porosity, dimensionless
ϕ	porosity, dimensionless
ϕ	contact angle, dimensionless
ϕ	volume fraction, dimensionless
ϕ	Thiele modulus, dimensionless
η	effectiveness factor, dimensionless
γ_j	activity coefficient of component j, dimensionless
Λ	residual life, s
μ	viscosity, Pa s
μ_i	ith moment of the residence time distribution, dimensionless
ν	Stoichiometric coefficient, negative for reactants, positive for products
ν	normalized moments of RTD function, dimensionless
θ	normalized residence time, dimensionless

θ_i	fractional coverage of surface sites by species i, dimensionless
θ_V	fraction of vacant sites, dimensionless
ρ	density, kg/m^3
ρ_B	bulk density, catalyst mass divided by reactor volume, kg/m^3
ρ_C	catalyst density, kg/m^3
σ	surface tension, N/m
σ^2	variance of the residence time distribution, dimensionless
τ	space time, s

Abbreviations

AES	Auger electron spectroscopy
BET	Brunauer–Emmett–Teller
CSTR	Continuous stirred tank reactor
CVBR	Constant volume batch reactor
GHSV	Gas hourly space velocity
HTF	Heat transfer fluid
LHHW	Langmuir–Hinshelwood–Hougen–Watson
LHSV	Liquid hourly space velocity
PFR	Plug flow reactor
RTD	Residence time distribution
SAM	Scanning Auger microscopy
SEM	Scanning electron microscopy
SV	Space velocity
TEM	Transmission electron microscopy
XPS	X-ray photoelectron spectroscopy

Authors

R.E. Hayes is a professor of chemical engineering in the Department of Chemical and Materials Engineering at the University of Alberta in Edmonton, Alberta, Canada. He received his undergraduate education in chemical engineering at Dalhousie University in Halifax, Nova Scotia, followed by a PhD in chemical engineering at the University of Bath, UK. He is the author or coauthor of numerous research articles in the area of reaction engineering and is a coauthor of the book *Introduction to Catalytic Combustion* (Gordon and Breach, 1997). He is a Fellow of the Canadian Society of Chemical Engineering and a registered professional engineer with the Association of Professional Engineers, Geologists, and Geophysicists of Alberta.

J.P. Mmbaga is a senior research associate in the Department of Chemical and Materials Engineering at the University of Alberta. He was formerly a senior lecturer in chemical engineering at the University of Dar es Salaam, Tanzania. He received his undergraduate education in chemical and process engineering from University of Dar es Salaam, Tanzania, and masters in chemical engineering from New Jersey Institute of Technology. He obtained his doctorate in chemical engineering from the University of British Columbia, Canada. He has spent more than 25 years in teaching, research and process consulting in oil and gas processing, energy, and other chemical process industries in Alberta and Tanzania. He has authored and coauthored a number of technical publications and is a member of the Canadian Society of Chemical Engineering and the Association of Professional Engineers, Geologists, and Geophysicists of Alberta.

1

Introduction

The analysis and design of chemical reactors is an area that is a distinguishing feature of chemical engineering. Although some other engineering disciplines deal with chemically reacting systems, this domain is usually reserved for the chemical engineering profession. As will be seen throughout this text, it is an area that relies extensively on all of the basic building blocks of the chemical engineering curriculum. A well-designed reactor is one that produces a desired product safely, without adverse environmental effects, in an economical manner, and to a desired specification.

Chemical reaction engineering involves the application of basic chemical engineering principles to the analysis and design of chemical reactors. Chemical reactors are the heart of the majority of industrial chemical plants, including those based on the exploitation of biotechnology. The understanding, analysis, and design of chemical reactors utilize all of the concepts taught in a classical chemical engineering program, and, as such, can be said to be at the apex of a chemical engineer's education. Many of the operations in a chemical plant can be viewed as supporting the reactor. For example, heat exchange, separations processes, and so on may be used to pretreat a reactor feed, and then to separate the reactor effluent into its constituent parts. Process control may be used to maintain the reactor operation at the desired set point. The development of a good level of proficiency in reactor analysis therefore requires a good understanding of all of the basic chemical engineering principles.

The importance and versatility of reaction engineering in the manufacture of useful products can be illustrated by considering some of the possible products that can be made from a natural gas feedstock. An overview of some of the myriad possibilities is illustrated in Figure 1.1. Natural gas consists primarily of methane and ethane, with some higher light hydrocarbons (sometimes referred to as condensate). Typically, natural gas is separated into three streams, consisting of methane, ethane, and other hydrocarbons. The ethane can be fed to a chemical reactor where it is dehydrogenated to form a mixture of ethene and hydrogen. After removing the hydrogen from the effluent, the ethene may then be reacted in different reactors under different operating conditions to form various types of polyethylene, such as low-density polyethylene (LDPE) or high-density polyethylene (HDPE). Polyethylene is used in a wide variety of consumer products. Ethane can also be partially oxidized to produce ethylene oxide, which may then be reacted with water to produce ethylene glycol. Ethylene glycol is widely used as a coolant.

Methane gas can be burned with air in a chemical reaction to provide thermal energy for heating applications. Alternatively, methane may be mixed with steam and fed to a catalytic reactor to produce a mixture of carbon monoxide and hydrogen, often referred to as synthesis gas (or colloquially, syngas). Synthesis gas may be used as feed to another reactor to produce methanol, which is used in many consumer products, for example, deicing fluid and gasoline additives. Methanol may also be reacted with oxygen to form formaldehyde. Methanol and butene can be combined in a different type of reactor to form methyl-*tert*-butyl-ether (MTBE), which is used as a gasoline additive, among other things. Methanol can also be reacted over yet another type of catalyst to form gasoline and water.

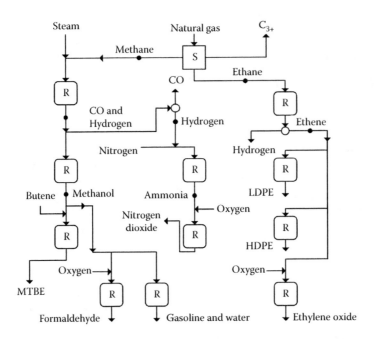

FIGURE 1.1

Illustration of some possible products that can be made from natural gas using chemical reactors. The symbol R in a box denotes a reactor and S denotes a separator. Each reactor is operated differently. Note that not all reactants and/or products present are shown in each case.

In another process, the carbon monoxide is removed from the syngas to leave an essentially pure hydrogen stream. The hydrogen is mixed with nitrogen, and then passed through a high-pressure reactor to produce ammonia. Ammonia is used directly in the manufacture of many items. Ammonia can also be oxidized to form nitrogen dioxide (NO_2), which, when mixed with water, forms nitric acid. Note that the heart of each of these processes is the chemical reactor. The choice of the operating conditions of the reactor, including the catalyst type, determines the product produced.

Chemical reaction engineering is involved in both the analysis of existing systems and the development of new processes, and both roles may be encountered by the active chemical engineer.

1.1 Process Development

Chemical reaction engineering plays a role at all levels of process design and development. At each step, different methodologies are employed, as each step in the process has different objectives. The steps in the development of a process from idea to production plant may be broadly outlined as:

- *Discovery*—A new reaction is usually discovered and developed in a research laboratory, either at a university or in an industry. The discovery may be the result of fundamental or exploratory research, conducted solely to "see what happens if …."

Such research is often conducted in university laboratories. Alternatively, there may be a perceived need for a product, and a research program is designed to find a way of producing that product economically.

- *Technical and economic feasibility*—Once a new discovery has been made, analysis is usually performed to determine its economic viability. The discovery must offer the promise of a new product, or a new or improved process for manufacturing an existing product.

- *Commercial reactor development*—Once it has been determined that a process offers promise, a reactor must be design. The design will include a determination of the type of reactor to be used and the operating conditions. The design of a reactor requires that the kinetics of the reaction be known, which, in turn, will require the experimental determination of reaction rate parameters. The development phase will include much bench-scale and pilot-scale work, and comprehensive reactor modeling before a full-scale reactor can be built.

- *Process or plant development*—The final phase of the analysis is the development of the complete chemical plant or process unit. This phase involves the design of feed and product stream processing equipment, including separations units, heat exchangers, and so on.

1.2 Basic Building Blocks of Chemical Reaction Engineering

It was stated that chemical reaction engineering is at the apex of chemical engineering, using as it does all of the principles of chemical engineering. This section briefly outlines the importance of each major engineering concept and its contribution to the area of chemical reaction engineering. Furthermore, the difference between the three fundamental parts of chemical reaction engineering—*chemical kinetics, chemical reactor analysis,* and *chemical reactor design*—is explained. Figure 1.2 shows a simplified diagram of the relationship among some of the key chemical engineering areas.

Chemical kinetics deals with the velocity of reactions: broadly speaking, it tells us how long it will take to achieve a specified level of conversion and what products will be formed. Transport phenomena are key elements of chemical engineering, and hence chemical reaction engineering. The three transport phenomena are the transport of momentum (fluid flow), the transport of energy, and the transport of mass. These three phenomena are usually covered in courses on fluid mechanics, heat transfer, and mass transfer (including separations processes) in the chemical engineering curriculum. Thermodynamics plays a role in the prediction of equilibrium states of reactions, energy changes on reaction, and, in some cases, vapor–liquid equilibrium. Combining kinetics, transport equations, and thermodynamics, reactor analysis is used to predict concentrations and temperatures in a given type of reactor. Reactor analysis involves the solution of systems of complex equations that govern the transport, thermodynamics, and kinetics of the reaction. The solution of these equations usually requires equally complex mathematical methods, involving the numerical solution of systems of equations on a computer. Therefore, both mathematical and computer methods are used extensively. Reactor analysis is used to predict temperature and concentration distributions, and hence the rate of conversion of reactants into products, in a given reactor type with specified operating conditions. It does not deal with topics such as

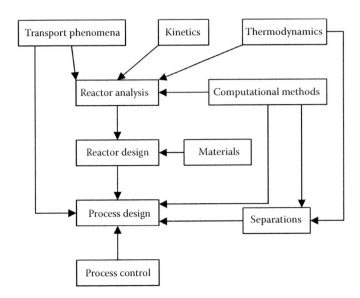

FIGURE 1.2
Illustration of the relationship of kinetics, reactor analysis, and reactor design to the main areas of the chemical engineering curriculum.

vessel wall thickness or other mechanical design considerations. The complete topic of reactor design, as opposed to reactor analysis, includes these considerations, and leads to a working reactor concept. Reactor design will thus require knowledge in areas such as strength of materials and other aspects of materials science. This book deals exclusively with reactor *analysis*, and not reactor *design*. Finally, once a reactor design is selected, the ancillary equipment, such as separations units, heat exchangers, and so on, is included to give a complete process design. The development of a final design usually requires an iterative procedure. For example, a preliminary reactor design may require the use of less than ideal separations equipment, and therefore the engineer designing the process may return to the reactor design and analysis stages to make needed changes to the final design.

1.3 Outline of the Book

The remainder of this book covers the material required for basic reactor analysis. The rest of this chapter is devoted to an introduction to kinetics, reactor types, and some commonly used terms. Chapter 2 provides a review of chemical engineering thermodynamics. Chapters 3 through 6 cover homogeneous reactions and homogeneous reactors. Chapter 3 covers mole balances in ideal reactors for three common reactor types. Chapter 4 covers energy balances in ideal reactors. Chapter 5 covers the area of chemical reaction kinetics, including mechanisms of reaction and rate expressions. Experimental methods are also explained in this chapter. Chapter 6 provides an introduction to nonideal reactors. Chapters 7 through 11 deal with material on catalytic systems, both kinetics and reactors. Chapter 7 is an introduction and overview of catalytic systems. Chapter 8 covers catalytic kinetics, including adsorption. Chapter 9 deals with the transport processes that occur in catalytic systems. Chapter 10 explores some of the common catalytic reactor designs, with an emphasis on the fixed bed reactor for gas-phase reactants. Finally, Chapter 11 explains some of the experimental methods used in catalytic reactions.

1.4 Introduction to Chemical Reactions

A chemical reaction involves the rearrangement of the chemical bonds in the molecules that participate in the reaction. This rearrangement of bonds may involve a single molecule that can undergo an internal rearrangement to form an *isomer*, or one in which the molecule can split into two or more fragments. More commonly, the reaction proceeds between two or more molecules, with a complex restructuring of these molecules. For example, the combustion of methane, CH_4, is represented by the overall reaction

$$CH_4 + 2O_2 \rightarrow CO_2 + 2H_2O \tag{1.1}$$

The *stoichiometry* of the reaction refers to the number of reactant and product molecules that participate in the reaction. In the reaction of Equation 1.1, one molecule of methane reacts with two molecules of oxygen to give one molecule of carbon dioxide and two molecules of water. Typically, methane is mixed with air and then burned: the air provides a source of oxygen. In addition, air contains nitrogen that does not participate in the combustion reaction: species present in a reacting mixture that do not react are referred to as *inert*. Furthermore, if methane is burned in air, there are usually more than 2 mol of oxygen present for each mole of methane present. Oxygen would therefore be considered an *excess* reactant, as there is more than the amount required by the reaction stoichiometry. Methane would be the *limiting reactant* because, if the reaction proceeds to completion, methane will be the first reactant to be totally consumed.

The reader may have noticed that the reaction representing the combustion of methane was written with an arrow pointing from the reactants (methane and oxygen) to the products (water and carbon dioxide). Before proceeding further, we introduce the concept of reversible and irreversible reactions. Consider, for illustration purposes, the reactions involving carbon monoxide, water, hydrogen, and methane. It was stated in the introduction that a mixture of methane and water can be passed over a catalyst to produce carbon monoxide and hydrogen. This reaction can be written as

$$CH_4 + H_2O \rightarrow CO + 3H_2 \tag{1.2}$$

However, if carbon monoxide and hydrogen are passed over a nickel catalyst under the appropriate conditions of temperature and pressure, methane and water will be produced. This reaction is

$$CO + 3H_2 \rightarrow CH_4 + H_2O \tag{1.3}$$

It can be seen that the reaction of Equation 1.3 is simply the reverse reaction of Equation 1.2. The reaction among these species is therefore called a *reversible reaction* and the reaction is typically represented in the following manner:

$$CO + 3H_2 \rightleftharpoons CH_4 + H_2O \tag{1.4}$$

The double arrow between the reactants and products shows that the reaction can proceed in either direction. The distribution between the reactants and the products depends on the reactor temperature and pressure, the catalyst (if present), and the value of the equilibrium constant (equilibrium considerations are discussed in Chapter 2). When a reaction can easily proceed in either direction, particular attention often must be paid to the reactor operating conditions. In contrast to reversible reactions, there are many reactions that may, for all practical purposes, be considered *irreversible*. A reaction is usually called irreversible if the equilibrium constant is so large that the reaction can proceed essentially to completion, that is, the conversion of reactants into products is essentially complete, and little if any of the limiting reactant remains (assuming sufficient residence time in the reactor). Such reactions are usually written with a single arrow between reactants and products, as was done with the combustion of the methane example cited earlier in this section.

In common with other disciplines, the study of chemical reactions can be codified by classifying reactions according to their important features. Although there are countless chemical reactions, all of them can be divided into *homogeneous* and *heterogeneous* reactions.

- *Homogeneous reaction*—In a homogeneous system, all of the species involved in the reaction are present in a single phase. For example, the combustion of methane and air in a burner where a flame is present is a homogeneous reaction. Many industrial reactions are carried out in a single phase. In the illustration of reactions from natural gas discussed previously, the dehydrogenation of ethane to produce hydrogen and ethene is a homogeneous reaction.

- *Heterogeneous reaction*—In a heterogeneous system, more than one phase is involved in the reaction. An example is the combustion of a solid fuel, such as wood being burned, where both solid and gas phases are present. Many industrial processes use solid catalysts. In such a system, the reactants and products are either gases or liquids, so such systems are also heterogeneous.

1.4.1 Rate of Reaction

In reactor analysis, the *rate of reaction* is of primary interest. The rate of reaction is a measure of how quickly a reaction occurs. In other words, for specified conditions of temperature and reactant concentration, the rate of reaction determines how long it will take to achieve a given conversion of reactants into products. The rate of reaction may be expressed in terms of the rate of disappearance of a reactant, or the rate of formation of a product. The rate of reaction may have a variety of units, the common one is mol/s. Usually, the reaction rate is expressed in terms of reactor volume (e.g., $mol/m^3 \cdot s$), mass of catalyst (e.g., $mol/kg \cdot s$), or catalyst surface area (e.g., $mol/m^2 \cdot s$).

Chemical reactions may be conducted in the presence of a *catalyst*. A catalyst is a substance which increases the rate of a reaction by providing an alternative reaction pathway between reactants and products. This path has a smaller activation energy than the uncatalyzed reaction. The catalyst participates in the chemical reaction but is not ultimately changed (in the ideal case), although the catalyst may lose its activity over time. Catalysts can be divided into *homogeneous catalysts*, in which the catalyst and the reactants are in the same phase, and *heterogeneous catalysts*, in which the catalyst is in a different phase than the reactants.

1.5 Classification and Types of Reactors

Chemical reactors have many designs and configurations. Although there is a danger of oversimplification when attempting to classify reactors, there are nonetheless a number of factors that are typically used to distinguish among reactor types.

The presence of flow streams during reaction is one important criterion for reactor classification. *Batch reactors* do not have inlet and outlet flow streams during operation. In contrast, *continuous flow* reactors are operated with inlet and outlet process streams. The level of mixing, which is related to the flow pattern, is also important, and reactors can be operated in any region from perfect mixing to negligible mixing. The number of phases (solid, liquid, or vapor) in the reactor is also important. *Homogeneous reactors* contain a single phase, and *heterogeneous reactors* (also called multiphase reactors) contain more than one phase.

1.5.1 Batch and Semibatch Operation

One of the most fundamental classifications of chemical reactors distinguishes between those reactors in which an initial quantity of material is reacted without the addition or removal of material, and those that operate with inlet and outlet flows. Those without flows are called *batch* reactors, and form an important area of reactor analysis.

1.5.1.1 Batch Reactor

A batch reactor has no inlet or outlet flows. In operation, the reactor is charged with reactants, the reaction proceeds, and then the reactor is emptied. Batch reactors can have constant or variable volume. They are often used where small quantities of very expensive products are produced, for example, in the pharmaceutical industry to produce drugs. In addition, many fermentation reactors (e.g., those used in the production of beer or ale) are operated as batch reactors. Batch reactors are also used in kinetic analysis, a topic discussed in detail in Chapter 5.

A typical batch reactor consists of a stirred tank. Figure 1.3 shows a tank that is stirred by a single impeller located along the axis of the tank. Typically, baffles are used at the

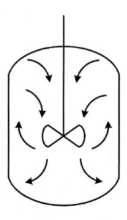

FIGURE 1.3
A typical batch reactor that may consist of a stirred tank containing the reactants. There are no inlet or outlet streams present during operation.

tank wall to ensure a flow pattern conducive to good mixing of the reactor contents. Note that the reactor illustrated is just one possible arrangement, and many configurations are used in practice. Batch reactors may be used for reactions with single or multiple phases, and for both catalytic and noncatalytic reactions. Furthermore, batch reactors may be well mixed, in which case the temperature and concentrations of all species are uniform throughout the reaction vessel, or they may contain significant gradients of temperature and/or concentration.

1.5.1.2 Semibatch Reactor

The term "semibatch reactor" is often applied to reactors that operate in neither fully continuous nor batch mode. A typical example is shown in Figure 1.4. The reactor may initially contain a mass of material, and, during the course of reaction, material is added but none is removed. After some time, the addition might be stopped and the products removed. Alternatively, once the addition of material stops, the reactor may continue to operate as a batch reactor. The semibatch reactor is a transient reactor, and cannot operate at steady state.

Semibatch operation can also be used to describe the start-up of a continuous flow reactor (see Section 1.6.2). In this scenario, once the reactor is full, an effluent stream from the reactor commences, and the reactor volume may remain constant during subsequent operation.

1.5.2 Continuous Flow Reactors

Continuous reactors have inlet and/or outlet streams. As with batch reactors, they may be homogeneous or heterogeneous reactors. Continuous reactors are often operated at steady state, that is, where the mass flow rate into the reactor is equal to the mass flow rate out of the reactor, and the temperatures and concentrations at all points in the reactor do not change with time. Unsteady-state operation may be used under some circumstances and will prevail during start-up, shutdown, or after a change in one of the process variables. Continuous reactors vary widely in configuration and are often classified according to

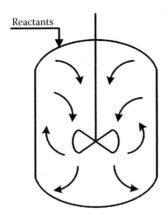

FIGURE 1.4
A semibatch reactor that may be a well-stirred tank reactor to which material is added during the operation. For a liquid-phase reaction with no effluent stream, the reacting volume will increase with time.

their internal flow patterns. In common with many chemical engineering applications, much use is made of "ideal" or model flow patterns. A large amount of reactor analysis is based on these ideal flow patterns. The two commonly encountered ideal reactor types are the perfectly mixed and the plug flow reactors.

1.5.2.1 Perfectly Mixed Flow Reactor

In an ideal perfectly mixed reactor, the mixing inside the reactor is assumed to be complete; thus, both the temperature and the concentration of all of the species are the same at every point in the reactor. A consequence of this uniform composition is that the temperature and composition of the effluent stream is the same as that in the reactor. Although many reactor configurations are possible, a perfectly mixed flow reactor usually consists of a stirred tank, as seen in Figure 1.5.

The reactor vessel usually contains some baffles at the wall to enhance turbulence and hence the level of mixing. A variety of impellers designs have been used in commercial reactors: two common examples are Rushton turbines and propellers. With a well-designed reactor, nearly perfect mixing can be achieved under some operating conditions. Typically, perfect mixing can be closely approximated in liquid-phase reactions where the fluid has a low viscosity and simple rheological behavior. Reactions involving highly viscous non-Newtonian fluids usually require special consideration, and perfect mixing is rarely achieved with these fluids.

The sizing of the tank and impeller owes a lot to experience and empirical data. As mentioned, stirred tank reactors are most commonly used for liquid-phase reactions, although some gas-phase reactions are also conducted in stirred tank reactors.

1.5.2.2 Plug Flow Reactor

A plug flow reactor consists of a vessel (usually a tube) through which the reacting fluid flows. It is assumed that there is no mixing in the direction of the flow and complete mixing in the direction transverse to the flow. No reactor can operate in perfect plug flow, but in many cases the operation is close enough for the assumption of plug flow to be valid.

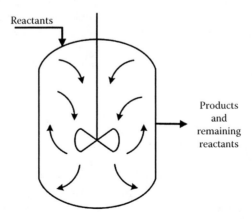

FIGURE 1.5
Illustration of a typical well-mixed continuous flow reactor. When the fluid in the reactor is perfectly mixed, the temperature and concentration are the same at every point in the reactor.

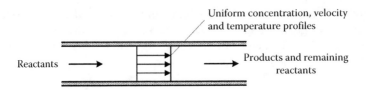

FIGURE 1.6
Velocity pattern in a plug flow reactor. The velocity is uniform across the diameter. Concentration and temperature are also assumed to be constant across the diameter.

In a plug flow reactor, concentration, temperature, and velocity gradients can therefore exist in the axial direction, but not in the radial direction. The plug flow reactor flow pattern is illustrated in Figure 1.6.

Perfect mixing and plug flow are the two extremes of mixing which can occur in flow reactors. Most flow reactors operate with intermediate levels of mixing, and very sophisticated models have been developed to explain mixing in reactors. Plug flow and perfectly mixed reactors are treated in detail in Chapter 3 (isothermal case) and Chapter 4 (nonisothermal case), while some of the methods used to analyze nonideal reactors are introduced in Chapter 6.

1.5.3 Classification by Number of Phases

One of the principal methods of classifying reactors uses the number of phases (solid, liquid, and/or gas) that are present in the reactor. The reactants, products, and catalyst (if present) may exist in more than one phase, which gives rise to the distinction made between *homogeneous* and *heterogeneous* reactors.

- *Homogeneous reactors*—The contents of the reactor are all in a single phase. In other words, the reactants, products, and any catalyst present are all of the same phase. In practical chemical engineering systems, this single phase is either a liquid or a gas.
- *Heterogeneous reactors*—In a heterogeneous reactor, two or more phases are present. In such a reactor, there could be solid, liquid, or gaseous reactants and products, as well as a catalyst. A reactor in which all of the reactants and products exist in a single phase (liquid or gas), but which contains a solid catalyst is a heterogeneous reactor. The transport of mass and energy in multiphase reactors is more complex than in single-phase reactors; hence, these reactors are more difficult to analyze and model.

1.5.4 Catalytic Reactors

In Section 1.1, the concept of the catalyst was introduced. It was stated that a catalyst is a substance that can increase the rate of reaction. The role of the catalyst and its influence on reactor analysis is discussed in detail in Chapters 7 through 11: suffice to say at this point that a catalyst can be an extremely useful tool for the process design engineer. Although catalysts and catalytic reactors have a multitude of forms, catalytic reactors in which a solid catalyst is used comprise a class of heterogeneous reactors that are extremely important in the chemical process industries. Reactors with solid catalysts have many designs, which are introduced in Chapter 7 and discussed in more detail in Chapter 10.

In this book, both homogeneous and heterogeneous reactors are considered. Because homogeneous reactors are easier to analyze than heterogeneous reactors, a "pseudo-homogeneous" model is often used for heterogeneous reactors. In such a model, the presence of the catalyst is not accounted for in a rigorous way.

1.6 Reactor Performance Measures

When considering the design or operation of a chemical reactor, a quantifiable means of identifying the performance or behavior of a reactor is needed. The performance of a chemical reactor is often related to the extent of reaction that occurs in it, or by the amount of feed that is processed in a given time for a given reactor volume. The first measure gives an indication of the fraction of reactants that are converted into products, while the second gives an indication of the total production rate of the desired product. The following terms are commonly used.

1.6.1 Conversion

The conversion (or fractional conversion), denoted X, is a frequently used measure of the degree of reaction. It is defined as

$$X \equiv \frac{\text{moles of a species that have reacted}}{\text{moles of same species initially present}} \tag{1.5}$$

This definition will be applied differently for batch and flow reactors. For a batch reactor, the conversion of reactant A is defined in terms of the number of moles of reactant initially present, N_{A0}, and the number of moles present after reaction, N_{Af}:

$$X_A = \frac{N_{A0} - N_{Af}}{N_{A0}} \tag{1.6}$$

In a flow reactor, the fractional conversion is often expressed in terms of the molar flow rate of the reactant at the reactor inlet, F_{A0}, and the reactor outlet, F_{AE}:

$$X_A = \frac{F_{A0} - F_{AE}}{F_{A0}} \tag{1.7}$$

Sometimes the conversion may be written using concentration. This definition is only valid for a constant-density system, and should not be used where the mass density of the mixture changes.

$$X_A = \frac{C_{A0} - C_{Af}}{C_{A0}} \tag{1.8}$$

An example of a constant-density system is a constant-volume batch reactor. In such a system, both mass and volume are constant; hence, so is the density.

Example 1.1

Consider the reaction represented by the overall expression

$$A + B \rightleftarrows C \qquad (1.9)$$

Initially, 50 mol of A and 25 mol of B are present. Calculate the fractional conversion of A and B after 10 mol of C are formed.

SOLUTION

The formation of 10 mol of C requires the reaction of 10 mol each of A and B. Therefore, 40 mol of A and 15 mol of B are left, and the conversions are

$$X_A = \frac{N_{A0} - N_{Af}}{N_{A0}} = \frac{50 - 40}{50} = 0.2 \qquad (1.10)$$

$$X_B = \frac{N_{B0} - N_{Bf}}{N_{B0}} = \frac{25 - 15}{25} = 0.4 \qquad (1.11)$$

Comments: The fractional conversion of A is 1/2 that of B. If all of the B reacted, 25 mol of A would still be left, so the maximum fractional conversion of A is 0.50. In this example, A is present in excess, and B is the limiting reactant.

1.6.2 Space Velocity

Space velocity is a term frequently used in reactor analysis. It is defined as the ratio of the volumetric feed flow rate and the reactor volume. The space velocity has units of inverse time. The definition is given mathematically as

$$SV = \frac{Q}{V} \qquad (1.12)$$

The volumetric feed flow rate, denoted Q, is specified at some arbitrary conditions of pressure and temperature, which are sometimes but not always the reactor inlet conditions. Because the flow rate is usually a function of temperature and pressure, a space velocity specified without T and P given should be treated with caution. Space velocity can be a useful measure when scaling reactors, that is, when designing a reactor for a larger throughput using information obtained on smaller reactors. Some examples of space velocities are given in the following paragraphs.

1.6.3 Liquid Hourly Space Velocity

To compute liquid hourly space velocity (LHSV), the volumetric feed flow rate to the reactor is specified as a liquid at some temperature, for example, 20°C. LHSV is often used when the reactants are liquid at room temperature. Note that this does not imply that the reactants are fed to the reactor in liquid form. If they are preheated, they may partially or completely vaporize. A good example is the hydrotreating of middle distillate, where the distillate is a liquid at ambient temperature. When preheated to reactor inlet temperature, some of the feed vaporizes; the amount depends on the distillate composition. The unit of LHSV is h^{-1}.

1.6.4 Gas Hourly Space Velocity

The gas hourly space velocity (GHSV) is used for gaseous feed streams with the volumetric flow rate expressed in terms of volume per hour; thus, the GHSV has a unit of h^{-1}. The conditions may be standard temperature and pressure (STP), reactor inlet conditions, or other specified conditions. It is important that the temperature and pressure of the feed be stated because the volumetric flow rate of a gas will change significantly if it is subjected to temperature and pressure changes.

1.6.4.1 Space Time

The space time is closely related to the space velocity, and is defined as

$$\tau = \frac{V}{Q} \tag{1.13}$$

From Equation 1.13 it is seen that the space time is essentially the reciprocal of the space velocity. However, it must be emphasized that for the space time to be equal to the reciprocal of the space velocity, the conditions of temperature and pressure under which the feed flow rate is specified must be the same. Space time data are frequently given with the feed conditions at the reactor inlet temperature and pressure, although they are sometimes quoted at other conditions. It is necessary to be vigilant when using either space time or space velocity.

1.6.4.2 Residence Time

The residence time is the length of time that species spend in the reactor. Depending on the reactor type, not all molecules that enter the reactor spend the same length of time in the reactor, and thus for any given reactor there will exist a distribution of residence times. The distribution of residence times is called (appropriately!) the residence time distribution (RTD). The average length of time that molecules spend in the reactor is called the mean residence time, t_m. The mean residence time is not equal to the space time except for constant-density fluids. The concept of residence time is explored more fully in Chapter 6.

1.7 Introduction to Rate Function

When analyzing a reactor, it is necessary to quantify the reaction rate. To this end, the rate of reaction is usually expressed as a function of the temperature and the concentrations of the reactants and products. The *rate function*, *rate equation*, and *rate expression* are all terms for the relationship among the rate of reaction, the concentration of species in the mixture, and the temperature. This concept is introduced in this section and considered in further detail in Chapter 5. Consider a general reversible reaction, represented by the following stoichiometry:

$$aA + bB + \cdots \rightleftharpoons rR + sS + \cdots \tag{1.14}$$

The rate expression is written in the following generic functional form:

$$(-r_A) = f(T, C_A, C_B, \ldots, C_R, C_S, \ldots) \tag{1.15}$$

where $(-r_A)$ is the rate of disappearance of species A. The concentrations of both reactants and products may appear in the function because either the reaction is reversible or the products may inhibit the forward reaction in some manner. In Equation 1.15, the rate is expressed in terms of reactant A and the rates of reaction of the different components in the mixture are related to each other by the stoichiometry of the reaction. For example, for the reaction given by Equation 1.14, the rate of reaction of B is related to the rate of reaction of A by

$$(-r_A) = \frac{a}{b}(-r_B) \tag{1.16}$$

The general relationship among the rates of reaction for this reaction can be written as

$$\frac{1}{a}(-r_A) = \frac{1}{b}(-r_B) = \frac{1}{r}r_R = \frac{1}{s}r_S, \quad \text{and so on} \tag{1.17}$$

When referring to the rate of reaction, it is important to know to which component the rate refers. The rate equation can take many forms. Consider, for example, a simple irreversible reaction

$$A \rightarrow R \tag{1.18}$$

The rate equation might have the form of a simple power law model, such as

$$(-r_A) = kC_A \tag{1.19}$$

Alternatively, the rate equation may have a complex nonlinear form, for example, a ratio like

$$(-r_A) = \frac{k_1 C_A}{(1 + k_2 C_A)^2} \tag{1.20}$$

where k, k_1, and k_2 are constants that are functions of temperature.

Note that the value of the rate constant(s) may vary, depending on the stoichiometry and the way in which the rate is expressed. For example, consider the irreversible reaction

$$A + 2B \rightarrow \text{products}$$

Note that 2 mol of B reacts for each mole of A. For illustration purposes, suppose that the rate equations for the rates of disappearance of A and B might respectively be written as

$$(-r_A) = k_1 \, C_A \, C_B \qquad (1.21)$$

$$(-r_B) = k_2 \, C_A \, C_B \qquad (1.22)$$

The rates of disappearance of A and B must both have the same functional form. Hence, the concentration dependence in both rate equations is the same. Note, however, that because 2 mol of B reacts for each mole of A, the rate of disappearance of B must be twice the rate of disappearance of A. The value of the rate constant is different in each equation, reflecting the stoichiometry of the reaction. In short

$$k_2 = 2k_1 \qquad (1.23)$$

It is very important that when numerical values of the rate constant are quoted for a given reaction, the basis of the value is also given. Otherwise, the risk of error is greatly increased.

A wide variety of functional forms for the rate equation are possible. Although the design engineer is often interested in a simple expression for computational simplicity, such a form is not possible for every reaction of industrial significance.

1.8 Transport Phenomena in Reactors

Reactor analysis relies on the conservation equations for momentum, mass, and energy. These balances are discussed briefly in the following sections and in detail in Chapters 3 and 4.

1.8.1 Material Balances

Material balances are important in reactor analysis because they provide information about the composition of the species and the variation of concentration within a reactor. The distribution of reactants, intermediates, and products may all be important. Material balances enable the reaction rate to be quantified and hence the transfer of energy from one form to another. Material balances are normally performed by following the molar flow rate of a species. For example, if methane was the reactant, a material balance performed for a control volume in an open system would be

$$
\begin{bmatrix} \text{Molar flow} \\ \text{of methane} \\ \text{into} \\ \text{the system} \end{bmatrix}
-
\begin{bmatrix} \text{Molar flow} \\ \text{of methane} \\ \text{out of} \\ \text{the system} \end{bmatrix}
=
\begin{bmatrix} \text{Disappearance} \\ \text{of methane} \\ \text{by reaction} \end{bmatrix}
+
\begin{bmatrix} \text{Rate of} \\ \text{accumulation} \\ \text{of methane} \\ \text{in the system} \end{bmatrix}
$$

1.8.2 Energy Balance

In most reactors, the energy balance is of paramount importance. The conservation of energy equation enables the calculation of things such as the temperature changes that occur as the reaction proceeds and tells us how much energy is transferred to or from a system. In particular, we are interested in knowing how much thermal energy is released or absorbed for a given feed composition and the resulting temperature of the process stream. The temperature of the reacting stream governs, among other things, the rate of reaction. The general energy balance equation for a system is

$$
\begin{bmatrix} \text{Rate of} \\ \text{energy flow} \\ \text{into the} \\ \text{system} \end{bmatrix} - \begin{bmatrix} \text{Rate of} \\ \text{energy flow} \\ \text{out of} \\ \text{the system} \end{bmatrix} = \begin{bmatrix} \text{Rate of} \\ \text{accumulation} \\ \text{of energy} \\ \text{in the system} \end{bmatrix}
$$

In addition to the energy flows, we are also interested in changes between different forms of energy. For example, during the course of a chemical reaction, a change in composition of the reacting mixture occurs. When a composition change occurs, the internal energy of the molecules of the process stream also changes. The energy gained or lost by the molecules must be balanced by a corresponding energy change of the system or surroundings so as to preserve the continuity of energy principle. Usually, the counterbalancing energy change is a change in the amount of thermal energy present, which manifests itself as a temperature change of the system or requires some heat transfer with the surroundings. Energy changes in systems are discussed in the thermodynamics review of Section 2.2 and the analysis of energy changes in reactors is developed in Chapter 4.

1.8.3 Momentum Balances

Momentum balances can be important in reactor analysis because they provide information about the pressure drop across the unit, and the variations in velocity and pressure profiles within the reactor. The relative importance of these terms varies with the application. In some cases, it may suffice to have an estimate of the overall pressure drop across the reactor; in others, detailed information on velocity and pressure profiles may be important, for example, to calculate flow patterns or developing flow profiles. The momentum balance for a control volume in an open system is

$$
\begin{bmatrix} \text{Rate of} \\ \text{momentum} \\ \text{into the} \\ \text{system} \end{bmatrix} - \begin{bmatrix} \text{Rate of} \\ \text{momentum} \\ \text{out of} \\ \text{the system} \end{bmatrix} + \begin{bmatrix} \text{Sum of} \\ \text{forces acting} \\ \text{on the} \\ \text{system} \end{bmatrix} = \begin{bmatrix} \text{Rate of} \\ \text{momentum} \\ \text{accumulation} \\ \text{in the system} \end{bmatrix}
$$

In a flow reactor, forces such as friction and pressure act on the system. The velocity depends on fluid density and hence temperature and pressure. As reactions proceed, the temperature of the fluid may change which affects the momentum balance.

1.8.4 Coupling of Material, Energy, and Momentum Balances

In a reactor, the material, energy, and momentum balances are all interrelated and should (strictly speaking) be solved simultaneously. It is clear that as the temperature has a very strong effect on the rate of reaction, then, at a minimum, material and energy balances would normally need to be solved together. The inclusion of the momentum balance may not be necessary, however, because pressure drops across many reactors are low and hence the effect that a change in pressure has on the solution of the material and energy balances is negligible. Changes in velocity as a result of density or temperature variations can often be monitored with a simple expression. However, depending on the nature of the application modeled (e.g., the rate expression is strongly dependent on pressure), it may be necessary to couple the momentum balance equations.

Most of the analysis in this text is based on the ideal reactor assumption, in which the momentum balance equation is not used. However, the reader should not always assume that such a proposition is valid, even though a very large number of practical problems can be solved to a satisfactory level of accuracy with this assumption.

1.9 Numerical Methods

Many of the problems encountered in chemical reactor analysis do not readily yield to analytical (exact) solutions. Difficulties arise because of the nonlinear nature of chemical reaction phenomena, especially in the exponential temperature dependence of the rate constants (see Chapters 4 and 5 for details). The problems in this book require the ability to solve systems of nonlinear algebraic equations as well as systems of nonlinear ordinary differential equations (ODE). A wide variety of methods exist for the solutions of such systems. For example, systems of nonlinear equations can be solved using Gauss–Seidel or Newton–Raphson methods, while systems of ODE (initial value problems) can be solved using one of the variants of the Runga–Kutta method. These methods are explained in various books on numerical analysis, for example, Conte and de Boor (1980) and Faires and Burden (1993). It is assumed that the reader is able to use an appropriate numerical method where necessary to solve such systems.

Many software packages exist which can be used to solve equations numerically. These packages, when used appropriately, can offer a considerable saving of time in problem solution, and the reader is encouraged to explore such tools.

Appendix 1 gives a good primer of the numerical techniques required for chemical reactor analysis.

References

Conte, S.D. and C. de Boor, 1980, *Elementary Numerical Analysis: An Algorithmic Approach*, 3rd Ed., McGraw-Hill, New York.
Faires, J.D. and R.L. Burden, 1993, *Numerical Methods*, PWS-Kent, Boston.

2

Thermodynamics of Chemical Reactions

Thermodynamics was identified in Chapter 1 as one of the key pillars in the field of chemical reactor analysis. Indeed, few reactor designs can be successfully executed in the absence of comprehensive thermodynamic calculations or, at the least, making some key thermodynamic assumptions. Several areas of chemical engineering thermodynamics are of major importance in chemical reaction engineering. They include, but are not limited to, the following:

1. *Enthalpy change of reaction and enthalpy changes owing to temperature changes.* Knowledge of these factors enables the prediction of temperature changes in the reactor and the calculation of heating/cooling requirements.
2. *Chemical reaction equilibrium* calculations enable the prediction of the maximum theoretical conversion at a given temperature.
3. *Vapor liquid equilibrium* is important in systems with both gaseous and liquid species. In such systems, a combined chemical reaction and phase equilibrium may be present.

2.1 Basic Definitions

The following sections give a brief introduction to the basic thermodynamic terms. A more detailed presentation of the enthalpy changes during reactions is then given. Finally, the calculation of reaction equilibrium is reviewed.

2.1.1 Open and Closed Systems

Thermodynamics divides the universe into a system and its surroundings. The system is the portion of the universe (e.g., a chemical reactor) that is to be analyzed while the remainder of the universe is the surroundings. The system boundary separates the system from the surroundings. The definition of the system is arbitrary, and depends simply on where one draws the boundary.

Chemical reactors may be either open or closed systems. In a closed system, the mass of the total material in the reactor remains constant and no mass enters or leaves the system. The volume of a closed system may change, and energy in the form of work or heat can enter or leave the system. The batch reactor is thus a closed system. In an open system, material is allowed to cross the system boundary. The mass contained within the system may remain constant with time (inflow equals outflow) or it may change with time (inflow not equal to outflow). Flow reactors are all examples of open systems, as material enters and leaves these systems. When deriving the material and energy balance equations for systems, it is important to know whether they are open or closed.

2.1.2 Thermodynamic Standard State

The thermodynamic state of a system is the condition in which that system exists. Associated with a state are values of temperature, pressure, volume, energy, enthalpy, entropy, and so on. Thermodynamics is concerned with changes that occur in system variables when the system moves from one state to another along a specified path. A thermodynamic convenience commonly used is the *standard state*. It is used, for example, in calculating enthalpy changes in reactions. For gaseous systems, the standard state is usually taken as the pure component at ideal gas conditions (see Section 2.1.3.1), 1 bar pressure (100 kPa), and the temperature of the system. Prior to 1982, the official *standard state pressure* for a gas was 1 atm (101.325 kPa) and many texts still use this value as the standard state. The differences between the two states are minor. For liquid systems, the standard state is taken as the pure liquid at the temperature of the system and 1 bar pressure.

2.1.3 Equations of State

The behavior of a fluid is governed by an *equation of state*. This equation gives the relationship among pressure, volume, and temperature (*PVT*). These variables are all easily measured quantities, so it is not surprising that they have been selected as the variables used to characterize the behavior of fluids. Changes in other thermodynamic quantities are often expressed in terms of changes in pressure, volume, and temperature for the same reason.

2.1.3.1 Ideal Gas Law

The simplest equation of state for a gas is the *ideal gas law*: although it is an approximation and involves a number of simplifying assumptions, it is widely used for gases at low-to-moderate pressure (up to a few bars for most gases). The equation is

$$PV = NR_gT \tag{2.1}$$

where N is the number of moles. The value of the gas constant, R_g, is 8.314 J/mol·K. The ideal gas law involves assumptions and should be used carefully. It works best for gases at high temperature and low pressure. For example, for simple gases and mixtures such as air, it is accurate to within 1% for temperatures from 300 to 1800 K and pressures in the range of 1–20 bar (100–2000 kPa).

2.1.3.2 Nonideal Gas Behavior

At some temperatures or pressures, gases will not behave according to the ideal gas law because at those conditions the assumptions made in developing the ideal gas law are not valid. In such a case, the gas is said to be nonideal. Many equations have been developed to model the behavior of nonideal gases, with the simplest equation using a *compressibility factor*, Z, to account for the deviation from the ideal. The equation is

$$PV = ZNR_gT \tag{2.2}$$

If the compressibility factor equals one, the gas behaves as an ideal gas. The computation of Z is relatively straightforward. It is usually calculated using the *reduced temperature and pressure*, which are in turn defined as the temperature and pressure divided by the *critical temperature and pressure*, respectively. See, for example, Smith et al. (1996) or Sandler (1999).

2.1.4 Multicomponent Mixtures

Chemical reactions always involve multicomponent mixtures, and the properties of each component can be related to the average properties of the mixture. The properties may be based on the mass of the system or on the number of moles in it. Let m_j denote the mass of component j in a mixture of n components: hence m_1, m_2, m_3, and so on. The total system mass is therefore

$$m = \sum_{j=1}^{n} m_j \tag{2.3}$$

For a mixture of volume V, the mixture mass density and the species mass density are given as a ratio of mass to volume, as follows:

$$\rho = \frac{m}{V}, \quad \rho_j = \frac{m_j}{V} \tag{2.4}$$

The mass fraction, w_j, is defined as

$$w_j = \frac{m_j}{m} = \frac{\rho_j}{\rho} \tag{2.5}$$

When analyzing reacting systems, it is most common to work with moles and molar concentrations, rather than mass and mass fractions, because rate expressions are usually expressed in terms of molar concentrations. If M_j denotes the molar mass of species j, the number of moles, N_j, is

$$N_j = \frac{m_j}{M_j} \tag{2.6}$$

The molar density (or concentration), C_j, is given by

$$C_j = \frac{N_j}{V} = \frac{\rho_j}{M_j} \tag{2.7}$$

The mole fraction of component j in a gaseous mixture is denoted y_j and is

$$y_j = \frac{C_j}{C} \tag{2.8}$$

The partial pressure of a component in an ideal gas mixture is defined as the product of mole fraction and the total pressure as follows:

$$P_j = y_j P \tag{2.9}$$

Equation 2.9 is only valid for an ideal gas. The average molar mass of a mixture, M_m, is

$$M_m = \sum_{j=1}^{n} y_j M_j \qquad (2.10)$$

The use of Equation 2.10 is illustrated in Example 2.1.

Example 2.1

Calculate the molar mass of air composed of 78.08 mol% N_2, 20.95 mol% O_2, and 0.97 mol% Ar.

SOLUTION

The average molar mass of the mixture is the weighted average of the molar masses of the components. The molar mass of N_2 is 28.01, that of O_2 is 32, and Ar is 39.9. Therefore, the molar mass of air is

$$28.01 \times 0.7808 + 32.00 \times 0.2095 + 0.0097 \times 39.9 = 28.96 \text{ g/mol}$$

Therefore, 28.96 kg of air is equivalent to 1000 mol of air.

The volumetric flow rate, Q, of an ideal gas is related to the total molar flow rate, F_t:

$$PQ = F_t R_g T \qquad (2.11)$$

The use of Equation 2.11 is illustrated in Example 2.2.

Example 2.2

A process stream has a pressure of 20 bar and a temperature of 800 K. The stream is a mixture of methane and air. The mass flow rate of air is 10 kg/s and the mass flow rate of methane is 0.25 kg/s. Calculate the mole fraction of methane in the feed and the volumetric flow rate.

SOLUTION

The mass flow rates are converted to molar flow rates by dividing by the molar mass; therefore, the molar flow rate of methane is

$$F_{CH4} = \frac{\dot{m}_{CH4}}{M_{CH4}} = \frac{0.25 \text{ kg/s}}{16.03 \text{ kg/kmol}} = 0.0156 \text{ kmol/s} \equiv 15.6 \text{ mol/s}$$

The molar flow rate for air is

$$F_{air} = \frac{\dot{m}_{air}}{M_{air}} = \frac{10 \text{ kg/s}}{28.96 \text{ kg/kmol}} = 0.3453 \text{ kmol/s} \equiv 345.3 \text{ mol/s}$$

The mole fraction is the moles of a substance divided by the total number of moles. The number of moles can be replaced by the molar flow rates, therefore

$$y_{CH4} = \frac{F_{CH4}}{F_{CH4} + F_{air}} = \frac{15.6}{15.6 + 345.3} = 0.0432$$

The sum of the mole fraction of air and methane must be one, therefore

$$y_{air} = 1 - y_{CH4} = 1 - 0.0432 = 0.9568$$

To calculate the volumetric flow rate, substitute the values of molar flow rate, temperature, pressure, and the gas constant into the ideal gas law for a stream to obtain

$$Q = \frac{F_t R_g T}{P} = \frac{(15.6 + 345.3) \text{ mol/s} \times 8.314 \text{ J/mol} \cdot \text{K} \times 800 \text{ K}}{20 \times 10^5 \text{ Pa}} = 1.20 \text{ m}^3/\text{s}$$

When canceling units in this equation, note that the unit of energy, J, is equivalent to units of $Pa \cdot m^3$.

2.2 Energy Changes in Systems

It is necessary to analyze the energy changes that accompany a chemical reaction. The energy of a system is composed of three parts: the kinetic energy that depends on the velocity of a flowing stream, the potential energy resulting from the position of the system relative to a reference point, and the internal energy, U. The latter quantity is a measure of the molecular and atomic motion within the fluid. In most chemical reactors, changes in kinetic and potential energy are assumed to be minor, and are usually ignored. Internal energy changes are discussed in detail in Chapter 4.

2.2.1 Enthalpy

In many chemical reactors, the pressure of the system is assumed to be constant, or approximately constant. In a constant-pressure system, it is convenient to work in terms of the enthalpy of the system, and most energy balances performed on reacting systems are in fact enthalpy balances. The enthalpy of a system is defined as

$$H = U + PV \tag{2.12}$$

We are concerned with enthalpy changes that occur because of temperature changes, composition changes owing to chemical reaction, and, to a lesser extent, pressure changes. These effects are discussed below.

2.2.2 Energy Change Resulting from Temperature Change: Heat Capacity

The enthalpy and internal energy change that occurs because of a temperature change is quantified using the *heat capacity*. Two heat capacities are used: the constant-volume heat capacity, denoted C_V, and the constant-pressure heat capacity, denoted C_P.

The constant-volume heat capacity is defined as the rate of change in internal energy with temperature when the volume of the system is held constant, or

$$C_V = \left(\frac{\partial U}{\partial T} \right)_V \tag{2.13}$$

The constant-pressure heat capacity is defined as the rate of change in enthalpy with temperature when the pressure of the system is held constant, or

$$C_P = \left(\frac{\partial H}{\partial T} \right)_P \tag{2.14}$$

These equations are written in partial derivative form to indicate that although the internal energy, U, is usually a function of temperature and density (or specific volume), the heat capacity, C_V, has been measured along a path of constant volume. Likewise, C_P has been evaluated when the pressure has been held constant. The heat capacity is a strong function of temperature and a weak function of pressure. Usually the heat capacity of a gas is expressed as a polynomial function, such as

$$C_P = a + bT + cT^2 + dT^3 \tag{2.15}$$

This heat capacity is valid for the ideal gas state, that is, where the ideal gas law is being used (i.e., when the compressibility factor is one). At very high pressures in critical applications, the pressure dependency of heat capacity may be important. See Reid et al. (1987) for calculation methods. For ideal gas mixtures, the molar heat capacity of the mixture is the sum of the heat capacities of the components multiplied by their respective mole fractions. For an ideal gas, the heat capacities are related by the universal gas constant:

$$C_P = C_V + R_g \tag{2.16}$$

For liquids and solids, the two heat capacities are approximately equal. Constants in the heat capacity polynomial are given in Appendix 2 for some simple compounds.

Example 2.3

A gas mixture is made by mixing 2 mol% CH_4, 1 mol% CO_2, 1 mol% H_2O, and 96 mol% air. Use the data in Appendix 2 to perform the following calculations:

   a. Derive an expression for C_P of this mixture as a function of temperature.
   b. Calculate the value of C_P at 500 K.
   c. Determine the enthalpy change per mole for an increase in temperature of this mixture from 500 to 800 K at 1 bar (100 kPa) pressure.

SOLUTION

   a. The heat capacity of the mixture is a weighted average of the heat capacities of the components. From Appendix 2, we have the values for the coefficients in the heat capacity polynomial.

	a	b	c	d
CH_4	19.86	5.016×10^{-2}	1.267×10^{-5}	-10.99×10^{-9}
CO_2	22.22	5.9711×10^{-2}	-3.495×10^{-5}	7.457×10^{-9}
H_2O	32.19	0.1920×10^{-2}	1.054×10^{-5}	-3.589×10^{-9}
Air	28.09	0.1965×10^{-2}	0.4799×10^{-5}	-1.965×10^{-9}

The heat capacity of the mixture expressed as a function of temperature is a polynomial in which the coefficients (i.e., the *a*, *b*, *c*, and *d*) are the weighted averages of the *a*, *b*, *c*, and *d* of the components. Therefore, the value of *a* in the polynomial for the mixture is

$$a = (19.86)(0.02) + (22.22)(0.01) + (32.19)(0.01) + (28.09)(0.96) = 27.91$$

Similarly,

$$b = 0.3506 \times 10^{-2}, \ c = 0.4616 \times 10^{-5}, \ d = -2.068 \times 10^{-9}$$

These values are substituted into the heat capacity polynomial to give the temperature function for the heat capacity of 1 mol of the mixture:

$$C_P = 27.91 + 3.506 \times 10^{-3}T + 4.616 \times 10^{-6}T^2 - 2.068 \times 10^{-9}T^3 \ J/mol \cdot K$$

b. Substituting 500 K into the preceding equation gives a value of

$$C_P = 30.56 \ J/mol \cdot K$$

c. For a constant-pressure process, the heat transfer to the system equals the enthalpy change:

$$\Delta H = \int_{T_1}^{T_2} C_P \, dT$$

Substituting the polynomial expression for heat capacity and the numerical values for the constants gives the integral equation

$$\Delta H = \int_{500}^{800} (27.91 + 3.506 \times 10^{-3}T + 4.616 \times 10^{-6}T^2 - 2.068 \times 10^{-9}T^3) dT$$

Performing the integration and substituting the numbers gives

$$\Delta H = 27.91(800 - 500) + \frac{3.506 \times 10^{-3}}{2}(800^2 - 500^2) + \frac{4.616 \times 10^{-6}}{3}(800^3 - 500^3)$$

$$- \frac{2.068 \times 10^{-9}}{4}(800^4 - 500^4) = 9473 \ J/mol$$

A common engineering approximation is to compute the enthalpy change by using the average of the heat capacities evaluated at 500 and 800 K. From part (b)

$$C_P = 30.56 \ J/mol \cdot K \ at \ 500 \ K \quad and \quad C_P = 32.61 \ J/mol \cdot K \ at \ 800 \ K$$

The average of the two values is 31.58 J/mol·K. Therefore,

$$\Delta H \approx (31.58 \text{ J/mol} \cdot \text{K})(800 - 500)\text{K} = 9475 \text{ J/mol}$$

In this case, it is a good approximation to use the average value of heat capacity to calculate the enthalpy change.

2.2.3 Enthalpy Change with Pressure

The enthalpy change with pressure at constant temperature can be defined in terms of PVT (details are beyond the scope of this book; further information is available in Sandler, 1999). It can be shown that the enthalpy change with pressure at constant temperature and composition is given by

$$\left(\frac{\partial H}{\partial P}\right)_T = V - T\left(\frac{\partial V}{\partial T}\right)_P = V - TV\beta \tag{2.17}$$

where β is the coefficient of thermal expansion. The advantage of using Equation 2.17 to compute changes in enthalpy with pressure is that the calculation can be done using easily measurable PVT data. For example, the enthalpy change with pressure for an ideal gas is found by differentiating the ideal gas law. Take the partial derivative of V with respect to T and constant P to give

$$\left(\frac{\partial V}{\partial T}\right)_P = \frac{NR_g}{P} \tag{2.18}$$

Substitution of Equation 2.18 into Equation 2.17 gives

$$\left[V - T\left(\frac{\partial V}{\partial T}\right)_P\right] = \left[V - T\left(\frac{NR_g}{T}\right)\right] = [V - V] = 0 \tag{2.19}$$

It is seen that this relationship reduces to zero: in other words, as long as ideal gas behavior can be assumed for the gas mixture under study, pressure changes will not affect the enthalpy of the gas. In cases where real gas behavior must be used in the reactor analysis, the equation of state can be used to evaluate the change in enthalpy that occurs when the pressure changes. As always, when analyzing reacting systems, it is necessary to decide whether the increased complexity of using a real gas model, rather than assuming ideal behavior, is justified in terms of the improved accuracy of the solution.

2.2.4 Energy Change owing to Composition Change: Heat of Reaction

Energy changes that accompany chemical reactions are very important. Two different quantities are commonly used to quantify energy changes in reactors, and the choice of which to use depends on the type of reactor being analyzed. The two quantities are the *internal energy change of reaction* and the *enthalpy change of reaction*. The enthalpy change of reaction is often referred to as the *heat of reaction*. These two energy changes can be defined as follows:

$\Delta U_R \equiv$ internal energy change of reaction: the difference in internal energy between reactants and products.

$\Delta H_R \equiv$ enthalpy change of reaction: the difference in enthalpy between reactants and products owing to reaction. Sometimes, this quantity is called the "heat of reaction."

If the energy change of reaction is negative, the reaction is *exothermic*, and heat is released. If it is positive, the reaction is *endothermic* and heat is absorbed. A complete description of ΔU_R and ΔH_R includes the temperature and pressure at which the value is applicable. Both quantities depend on temperature and pressure, although the pressure dependence is often ignored.

To standardize calculation and reporting methods, the standard state is usually used. Recall that the standard state is defined as:

1. For gases—ideal gas state at 1 bar pressure
2. For liquids—pure liquid at 1 bar pressure

and the temperature of the system. The standard state is commonly denoted with a superscript "∘"; therefore, ΔH_R° and ΔU_R° denote the standard enthalpy change and standard internal energy change, respectively. Because the standard state is specified at the temperature of the system, both ΔH_R° and ΔU_R° are functions of temperature. Data are usually tabulated at a reference temperature of 298 K; thus, $\Delta H_{R,298}^{\circ}$ denotes the standard enthalpy change of reaction at 298 K. $\Delta U_{R,298}^{\circ}$ can be calculated from enthalpy of formation data. The standard heats of formation are defined as the enthalpy change that occurs when compounds are formed from the elements with both reactants and products in their standard state (1 bar pressure and the temperature of the system). They are usually tabulated at 298.15 K and some values for some common compounds are given in Appendix 2.

Consider the hypothetical reversible reaction

$$a\mathrm{A} + b\mathrm{B} \rightleftharpoons c\,\mathrm{C} + d\mathrm{D} \tag{2.20}$$

where a, b, c, and d are the stoichiometric coefficients. Let $\Delta H_{f,298}^{\circ}$ denote the standard enthalpy of formation. Then the enthalpy of reaction is given by

$$\Delta H_{R,298}^{\circ} = c\left(\Delta H_{f,298}^{\circ}\right)_{\mathrm{C}} + d\left(\Delta H_{f,298}^{\circ}\right)_{\mathrm{D}} - a\left(\Delta H_{f,298}^{\circ}\right)_{\mathrm{A}} - b\left(\Delta H_{f,298}^{\circ}\right)_{\mathrm{B}}$$

Usually, the value is normalized with respect to one of the components. Define α_j as the stoichiometric coefficient for any component j in the reaction, positive for both reactants and products. The enthalpy of reaction per mole of species k is

$$\Delta H_{R,298}^{\circ} = \frac{1}{\alpha_k}\left[\sum_{\text{products}} \alpha_j \Delta H_{f,298}^{\circ} - \sum_{\text{reactants}} \alpha_j \Delta H_{f,298}^{\circ}\right] \tag{2.21}$$

For example, for our hypothetical reaction, the $\Delta H_{R,298}^{\circ}$ per mole of species A that reacts is

$$\Delta H_{R,298}^{\circ} = \frac{1}{a}\left[c\left(\Delta H_{f,298}^{\circ}\right)_{\mathrm{C}} + d\left(\Delta H_{f,298}^{\circ}\right)_{\mathrm{D}} - a\left(\Delta H_{f,298}^{\circ}\right)_{\mathrm{A}} - b\left(\Delta H_{f,298}^{\circ}\right)_{\mathrm{B}}\right] \tag{2.22}$$

Example 2.4

Use heat of formation data to calculate the standard heat of reaction for the combustion of 1 mol of methane. The overall reaction is

$$CH_4 + 2O_2 \rightarrow CO_2 + 2H_2O$$

SOLUTION

From Appendix 2, the relevant heats of formation at 298.15 K are

Methane	−74.90 kJ/mol
Oxygen	0.0
Carbon dioxide	−393.8 kJ/mol
Water vapor	−242.0 kJ/mol

The standard heat of reaction is the sum of the heats of formation of the products minus the heats of formation of the reactants. This calculation must take into account the stoichiometry of the reaction. If we let α_j represent a stoichiometric coefficient (which is defined here as a positive number for both reactants and products), the generic formula for computing a standard heat of reaction is

$$\Delta H_{R,298}^{\circ} = \frac{1}{\alpha_k} \left[\sum_{products} \alpha_j \Delta H_{f,298}^{\circ} - \sum_{reactants} \alpha_j \Delta H_{f,298}^{\circ} \right] \qquad (2.23)$$

Equation 2.23 gives the standard heat of reaction per mole of species k, which could be either a reactant or a product. For the combustion of methane, we have

$$\Delta H_{R,298}^{\circ} = \left(\Delta H_{f,298}^{\circ}\right)_{CO_2} + 2\left(\Delta H_{f,298}^{\circ}\right)_{H_2O} - 2\left(\Delta H_{f,298}^{\circ}\right)_{O_2} - \left(\Delta H_{f,298}^{\circ}\right)_{CH_4}$$

Note that the heats of formation are multiplied by the stoichiometric coefficients of the reaction. Substitution of the numerical values gives

$$\Delta H_{R,298}^{\circ} = (-393.8) + (2)(-242.0) - (2)(0) - (-74.90) = -802.9 \text{ kJ/mol}$$

This value is the heat of reaction for 1 mol of methane.

A negative value of $\Delta H_{R,298}^{\circ}$ indicates an exothermic reaction, while a positive value denotes an endothermic reaction. Thus, the combustion of methane occurs with a release of thermal energy.

2.2.4.1 Computing the Enthalpy of Reaction at Other Temperatures

Chemical reactions are rarely carried out at the standard state and 298 K. To use data at 298 K, we make use of the fact that ΔH_R° is a *state function*, that is, its value does not depend on the path, only on the initial and final state of the system.

Consider a system compound of reactants at some temperature T_1 and pressure P_1. We wish to compute $\Delta H_R^{\circ} (T_1, P_1)$ for converting these reactants to products. This calculation may be done by defining an alternate path, such that the reaction is carried out at standard conditions

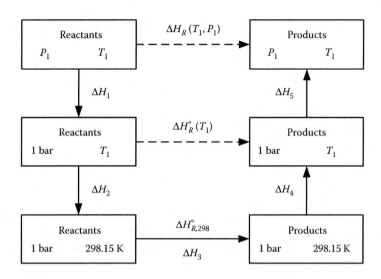

FIGURE 2.1
Alternate calculation pathway used to compute the enthalpy change of reaction at different temperatures and pressures.

and 298 K, at which conditions ΔH_R° may be readily computed from enthalpy of formation data. This path is a five-step process, as shown in Figure 2.1 and summarized in the following:

Step 1: Take the reactants at T_1 and P_1 to the ideal gas state at 1 bar and T_1. The enthalpy change associated with this step is ΔH_1.

Step 2: Take the reactants from T_1 and 1 atm to 298 K and 1 bar. This enthalpy change is ΔH_2.

Step 3: Convert the reactants in the standard state at 298 K to products in the standard state at 298 K. This enthalpy change is ΔH_3.

Step 4: Take the products from the standard state at 298 K to standard state and temperature T_1. Call this enthalpy change ΔH_4.

Step 5: Take the products to temperature T_1 and pressure P_1. Call this ΔH_5.

The total enthalpy change is the sum of the changes that occur in these five steps, or

$$\Delta H_R\left(T_1, P_1\right) = \Delta H_1 + \Delta H_2 + \Delta H_3 + \Delta H_4 + \Delta H_5 \tag{2.24}$$

Note that $\Delta H_3 = \Delta H_{R,298}^{\circ}$, the standard heat of reaction at 298 K, and is computed from the values of standard heats of formation. To evaluate temperature and pressure effects, we recall

$$dH = \left(\frac{\partial H}{\partial T}\right)_P dT + \left(\frac{\partial H}{\partial P}\right)_T dP \tag{2.25}$$

Furthermore, recall that $(\partial H/\partial T)_P = C_P$, the constant-pressure heat capacity. ΔH_2 is computed as

$$\Delta H_2 = \int_{T_1}^{298} C_P \, dT \tag{2.26}$$

The C_P of the reactants is based on the stoichiometric amounts. ΔH_4 is computed as

$$\Delta H_4 = \int\limits_{298}^{T_1} C_P \, dT \tag{2.27}$$

The C_P is for the products, based on the stoichiometric amount. To compute the pressure effect, we use the following thermodynamic relationship:

$$\left(\frac{\partial H}{\partial P} \right)_T = V - T \left(\frac{\partial V}{\partial T} \right)_P \tag{2.28}$$

If the equation of state is known, the term $(\partial V / \partial T)_P$ can be computed and thus $(\partial H / \partial P)_T$ obtained. Alternatively, the enthalpy departure curves, derived from the theory of corresponding states, may be used to calculate ΔH_1 and ΔH_5. Example 2.5 illustrates the calculation of ΔH_R° at temperatures other than 298 K.

Example 2.5

Consider the oxidation of methane, represented by the reaction

$$CH_4 + 2O_2 \rightarrow CO_2 + 2H_2O$$

Derive an expression for the standard heat of reaction as a function of temperature and use it to calculate the standard heat of reaction at 800 K.

SOLUTION

Calculate the standard heat of reaction, that is, the heat of reaction at 1 bar pressure. The heat of reaction at a temperature T is calculated by defining a path as follows:

a. Change the temperature of the reactants from T to 298.15 K.
b. Carry out the reaction at 298.15 K (in the previous example $\Delta H_{R,298}^{\circ}$ for this reaction was computed to be −802.9 kJ/mol).
c. Take the temperature of the products to the (same) temperature T.

The enthalpy changes which accompany the temperature changes can be computed using the heat capacities of the respective mixtures. It is useful to take a basis for calculation purposes, and therefore we take the basis of 1 mol of methane and 2 mol of oxygen. The resulting heat of reaction will therefore be the value per mole of methane that reacts. The heat of reaction at T is the sum of the enthalpy changes for these steps. The generic formula for this operation is

$$\Delta H_{R,T}^{\circ} = \frac{1}{\alpha_k} \left[\sum_{\text{reactants}} \int\limits_T^{298} \alpha_j C_P \, dT \right] + \Delta H_{R,298}^{\circ} + \frac{1}{\alpha_k} \left[\sum_{\text{products}} \int\limits_{298}^{T} \alpha_j C_P \, dT \right]$$

Note that the result is normalized to species k. Rearrangement gives

$$\Delta H_{R,T}^{\circ} = \Delta H_{R,298}^{\circ} + \frac{1}{\alpha_k} \left[\int\limits_{298}^{T} \left(\sum_{\text{products}} \alpha_j C_P - \sum_{\text{reactants}} \alpha_j C_P \right) dT \right] \tag{2.29}$$

The difference in heat capacity between products and reactants is defined as

$$\Delta C_P = \left[\sum_{products} \alpha_j C_P - \sum_{reactants} \alpha_j C_P \right] \tag{2.30}$$

If the heat capacity is expressed as a polynomial in temperature, then

$$\Delta C_P = \Delta a + \Delta b T + \Delta c T^2 + \Delta d T^3 \tag{2.31}$$

The coefficients are defined in terms of the reactants and products, that is

$$\Delta a = \left[\sum_{products} \alpha_j a_j - \sum_{reactants} \alpha_j a_j \right] \tag{2.32}$$

Similar definitions apply for Δb, Δc, and Δd. The values for the coefficients in the heat capacity polynomials are given in Appendix 2:

	a	b	c	d
CH_4	19.86	5.016×10^{-2}	1.267×10^{-5}	-10.99×10^{-9}
CO_2	22.22	5.9711×10^{-2}	-3.495×10^{-5}	7.457×10^{-9}
H_2O	32.19	0.1920×10^{-2}	1.054×10^{-5}	-3.589×10^{-9}
O_2	25.44	1.518×10^{-2}	-0.7144×10^{-5}	1.310×10^{-9}

Using the basis of 1 mol of methane and 2 mol of oxygen, the constants in the heat capacity polynomial for the mixture are computed, that is, the resulting constants are the values for the mixture of 1 mol of methane and 2 mol of oxygen. The values are

$$a = 1 \times 19.86 + 2 \times 25.44 = 70.74$$

$$b = 1 \times 5.016 \times 10^{-2} + 2 \times 1.518 \times 10^{-2} = 8.052 \times 10^{-2}$$

$$c = 1 \times 1.267 \times 10^{-5} + 2 \times -0.7144 \times 10^{-5} = -1.618 \times 10^{-6}$$

$$d = 1 \times -10.99 \times 10^{-9} + 2 \times 1.310 \times 10^{-9} = -8.37 \times 10^{-9}$$

When using these data, it should be remembered that the coefficients apply to an equation for C_p that has units of J/mol·K. The coefficients of the product mixture (1 mol of carbon monoxide and 2 mol of water) are computed in a similar manner:

$$a = 86.60$$

$$b = 6.355 \times 10^{-2}$$

$$c = -1.387 \times 10^{-5}$$

$$d = 2.79 \times 10^{-10}$$

Note: These are the heat capacities of an amount of mixture based on 1 mol of methane, not 1 mol of mixture. Therefore, the difference in each constant value between the products and the reactants is computed as

$$\Delta a = 86.60 - 70.74 = 15.86$$

$$\Delta b = 6.355 \times 10^{-2} - 8.052 \times 10^{-2} = -1.697 \times 10^{-2}$$

$$\Delta c = -1.387 \times 10^{-5} - (-1.618 \times 10^{-6}) = -1.189 \times 10^{-5}$$

$$\Delta d = 2.79 \times 10^{-10} - (-8.37 \times 10^{-9}) = 8.649 \times 10^{-9}$$

The difference in heat capacity between the products and the reactants is therefore

$$\Delta C_p = 15.86 + -1.697 \times 10^{-2} T - 1.225 \times 10^{-5} T^2 + 8.649 \times 10^{-9} T^3$$

The integral of this quantity can be obtained by substitution and integration:

$$\int_{298}^{T} \Delta C_p \, dT = -3886 + 15.86T - 8.485 \times 10^{-3} T^2 - 3.963 \times 10^{-6} T^3 + 2.16 \times 10^{-9} T^4$$

We divide by 1000 (to convert to kJ/mol·K) and substitute values to give

$$\Delta H_R^\circ = -806.9 + 1.586 \times 10^{-2} T - 8.485 \times 10^{-6} T^2 - 3.963 \times 10^{-9} T^3 + 2.16 \times 10^{-12} T^4$$

This equation is a general equation for the standard heat of reaction in kJ/mol. Substituting 800 K into the equation gives a value of −800.9 kJ/mol. This is within 1% of the value at 298 K.

Note: This value for the heat of reaction is valid when water is present as a vapor, which is usually the case in combustion calculations when the products have a high temperature. If water will be present as a liquid during the reaction, the latent heat of vaporization of water must be considered. The value of the heat of reaction in that case would be altered by the value of the latent heat of vaporization of water, and the heat of combustion of methane would have a larger negative value.

2.2.4.2 Generalized Stoichiometric Coefficient

The calculations with the enthalpy of reaction shown in this section have adopted the convention that the stoichiometric coefficients are positive for both reactants and products. Although this convention is widely used, and can make grasping the concepts and calculation methodology easier, a more generalized set of equations can also be written using another convention. In this alternate convention, the stoichiometric coefficients are defined as negative for reactants and positive for products. These alternate stoichiometric coefficients will be denoted with the symbol ν. For the same hypothetical reaction considered previously

$$a\text{A} + b\text{B} \rightleftharpoons c\text{C} + d\text{D} \tag{2.33}$$

The stoichiometric coefficients a and b have negative values, while the values c and d have positive values. Using generalized coefficients, the enthalpy of reaction per mole of species k can be written in terms of a single summation:

$$\Delta H_{R,298}^\circ = \frac{1}{|\nu_k|} \left[\sum_{\text{all species}} \nu_j \Delta H_{f,298}^\circ \right] \tag{2.34}$$

The absolute value must be used on the stoichiometric coefficient of the normalizing component. The enthalpy of the reaction in Equation 2.33 is obtained by substitution into Equation 2.34:

$$\Delta H^{\circ}_{R,298} = c\left(\Delta H^{\circ}_{f,298}\right)_C + d\left(\Delta H^{\circ}_{f,298}\right)_D - a\left(\Delta H^{\circ}_{f,298}\right)_A - b\left(\Delta H^{\circ}_{f,298}\right)_B$$

It is easily seen that this result is the same as before. Because it is advantageous to use Equation 2.34 when programming general solutions, this second convention is also widely used. The second convention is also used in developing the equations that govern chemical reaction equilibrium, a topic that is considered in Section 2.3. As always, when referring to work in other books, and using literature results, it is essential to take care when interpreting results and the explanations offered there.

2.3 Chemical Reaction Equilibrium

The equilibrium state for a system is defined as the state in which a system exists when the free energy of the system is at a minimum. For a nonreacting system, it is relatively easy to determine a minimum free energy for the system at any given temperature and pressure; for a system in which chemical reactions occur, it is more difficult. A chemical reaction results in a change in free energy. For any given set of reactants, there are often many possible products, and the free energy changes therefore depend on what reactions actually occur (i.e., the minimum free energy for a system can change if different reactions are allowed to occur). For example, consider the combustion of methane. This reaction can be represented by

$$CH_4 + 2O_2 \rightleftharpoons CO_2 + 2H_2O$$

The equilibrium composition for this reaction can be calculated (using the methods described in this section) assuming that only the two reactants and two products are present. At any given temperature and pressure, the equilibrium composition will have a unique value corresponding to the equilibrium. However, the reaction scheme as written may not accurately represent what is actually occurring in the reactor. For example, it may be more realistic to represent the combustion of methane by a two-step process in which CO is present as follows:

$$CH_4 + \tfrac{3}{2}O_2 \rightleftharpoons CO + 2H_2O$$

$$CO + \tfrac{1}{2}O_2 \rightleftharpoons CO_2$$

In this scheme, an additional species, CO, is allowed to be present. The presence of CO changes the composition of the system at which the free energy minimum occurs. It follows therefore that when performing equilibrium calculations on reacting mixtures it is necessary to specify what products are to be permitted in the system. Obviously, this selection should reflect what actually occurs in the reactor. Such an equilibrium is referred to as a constrained equilibrium.

This concept can be illustrated on a more complex system. Consider a mixture of synthesis gas, CO and H_2. This mixture can be reacted over various catalysts to form different products: for example, methanol, hydrocarbons and water (Fischer–Tropsch synthesis), or methane and water. With such a wide variety of possible products, it is very important to identify carefully what species will be present at equilibrium. The equilibrium calculations are restricted to the products that are observed to form in the reactor under the operating conditions of interest.

2.3.1 Derivation of Equilibrium Constant

We now derive the thermodynamic statement of reaction equilibrium. Consider the reaction

$$bB + cC \rightleftharpoons dD + eE \tag{2.35}$$

The standard free energy change for this reaction, denoted with the superscript "∘," is defined as the free energy difference between the reactants and products with all species in their standard state. For the model reaction, this quantity is

$$\Delta G_R^\circ = d\bar{G}_D^\circ + e\bar{G}_E^\circ - b\bar{G}_B^\circ - c\bar{G}_C^\circ \tag{2.36}$$

where $\bar{G}_i$ is the partial molar free energy of species i. For any arbitrary isothermal process, the difference in free energy between the reactants and products is

$$\Delta G_R = d\bar{G}_D + e\bar{G}_E - b\bar{G}_B - c\bar{G}_C \tag{2.37}$$

Combining Equations 2.36 and 2.37 gives

$$\Delta G_R - \Delta G_R^\circ = d\left(\bar{G}_D - \bar{G}_D^\circ\right) + e\left(\bar{G}_E - \bar{G}_E^\circ\right) - b\left(\bar{G}_B - \bar{G}_B^\circ\right) - c\left(\bar{G}_C - \bar{G}_C^\circ\right) \tag{2.38}$$

A change in partial molar free energy for an isothermal process is related to a change in the fugacity of the component:

$$\left(d\bar{G}_j = R_g T d\ln \hat{f}_j\right)_T \tag{2.39}$$

where $\hat{f}_j$ is the fugacity of component j in the mixture. If we integrate Equation 2.39 between the standard state and some arbitrary state, we obtain

$$\bar{G}_j = \bar{G}_j^\circ + R_g T \ln\left(\frac{\hat{f}_j}{f_j^\circ}\right) \tag{2.40}$$

The ratio of fugacities is a property called the activity, denoted $\hat{a}_i$; therefore

$$\bar{G}_j - \bar{G}_j^\circ = R_g T \ln \hat{a}_j \tag{2.41}$$

Substituting Equation 2.41 into Equation 2.38 gives

$$\Delta G_R - \Delta G_R^\circ = dR_g T \ln \hat{a}_D + eR_g T \ln \hat{a}_E - bR_g T \ln \hat{a}_B - cR_g T \ln \hat{a}_C \qquad (2.42)$$

Group the terms containing the ln into a single term

$$\Delta G_R - \Delta G_R^\circ = R_g T \ln \left(\frac{\hat{a}_D^d \hat{a}_E^e}{\hat{a}_B^b \hat{a}_C^c} \right) \qquad (2.43)$$

The criterion for chemical reaction equilibrium is that the free energy difference between reactants and products is zero, that is, $\Delta G_R = 0$. Substituting $\Delta G_R = 0$ into Equation 2.43 gives

$$-\frac{\Delta G_R^\circ}{T} = R_g \ln \left(\frac{\hat{a}_D^d \hat{a}_E^e}{\hat{a}_B^b \hat{a}_C^c} \right)_{\text{equilibrium}} = R_g \ln K \qquad (2.44)$$

The equilibrium constant, K, relates the values of the reactant and product activities when the system is at equilibrium. The constant K is dimensionless because the activities are dimensionless, and its value can be computed from the standard free energy change of reaction. K can be expressed in a general form. Let ν_j be the stoichiometric coefficient of component j, negative for reactants and positive for products. The expression for K is then

$$K = \prod_{j=1}^n \hat{a}_j^{\nu_j} \qquad (2.45)$$

Note that the equilibrium constant includes the stoichiometric coefficients. Thus, the numerical value of K for the reaction

$$N_2 + 3H_2 \rightleftharpoons 2NH_3$$

would be different from the value for the same reaction written as

$$\tfrac{1}{2} N_2 + \tfrac{3}{2} H_2 \rightleftharpoons NH_3$$

When using K values, it is essential to know the stoichiometric coefficients that are to be used.

2.3.2 Computing Standard Free Energy Change

The equilibrium constant is defined in terms of the standard free energy change and thus its numerical value can be calculated if the standard free energy change is known. In most chemical engineering calculations, the standard state is taken as 1 bar pressure and the temperature of the system (the same standard state as used for ΔH_R°). Standard free energy changes of reaction can be computed from standard free energies of formation. These values are usually tabulated at a temperature of 298.15 K. The standard free energy of reaction is given by

$$\Delta G_{R,298}^{\circ} = \sum_{products} \Delta G_{f,298}^{\circ} - \sum_{reactants} \Delta G_{f,298}^{\circ} \tag{2.46}$$

Equation 2.46 can be used to calculate the free energy change for the reaction at 298 K, and hence the value of the equilibrium constant at 298 K. Because reactions are rarely carried out at 298 K, it is necessary to develop methods for calculating the free energy change at other temperatures.

2.3.3 Temperature Dependence of Equilibrium Constant

The equilibrium constant is defined at the standard state, which is 1 bar pressure and the temperature of the system. Because the standard free energy of reaction is defined at 1 bar, it follows that the numerical value of the equilibrium constant is not a function of pressure. However, as the standard state is defined at the temperature of the system, the standard free energy change of reaction depends on the temperature, and hence so does the equilibrium constant. The standard free energy change of reaction at other temperatures may be computed provided that the entropy change of reaction, denoted ΔS_R°, and ΔH_R° are available. The following equation provides a relationship among these variables:

$$\frac{\Delta G_R^{\circ}}{T} = \frac{\Delta H_R^{\circ}}{T} - \Delta S_R^{\circ} \tag{2.47}$$

Both ΔS_R° and ΔH_R° are functions of temperature, as follows:

$$\frac{\partial \left(\Delta H_R^{\circ} \right)}{\partial T} = \Delta C_P \quad \text{and} \quad \frac{\partial \Delta S_R^{\circ}}{\partial T} = \frac{\Delta C_P}{T}$$

where ΔC_P is the difference in heat capacity between the products and reactants. Equation 2.47 can be used directly to compute the free energy change at other temperatures, or, alternatively, the equation can be manipulated to give an explicit equation in terms of the equilibrium constant. We start by differentiating Equation 2.47 at constant P to give

$$\left(\frac{\partial \left(\frac{\Delta G_R^{\circ}}{T} \right)}{\partial T} \right)_P = \frac{1}{T} \left(\frac{\partial \Delta H_R^{\circ}}{\partial T} \right)_P - \frac{\Delta H_R^{\circ}}{T^2} - \left(\frac{\partial \Delta S_R^{\circ}}{\partial T} \right)_P \tag{2.48}$$

The change in the values of ΔS_R° and ΔH_R° with temperature is related to the difference in the heat capacity between the products and the reactants, as follows:

$$\frac{\partial \left(\Delta H_R^{\circ} \right)}{\partial T} = T \frac{\partial \Delta S_R^{\circ}}{\partial T} = \Delta C_P \tag{2.49}$$

Substitute the terms of Equation 2.49 into Equation 2.48 and simplify:

$$-\frac{\partial\left(\dfrac{\Delta G_R^{\circ}}{T}\right)}{\partial T} = \frac{\Delta H_R^{\circ}}{T^2} \tag{2.50}$$

Recall the definition of the equilibrium constant:

$$-\frac{\Delta G_R^{\circ}}{T} = R_g \ln K \tag{2.51}$$

Equation 2.51 can be differentiated:

$$\partial\left(\frac{\Delta G_R^{\circ}}{T}\right) = -R_g \partial\left(\ln K\right) \tag{2.52}$$

Substitution of Equation 2.52 into Equation 2.50 gives the final result

$$\frac{\partial \ln K}{\partial T} = \frac{\Delta H_R^{\circ}}{R_g T^2} \tag{2.53}$$

Equation 2.53 is the van't Hoff equation. Remember, the value of ΔH_R° depends on the temperature; hence, the usual procedure is to develop an equation for ΔH_R° as a function of temperature, which is then substituted into Equation 2.53 and the result integrated. Provided that K at one temperature is known, the value of K at other temperatures may be determined.

An approximation that is often made when calculating equilibrium constants is to assume a constant value for ΔH_R°. The integrated form of the van't Hoff equation may then be written as

$$\ln\left(\frac{K_2}{K_1}\right) = \frac{\left(-\Delta H_R^{\circ}\right)}{R_g}\left(\frac{1}{T_2} - \frac{1}{T_1}\right) \tag{2.54}$$

The best value of ΔH_R° to use is the value calculated at the average of T_1 and T_2.

2.3.4 Computing Equilibrium Composition

The quantity usually desired in an equilibrium calculation is the mixture composition when equilibrium is achieved at a given temperature and pressure. This calculation has minor variations for liquid or gas mixtures, but the principles are the same in both cases. Consider the generic form of the equation for the equilibrium constant developed in Section 2.3.1:

$$K = \prod_{j=1}^{n} \hat{a}_j^{\nu_j} \tag{2.55}$$

Recall that v_j is negative for reactants and positive for products. Also recall that the activity, $\hat{a}_j$, is defined as the fugacity ratio $\hat{f}_j/f_j^\circ$. The fugacity of component j in solution, $\hat{f}_j$, can be defined in terms of the pure component fugacity, f_j, and an activity coefficient, γ_j, that is

$$\hat{f}_j = \gamma_j z_j f_j \tag{2.56}$$

where z_j is the mole fraction of j in the solution. The activity coefficient may be a function of both temperature and composition and, in the general case, is usually determined experimentally. Therefore, the activity of component j in solution can be written as

$$\hat{a}_j = \frac{\gamma_j z_j f_j}{f_j^\circ} \tag{2.57}$$

Thus, the equilibrium constant is now written explicitly in terms of the mixture mole fractions:

$$K = \prod_{j=1}^{n} \left(\frac{\gamma_j z_j f_j}{f_j^\circ} \right)^{v_j} \tag{2.58}$$

Provided that K, γ_j, and f_j are known, then z_j, the mole fraction at equilibrium, can be computed. The difference for liquid-, gas-, and solid-phase reactions are given below.

2.3.4.1 Liquid-Phase Reactions

For liquid-phase reactions, the usual convention is to denote liquid-phase mole fractions as x_j, that is, set $z_j = x_j$. For most liquids, the fugacity is not a strong function of pressure and therefore the fugacity ratio is approximately written as

$$\frac{f_j}{f_j^\circ} \approx 1 \tag{2.59}$$

For ideal solutions, the activity coefficient has a value of one, $\gamma_j = 1$. For nonideal solutions, $\gamma_j \neq 1$ and furthermore it is a strong function of x_j. Its value is computed from relationships such as the van Laar, Margules, and Wilson equations. For nonideal solutions, therefore

$$a_j = \gamma_j x_j \tag{2.60}$$

and

$$K = \prod_{j=1}^{n} (x_j \gamma_j)^{v_j} \tag{2.61}$$

The calculation of the equilibrium composition for a liquid reacting mixture requires equations for the activity coefficients of every species present.

2.3.4.2 Gas-Phase Reactions

For gaseous reactions, the usual convention is to denote gas-phase mole fractions as y_j; thus, $z_j = y_j$. The fugacity of component j in solution, $\hat{f}_j$, can be written as

$$\hat{f}_j = \gamma_j y_j f_j = \gamma_j y_j \left(\frac{f_j}{P}\right) P \tag{2.62}$$

where P is the total pressure. The fugacity coefficient, ϕ_j, is defined as $\phi_j = f_j/P$ and may be determined from an equation of state (e.g., Peng–Robinson or Redlich–Kwong equations). Thus, the fugacity of component j in solution is

$$\hat{f}_j = \gamma_j y_j \phi_j P \tag{2.63}$$

Most gases form ideal solutions for which $\gamma_j = 1$, so

$$K = \prod_{j=1}^{n} \left(\frac{y_j \phi_j P}{f_j^\circ}\right)^{v_j} \tag{2.64}$$

Equation 2.64 can be written as

$$K = \frac{K_\phi K_y}{K_{f^\circ}} P^{\sum v_j} \tag{2.65}$$

where

$$K_\phi = \prod_{j=1}^{n} \phi_j^{v_j} \quad K_y = \prod_{j=1}^{n} y_j^{v_j} \quad K_{f^\circ} = \prod_{j=1}^{n} \left(f_j^\circ\right)^{v_j}$$

When f_j° is equal to 1 bar, then $K_{f^\circ} = 1$. Note that for this definition of standard state, the units of pressure in the previous equations must be in bar. If P and T are known, K and K_ϕ can be computed; hence K_y can be calculated and thereby the composition at equilibrium. For the special case of an ideal gas, $\phi_j = 1$, $P_j = \hat{f}_j = y_j P$ and thus $K_\phi = 1$. Then

$$K = K_y P^{\sum v_j} = \prod_{i=1}^{n} P_j^{v_j} = K_P \tag{2.66}$$

Many gas-phase reactions in industrial reactors are carried out at pressures near atmospheric, and in such cases, assuming that the gas behaves like an ideal gas is often appropriate.

Example 2.6

Consider the gas-phase reaction of CO and H_2 to form methanol:

$$CO + 2H_2 \rightleftharpoons CH_3OH$$

For a stoichiometric feed composition, perform the following calculations:

a. Find the equilibrium conversion at 298 K and 1 bar, assuming ideal gases.
b. Find the equilibrium conversion at 515 K and 100 bar, assuming ideal gases.
c. Find the equilibrium conversion at 515 K and 100 bar, assuming real gases. For the real gas case, we must compute ϕ using, for example, the method of corresponding states. The critical properties, reduced properties, and values of ϕ are given in the following table:

	T_c (K)	P_c (atm)	T_R (K)	P_R (atm)	ϕ
CH_3OH	512.6	81.0	1.00	1.235	0.54
CO	132.9	35.0	3.88	2.86	0.98
H_2	33.2	13.0	15.5	7.69	1.05

SOLUTION

With $K_f^o = 1$, Equation 2.65 is written as

$$K = K_y K_\phi P^{\sum v_j}$$

The pressure P is in bar, and

$$K_y = \prod_{j=1}^{n} y_j^{v_j} \quad K_\phi = \prod_{j=1}^{n} \phi_j^{v_j}$$

To calculate the value of the equilibrium constant, first compute the free energy change of reaction. Using the information in Appendix 1, the following values can be calculated:

$$\Delta G_{R,298}^{\cdot} = -25,200 \text{ J/mol} \quad \text{and} \quad \Delta H_{R,298}^{\cdot} = -90,700 \text{ J/mol}$$

The equilibrium constant is calculated from Equation 2.51:

$$\Delta G_R^{\cdot} = -R_g T \ln K$$

Therefore

$$-25,200 = -(8.314)(298)\ln K$$

which gives

$$K_{298} = 2.614 \times 10^4$$

This value is the equilibrium constant at 298 K. To perform the equilibrium calculations, it is necessary to take a basis for the reaction. Let the basis be a starting mixture of 1 mol of CO and 2 mol of H_2. At 298 K and 1 bar, the mixture is given as an ideal gas and therefore $K_\phi = 1$. The summation of the stoichiometric coefficients is

$$\sum_{j=1}^{n} v_j = 1 - 2 - 1 = -2$$

The equilibrium constant can thus be expressed as

$$K = \frac{y_{CH_3OH}}{y_{CO} y_{H_2}^2} P^{-2}$$

Taking the starting basis of 1 mol of CO, let ε moles react. The number of moles of each component at equilibrium can then be expressed by

$$N_{CO} = 1 - \varepsilon$$

$$N_{H_2} = 2(1 - \varepsilon)$$

$$N_{CH_3OH} = \varepsilon$$

$$N_t = 3 - 2\varepsilon$$

Substitution of these values into the expression for the equilibrium constant gives

$$K = \frac{\varepsilon}{4(1-\varepsilon)^3}(3-2\varepsilon)^2 P^{-2} \quad \text{or} \quad 2.614 \times 10^4 = \frac{\varepsilon(3-2\varepsilon)^2}{4(1-\varepsilon)^3}$$

Solve by iteration for the unknown value to get $\varepsilon = 0.9783$. The conversion of CO is 97.83%. This value is the solution to part (a).

To solve part (b), it is first necessary to calculate the value of the equilibrium constant at 515 K. To do this calculation, Equation 2.53 is used.

$$\frac{\partial(\ln K)}{\partial T} = \frac{\Delta H_R^\circ}{R_g T^2}$$

The first step is compute an equation for ΔH_R°. Recall that the temperature dependence is given by Equation 2.49:

$$\frac{\partial(\Delta H_R^\circ)}{\partial T} = \Delta C_P$$

The change in heat capacity between reactants and products is

$$\Delta C_P = \sum_{j=1}^{n} v_j C_{P,j}$$

From Appendix 1, the following values are obtained for the heat capacity polynomial:

$$C_P = a + bT + cT^2 + dT^3$$

Compound	a	b	c	d
CO	28.11	0.1675×10^{-2}	0.5363×10^{-5}	-2.218×10^{-9}
H_2	29.06	-0.1913×10^{-2}	0.3997×10^{-5}	-0.869×10^{-9}
CH_3OH	19.02	9.137×10^{-2}	-1.216×10^{-5}	-8.030×10^{-9}

Substitution of these coefficients gives the equation

$$\Delta C_P = -67.21 + 9.352 \times 10^{-2}T - 2.552 \times 10^{-5}T^2 - 4.074 \times 10^{-9}T^3$$

The general equation for ΔH_R° can be written as

$$\Delta H_{R,T}^\circ = \Delta H_{R,298}^\circ + \int_{298}^{T} \Delta C_P \, dT \tag{2.67}$$

Substitution of the heat capacity and enthalpy of formation data, followed by integration gives

$$\Delta H_{R,T}^\circ = -66,326 - 67.21T + 4.676 \times 10^{-2}T^2 - 8.506 \times 10^{-6}T^3 - 1.018 \times 10^{-9}T^4 \tag{2.68}$$

Equation 2.68 for ΔH_R° can now be used to develop an equation for $\ln K$ using the van't Hoff equation (Equation 2.53):

$$\ln K_T = \ln K_{298} + \int_{298}^{T} \frac{\Delta H_R^\circ}{R_g T^2} \, dT \tag{2.69}$$

Substitution of the numerical values and integration gives

$$\ln(K_T) = 27.74 + \frac{7978}{T} - 8.084 \ln T + 5.624 \times 10^{-3}T - 5.115 \times 10^{-7}T^2 - 4.081 \times 10^{-11}T^3 \tag{2.70}$$

At a temperature of 515 K, Equation 2.70 gives a value for the equilibrium constant of $K_{515} = 1.120 \times 10^{-2}$. Therefore, substitution gives the following expression for the conversion at equilibrium:

$$1.120 \times 10^{-2} = \frac{\varepsilon(3 - 2\varepsilon)^2}{4(1 - \varepsilon)^3}(100)^{-2} \tag{2.71}$$

Solving Equation 2.71 by trial and error gives the result $\varepsilon = 0.8529$. By increasing T and P, we reduce equilibrium yield to 85.29%. However, this value assumes that the compounds are ideal gases, which is not true at this temperature and pressure.

For part (c), it is necessary to add the effect of the fugacity coefficients of the components. For each component, the value of ϕ is given. These values were computed using the method of corresponding states. Substituting these values gives

$$K_\phi = \frac{0.54}{(0.98)(1.05)^2} = 0.500 \tag{2.72}$$

When Equation 2.72 is included in the equilibrium calculation, we obtain the expression

$$1.120 \times 10^{-2} = \frac{\varepsilon(3-2\varepsilon)^2}{4(1-\varepsilon)^3} \times 0.500 \times 100^{-2} \tag{2.73}$$

Solution of Equation 2.73 gives $\varepsilon = 0.8857$. The conversion at this T and P is 88.57% using the real gas assumption.

2.3.4.3 Solid-Phase Reactions

When a component in reacting mixture is present as a pure solid, its activity has a value of one at low-to-moderate pressure. Therefore, if a mixture of gases and pure solids are involved in a reaction, the presence of the solid does not affect the overall gas-phase composition at equilibrium. Consider the reaction between ferrous oxide and carbon monoxide to produce metallic iron and carbon dioxide. The overall reaction is given by the following equation, where the subscript (s) denotes a solid and the subscript (g) denotes a gaseous component:

$$FeO_{(s)} + CO_{(g)} \rightleftharpoons Fe_{(s)} + CO_{2(g)} \tag{2.74}$$

The equilibrium constant can be written in terms of activities at equilibrium:

$$K = \left(\frac{\hat{a}_{Fe}\hat{a}_{CO_2}}{\hat{a}_{FeO}\hat{a}_{CO}} \right)_{equilibrium} \tag{2.75}$$

If both FeO and Fe are present as pure components (i.e., they do not form a solution), their activities are both one. For ideal gases, the activities depend on the partial pressures. The equilibrium composition can thus be written in terms of the gas-phase partial pressures, that is

$$K = \frac{P_{CO_2}}{P_{CO}} \tag{2.76}$$

Because the equilibrium composition does not include the solid activities, the amount of FeO and Fe present at equilibrium cannot be determined directly from the equilibrium constant alone, but rather requires a knowledge of the starting composition as well. The ratio of Fe and FeO in the final mixture can be calculated from a mole balance, as illustrated in Example 2.7.

Example 2.7

Consider the reduction of FeO by CO gas. The overall reaction is given as follows, where the subscript (s) denotes a solid and the subscript (g) denotes a gaseous component:

$$FeO_{(s)} + CO_{(g)} \rightleftharpoons Fe_{(s)} + CO_{2(g)} \tag{2.77}$$

At 1273 K, the equilibrium constant for the reaction has a value of 0.4. At this temperature, pure CO is added to FeO and the mixture is allowed to reach equilibrium at a total pressure of 1 bar. Calculate the number of moles of Fe formed for each mole of CO initially in the mixture.

SOLUTION

The assumption is made that the gases are ideal. Take the initial total pressure as 1 bar of CO, and let x denote the partial pressure of CO at equilibrium. The equation that describes the equilibrium composition of the gas phase is then

$$K = \frac{P_{CO2}}{P_{CO}} = \frac{1-x}{x} = 0.4 \tag{2.78}$$

Solving for x gives a value of $x = 0.714$; therefore, the final equilibrium partial pressure of CO is 0.714 bar. The fraction of the CO converted is equal to $(1 - x)$, or 0.286. The partial pressure is directly proportional to the number of moles, and therefore at equilibrium, each mole of CO initially present will react to leave 0.714 mol of CO and 0.286 mol of CO_2. Thus, from the reaction stoichiometry, for each mole of CO initially present, 0.286 mol of Fe will be formed.

   Note that the preceding calculation gives no indication of the actual amounts of either FeO or Fe present at equilibrium. Furthermore, it is assumed in the calculation that sufficient FeO is initially present so that, when equilibrium is reached, both FeO and Fe are present. Suppose, for example, that the initial mixture contained 1 mol of CO and 0.5 mol of FeO. The calculation tells us that 0.286 mol of Fe is formed for each mole of CO initially present, and thus it is evident that the final solid composition will be 0.286 mol of Fe and 0.214 mol of FeO. If, on the other hand, 1 mol of CO was to be added to 0.25 mol of FeO, then equilibrium cannot be established because <0.285 mol of FeO is available. In this case, all of the FeO would react to Fe and the final gas-phase composition would be 0.75 mol of CO and 0.25 mol of CO_2.

2.3.5 Equilibrium Compositions with Multiple Reactions

When more than one reaction occurs in a mixture, the computation of the equilibrium composition is more complex than the single reaction case. In the multiple reaction case, all of the possible reactions must simultaneously be in equilibrium. Calculations of equilibrium compositions in such systems can be quite complex, and require a knowledge of which products can reasonably be expected to occur. For example, in the illustration used at the beginning of the chapter, in which the possible reactions of synthesis gas were discussed, it was seen that a myriad of possible products could be formed. Indeed, in this example, the possibilities are literally almost infinite. Two methods are commonly used to solve multireaction equilibrium problems. In the first method, the minimum number of independent reactions required to describe the system is written, then an equilibrium expression is written for each one. The resulting set of nonlinear equations is then solved numerically. The second method is based on the minimization of the Gibbs free energy for the system as a whole. Details of these methods can be found in, for example, Smith et al. (1996) and Sandler (1999).

2.4 Summary

Chemical reaction engineering encompasses a broad range of practical applications which have seen the development of a wide variety of reactor types. The analysis of chemical reactors draws on many of the fundamentals of chemical engineering, and, as such, is often represented as being the pinnacle of the education of a chemical engineer. Although there are many types of chemical reactor, a solid foundation in the understanding of the analysis of the basic ideal reactor provides a good starting point for the analysis of more complex types.

One of the more important underpinnings of chemical reactor analysis is the science of thermodynamics. Thermodynamics is employed in calculating enthalpy changes in reactors as well as equilibrium compositions.

More detailed information on thermodynamics can be found in the references given in the "Recommended Readings" in frontmatter.

PROBLEMS

2.1 Ammonia is an important raw material used in the manufacture of, among other things, fertilizers. Ammonia can be produced by the catalytic reaction between nitrogen and hydrogen. The overall reaction is

$$\tfrac{1}{2}N_2 + \tfrac{3}{2}H_2 \rightleftharpoons NH_3$$

Industrially, this reaction is carried out at elevated pressures where the gases are not ideal. Calculate the equilibrium composition for this operation at a reactor pressure of 1000 bar and temperature of 723 K. The feed consists of 21 mol% N_2, 63 mol% H_2, and 16 mol% inert Ar. The fugacities of pure hydrogen, nitrogen, and ammonia are 1350, 1380, and 860 bar, respectively. Other thermodynamic data are available in Appendix 1.

2.2 Ethene (ethylene) is used in the production of many useful chemicals, including polyethylene. Ethene is produced by the thermal cracking of ethane, which desaturates the ethane molecule, producing ethene and hydrogen. The overall reaction is

$$C_2H_6 \rightleftharpoons C_2H_4 + H_2$$

a. Calculate the equilibrium fractional conversion of ethane cracked for a reactor feed of pure ethane at a pressure of 120 kPa and a temperature of 1023 K. Use the data in Appendix 1. At this temperature and pressure, the fugacity coefficients of the three components are all equal to one.

b. Use the data in Appendix 1 to calculate the amount of heat transfer required per mole of ethane cracked for the operation in part (a) to maintain the temperature at 1023 K. Is the heat transferred to or from the reactor?

2.3 The manufacture of sulfuric acid required sulfur trioxide. In this process, sulfur is first burned with air to produce sulfur dioxide. The sulfur dioxide is then oxidized in a catalytic reactor to produce sulfur trioxide. The gas from a sulfur burner consists of 8 mol% SO_2, 11 mol% O_2, and 81 mol% N_2. This gaseous mixture is passed to a catalytic reactor where SO_2 is oxidized to SO_3. The reactor exit temperature is 500°C and the pressure is 1 bar. The equilibrium constant at 500°C is 85 and fugacity coefficients are essentially equal to one. Calculate the reactor exit conversion of SO_2 if equilibrium is established. The reaction is

$$SO_2 + \tfrac{1}{2}O_2 \rightleftharpoons SO_3$$

Repeat the calculation for a total reactor pressure of 2 bar.

2.4 Consider the oxidation of sulfur dioxide to sulfur trioxide:

$$SO_2 + \tfrac{1}{2}O_2 \rightleftharpoons SO_3$$

The equilibrium constant is 85 at 500°C. The enthalpy of reaction is −98.2 kJ/mol of SO_2 converted, and may be assumed to be constant. The heat capacities of the species may be assumed to have constant values of

N_2	28.8 J/mol	O_2	32.6 J/mol
SO_2	47.7 J/mol	SO_3	64.0 J/mol

A mixture of SO_2, O_2, and N_2 enters an adiabatic oxidation reactor. The mixture reacts and comes to equilibrium. Calculate the final temperature of the mixture if the initial temperature is 400°C. The initial composition of the mixture is 8 mol% SO_2, 11 mol% O_2, and 81 mol% N_2. The total pressure is 100 kPa (1 bar).

2.5 Palladium catalyst is used in the catalytic oxidation of methane. The active form of palladium is the oxide form, which can decompose to palladium metal and oxygen gas. The overall reaction is

$$PdO \rightleftharpoons Pd + \tfrac{1}{2}O_2$$

Palladium oxide and palladium metal are solids. The free energy change of reaction is

$$\frac{\Delta G_R^\circ}{R_g} = 11{,}273 + T[2.89\log_{10}(T) - 18.57]$$

If the oxygen partial pressure above the catalyst is 0.2 bar, calculate the temperature at which the PdO catalyst will decompose to Pd. Repeat the calculation for an oxygen pressure of 1 bar.

2.6 The steam reforming reaction, that is, the reaction between water vapor and methane, typically involves multiple reactions. For example, the following two reactions can be used to describe the process:

$$CH_4 + H_2O \rightleftharpoons CO + 3H_2$$

$$CH_4 + 2H_2O \rightleftharpoons CO_2 + 4H_2$$

Assume that both of these reactions achieve equilibrium at 593 K. The equilibrium constants at this temperature for the two reactions are 0.41 and 1.09, respectively. Calculate the equilibrium composition if the starting composition is 5 mol of steam and 1 mol of methane at a pressure of 2 bar.

2.7 Many solids decompose to yield another solid and a gas. Consider the decomposition of limestone (calcium carbonate), which produces lime (calcium oxide) and carbon dioxide gas. The overall reaction is

$$CaCO_{3(s)} \rightleftharpoons CaO_{(s)} + CO_{2(g)}$$

The decomposition reaction will only occur if the activity of the gas in contact with the solid has a value lower than the equilibrium value. The pressure of the gas at which the decomposition of the solid occurs is called the decomposition pressure.

a. Calculate the decomposition pressure of limestone at 1000 K.

b. Calculate the temperature that corresponds to a decomposition pressure of 1 bar.

The standard free energy change of the reaction at 298.15 K is 134.3 kJ/mol. The standard enthalpy change of reaction at 298.15 K is 182.1 kJ/mol. The heat capacities are

$$\left(C_P\right)_{CaCO_3} = 82.26 + 4.97 \times 10^{-2}T - \frac{1.286 \times 10^6}{T^2} \, \text{J/mol} \cdot \text{K}$$

$$\left(C_P\right)_{CaO} = 41.8 + 2.02 \times 10^{-2}T - \frac{4.51 \times 10^5}{T^2} \, \text{J/mol} \cdot \text{K}$$

References

Reid, R.C., J.M. Prausnitz, and B.E. Poling, 1987, *The Properties of Gases and Liquids*, 4th Ed., McGraw-Hill, New York.

Sandler, S.I., 1999, *Chemical and Engineering Thermodynamics*, 3rd Ed., John Wiley and Sons, Toronto.

Smith, J.M., H.C. van Ness, and M.M. Abbot, 1996, *Introduction to Chemical Engineering Thermodynamics*, 5th Ed., McGraw-Hill, New York.

3

Mole Balances in Ideal Reactors

Chapter 1 gave an introduction and overview of chemical reaction engineering and presented an introduction to some of the common reactor types. This chapter begins the development of the methodology required for the analysis of chemically reacting systems. The material presented in this chapter describes the development of the mole balance equations required for the analysis of three types of ideal reactors. To preserve simplicity at this stage in the development, temperature effects are not considered in this chapter. Three conditions must therefore be satisfied. First, the reacting fluid must have a uniform temperature at all locations in the reactor. Second, the temperature at every point in the reactor must stay constant with time. Third, when a flow reactor is considered, the inlet and outlet streams must also have the same temperature. When these three constraints are met, it is not necessary to solve an energy balance equation to calculate the extent of reaction because the temperature is known for the entire system.

The discussion is furthermore limited to ideal reactors. These reactors have a well-defined flow pattern and a known level of mixing. One of two extremes of mixing, either perfect mixing or zero mixing, is assumed. These mixing patterns were introduced in Chapter 1 as the perfect mixing and plug flow assumptions. In this chapter, these flow patterns are explained in more detail and the effect that each assumption has on the development of the mole balance is shown. When an ideal reactor is assumed, pressure drops across the reactor are ignored and because the flow pattern within the reactor is also assumed to be known, it is not necessary to solve the momentum balance equation to deduce the velocity and pressure distribution in the reactor.

With the temperature, pressure, and velocity patterns known, the only remaining variable in the reactor is the concentration distribution of the reactants and products. In the following sections, the mole balance equations are derived for the three major types of ideal reactors. These three reactors were introduced in Chapter 1 and are the perfectly mixed batch reactor, the perfectly mixed continuous stirred tank reactor, and the plug flow reactor (PFR). We start by introducing the general mole balance equation, and then show how variants of the general equation are used to describe the different reactor types. The mole balance equations are developed for single- and multiple-reactor systems, and for single and multiple reactions. Both steady-state and transient systems are considered in this chapter.

3.1 General Mole Balance Equation

In chemical reaction analysis, the conservation equations are usually based on mole balances rather than mass balances. Although mass can be used as a basis for calculations involving reactors, it is usually considered to be more convenient to work in terms of moles. Mass is always conserved in a chemical reaction, but the total number of moles may change. When the mole balance is performed in reactor analysis, it is usually done with respect to a single species.

We start the derivation of the mole balance equation by developing the general mole balance over an incremental reactor volume element. Consider an incremental volume element of a chemical reactor, ΔV, over an incremental time, Δt. Let j be some species in the reactor. The species j flows into the volume element at some rate and flows out at another rate. Furthermore, some moles may be created or destroyed by reaction inside the volume element and, in the transient case, some accumulation of moles may occur as well. The mole balance for some component j over the volume increment can be expressed in words by

$$\begin{bmatrix} \text{Molar flow} \\ \text{of } j \\ \text{into } \Delta V \end{bmatrix} - \begin{bmatrix} \text{Molar flow} \\ \text{of } j \text{ out} \\ \text{of } \Delta V \end{bmatrix} + \begin{bmatrix} \text{Rate of formation} \\ \text{of } j \text{ by reaction} \\ \text{in } \Delta V \end{bmatrix} = \begin{bmatrix} \text{Rate of} \\ \text{accumulation} \\ \text{of } j \text{ in } \Delta V \text{ during } \Delta t \end{bmatrix}$$

Let F_j represent the molar flow rate of component j, and r_j the rate of formation of component j. The rate of formation of j is expressed as a rate per volume, so the rate of formation of j in the volume ΔV is the product of the rate and the volume. The rate of accumulation of moles in the volume element can be expressed in terms of the number of moles of j, denoted N_j, at the beginning and end of a time increment, Δt. Therefore, the mole balance equation can be represented mathematically as

$$F_j\big|_{in} - F_j\big|_{out} + r_j \Delta V = \frac{\left[(N_j)_{t+\Delta t} - (N_j)_t \right]}{\Delta t} = \frac{\Delta N_j}{\Delta t} \tag{3.1}$$

Equation 3.1 represents a general mole balance over a discrete volume element. Depending on the type of reactor to which Equation 3.1 is applied, different mole balance equations result. The final form depends on whether or not the concentration in the reactor varies with position or is uniform throughout the entire volume. The following sections show how the specific mole balances for the perfectly mixed batch reactor, the PFR, and the perfectly mixed continuous reactor are developed.

3.2 Perfectly Mixed Batch Reactor

A batch reactor does not have any inlet or outlet streams during operation. Material is added to the vessel, the reaction occurs, then the vessel is emptied: therefore, $F_j = 0$ for both inlet and outlet. Batch reactors can be operated in a variety of ways; in the following analysis, it is assumed that the reaction mixture is homogeneous, that is, all the contents are in a single phase. Perfect mixing is also assumed, which means that there are no spatial variations in temperature and composition. Eliminating the flow terms from Equation 3.1 gives

$$r_j \Delta V = \frac{\left[(N_j)_{t+\Delta t} - (N_j)_t \right]}{\Delta t} = \frac{\Delta N_j}{\Delta t} \tag{3.2}$$

If the concentration of all species is the same at every point in the reactor volume, the incremental volume can be replaced by the total volume. Furthermore, the discrete time change can be replaced by the differential time element by taking the limit of an infinitesimally small time step. With these changes, Equation 3.2 is written as a differential equation:

$$r_j = \frac{1}{V} \frac{dN_j}{dt} \tag{3.3}$$

Equation 3.3 is valid for all homogeneous perfectly mixed batch reactors. The volume, V, is the total volume of the reaction mixture and may change during the course of the reaction.

Note that r_j is the *rate of formation* of component j. Equation 3.3 can also be (and quite often is) written in terms of the *rate of disappearance* of a reactant, denoted A. The rate of disappearance of a species is written as the minus of the rate of formation, that is as $(-r_A)$. Note that if reactant A is being consumed in a reaction, its rate of disappearance, $(-r_A)$, is a *positive* number. The mole balance for species A in terms of its rate of disappearance is then

$$(-r_A) = -\frac{1}{V} \frac{dN_A}{dt} \tag{3.4}$$

The following sections show how the batch reactor mole balance is applied to both constant- and variable-volume batch reactors.

3.2.1 Constant-Volume Batch Reactor

The constant-volume batch reactor is perhaps the most widely used type of batch reactor. In this mode of operation, the volume of the reaction mixture does not change with time. Most liquid-phase reactions do not undergo significant volume changes with reaction, although there are some exceptions. A gas-phase reaction that occurs in a reactor with fixed walls will also be constant in volume. If the moles change on reaction for a gas-phase reaction, the pressure inside the reactor will change to compensate for the change in number of moles.

In a constant-volume batch reactor, it is common to perform the mole balance calculations in terms of concentrations, rather than the number of moles. The moles of species A is simply the product of the volume and the concentration, that is

$$N_A = C_A V \tag{3.5}$$

Substitution of Equation 3.5 into Equation 3.4, and assuming constant volume, gives

$$(-r_A) = -\frac{1}{V} \frac{d(VC_A)}{dt} = -\frac{dC_A}{dt} \tag{3.6}$$

It is important to emphasize that Equation 3.6 is only valid for a constant-volume batch reactor, is not generally applicable to all reactors, and is *not* a definition of reaction rate.

Because the volume, and hence the mass density, is constant, the conversion of a reactant A may be defined using the concentrations. The fractional conversion of A for a constant-volume process is defined as

$$X_A = \frac{C_{A0} - C_A}{C_{A0}} \tag{3.7}$$

In Equation 3.7, C_{A0} is the initial concentration of A in the reactor, and C_A is the concentration at time t. Equation 3.7 can be rearranged to give an explicit equation for C_A:

$$C_A = C_{A0}(1 - X_A) \tag{3.8}$$

Both sides of Equation 3.8 can be differentiated with respect to time to give

$$-\frac{dC_A}{dt} = C_{A0}\frac{dX_A}{dt} \tag{3.9}$$

Substitution of Equation 3.9 into Equation 3.6 gives an expression for the reaction rate in terms of the fraction conversion of reactant A:

$$(-r_A) = -\frac{dC_A}{dt} = C_{A0}\frac{dX_A}{dt} \tag{3.10}$$

Equation 3.10 is the basic mole balance for the constant-volume batch reactor. The analysis of a constant-volume batch reactor is illustrated in Example 3.1 following Section 3.2.2.

3.2.2 Variable-Volume Batch Reactor

In a variable-volume batch reactor, the volume of the reacting mixture changes with time. This change can result from an imposed volume change by an external force. For example, a gas-phase reaction might be conducted in a cylinder fitted with a piston. During operation, the piston might be moved by the application of an external force. Alternatively, during the course of reaction, a gas-phase reaction might result in a change in the number of moles present in the reactor. If the reactor is operated in such a way that the pressure of the reactor is held constant, the volume of the reactor will adjust. In a variable-volume reactor, the conversion cannot be easily defined in terms of the concentration because a change in the volume yields a change in concentration even if no reaction has occurred. The conversion must be defined in terms of the number of moles. For reactant A

$$X_A = \frac{N_{A0} - N_A}{N_{A0}} \tag{3.11}$$

Equation 3.11 can be rearranged to the form

$$N_A = N_{A0}(1 - X_A) \tag{3.12}$$

The mole balance equation for the variable-volume well-mixed batch reactor is simply the general equation developed earlier. That is, in terms of the rate of disappearance of species A, the mole balance is

$$\left(-r_A\right) = -\frac{1}{V}\frac{dN_A}{dt} \tag{3.13}$$

The reactor volume, V, must be related to either time or the moles of A present at any time. The use of the batch reactor mole balance for the variable volume case is illustrated in Example 3.1.

Example 3.1

An irreversible gas-phase reaction is carried out in a batch reactor at 25°C. The reaction is

$$A + B \rightarrow C$$

The rate of disappearance of reactant A is given by the following function:

$$\left(-r_A\right) = kC_A C_B \, \text{mol/m}^3\text{s}, \quad \text{where} \quad k = 3.5 \times 10^{-5} \, \text{m}^3/\text{mol s} \tag{3.14}$$

The reactor is filled with an equal number of moles of A and B. The initial concentration of A is $C_{A0} = 50.0 \, \text{mol/m}^3$. Calculate the fractional conversion of A after 500 s for the following three types of batch reactors. Assume ideal gases.

    a. Constant-volume reactor
    b. Constant-pressure reactor
    c. Variable-volume reactor where $V = V_0\,(1 + 0.001t)$

SOLUTION
The initial concentrations of A and B are equal, and 1 mol of A reacts for each mole of B that reacts. It follows that the concentrations of A and B are equal at all times and we may write the rate function as

$$\left(-r_A\right) = kC_A^2, \quad \text{where} \quad C_A = \frac{N_A}{V} \tag{3.15}$$

The solution for each of the three cases is obtained by application of the appropriate form of the mole balance equation.

a. Constant-Volume Batch Reactor (CVBR)
The first step is to substitute the rate function, Equation 3.15, into the mole balance equation. As the volume of the reactor is constant, Equation 3.6 is used:

$$\left(-r_A\right) = -\frac{dC_A}{dt} = kC_A^2 \tag{3.16}$$

Equation 3.16 contains one dependent variable, the concentration of A. Equation 3.16 can be integrated over a time interval of 0 to t, during which the concentration changes from C_{A0} to C_{Af}. The integral equation is

$$\int_0^t k\,dt = -\int_{C_{A0}}^{C_{Af}} \frac{dC_A}{C_A^2} \tag{3.17}$$

Integrating Equation 3.17 between the limits gives

$$kt = \frac{1}{C_A}\Big|_{C_{A0}}^{C_{Af}} = \frac{1}{C_{Af}} - \frac{1}{C_{A0}} \tag{3.18}$$

Substitution of the numerical values of the variables leads to

$$\left(3.5 \times 10^{-5}\right)\mathrm{m}^3/\mathrm{mol}\,\mathrm{s}(500)\mathrm{s} = \frac{1}{C_{Af}} - \frac{1}{50}\,\mathrm{m}^3/\mathrm{mol} \tag{3.19}$$

Note that the units are consistent. Solving Equation 3.19 for the final concentration gives a value of $C_{Af} = 26.66\ \mathrm{mol/m}^3$. The final conversion can be computed from this value as follows:

$$X_A = \frac{C_{A0} - C_{Af}}{C_{A0}} = \frac{50 - 26.66}{50} = 0.467$$

Note that this definition of conversion is only valid because the volume of the reactor is constant.

b. Constant-Pressure Batch Reactor (CPBR)

The reaction being considered is a gas-phase reaction which undergoes a change in moles on reaction. If both the pressure and the temperature of the reactor are to be held constant, the volume of the reactor must change as the reaction proceeds. The mole balance equation to use is therefore Equation 3.13. Substitute the rate expression, Equation 3.15, into Equation 3.13, to give

$$-\frac{1}{V}\frac{dN_A}{dt} = kC_A^2 \tag{3.20}$$

Equation 3.20 contains three independent variables: N_A, C_A, and V. The equation must be expressed in terms of one of these variables only. It is best to write the mole balance in terms of the number of moles of A. The concentration and the number of moles are related by the volume of the reaction mixture; hence

$$C_A = \frac{N_A}{V} \quad \text{which gives} \quad -\frac{1}{V}\frac{dN_A}{dt} = \frac{kN_A^2}{V^2} \tag{3.21}$$

The volume occupied by a gas is a function of the total number of moles present, the temperature, and the pressure. The total number of moles present is simply the sum of all the moles of the different species in the system:

$$N_t = \sum N_i = N_A + N_B + N_C \tag{3.22}$$

The number of moles of A, B, and C are related by the reaction stoichiometry. We want to express the total number of moles in terms of the number of moles of A only. The easiest way to represent this is to build a stoichiometric table, as follows:

Species	Initial Moles	Moles after Reaction
A	N_{A0}	N_A
B	$N_{B0} = N_{A0}$	$N_B = N_{B0} - (N_{A0} - N_A) = N_A$
C	$N_{C0} = 0$	$N_C = N_{A0} - N_A$
Total	$N_t = 2N_{A0}$	$N_t = N_{A0} + N_A$

If we assume that the mixture behaves like an ideal gas, the volume, pressure, and the total number of moles can be related using the ideal gas law:

$$V = N_t \frac{R_g T}{P}$$

The initial volume, V_0, and the volume for any given extent of reaction, V, are therefore, respectively, given by

$$V_0 = 2N_{A0} \frac{R_g T}{P} \quad \text{and} \quad V = \left(N_{A0} + N_A\right) \frac{R_g T}{P}$$

Taking the ratio of the two volumes yields an expression for the volume as a function of the extent of reaction:

$$\frac{V}{V_0} = \left(\frac{N_{A0} + N_A}{2N_{A0}}\right) \quad \text{or} \quad V = V_0 \left(\frac{N_{A0} + N_A}{2N_{A0}}\right) \tag{3.23}$$

Substituting Equation 3.23 into Equation 3.21 gives

$$-\frac{dN_A}{dt} = \frac{2N_{A0}kN_A^2}{V_0\left(N_{A0} + N_A\right)} \tag{3.24}$$

Equation 3.24 can be rearranged and expressed in integral form:

$$\frac{2N_{A0}k}{V_0} \int_0^t dt = \int_{N_{A0}}^{N_{Af}} \left(\frac{N_{A0} + N_A}{N_A^2}\right) dN_A = \int_{N_{A0}}^{N_{Af}} \left(\frac{N_{A0}}{N_A^2} + \frac{1}{N_A}\right) dN_A$$

These integrals have analytical solutions:

$$-2\frac{N_{A0}k}{V_0}t \Big|_0^t = N_{A0}\left[-\frac{1}{N_A}\right]_{N_{A0}}^{N_{Af}} + \ln N_A \Big|_{N_{A0}}^{N_{Af}} \tag{3.25}$$

Substitution of the limits in Equation 3.25 gives

$$-2\frac{N_{A0}k}{V_0}t = -2kC_{A0}t = N_{A0}\left[\frac{1}{N_{A0}} - \frac{1}{N_{Af}}\right] + \ln\left(\frac{N_{Af}}{N_{A0}}\right) \tag{3.26}$$

Finally, substitution of the numerical values into Equation 3.26 gives

$$(-2)(3.5 \times 10^{-5})(50)(500) = 1 - \frac{N_{A0}}{N_{Af}} - \ln\left(\frac{N_{A0}}{N_{Af}}\right)$$

Solving for (N_{Af}/N_{A0}) by iteration gives a value of $(N_{Af}/N_{A0}) = 0.491$. This ratio is related to the fractional conversion as follows:

$$X_A = \frac{N_{A0} - N_{Af}}{N_{A0}} = 1 - \frac{N_{Af}}{N_{A0}} = 1 - 0.491 = 0.509$$

Note that the volume decreases during the course of the reaction, which gives a higher effective concentration, compared to the constant volume case, and therefore the average rate of reaction is higher. Ultimately, a higher fractional conversion is obtained for the same reaction time, compared to the constant volume case.

c. Imposed Volume Change with Time

In the final case, the reactor volume is changed using an imposed external means, so that the volume varies according to the following function in time:

$$V = V_0(1 + 0.001t) \tag{3.27}$$

In this case, the volume change is imposed in a defined manner, so both the reactor volume and pressure must change with time. The mole balance for this reactor can be obtained by substitution of Equation 3.27 for the reactor volume into the batch reactor mole balance, Equation 3.13, to get

$$\frac{dN_A}{dt} = -\frac{kN_A^2}{V_0(1 + 0.001t)} \tag{3.28}$$

Equation 3.28 can be written in integral form as

$$\int_{N_{A0}}^{N_{Af}} \frac{dN_A}{N_A^2} = -\frac{k}{V_0}\int_0^t \frac{dt}{(1 + 0.001t)} \tag{3.29}$$

Performing the integration

$$-\frac{1}{N_A}\Bigg|_{N_{A0}}^{N_{Af}} = -\frac{k}{V_0}\left(\frac{1}{0.001}\right)\ln(1 + 0.001t)\Big|_0^t \tag{3.30}$$

Substituting the limits into Equation 3.30 gives

$$\frac{1}{N_{Af}} - \frac{1}{N_{A0}} = \frac{k}{0.001V_0}\ln(1 + 0.001t) \tag{3.31}$$

Rearranging Equation 3.31

$$\frac{N_{A0}}{N_{Af}} - 1 = \frac{k}{0.001} \frac{N_{A0}}{V_0} \ln\left(1 + 0.001t\right) = \frac{kC_{A0}}{0.001} \ln\left(1.5\right) \tag{3.32}$$

Performing the final substitutions and solving give the final results

$$\frac{N_{A0}}{N_{Af}} = 1.7096 \quad \text{or} \quad \frac{N_{Af}}{N_{A0}} = 0.585 \quad \text{and so } X_{A0} = 0.415$$

Summary of the Three Cases

The results of the three cases are summarized below. Note that the highest conversion is achieved when the volume decreases during the reaction. A decreasing volume effectively results in a higher average concentration during the reaction, which gives a higher average rate of reaction.

a. Constant volume	$X_A = 0.467$
b. Constant pressure	$X_A = 0.509$
c. Variable volume and pressure	$X_A = 0.415$

This section has introduced the concept of the well-mixed batch reactor. With the assumptions used, it was seen that the mole balance is an ordinary first-order differential equation that relates the moles of the species in the reactor to time using a rate of reaction for those species. It is important to appreciate the differences that are obtained when a reactor is operated with either constant or variable volume. These differences are especially important when defining the fractional conversion of a reactant. Generally speaking, liquid-phase reactions are usually assumed to occur in a constant-volume environment, while gas-phase reactions may have either constant or variable volume. If the reactor walls are fixed and the reacting mixture fills the vessel entirely, the volume must be constant by definition.

3.3 Plug Flow Reactor

This section introduces the mole balance equation for the PFR operating at steady state. The PFR usually consists of a tube or a duct through which the reacting fluid flows. Although PFRs can be of any cross-sectional shape, they are commonly comprised of a circular tube of constant cross-sectional area. This version is often called a *tubular reactor*, although caution should be used with these labels. The term "tubular reactor" may also be used for a reactor of circular cross section which does not operate in plug flow.

3.3.1 Plug Flow Assumptions

The plug flow assumptions refer to the assumed nature of the fluid flow within the tube. The assumptions are described in the following:

1. Flat velocity profile along the reactor radius. When a fluid flows down a tube, the wall exerts a drag force on the fluid. This drag force is passed by adjacent fluid layers into the bulk of the fluid. The result is a velocity profile along the radius of

the tube, with the fluid velocity equal to zero at the wall and having a maximum value at the centerline. This profile is most pronounced in laminar flow, where the fully developed velocity profile in a circular duct is parabolic. As the flow becomes increasingly turbulent, the velocity profile will tend to become flatter at the center of the pipe, with a steeper drop off to a zero velocity at the wall. Therefore, the higher the mean velocity in the reactor, the flatter the velocity profile will become. Plug flow assumes that the velocity profile is completely flat, that is, there are no radial variations in the velocity. In other words, the velocity at any radial position is equal to the average velocity of the fluid. This assumption will cause the least error in fully developed turbulent flow, that is, when the Reynolds number is greater than 10,000.

2. No radial variations in concentration or temperature. In addition to assuming a flat velocity profile, it is also assumed that there is perfect mixing in the radial direction. In other words, the concentration and temperature along a radius at any given axial position are constant. This assumption is best when applied to the fully developed turbulent flow case.

3. No axial mixing. A key assumption in plug flow is that there is no mixing along the axial direction as a result of either molecular diffusion or dispersive mixing processes. In reality, the concentration of the reacting stream will change as it flows through the reactor, creating axial gradients in concentration. These gradients will cause diffusion to occur. The plug flow model assumes that these diffusion processes are minor compared to the bulk movement of fluid, and can be ignored.

3.3.2 Plug Flow Mole Balance

Consider a volume element within a pipe through which the reacting fluid is flowing, as represented in Figure 3.1. The mole balance over this volume element for a time interval Δt may be written in terms of the molar flow rate of species j into and out of the volume element, and the change in concentration of j within the element over the time interval.

$$F_j\big|_V - F_j\big|_{V+\Delta V} + r_j \Delta V = \Delta V \frac{C_j\big|_{t+\Delta t} - C_j\big|_t}{\Delta t} \tag{3.33}$$

At steady state, the right-hand side of Equation 3.33 is equal to zero and

$$F_j\big|_V - F_j\big|_{V+\Delta V} + r_j \Delta V = 0 \tag{3.34}$$

FIGURE 3.1
Volume element of a plug flow reactor used to develop the mole balance equation.

Equation 3.34 can be rearranged to give an expression in terms of reaction rate:

$$r_j = \frac{F_j\big|_{V+\Delta V} - F_j\big|_V}{\Delta V} \tag{3.35}$$

In the limit, ΔV tends to the differential volume element dV which gives a differential equation

$$r_j = \frac{dF_j}{dV} \tag{3.36}$$

Equation 3.36 is the steady-state PFR mole balance equation. The same expression written in terms of the rate of disappearance of a reactant A is

$$(-r_A) = -\frac{dF_A}{dV} \tag{3.37}$$

The mole balance equation can also be written in terms of the conversion of a reactant, denoted A. The conversion in a steady-state flow reactor is defined in terms of a change in molar flow rate of the species of interest, as follows:

$$X_A = \frac{F_{A0} - F_A}{F_{A0}} \quad \text{or} \quad F_A = F_{A0}(1 - X_A)$$

Substitution of the conversion of A into Equation 3.37 gives

$$(-r_A) = F_{A0} \frac{dX_A}{dV} \tag{3.38}$$

Equation 3.38 can be integrated over a volume, V, between the inlet and outlet molar flow rates, denoted F_{A0} and F_{AE}. The integral form of the PFR mole balance, Equation 3.37, is

$$\int_0^V dV = V = -\int_{F_{A0}}^{F_{AE}} \frac{dF_A}{(-r_A)} \tag{3.39}$$

The integral form of Equation 3.38 is

$$\frac{V}{F_{A0}} = \int_0^{X_{AE}} \frac{dX_A}{(-r_A)} \tag{3.40}$$

3.3.3 Space Time in Plug Flow Reactor

The space time and space velocity were introduced in Chapter 1 as useful methods of quantifying reactor throughput. Recall that the space time in a flow reactor is typically defined as the ratio of the reactor volume to the volumetric flow rate of the inlet stream measured at the reactor inlet conditions. The inlet molar flow rate in the PFR mole balance can be expressed in terms of the inlet volumetric flow rate, Q_0. The relationship between molar and volumetric flow rates is

$$F_{A0} = Q_0 C_{A0} \tag{3.41}$$

Equation 3.40 can then be written as

$$\frac{V}{Q_0} = \tau = C_{A0} \int_0^{X_{AE}} \frac{dX_A}{(-r_A)} \tag{3.42}$$

In Equation 3.42, τ is the space time of the reactor.

3.3.4 Mean Residence Time in Plug Flow Reactor

The mean residence time was also introduced in Chapter 1 as a measure of the average amount of time that molecules can spend in the reactor. A higher mean residence time implies a higher conversion of reactants, provided that equilibrium is not reached. The mean residence time, t_m, for a PFR is defined as

$$t_m = \int_0^V \frac{dV}{Q} \tag{3.43}$$

The volumetric flow rate in the reactor changes with position if the mass density of the fluid changes, either as a result of temperature changes or if the number of moles changes in a gas-phase reaction. When the mass density of the fluid is constant, the following relationship is true:

$$t_m = \int_0^V \frac{dV}{Q} = \frac{V}{Q_0} = \tau \tag{3.44}$$

The mean residence time is equal to the space time, if, and only if, the density is constant and the space time is based on the conditions at the reactor inlet.

The application of the plug flow mole balance is illustrated in Example 3.2.

Example 3.2

Consider the irreversible ideal gas-phase reaction as follows:

$$A + B \rightarrow C$$

The reaction is carried out in a PFR at a constant 25°C, with a space time of 500 s. The feed contains only reactants A and B in equal concentrations of 50 mol/m³. The rate of reaction of A is given by

$$(-r_A) = kC_A C_B, \quad \text{where } k = 3.5 \times 10^{-5}\ \text{m}^3/\text{mol} \cdot \text{s} \tag{3.45}$$

This reaction is the same one used in Example 3.1. Calculate the exit fractional conversion of A and the mean residence time for the reactor.

SOLUTION

The fractional conversion is calculated directly from the plug flow mole balance equation. The mole balance equation for the PFR is

$$\left(-r_A\right) = -\frac{dF_A}{dV} = kC_A C_B \tag{3.46}$$

In Equation 3.46, both F_A and C_A are present as dependent variables. Only one of these variables may be used. In this example, we will work in terms of F_A because the reaction is a gas-phase reaction in which moles are not conserved. The concentration of the species thus changes as a result of density changes that occur as the reaction proceeds, which make the direct use of concentrations problematic. The relationship between F_A and C_A depends on the volumetric flow rate, Q:

$$QC_A = F_A \tag{3.47}$$

If we assume that the gases obey ideal mixing and the pressure is constant, then Q is proportional to F_t, the total molar flow rate. In other words, Q will change along the reactor length. The ideal gas law gives the relationship between Q and F_t:

$$P_t Q = F_t R_g T \quad \text{and} \quad P_t Q_0 = F_{t0} R_g T \tag{3.48}$$

We take the ratio of the conditions at any point in the reactor and at the inlet conditions to give

$$\frac{Q}{F_t} = \frac{Q_0}{F_{t0}} \quad \text{or} \quad Q = \frac{Q_0}{F_{t0}} F_t \tag{3.49}$$

The overall reaction is

$$A + B \rightarrow C$$

The molar flow rate of each component can be expressed in terms of F_A. Note that the inlet molar flow rates of A and B are the same; therefore, the flow rate of A and B (and their concentrations) will be equal at any point in the reactor.

Species	Inlet Flow Rate	Flow Rate after Reaction
A	F_{A0}	F_A
B	$F_{B0} = F_{A0}$	$F_B = F_{B0} - (F_{A0} - F_A) = F_A$
C	$F_{C0} = 0$	$F_C = F_{A0} - F_A$
Total	$F_t = 2F_{A0}$	$F_t = F_{A0} + F_A$

The volumetric flow rate can then be expressed as

$$Q = Q_0 \frac{\left(F_{A0} + F_A\right)}{2F_{A0}} \tag{3.50}$$

The concentration terms in the rate expression are now replaced by flow rates (and noting that the concentrations of A and B are equal) to give

$$(-r_A) = kC_AC_B = kC_A^2 = k\frac{F_A^2}{Q^2} \tag{3.51}$$

Substituting the volumetric flow rate from Equation 3.50 and simplifying give

$$(-r_A) = \frac{kF_A^2}{(Q_0/2F_{A0})^2(F_{A0} + F_A)^2} = k\left(\frac{2F_{A0}}{Q_0}\right)^2\frac{F_A^2}{(F_{A0} + F_A)^2} \tag{3.52}$$

Equation 3.52 can be expressed in terms of the inlet concentration of A:

$$(-r_A) = 4kC_{A0}^2\left(\frac{F_A}{F_{A0} + F_A}\right)^2 \tag{3.53}$$

The mole balance expressed in terms of F_A is therefore written as

$$-\frac{dF_A}{dV} = 4kC_{A0}^2\left(\frac{F_A}{F_{A0} + F_A}\right)^2 \tag{3.54}$$

Equation 3.54 may be rearranged and written in integral form as

$$4kC_{A0}^2\int_0^V dV = -\int_{F_{A0}}^{F_{AE}}\left(\frac{F_{A0} + F_A}{F_A}\right)^2 dF_A \tag{3.55}$$

Integrating the left-hand side of Equation 3.55 and expanding the integrand on the right-hand side give

$$4kC_{A0}^2V = -\int_{F_{A0}}^{F_{AE}}\left[\frac{F_{A0}^2 + 2F_{A0}F_A + F_A^2}{F_A^2}\right] dF_A \tag{3.56}$$

Simplifying the integrand then gives a simple form for analytical solution:

$$4kC_{A0}^2V = -\int_{F_{A0}}^{F_{AE}}\left[\left(\frac{F_{A0}}{F_A}\right)^2 + \frac{2F_{A0}}{F_A} + 1\right] dF_A \tag{3.57}$$

Integrating from the inlet to the outlet flow rate gives

$$-4kC_{A0}^2V = F_{A0}^2\left[\frac{1}{F_{A0}} - \frac{1}{F_{AE}}\right] + 2F_{A0}\ln\left(\frac{F_{AE}}{F_{A0}}\right) + F_{AE} - F_{A0} \tag{3.58}$$

Divide both sides of Equation 3.58 by F_{A0} and simplify to give

$$-4kC_{A0}^2 \frac{V}{F_{A0}} = 1 - \frac{F_{A0}}{F_{AE}} + 2\ln\left(\frac{F_{AE}}{F_{A0}}\right) + \frac{F_{AE}}{F_{A0}} - 1 \tag{3.59}$$

Consider the left-hand side of Equation 3.59. The molar flow rate is given by

$$F_{A0} = Q_0 C_{A0} \tag{3.60}$$

Substitute Equation 3.60 into the left-hand side of Equation 3.59 to obtain

$$-4kC_{A0}^2 \frac{V}{F_{A0}} = -4kC_{A0}^2 \frac{V}{Q_0 C_{A0}} = -4kC_{A0}\tau \tag{3.61}$$

Substitute the numerical values for the known parameters

$$-4kC_{A0}\tau = (-4)(3.5 \times 10^{-5})(50)(500) = -3.5 \tag{3.62}$$

Equation 3.59 is therefore written as

$$\frac{F_{AE}}{F_{A0}} - \frac{F_{A0}}{F_{AE}} + 2\ln\left(\frac{F_{AE}}{F_{A0}}\right) = -3.5 \tag{3.63}$$

Equation 3.63 can be solved by iteration to give a value of $(F_{AE}/F_{A0}) = 0.438$. In terms of conversion, we can write

$$X_A = \frac{F_{A0} - F_{AE}}{F_{A0}} = 1 - \frac{F_{AE}}{F_{A0}} = 0.562 \tag{3.64}$$

The conversion in the reactor is 56.2%.

Mean Residence Time

The mean residence time, t_m, for a PFR is given by

$$t_m = \int_0^V \frac{dV}{Q} \tag{3.65}$$

The relationship between the molar flow rate and the volumetric flow rate for this reaction is given by Equation 3.50 as

$$Q = \frac{Q_0}{2F_{A0}}(F_{A0} + F_A) = \frac{1}{2C_{A0}}(F_{A0} + F_A) \tag{3.66}$$

It is necessary to change the variables in Equation 3.65 from dV to dF_A. The mole balance, Equation 3.54, can be used for this purpose:

$$-\frac{dF_A}{dV} = 4kC_{A0}^2\left(\frac{F_A}{F_{A0} + F_A}\right)^2 \tag{3.67}$$

Equation 3.67 may be rearranged to give an explicit expression for dV:

$$dV = -dF_A \left\{ \left[\frac{F_A^2}{(F_{A0} + F_A)^2} \right] 4kC_{A0}^2 \right\}^{-1} = -\frac{(F_A + F_{A0})^2}{4kC_{A0}^2 F_A^2} dF_A \tag{3.68}$$

Substitute dV from Equation 3.68 and Q from Equation 3.66 into Equation 3.65, the residence time definition

$$t_m = \int_0^V \frac{dV}{Q} = -2C_{A0} \left(\frac{1}{4kC_{A0}^2} \right) \int_{F_{A0}}^{F_{AE}} \frac{(F_{A0} + F_A)}{F_A^2} dF_A$$

$$= -\frac{1}{2kC_{A0}} \int_{F_{A0}}^{F_{AE}} \left(\frac{F_{A0}}{F_A^2} + \frac{1}{F_A} \right) dF_A \tag{3.69}$$

Integrating Equation 3.69 gives

$$t_m = -\frac{1}{2kC_{A0}} \left[F_{A0} \left(\frac{1}{F_{A0}} - \frac{1}{F_{AE}} \right) + \ln \left(\frac{F_{AE}}{F_{A0}} \right) \right] = -\frac{1}{2kC_{A0}} \left[1 - \frac{F_{A0}}{F_{AE}} + \ln \left(\frac{F_{AE}}{F_{A0}} \right) \right] \tag{3.70}$$

The mean residence time is desired for an operating condition that gives a value of $(F_{AE}/F_{A0}) = 0.438$. Therefore, we substitute this value into Equation 3.70 and solve to obtain $t_m = 602$ s. Recall that the space time given for this reactor was 500 s, and therefore, the mean residence time is not equal to the space time.

Comment: In the constant-pressure batch reactor example, Example 3.1, the fractional conversion of A was equal to 0.5093. This value corresponds to $(F_{AE}/F_{A0}) = 0.4907$. If we substitute this value into Equation 3.70, we obtain a mean residence time of 500 s. Thus, when the mean residence time in a PFR equals the mean residence time (reaction time) in a constant-pressure batch reactor, the conversion is the same, provided that the reactor is isothermal.

The plug flow mole balance for the steady-state reactor yields an ordinary differential equation, the integration of which gives the outlet molar flow rates. When using the mole balance equation, one of the most important questions that must be answered first is whether the volumetric flow rate changes as the process stream moves along the reactor. If it does, then care should be taken in performing the mole balance, and the use of concentrations is best avoided in favor of molar flow rates.

3.4 Continuous Stirred Tank Reactor

We now consider the second of the ideal flow reactors, the continuous stirred tank reactor (CSTR). The CSTR usually consists of a baffled tank with mixing induced by an impeller, as illustrated in Figure 3.2. The steady-state reactor shall be considered first, followed by the transient case. At steady state, the mass flow rate into the tank equals the mass flow rate out of the tank, and the feed and product properties are not functions of time.

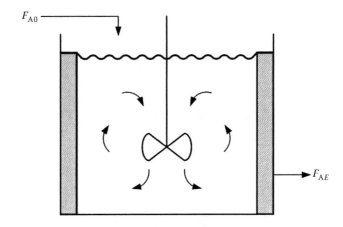

FIGURE 3.2
A CSTR typically consists of a fluid-filled tank which is stirred by an impeller. Baffles are usually used to ensure good mixing.

Furthermore, the reactor volume is constant. For a well-mixed reactor, the composition and temperature of the fluid in the reactor are uniform over the entire volume, that is, they do not vary with position. It follows that the temperature and concentration in the reactor are the same as the exit temperature and concentration. Good mixing may be achieved *fairly* easily for liquid-phase homogeneous reactions where the viscosity of the solution is less than about 100 cP and the fluid is Newtonian. It is more difficult to achieve CSTR behavior with gas-phase reactions, although not impossible.

Recall the general mole balance from Section 3.1.

$$F_j\big|_{\text{in}} - F_j\big|_{\text{out}} + r_j\Delta V = \frac{\left[\left(N_j\right)_{t+\Delta t} - \left(N_j\right)_t\right]}{\Delta t} = \frac{\Delta N_j}{\Delta t} \tag{3.71}$$

At steady state, the right-hand side equals zero. Furthermore, because the reactor contents have a uniform composition, the incremental volume is replaced by the total volume. The steady-state mole balance is thus

$$F_{j0} - F_{jE} + r_j V = 0 \tag{3.72}$$

Equation 3.72 can be rearranged explicitly in terms of the reaction rate

$$\left(-r_j\right) = \frac{F_{j0} - F_{jE}}{V} \tag{3.73}$$

Note that Equation 3.73 is an algebraic, not a differential, equation. Equation 3.73 may be written in terms of the rate of disappearance of reactant A:

$$\left(-r_A\right) = \frac{F_{A0} - F_{AE}}{V} \tag{3.74}$$

The conversion of reactant A is defined in terms of the difference between the inlet and outlet molar flow rates as

$$X_A = \frac{F_{A0} - F_{AE}}{F_{A0}} \tag{3.75}$$

Substitute Equation 3.75 into Equation 3.74 and simplify to give

$$(-r_A) = \frac{F_{A0} X_A}{V} \tag{3.76}$$

Equations 3.74 and 3.76 are both common forms of the mole balance equation for a CSTR.

3.4.1 Space Time and Mean Residence Time in CSTR

The mean residence time in a CSTR is defined as the reactor volume divided by the outlet volumetric flow rate:

$$t_m = \frac{V}{Q_E} \tag{3.77}$$

The space time is defined as the reactor volume divided by the inlet volumetric flow rate:

$$\tau = \frac{V}{Q_0} \tag{3.78}$$

Therefore, Equation 3.76 written in terms of the space time as

$$(-r_A) = \frac{F_{A0} X_A}{Q_0 \tau} = \frac{C_{A0}}{\tau} X_A \tag{3.79}$$

Note that because reactor concentration equals exit concentration, the rate occurs at the exit concentration and $(-r_A)$ is evaluated at C_{AE}.

Example 3.3

Consider the irreversible ideal gas-phase reaction given by

$$A + B \rightarrow C$$

The reaction is carried out in an isothermal CSTR. The feed is of a mixture of 50 mol% A and 50 mol% B, with concentrations of 50 mol/m³ each. The space time is 500 s and the reactor temperature is 25°C. The rate of disappearance of A is given by

$$(-r_A) = kC_A C_B, \quad \text{where } k = 3.5 \times 10^{-5} \, \text{m}^3/\text{mol} \cdot \text{s} \tag{3.80}$$

The pressure in the feed stream is the same as the pressure in the reactor. Calculate the exit fractional conversion of A and the mean residence time for the reactor. Note that this reaction is the same one used in Examples 3.1 and 3.2.

SOLUTION

The problem is solved by applying the CSTR mole balance equation. We start by substituting the rate expression, Equation 3.80, into the mole balance equation for the CSTR, Equation 3.74:

$$(-r_A) = \frac{F_{A0} - F_{AE}}{V} = kC_{AE}C_{BE} \tag{3.81}$$

Note that the reaction rate is evaluated at the exit concentration of A and B. In Equation 3.81, both F_A and C_A are present as dependent variables. Only one of these variables should be used. We will work in terms of F_A because this is a gas-phase reaction in which the number of moles changes with reaction. Therefore, the volumetric flow rate of the effluent is different from the inlet, and thus there is a concentration change owing to a volume change. The relationship between F_A and C_A is a function of the volumetric flow rate, Q, as follows:

$$QC_A = F_A \tag{3.82}$$

If we assume that the gases follow ideal mixing and that the pressure is constant, then Q is proportional to F_t, that is, Q at the reactor exit is different from the value at the inlet. The ideal gas law gives the relationship between Q and F_t for both the inlet and exit streams:

$$P_t Q_E = F_{tE} R_g T \quad \text{and} \quad P_t Q_0 = F_{t0} R_g T$$

We can take the ratio of the conditions at the reactor exit to those at the reactor inlet to give

$$\frac{Q_E}{F_{tE}} = \frac{Q_0}{F_{t0}} \quad \text{or} \quad Q_E = \frac{Q_0}{F_{t0}} F_{tE} \tag{3.83}$$

The overall reaction is

$$A + B \rightarrow C$$

The molar flow rate of each component can be expressed in terms of the molar flow rate of reactant A, F_A. Note that the inlet molar flow rates of A and B are the same; hence, the flow rate of A and B in the reactor effluent, and their concentrations at any point in the reactor, will be equal. Build the following mole balance table:

Species	Inlet Flow Rate	Flow Rate at Reactor Exit
A	F_{A0}	F_{AE}
B	$F_{B0} = F_{A0}$	$F_{BE} = F_{B0} - (F_{A0} - F_{AE}) = F_{AE}$
C	$F_{C0} = 0$	$F_{CE} = F_{A0} - F_{AE}$
Total	$F_t = 2F_{A0}$	$F_{tE} = F_{A0} + F_{AE}$

The volumetric flow rate at the reactor exit can be expressed as

$$Q_E = Q_0 \frac{(F_{A0} + F_{AE})}{2F_{A0}} \tag{3.84}$$

Substituting Equations 3.82 and 3.84 into the rate expression, Equation 3.80, gives an equation for the reaction rate in terms of the molar flow rate of A (noting that the concentrations of A and B are equal):

$$(-r_A)_E = kC_{AE}C_{BE} = kC_{AE}^2 = k\frac{F_{AE}^2}{Q_E^2} = \frac{kF_{AE}^2}{(Q_0/2F_{A0})^2(F_{A0} + F_{AE})^2}$$

$$= k\left(\frac{2F_{A0}}{Q_0}\right)^2 \frac{F_{AE}^2}{(F_{A0} + F_{AE})^2} \tag{3.85}$$

Equation 3.85 can be simplified to give

$$(-r_A) = 4k\left(\frac{F_{A0}}{Q_0}\right)^2\left(\frac{F_{AE}}{F_{A0} + F_{AE}}\right)^2 \tag{3.86}$$

Equation 3.86 can be substituted into the mole balance, Equation 3.81, to give

$$(-r_A) = \frac{F_{A0} - F_{AE}}{V} = 4k\left(\frac{F_{A0}}{Q_0}\right)^2\left(\frac{F_{AE}}{F_{A0} + F_{AE}}\right)^2 \tag{3.87}$$

Rearrange Equation 3.87 to give

$$F_{A0} - F_{AE} = 4k\left(\frac{F_{A0}}{Q_0}\right)\left(\frac{V}{Q_0}\right)F_{A0}\left(\frac{F_{AE}}{F_{A0} + F_{AE}}\right)^2 = 4kC_{A0}\tau F_{A0}\left(\frac{F_{AE}}{F_{A0} + F_{AE}}\right)^2 \tag{3.88}$$

Equation 3.88 can be further rearranged to give

$$4k\tau C_{A0} = \frac{F_{A0}(1 - (F_{AE}/F_{A0}))F_{A0}^2(1 + (F_{AE}/F_{A0}))^2}{F_{A0}^3(F_{AE}/F_{A0})^2}$$

$$= \frac{(1 - (F_{AE}/F_{A0}))(1 + (F_{AE}/F_{A0}))^2}{(F_{AE}/F_{A0})^2} \tag{3.89}$$

Substitute the numerical values into the left-hand side of Equation 3.89:

$$4k\tau C_{A0} = (4)(3.5 \times 10^{-5})(500)(50) = 3.50$$

Solving Equation 3.89 numerically gives a value of $(F_{AE}/F_{A0}) = 0.555$, which in turn gives a value of fractional conversion, $X_A = 0.445$. This conversion is lower than that observed in the PFR for the same space time.

Mean Residence Time

The mean residence time for a CSTR is governed by the volumetric flow rate at the reactor exit:

$$t_m = \frac{V}{Q_E} \tag{3.90}$$

The effluent flow rate can be expressed as, from Equation 3.84

$$Q_E = Q_0 \frac{(F_{A0} + F_{AE})}{2F_{A0}} = \frac{Q_0}{2}\left(1 + \frac{F_{AE}}{F_{A0}}\right) = \frac{Q_0}{2}(1 + 0.555) \tag{3.91}$$

The residence time is found by substituting Equation 3.91 into Equation 3.90:

$$t_m = \frac{2V}{Q_0 1.555} = \frac{2\tau}{1.555} = 643 \text{ s}$$

3.4.2 Summary of Mole Balance Equations for Three Ideal Reactors

The steady-state mole balance equations and the definition of conversion for each of the three basic reactor types are summarized in the following table:

Reactor	Conversion	Mole Balance Equation
Batch	$X_A = \dfrac{N_{A0} - N_A}{N_{A0}}$	$(-r_A) = -\dfrac{1}{V}\dfrac{dN_A}{dt} = \dfrac{N_{A0}}{V}\dfrac{dX_A}{dt}$
Plug flow	$X_A = \dfrac{F_{A0} - F_A}{F_{A0}}$	$(-r_A) = -\dfrac{dF_A}{dV} = F_A\dfrac{dX_A}{dV}$
CSTR	$X_A = \dfrac{F_{A0} - F_A}{F_{A0}}$	$(-r_A) = \dfrac{F_{A0} - F_{AE}}{V} = \dfrac{F_{A0} X_{AE}}{V}$

Observation: In the examples illustrated so far, the conversion was calculated for a given reactor volume or space time. Often the design problem is to determine the reactor volume or space time that is required to achieve a desired conversion. The calculation procedure is essentially the same, except that the unknown in the appropriate design equation is V (or τ) rather than X.

3.5 Reaction Rate in Terms of Catalyst Mass

The rate of reaction has been presented so far in terms of reactor volume, for example, mol/m^3 s. When a catalyst is used in the reactor (as in the case studies presented later in this chapter), it is common to quote the reaction rate in terms of the catalyst mass. Typical units are $mol/kg\text{-}cat \cdot s$. The following terms are commonly encountered in the analysis of catalytic reactors:

$$\text{Bulk density}, \rho_B = \frac{\text{Catalyst mass}}{\text{Reactor volume}} = \frac{W}{V} \tag{3.92}$$

$$\text{Catalyst density}, \rho_C = \frac{\text{Mass of a catalyst particle}}{\text{Volume of a catalyst particle}} \tag{3.93}$$

$$\text{Bed porosity}, \varepsilon = \frac{\text{Void volume in reactor}}{\text{Total reactor volume}} \tag{3.94}$$

The mass of catalyst in a reactor is denoted W: from Equation 3.92, $W = V\rho_B$. If the rate of reaction in terms of reactor volume is denoted $(-r_A)_V$ and the rate in terms of catalyst mass is denoted as $(-r_A)_W$, then

$$\rho_B (-r_A)_W = (-r_A)_V \tag{3.95}$$

The CSTR steady-state mole balance equation with the reaction rate expressed in terms of reactor volume (for the disappearance of reactant A) is

$$F_{A0} - F_{AE} - (-r_A)_V V = 0 \tag{3.96}$$

If the reaction rate is expressed in terms of the catalyst mass, the mole balance can be written as

$$F_{A0} - F_{AE} - (-r_A)_W \rho_b V = F_{A0} - F_{AE} - (-r_A)_W \frac{W}{V} V$$

$$= F_{A0} - F_{AE} - (-r_A)_W W = 0 \tag{3.97}$$

For the PFR, a similar logic can be applied. The mole balance equation becomes

$$-\frac{dF_A}{dV} = (-r_A)_V = \rho_b (-r_A)_W \tag{3.98}$$

Equation 3.98 can be rewritten as

$$-\frac{dF_A}{d(V\rho_b)} = -\frac{dF_A}{d\left(V \dfrac{W}{V}\right)} = -\frac{dF_A}{dW} = (-r_A)_W \tag{3.99}$$

It is always important to check that the units in the mole balance equation are consistent.

When the reactor mole balance for a catalytic reactor is expressed in terms of catalyst mass as represented by one of the preceding equations, mass transfer effects between the catalyst and the bulk fluid and within the catalyst particles themselves are being ignored. Such mass transfer effects may be important. The special issues arising from catalytic systems are discussed in detail in Chapters 7 through 11.

Case Study 3.1: Hydrogenation of Sulfur and Nitrogen Compounds in Oil

In this case study, the desulfurization (HDS) and denitrogenation (HDN) of oil will be considered. Most crude oil contain a number of different types of sulfur and nitrogen compounds. At some point in the oil refining process, these compounds must be eliminated. Typically, the oil stream containing these compounds is reacted with hydrogen over a catalyst to convert the sulfur into hydrogen sulfide and the nitrogen into ammonia. Hydrogen sulfide and ammonia are subsequently removed by stripping. For a given catalyst and gas oil combination, the rate of reaction of sulfur molecules was found to follow the rate expression:

$$(-r_S) = A_S \exp\left(\frac{-E_S}{R_g T}\right) C_S^{1.5} P_{H_2}^{0.8} \text{ mol of S/h gcat} \tag{3.100}$$

The rate of reaction of sulfur molecules is based on total sulfur concentration expressed as S, regardless of the molecular form of the sulfur. This type of kinetic expression is an average rate for the reaction of different types of sulfur-containing molecules. This expression is sometimes referred to as lumped kinetics. The rate expression for the nitrogen-containing compounds is

$$(-r_N) = A_N \exp\left(\frac{-E_N}{R_g T}\right) C_N P_{H_2}^{1.3} \text{ mol of N/h gcat} \tag{3.101}$$

Note that the rate is based on the mass of the catalyst, not the reactor volume. The unit gcat in Equations 3.100 and 3.101 means grams of catalyst. This representation is common when dealing with catalytic reactions. Refer to Section 3.5 for a discussion of rate expressions expressed in terms of catalyst mass. It is necessary to pay close attention to the units when solving this problem. The values of the kinetic parameters are

$$A_S = 5.22 \times 10^{11} \text{ cm}^{4.5}/\text{h gcat mol}^{0.5} \text{ MPa}^{0.8}, \quad \frac{E_S}{R_g} = 15,100 \text{ K}$$

$$A_N = 1.21 \times 10^7 \text{ cm}^3/\text{h gcat MPa}^{1.3}, \quad \frac{E_N}{R_g} = 12,300 \text{ K}$$

Note that the units of the rate constants are such that the rate will be in the units given in Equations 3.100 and 3.101, when the sulfur or nitrogen concentration is measured in mol/cm³ and the hydrogen pressure is measured in MPa. The oil contains 4.3 wt.% sulfur and its density is 0.93 g/cm³. We assume ideal, isothermal operation, constant density, and constant hydrogen partial pressure. This last assumption implies that the hydrogen is present in excess, and therefore its pressure does not change during the course of the reaction. In each of the following illustrations, the reactor contains 10 g of catalyst and is operated at 400°C with a hydrogen partial pressure of 13.9 MPa.

BATCH REACTOR OPERATION

In this step, we illustrate the reaction carried out in a constant-temperature batch reactor. We calculate the time required to react 95% of the sulfur present in 100 cm³ of oil. We also want to calculate the fraction of nitrogen that reacts during this time.

As a first step, we calculate the value of the rate constant at the given reactor temperature. The hydrogen pressure is a constant, and thus we will also include this value into a "pseudo" rate constant. Hence, define

$$k_1 = A_S \exp\left(\frac{-E_S}{R_g T}\right) P_{H_2}^{0.8} = 5.22 \times 10^{11} \exp\left(\frac{-15,100}{400 + 273}\right) 13.9^{0.8}$$

$$= 772.5 \, \text{cm}^{4.5}/\text{h gcat mol}^{0.5}$$

and

$$k_2 = A_N \exp\left(\frac{-E_N}{R_g T}\right) P_{H_2}^{1.3} = 1.21 \times 10^7 \exp\left(\frac{-12,300}{400 + 273}\right) 13.9^{1.3} = 4.28 \, \text{cm}^3/\text{h gcat}$$

The rate of reaction in terms of catalyst mass can be converted to a rate expressed in terms of liquid volume if it is multiplied by the ratio of catalyst mass in the reactor, W, and the reaction volume, V. The batch reactor mole balance equation can then be written as

$$-\frac{1}{V}\frac{dN_S}{dt} = k_1 C_S^{1.5}\frac{W}{V} \quad \text{or} \quad -\frac{1}{W}\frac{dN_S}{dt} = k_1 C_S^{1.5} = k_1 \frac{N_S^{1.5}}{V^{1.5}}$$

We will take the moles of sulfur as the dependent variable. Integrating the equation gives

$$-Wk_1 \int_0^t dt = V^{1.5} \int_{N_{S0}}^{N_S} \frac{dN_S}{N_S^{1.5}} \tag{3.102}$$

Integrating Equation 3.102 and solving for the moles of sulfur as a function of time, t, yields

$$t = \frac{V^{1.5}}{Wk_1}\frac{1}{0.5}\left(\frac{1}{N_S^{0.5}} - \frac{1}{N_{S0}^{0.5}}\right) \tag{3.103}$$

In Equation 3.103, the values of the variables are

W = mass of catalyst in reactor = 10 g
V = volume of liquid in reactor = 100 cm³ (assumed constant)
N_{S0} = initial moles of sulfur in gas oil
N_S = moles of sulfur in gas oil at time t

Initially there are 100 cm^3 of oil with a density of 0.93 g/cm^3, or 93 g of oil. The oil contains 4.3 wt.% sulfur, or $0.043 \times 93 = 4$ g of S. The molecular mass of S is 32 g/mol, and therefore there are 0.125 mol of sulfur initially present. If 95% is converted, then the final moles of sulfur will be 6.25×10^{-3} mol. If all the numerical values are substituted into the equation, we obtain

$$t = \frac{V^{1.5}}{Wk_1} \frac{1}{0.5} \left(\frac{1}{N_S^{0.5}} - \frac{1}{N_{S0}^{0.5}} \right) = \frac{100^{1.5}}{10 \times 772.5} \times \frac{1}{0.5} \times \left(\frac{1}{0.00625^{0.5}} - \frac{1}{0.125^{0.5}} \right)$$

$$= 2.54 \text{ h}$$

In other words, it takes 2.54 h to react 95% of the sulfur present.

We now want to calculate the extent of the nitrogen reaction during the 2.54 h reaction time. The mole balance equation for the rate of disappearance of nitrogen, expressed in terms of the catalyst mass, is

$$-\frac{1}{W} \frac{dN_N}{dt} = k_2 C_N = k_2 \frac{N_N}{V} \tag{3.104}$$

Integrating Equation 3.104 gives

$$-Wk_2 \int_0^t dt = V \int_{N_{N0}}^{N_N} \frac{dN_N}{N_N} \tag{3.105}$$

Integration of Equation 3.105 between the limits yields

$$\ln\left(\frac{N_{N0}}{N_N} \right) = \frac{Wk_2 t}{V} \tag{3.106}$$

where

N_{N0} = initial moles of nitrogen in gas oil
N_N = moles of nitrogen in gas oil at time t

Substituting 2.54 h into Equation 3.106 and solving give

$$\frac{N_{N0}}{N_N} = \exp\left[\frac{Wk_2 t}{V} \right] = \exp\left[\frac{10 \times 4.28 \times 2.54}{100} \right] = 2.966$$

This ratio can be converted to a fractional conversion

$$X_N = \frac{N_{N0} - N_N}{N_{N0}} = 1 - \frac{N_N}{N_{N0}} = 1 - \frac{1}{2.966} = 0.663$$

Hence, the fraction of nitrogen compounds converted is 0.663 or 66.3%. Note that the initial concentration of nitrogen is not required. For first-order reactions, fractional conversions can be calculated as a function of time without knowing the initial concentration. Note also that this example could have been solved using fractional conversion as the dependent variable.

Comments: The assumption of isothermal operation may not be very good because some reaction will occur as the reactor is slowly heated to the operating temperature. The assumption of constant volume might not be valid because the liquid volume may change as a result of thermal expansion and compositional changes. In addition, the hydrogen partial pressure can also vary significantly during the experiment due to the build-up of volatile products.

PLUG FLOW REACTOR

In the PFR case, we take the same feedstock and operating conditions, but now we wish to calculate the maximum feed rate of oil possible if 95% of the sulfur is to be reacted. Furthermore, we will calculate the fraction of nitrogen that reacts for this flow rate and the total amount of feedstock that can be processed at these above conditions in 2.54 h.

The rate equations can again be expressed in terms of the pseudo rate constants, whose values are the same as for the batch reactor case. It is also necessary to convert the reaction rate per unit mass to a rate per volume. The mole balance equation for the PFR in integral form can thus be written as

$$\frac{V}{F_{S0}} = \int_0^X \frac{1}{(-r_S)} \frac{V}{W} \, dX \quad \text{or} \quad \frac{W}{F_{S0}} = \int_0^X \frac{dX}{k_1 C_S^{1.5}} \tag{3.107}$$

The inlet molar flow rate of sulfur can be converted to a volumetric flow rate, and the result expressed in terms of the conversion. Note that

$$C_S = C_{S0}(1 - X) \quad \text{and} \quad F_{S0} = Q C_{S0}$$

Substituting into Equation 3.107 and simplifying give

$$\frac{k_1 C_{S0}^{0.5} W}{Q} = \int_0^X \frac{dX}{(1 - X)^{1.5}}$$

Integration and rearrangement to give a result in terms of the volumetric feed rate, Q, yield

$$Q = 0.50 W k_1 C_{S0}^{0.5} \frac{(1 - X)^{0.5}}{1 - (1 - X)^{0.5}}$$

where C_{S0}, the initial concentration of sulfur in the gas oil, is obtained from the batch reactor analysis section. Recall that there were 0.125 mol in 100 cm³ of oil, which gives an initial concentration of 0.00125 mol/cm³. The desired conversion of sulfur is 0.95; therefore, we can substitute to get the volumetric flow rate:

$$Q = 0.50 W k_1 C_{S0}^{0.5} \frac{(1-X)^{0.5}}{1-(1-X)^{0.5}} = 0.50 \times 10 \times 772.5 \times 0.00125^{0.5} \frac{(1-0.95)^{0.5}}{1-(1-0.95)^{0.5}}$$

$$= 39.33 \ \text{cm}^3/\text{h}$$

Applying the same logic to the nitrogen reaction gives

$$\frac{W}{F_{N0}} = \int_0^X \frac{dX}{k_2 C_N} \quad \text{or} \quad \frac{W}{Q C_{N0}} = \int_0^X \frac{dX}{k_2 C_{N0}(1-X)} \quad \text{or} \quad \frac{W}{Q} = \int_0^X \frac{dX}{k_2(1-X)}$$

Integration gives the following equation:

$$k_2 W = Q \int_0^X \frac{dX}{(1-X)} \quad \text{or} \quad -\ln[1-X] = \frac{W k_2}{Q}$$

with $Q = 39.33 \ \text{cm}^3/\text{h}$, $k_2 = 4.28 \ \text{cm}^3/\text{h gcat}$ and $W = 10 \ \text{g}$ solving for X gives $X = 0.663$. Note that this is the same conversion that was achieved in the batch reactor for 2.54 h of reaction time. The amount of material that can be processed in 2.54 h is

$$2.54 \times Q = 100 \ \text{cm}^3$$

Comment: It takes the same amount of time to process 100 cm^3 of oil in the batch reactor and the PFR to the same sulfur and nitrogen conversion.

CONTINUOUS STIRRED TANK REACTOR

Now take the same system but replace the PFR with a CSTR, all other conditions remaining the same. We want to calculate the fractional conversion of sulfur if the feed flow rate to the CSTR is the same as to the PFR. We start with the mole balance equation, and convert to an expression in terms of catalyst mass.

$$\frac{V}{F_{S0}} = \frac{X}{(-r_S)} \frac{V}{W} \quad \text{or} \quad \frac{W}{F_{S0}} = \frac{X}{k_1 C_S^{1.5}}$$

Writing in terms of fractional conversion and initial concentration of sulfur:

$$\frac{k_1 C_{S0}^{0.5} W}{Q} = \frac{X}{(1-X)^{1.5}} \tag{3.108}$$

Substituting $Q = 39.33 \ \text{cm}^3/\text{h}$, $k_1 = 772.5 \ \text{cm}^{4.5}/\text{h gcat mol}^{0.5}$ and $W = 10 \ \text{g}$ into Equation 3.108 and solving for X gives $X = 0.769$. Recall that for the PFR with the same feed flow rate, the conversion was 0.95. The nitrogen conversion is given by

$$\frac{k_2 W}{Q} = \frac{X}{(1-X)} \tag{3.109}$$

Substitute the numerical values into Equation 3.109 and solve for X to give a fraction conversion of nitrogen of $X = 0.521$.

Case Study 3.1 compared the performance of the three ideal reactor types for a constant-density fluid. It was seen that for the same reaction time, the batch and PFRs gave the same conversion, while the CSTR gave a lower conversion. If the reaction rate decreases as the concentration of reactants decreases, then for an isothermal reactor and the same operating conditions, the conversion in a CSTR is always lower than the conversion in a PFR, that is, backmixing causes decreases in conversion. The comparison of CSTR and PFR performance is considered in more detail in the following sections.

3.6 Comparison of PFR and CSTR Performance

A comparison of CSTR and PFR performance can be made graphically. Let the rate function be represented by a power law as follows:

$$(-r_A) = kC_A^n \quad n > 0$$

For this type of rate expression, the rate decreases as the concentration of A decreases, or as conversion increases. This type of rate/concentration dependence is sometimes referred to as "normal kinetics." We can prepare some generic plots illustrating the behavior of the rate function with conversion. We plot the ratio of the inlet molar flow rate and the reaction rate as a function of the conversion. Such a plot is shown in Figure 3.3.

We now use Figure 3.3 to make a comparison of the PFR and the CSTR. First, consider a PFR. The mole balance equation in terms of conversion of A is

$$V = F_{A0} \int_0^{X_{AE}} \frac{dX_A}{(-r_A)} \tag{3.110}$$

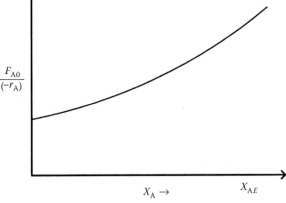

FIGURE 3.3

Representation of the rate behavior as a function of conversion for "normal" kinetics. The reaction rate decreases as conversion increases.

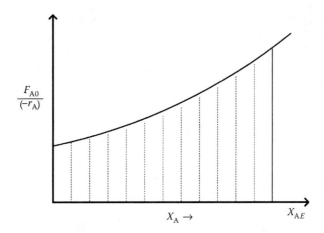

FIGURE 3.4
Reaction rate behavior verses conversion for "normal" kinetics. The hatched area corresponds to the volume of a PFR required to give the indicated conversion.

The right-hand side of Equation 3.110 is nothing more than the area under the curve in Figure 3.4. This area is in turn equal to V, the reactor volume, which is represented by the hatched area under the curve in Figure 3.4.

Next, consider the CSTR. The mole balance equation for a CSTR is

$$V = \frac{F_{A0}}{(-r_A)} X_{AE} \tag{3.111}$$

For the CSTR, the reactor volume is equal to the hatched area in Figure 3.5. From a comparison of the areas in Figure 3.5 with that in Figure 3.4 for the PFR, it is evident that for a given reaction, with equal conditions of feed composition, temperature and flow rate, and

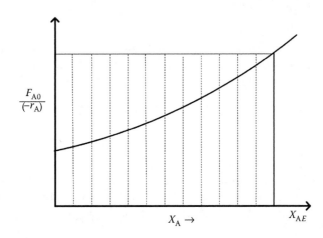

FIGURE 3.5
Reaction rate behavior verses conversion for "normal" kinetics. The hatched area corresponds to the volume of a CSTR required to give the indicated conversion.

specified outlet conversion, the volume of a CSTR will exceed the volume of a PFR. Also, for equal operating conditions and equivalent reactor volume, the conversion in a PFR will exceed that of a CSTR. It must be emphasized that this must be true for isothermal reactors in which a single reaction is occurring with "normal kinetics." It will be seen in Chapter 4 that this statement does not necessarily hold true for nonisothermal reactors, and it is possible for the volume of a CSTR to be less than that of a PFR for the same conversion with equal inlet conditions.

3.7 Multiple Reactions

In many industrial reactors, more than one chemical reaction occurs simultaneously. For example, in Case Study 3.1, both sulfur- and nitrogen-containing compounds were reacted with hydrogen. It is possible to envisage a variety of scenarios which could be loosely grouped into series reactions, parallel reactions, and series/parallel reactions. Some of the possible combinations are discussed in the following sections.

3.7.1 Series Reactions

A series reaction is generally considered to be one in which the products from one reaction can react further to produce other products. In many cases, it is the product from the first reaction that is desired, and the reactor is designed in an attempt to avoid the second reaction. An example of a series reaction is

$$A + B \rightarrow C + D \rightarrow R + S$$

In this example, the reaction products C and D from the first reaction can react further to form molecules of R and S.

3.7.2 Parallel Reactions

In a parallel reaction, a set of reactants may undergo individual reactions to form products. Two cases can be identified. In one case, two reactants may enter the reactor and undergo separate reactions, for example, reactants A and B might react according to

$$A \rightarrow C, \quad \text{where } (-r_A) = k_1 C_A$$

$$B \rightarrow D, \quad \text{where } (-r_B) = k_2 C_B$$

These two reactions are independent, or uncoupled, because the rates of disappearance of A or B depend only on their respective concentrations. In another case, reactant A might be involved in two competing reactions, for example

$$A \rightarrow B, \quad \text{where } (-r_A)_1 = k_1 C_A$$

$$A \rightarrow C, \quad \text{where } (-r_A)_2 = k_2 C_A$$

In this case, the rates of formation of B and C both depend on the concentration of A, so these reactions are said to be coupled. The selectivity of the reaction is defined as the rate of production of one of these products divided by the rate of production of the other.

3.7.3 Series–Parallel Reactions

In a combined series and parallel reaction, both series and parallel reactions occur. For example, the product from a reaction might react with one of the reactants to form additional products. An example of this type of scheme is represented by the following set of overall reactions:

$$A + B \rightarrow C + D, \quad \text{where } (-r_A)_1 = k_1 C_A C_B$$

$$D + A \rightarrow E + F, \quad \text{where } (-r_A)_2 = k_2 C_A C_D$$

In this scenario, the reaction of A in reaction (3.2) (the formation of products E and F) will not start until some product D has been formed in the first reaction.

Many combinations and permutations of these examples are possible. The complexity of the reaction scheme will increase with the number of reactants that are present. In some feed streams (e.g., process streams in oil refineries), literally thousands of species are present. In such a case there may be hundreds of reactions occurring; it is common in such cases to represent the kinetics in terms of an average rate and some model compounds.

Example 3.4 illustrates the analysis of PFR and CSTR with multiple reactions.

Example 3.4

Consider the following system of series–parallel liquid-phase reactions.

$$A + B \xrightarrow{k_1} C + D, \quad \text{where } (-r_A)_1 = k_1 C_A C_B \quad \text{(Reaction 1)}$$

$$A + C \xrightarrow{k_2} E, \quad \text{where } (-r_A)_2 = k_2 C_A C_C \quad \text{(Reaction 2)}$$

The rate constants have values of

$$k_1 = 20\,\text{m}^3/\text{mol} \cdot \text{s}, \quad k_2 = 10\,\text{m}^3/\text{mol} \cdot \text{s}$$

In this system, the reactant A reacts with one of the products, C, to form a further product, E. Consider a feed stream containing equal concentrations of A and B, such that

$$C_{A0} = C_{B0} = 1 \times 10^{-3}\,\text{mol}/\text{m}^3$$

The space time for the reactor is to be 100 s. Determine the outlet concentrations of all the species in:

a. A CSTR
b. A PFR

SOLUTION

a. CSTR

We begin by writing the mole balance equations for each species. The general mole balance equation for a CSTR may be written as

$$F_{j0} - F_{jE} = (-r_j)_E V \tag{3.112}$$

where $(-r_j)_E$ is the rate of disappearance of component j evaluated at the outlet concentration of the reactor. Note that subscript E is used here to emphasize the outlet condition. The rate of disappearance must account for all the reactions in which the species participates. We write the mole balance for each species present in the reactor. The mole balance for species A is

$$F_{A0} - F_{AE} - (-r_A)_{1,E} V - (-r_A)_{2,E} V = 0 \tag{3.113}$$

Molecules of A are consumed in both reactions, so the rates of each are included. The molar flow rates can be expressed in terms of the volumetric flow rate and the concentration, $F_A = C_A Q$. Because reaction occurs in the liquid phase, the density is assumed constant, and hence Q is also assumed to be constant. Substitute the rate expressions into the mole balance equation to get

$$Q C_{A0} - Q C_{AE} - k_1 C_{A,E} C_{B,E} V - k_2 C_{A,E} C_{C,E} V = 0 \tag{3.114}$$

Writing Equation 3.114 terms of the space time gives

$$\frac{1}{\tau} C_{A0} - \frac{1}{\tau} C_{AE} - k_1 C_{A,E} C_{B,E} - k_2 C_{A,E} C_{C,E} = 0 \tag{3.115}$$

Component B reacts only in reaction (3.1). The rate of disappearance of B is the same as the rate of disappearance of A in this reaction because 1 mol of B reacts for every mole of A that reacts. Therefore

$$F_{B0} - F_{BE} - (-r_A)_{1,E} V = 0 \tag{3.116}$$

or, in terms of concentration and space time

$$\frac{1}{\tau} C_{B0} - \frac{1}{\tau} C_{BE} - k_1 C_{A,E} C_{B,E} = 0 \tag{3.117}$$

C is formed in reaction (3.1) and consumed in reaction (3.2). The rate of formation of C in reaction (3.1) is equal to minus the rate of disappearance of A in reaction (3.1). The rate of disappearance of C in reaction (3.2) is the same as the rate of disappearance of A in reaction (3.2). The mole balance for C is therefore

$$F_{C0} - F_{CE} + (-r_A)_{1,E} V - (-r_A)_{2,E} V = 0 \tag{3.118}$$

In terms of concentration and space time

$$\frac{1}{\tau} C_{C0} - \frac{1}{\tau} C_{CE} + k_1 C_{A,E} C_{B,E} - k_2 C_{A,E} C_{C,E} = 0 \tag{3.119}$$

Component D is formed in reaction (3.1) at a rate equal to minus the rate of disappearance of A. The mole balance for component D is therefore

$$F_{D0} - F_{DE} + (-r_A)_{1,E}V = 0 \tag{3.120}$$

In terms of concentrations and space time

$$\frac{1}{\tau}C_{D0} - \frac{1}{\tau}C_{DE} + k_1 C_{A,E} C_{B,E} = 0 \tag{3.121}$$

Component E is formed in reaction (3.2) at a rate equal to minus the rate of disappearance of A in reaction (3.2). The mole balance for component E is therefore

$$F_{E0} - F_{EE} + (-r_A)_{2,E}V = 0 \tag{3.122}$$

In terms of concentrations and space time

$$\frac{1}{\tau}C_{E0} - \frac{1}{\tau}C_{EE} + k_2 C_{A,E} C_{C,E} = 0 \tag{3.123}$$

Equations 3.115, 3.117, 3.119, 3.121, and 3.123 are a system of five equations containing five unknowns. Substituting for the known parameters of space time, inlet concentrations, and rate constants gives a system of equations to solve:
From Equation 3.115

$$\frac{1}{100}1 \times 10^{-3} - \frac{1}{100}C_{AE} - 20C_{A,E}C_{B,E} - 10C_{A,E}C_{C,E} = 0 \tag{3.124}$$

From Equation 3.117

$$\frac{1}{100}1 \times 10^{-3} - \frac{1}{100}C_{BE} - 20C_{A,E}C_{B,E} = 0 \tag{3.125}$$

From Equation 3.119

$$-\frac{1}{100}C_{CE} + 20C_{A,E}C_{B,E} - 10C_{A,E}C_{C,E} = 0 \tag{3.126}$$

From Equation 3.121

$$-\frac{1}{100}C_{DE} + 20C_{A,E}C_{B,E} = 0 \tag{3.127}$$

From Equation 3.123

$$-\frac{1}{100}C_{EE} + 10C_{A,E}C_{C,E} = 0 \tag{3.128}$$

Equations 3.124 through 3.128 comprise a coupled system of five equations that can be solved by a suitable analytical or numerical method to give the following values of outlet concentration:

$$C_{AE} = 4.14 \times 10^{-4}\,\text{mol}/\text{m}^3, \quad C_{BE} = 5.47 \times 10^{-4}\,\text{mol}/\text{m}^3, \quad C_{CE} = 3.20 \times 10^{-4}\,\text{mol}/\text{m}^3,$$

$$C_{DE} = 4.53 \times 10^{-4}\,\text{mol}/\text{m}^3, \quad C_{EE} = 1.33 \times 10^{-4}\,\text{mol}/\text{m}^3$$

b. PFR

The PFR problem is handled in a manner similar to that used for the analysis of the CSTR. That is, the mole balance equations are written for each of the species, then the resulting system of equations is solved. In this case, however, we will have a system of differential equations to solve. The general mole balance equation for the PFR is

$$-\frac{dF_j}{dV} = (-r_j) \tag{3.129}$$

The reaction occurs in a liquid phase; therefore, we will assume that the volumetric flow rate is constant. The mole balance may be written in terms of the concentration and the space time.

$$-\frac{dF_j}{dV} = -Q\frac{dC_j}{dV} = -\frac{dC_j}{d\tau} = (-r_j) \tag{3.130}$$

The space time is selected as the independent variable because we are analyzing the system based on a known value of the space time. As in the CSTR case, the rate of reaction for each species in the system must account for its participation in each reaction. Species A reacts in reactions (3.1) and (3.2); therefore, the mole balance equation for A is

$$-\frac{dC_A}{d\tau} = k_1 C_A C_B + k_2 C_A C_C \tag{3.131}$$

B disappears by reaction (3.1) only:

$$-\frac{dC_B}{d\tau} = k_1 C_A C_B \tag{3.132}$$

C is formed in reaction (3.1) and disappears in reaction (3.2):

$$-\frac{dC_C}{d\tau} = -k_1 C_A C_B + k_2 C_A C_C \tag{3.133}$$

D is formed in reaction (3.1):

$$-\frac{dC_D}{d\tau} = -k_1 C_A C_B \tag{3.134}$$

E is formed in reaction (3.2):

$$-\frac{dC_E}{d\tau} = -k_2 C_A C_C \tag{3.135}$$

Equations 3.131 through 3.135 are a system of ordinary differential equations that must be solved simultaneously using a numerical method. The solution obtained gives the following values for the outlet concentrations of the five species:

$$C_{AE} = 2.25 \times 10^{-4}\, mol/m^3, \quad C_{BE} = 3.75 \times 10^{-4}\, mol/m^3, \quad C_{CE} = 4.75 \times 10^{-4}\, mol/m^3,$$

$$C_{DE} = 6.25 \times 10^{-4}\, mol/m^3, \quad C_{EE} = 1.50 \times 10^{-4}\, mol/m^3$$

Note the difference in the outlet product distribution compared with the CSTR. It is not *a priori* obvious which reactor type will give a better product distribution for complex reaction schemes.

3.8 Multiple-Reactor Systems

In some circumstances, it may be either desirable or necessary to use a system of reactors. Sometimes a sequence of the same type of reactor of equal volumes is used. In other cases, the reactors might be of different type or size. They might also be operated in series or in parallel. When reactors are operated in series, the effluent from any given reactor forms the feed stream to the next in the series. In a parallel system of reactors, a feed stream is divided into substreams, each of which is fed to a reactor. The effluent streams from all the reactors are then collected and mixed. Sometimes there are additional feed streams to some of the reactors in the system.

In the following section, various types of reactor systems are discussed. It is assumed in each case that all the reactors are operating at the same temperature and that no reaction occurs in any transfer piping between reactors.

3.8.1 Plug Flow Reactors in Series

Consider a system of a series of PFRs with no side streams between subsequent reactors, as shown in Figure 3.6. Three reactors are shown in this figure; in any given case, there might be fewer or more reactors. Denote with subscript j any reactor in the series, then the mole balance for reactor j in differential form is

$$-\frac{dF_A}{dV} = (-r_A) \tag{3.136}$$

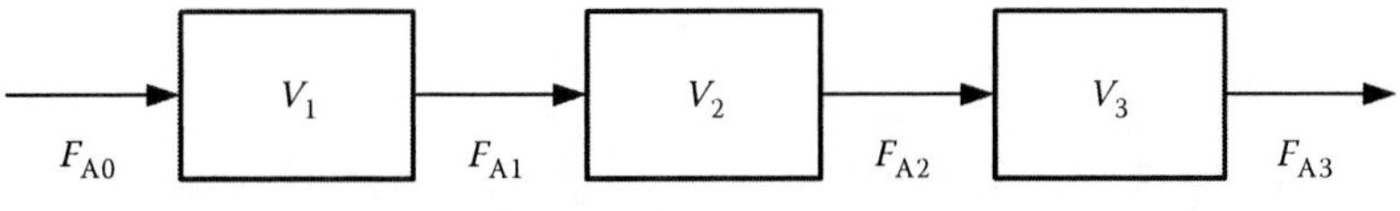

FIGURE 3.6
A system of plug flow reactors in series. The entire process stream is fed to the first reactor. The effluent from each reactor forms the feed stream to the next reactor.

The inlet molar flow rate to reactor j is the effluent from the previous reactor, denoted $j-1$. The input molar flow rate is therefore $F_{A,j-1}$ and the output molar flow rate is F_{Aj}. The integral form of the mole balance for reactor j is therefore

$$V_j = -\int_{F_{A,j-1}}^{F_{Aj}} \frac{dF_A}{(-r_A)} \tag{3.137}$$

The total volume, V_t, of a series of N reactors would be given by

$$V_t = \sum_{j=1}^{N} V_j \tag{3.138}$$

Equation 3.138 can be expressed in terms of the mole balance for each reactor

$$V_t = -\int_{F_{A0}}^{F_{A1}} \frac{dF_A}{(-r_A)} - \int_{F_{A1}}^{F_{A2}} \frac{dF_A}{(-r_A)} - \cdots - \int_{F_{A,j-1}}^{F_{Aj}} \frac{dF_A}{(-r_A)} - \int_{F_{Aj}}^{F_{A,j+1}} \frac{dF_A}{(-r_A)} - \cdots - \int_{F_{A,N-1}}^{F_{AN}} \frac{dF_A}{(-r_A)}$$

These integrals, however, may be added to give a single integral, as follows:

$$V_t = -\int_{F_{A0}}^{F_{AN}} \frac{dF_A}{(-r_A)} \tag{3.139}$$

In other words, the total volume of all the reactors is obtained by integrating the mole balance equation between the inlet to the first reactor and the effluent from the last one. This means that the series of PFR behaves in the same way as a single PFR of volume V_t.

The mole balance on any reactor may be calculated in terms of a fractional conversion of the feed to the first reactor. That is, define

$$X_j = \frac{F_{A0} - F_{Aj}}{F_{A0}} \tag{3.140}$$

The volume of any reactor in the series may be expressed in terms of this conversion as

$$V_j = F_{A0} \int_{X_{j-1}}^{X_j} \frac{dX_A}{(-r_A)} \tag{3.141}$$

3.8.2 Plug Flow Reactors in Parallel

In a parallel system of PFRs, the feed stream is divided into parts, each of which is fed to a single reactor, as shown in Figure 3.7. Each reactor can be analyzed individually, provided that the feed distribution among the various reactors is known. The overall conversion from

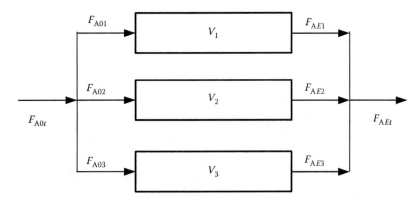

FIGURE 3.7
A system of plug flow reactors in parallel. The total molar flow rate is divided among the reactors in any desired manner.

the system of reactors can be determined by performing a mole balance on the point of convergence of the effluent streams. It may be shown that for a given temperature and total molar flow rate, the highest overall conversion is achieved with equal space velocity for each of the reactors. A common industrial arrangement is to have a multitubular reactor consisting of many (perhaps hundreds) parallel tubes of equal size. If each reactor has the same operating conditions, then any one tube may be considered to be representative of the reactor as a whole.

3.8.3 CSTR in Parallel

A parallel arrangement of CSTR is shown in Figure 3.8. The analysis of this system is similar to the analysis of the system of parallel PFR, as each reactor can be analyzed separately, provided that the flow distribution is known.

For a system of parallel CSTR, the highest overall conversion is achieved when the conversion is the same in each reactor. If the temperature in each reactor is the same, it follows that the space time τ is the same for all reactors. In other words, the total flow rate is divided among the reactors according to their volume. A system of N parallel CSTR of equal space times will give the same overall conversion as a single CSTR

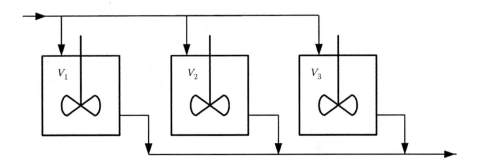

FIGURE 3.8
A system of CSTR in parallel. The total process stream is divided up among the reactors. The effluent streams from the reactors are then mixed.

with a volume (V_t) equal to the sum of the volumes of all the CSTR in parallel. In other words

$$V_t = \sum_{j=1}^{N} V_j \tag{3.142}$$

3.8.4 CSTR in Series

A system of CSTR in series is illustrated in Figure 3.9. In this arrangement, the effluent from one reactor forms the feed stream to the next in the series. In this section, we consider the case where there are no side streams between reactors. This latter case will be analyzed shortly.

The mole balance for reactant A can be written for any reactor in the series. For any reactor j, the mole balance is

$$(-r_A)_j = \frac{F_{A,j-1} - F_{A,j}}{V_i} \tag{3.143}$$

The fractional conversion of A, X_A, can be calculated on the basis of the molar flow rate to the first reactor, F_{A0}. The conversion from reactor j and reactor $j-1$ can thus be defined as

$$X_{A,j} = \frac{F_{A0} - F_{A,j}}{F_{A0}} \quad \text{and} \quad X_{A,j-1} = \frac{F_{A0} - F_{A,j-1}}{F_{A0}}$$

The molar flow rates from reactors j and $j-1$ in terms of this conversion are

$$F_{A,j} = F_{A0}\left(1 - X_{A,j}\right) \quad \text{and} \quad F_{A,j-1} = F_{A0}\left(1 - X_{A,j-1}\right)$$

The difference in molar flow rate of A between the inlet and outlet of any reactor is defined as

$$F_{A,j-1} - F_{A,j} = F_{A0}\left(X_{A,j} - X_{A,j-1}\right)$$

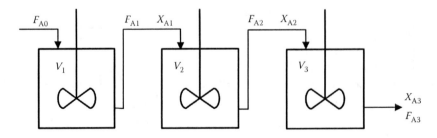

FIGURE 3.9
A system of CSTR in series. The entire process stream is fed to the first reactor. The effluent from each reactor forms the feed stream to the next reactor.

The mole balance equation for reactor j written in terms of the conversion and the molar flow rate to the first reactor is then written as

$$(-r_A)_j = \frac{F_{A0}(X_{A,j} - X_{A,j-1})}{V_j} = \frac{F_{A0}\Delta X}{V_j} \tag{3.144}$$

Equation 3.144 can be rearranged in terms of the conversion from reactor j:

$$X_{A,j} = X_{A,j-1} + \frac{(-r_A)_j V_j}{F_{A0}} \tag{3.145}$$

Equation 3.145 can be solved for each reactor sequentially to calculate the incremental conversion from each reactor in the series.

3.8.4.1 Minimum Total Reactor Volume for CSTR in Series

In this section, we examine a sequence of CSTR in series and determine the minimum total reactor volume required to achieve a given conversion for a given set of feed conditions to the first reactor. Consider the reactor system shown in Figure 3.10. The mole balance equation can be written for each CSTR, with the conversion in any reactor defined in terms of the molar flow rate of A to the first reactor:

Reactor 1

$$V_1 = \frac{F_{A0}}{(-r_A)_1} X_{A1}$$

Reactor 2

$$V_2 = \frac{F_{A0}}{(-r_A)_2}(X_{A2} - X_{A1})$$

The inlet molar flow rate divided by the reaction rate can be plotted as a function of conversion for this reactor system, as shown in Figure 3.11. Note that this type of behavior is valid for "normal" kinetics, that is, the rate of disappearance of reactant A decreases as the

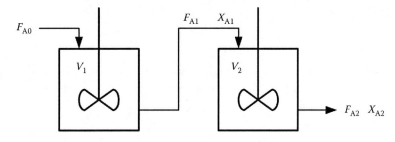

FIGURE 3.10
A system of two CSTR in series.

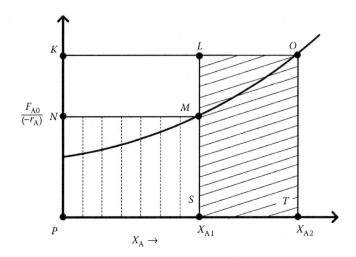

FIGURE 3.11
Conversion as a function of inverse reaction rate for two CSTR in series. The hatched areas are equal to the reactor volumes.

concentration of A decreases. On this plot, each reactor volume is equal to the hatched area of a rectangle. For example, rectangle *PSMN* is the volume of reactor 1 and rectangle *SLOT* is the volume of reactor 2. The rectangle *PKOT* is the volume of a single reactor required to achieve the same conversion as the two reactors in series. The area of rectangle *KLMN* is the reactor volume "saved" by using two reactors rather than one. Any sequencing and sizing of the reactors should aim to maximize the area *KLMN* to give the minimum total reactor volume, which is an optimization problem.

The minimization problem can be posed in the following terms. Find the value of X_{A1} that minimizes the total reactor volume. We first write the equation for the total reactor volume:

$$V_1 + V_2 = F_{A0} \frac{X_{A1}}{(-r_A)_1} + F_{A0} \frac{X_{A2} - X_{A1}}{(-r_A)_2} \tag{3.146}$$

The minimum total volume is found by taking the derivative of the total volume with respect to X_{A1}, and setting the result equal to zero. Thus, take the derivative of Equation 3.1 with respect to X_{A1}, and set the result equal to zero:

$$\left[\frac{d(V_1 + V_2)}{dX_{A1}} \right]_{X_{A2}} = 0 = \frac{F_{A0}}{(-r_A)_1} + F_{A0} X_{A1} \left(\frac{d(1/(-r_A)_1)}{dX_{A1}} \right) - \frac{F_{A0}}{(-r_A)_2} \tag{3.147}$$

Rearrange Equation 3.147 to give the condition that gives the minimum volume. The minimum volume for the reactor system is achieved when the following condition is satisfied:

$$\frac{d(1/(-r_A)_1)}{dX_{A1}} = \frac{(1/(-r_A)_2) - (1/(-r_A)_1)}{X_{A1}} \tag{3.148}$$

The methodology used above can be generalized to a system of N reactors in series. For a series of reactors, the outlet conversion from the last reactor in the series is fixed, while the other conversions can be varied to minimize the total volume. For example, for three reactors, X_{A1} and X_{A2} are adjustable; for four reactors, X_{A1}, X_{A2}, and X_{A3}, are adjustable, and so on. A global minimization is achieved if the rate of change of the total reactor volume with respect to a change in the conversion in every reactor is equal to zero. This concept can be expressed mathematically for reactor j as

$$\frac{d\left(1/(-r_A)_j\right)}{dX_{A,j}} = \frac{\left(1/(-r_A)_{j+1}\right) - \left(1/(-r_A)_j\right)}{X_{A,j} - X_{A,j-1}} \tag{3.149}$$

For a system of N reactors, Equation 3.149 is applied for reactors $1 < j < N - 1$. When all the equations are satisfied, the total reactor volume is a minimum. The use of Equation 3.149 is illustrated in Example 3.5.

Example 3.5

Consider a system of three CSTR in series, as shown in Figure 3.9. The following liquid-phase reaction occurs:

$$A + B \rightarrow C + D$$

The volumetric feed flow rate is 2.5×10^{-4} m³/s. The concentrations of A and B in the feed are 2.5×10^3 and 3.5×10^3 mol/m³, respectively. The rate of reaction is

$$(-r_A) = kC_A^{1.2}C_B \text{ mol/m}^3\text{s}$$

where

$$k = 1 \times 10^{-6} \left[\left(\frac{m^3}{mol}\right)^{1.2} \frac{1}{s}\right]$$

Calculate the volume of each reactor in the series when the outlet conversion from the third reactor is equal to 0.8266 and the total reactor volume is a minimum.

SOLUTION

It is desired to have a minimum total reactor volume with the conversion at the outlet of the third reactor equal to 0.95. The general condition for a minimum volume of the jth reactor is

$$\frac{d\left(1/(-r_A)_j\right)}{dX_{A,j}} = \frac{\left(1/(-r_A)_{j+1}\right) - \left(1/(-r_A)_j\right)}{X_{A,j} - X_{A,j-1}} \tag{3.150}$$

The reaction rate in terms of conversion, with substitution of the inlet concentrations, is

$$(-r_A) = 41.84(1 - X_A)^{1.2}\left(1 - \frac{5}{7}X_A\right) \tag{3.151}$$

It is necessary to take the derivative of the inverse rate.

$$\frac{d(1/(-r_A))}{dX_A} = \frac{1}{41.84} \frac{\left(1-(5/7)X_A\right)1.2\left(1-X_A\right)^{0.2} + (5/7)\left(1-X_A\right)^{1.2}}{\left[\left(1-X_A\right)^{1.2}\left(1-(5/7)X_A\right)\right]^2} \tag{3.152}$$

Equation 3.152 can be simplified

$$\frac{d\left(1/(-r_A)\right)}{dX_A} = \frac{1}{41.84}\left[\frac{1.914 - 1.571 X_A}{\left(1-X_A\right)^{2.2}\left(1-(5/7)X_A\right)^2}\right] \tag{3.153}$$

Equations 3.151 and 3.153 can be substituted into Equation 3.150 for reactors 1 and 2 to give the following two equations:

Reactor 1

$$\frac{1.914 - 1.571 X_{A1}}{\left(1-X_{A1}\right)^{2.2}\left(1-(5/7)X_{A1}\right)^2} =$$

$$\frac{\left[\left(1-X_{A2}\right)^{1.2}\left(1-(5/7)X_{A2}\right)\right]^{-1} - \left[\left(1-X_{A1}\right)^{1.2}\left(1-(5/7)X_{A1}\right)\right]^{-1}}{X_{A1}} \tag{3.154}$$

Reactor 2

$$\frac{1.914 - 1.571 X_{A2}}{\left(1-X_{A2}\right)^{2.2}\left(1-(5/7)X_{A2}\right)^2} =$$

$$\frac{\left[\left(1-X_{A3}\right)^{1.2}\left(1-(5/7)X_{A3}\right)\right]^{-1} - \left[\left(1-X_{A2}\right)^{1.2}\left(1-(5/7)X_{A2}\right)\right]^{-1}}{X_{A2} - X_{A1}} \tag{3.155}$$

The value of $X_{A3} = 0.8266$ is known; therefore, the numerical solution of Equations 3.154 and 3.155 gives values of

$$X_{A1} = 0.4819 \quad \text{and} \quad X_{A2} = 0.7080$$

These conversion values are substituted into the CSTR mole balance equation to obtain the reactor volumes.

$$V_1 = \frac{F_{A0}X_{A1}}{\left(-r_A\right)_1} = 2.42 \times 10^{-2}\,\text{m}^3$$

$$V_2 = \frac{F_{A0}\left(X_{A2} - X_{A1}\right)}{\left(-r_A\right)_2} = 2.99 \times 10^{-2}\,\text{m}^3$$

$$V_3 = \frac{F_{A0}\left(X_{A3} - X_{A2}\right)}{\left(-r_A\right)_3} = 3.54 \times 10^{-2}\,\text{m}^3$$

Comments: The total reactor volume is 8.95×10^{-2} m^3, compared with 9.00×10^{-2} m^3 for three equally sized reactors.

Note that for nth-order kinetics, that is, where $(-r_A) = kC_A^n$, we can make the following general observation. When a variety of differently sized CSTR are used in series, the ordering of the reactors to maximize conversion is

$n > 1$ order in increasing size will maximize conversion
$n = 1$ order has no effect on conversion
$n < 1$ order in decreasing size (large ones first) maximize conversion

If you are *designing* a system of series of CSTR with no side streams, the volume of all reactors is often made equal. Although this is not optimal from a point of view of minimizing the total reactor volume, there is often an advantage in having all equipment the same size.

3.8.5 CSTR–PFR in Series

In some cases, a combination of CSTR and PFR in series might be used. Depending on the reaction kinetics, the sequencing of the reactors may affect the conversion and product distribution. When analyzing a system of CSTR/PFR, it is possible to analyze each reactor by writing the mole balances around each reactor and solving for the outlet molar flow rates from each one in turn. We consider the example of a single PFR and a single CSTR in series, with a uniform temperature in both reactors. The case of a constant-density system is considered. The mole balances around each reactor system will be illustrated, with the objective of determining which sequence of reactors will give the highest conversion for a first-order isothermal reaction.

Case 1. CSTR followed by PFR

We consider first a case where the CSTR is placed first, and the PFR follows, as illustrated in Figure 3.12. Consider a first-order irreversible reaction with constant volumetric flow rate Q. The CSTR mole balance equation is

$$(-r_A) = \frac{F_{A0} - F_{A1}}{V_{CSTR}}, \quad \text{where } (-r_A) = kC_A \tag{3.156}$$

Substitute for a first-order reaction and write the equation in terms of fractional conversion:

$$k\frac{F_{A0}}{Q}(1 - X_{A1}) = \frac{F_{A0}X_{A1}}{V_{CSTR}} \tag{3.157}$$

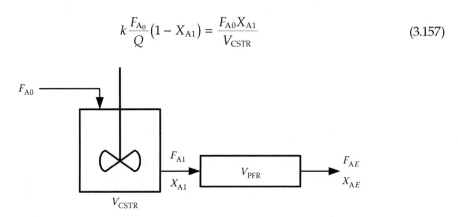

FIGURE 3.12
Representation of CSTR and a PFR in series, with the CSTR placed first.

Introducing the space time and rearranging give an explicit expression for the conversion in the CSTR. Note that nothing new has been introduced in this step, and the following equation is valid for any isothermal CSTR with a constant-density system and a first-order reaction.

$$X_{A1} = \frac{k\tau_{CSTR}}{1 + k\tau_{CSTR}} \tag{3.158}$$

Now we consider the PFR. The general mole balance equation for the PFR is

$$(-r_A) = -\frac{dF_A}{dV} \tag{3.159}$$

Substituting the rate equation for a first-order reaction and writing Equation 3.159 in terms of the fractional conversion of A, where the fractional conversion of A is based on the inlet molar flow rate to the first reactor (the CSTR), give

$$k\frac{F_{A0}}{Q}(1 - X_A) = F_{A0}\frac{dX_A}{dV} \tag{3.160}$$

The integral form of Equation 3.160 is

$$k\tau_{PFR} = \int_{X_{A1}}^{X_{AE}} \frac{dX_A}{(1 - X_A)} = -\ln(1 - X_A)\Big|_{X_{A1}}^{X_{AE}}$$

Substitution of the limits gives an expression for the conversion from the PFR:

$$X_{AE} = 1 - (1 - X_{A1})\exp(-k\tau_{PFR})$$

Substitute for X_{A1} from the CSTR solution, Equation 3.158, to obtain the expression for the overall conversion:

$$X_{AE} = 1 - \frac{\exp(-k\tau_{PFR})}{(1 + k\tau_{CSTR})} \tag{3.161}$$

Case 2. PFR followed by CSTR

Now let us reverse the order of the reactors. This sequence of reactors is shown in Figure 3.13.

The mole balance for the PFR is the same as for Case 1, except that the limits on the integral are different. Thus, write

$$k\tau_{PFR} = \int_{0}^{X_{A1}} \frac{dX_A}{(1 - X_A)} = -\ln(1 - X_A)\Big|_{0}^{X_{A1}} \tag{3.162}$$

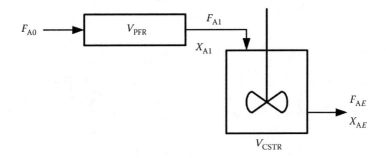

FIGURE 3.13
Representation of CSTR and a PFR in series, with the PFR placed first.

Substitution of the limits gives an expression for the conversion from the PFR:

$$X_{A1} = 1 - \exp(-k\tau_{PFR}) \tag{3.163}$$

The mole balance for the CSTR is

$$(-r_A) = \frac{F_{A1} - F_{AE}}{V_{CSTR}} \tag{3.164}$$

Substituting for the first-order rate expression and the conversion based on the feed flow rate to the first reactor gives

$$k\frac{F_{A0}}{Q}(1 - X_{AE}) = \frac{F_{A0}(X_{AE} - X_{A1})}{V_{CSTR}} \tag{3.165}$$

In terms of the space time, the conversion is

$$X_{AE} - X_{A1} = k\tau_{CSTR}(1 - X_{AE}) \tag{3.166}$$

The outlet conversion from the CSTR is thus given by

$$X_{AE} = \frac{X_{A1} + k\tau_{CSTR}}{1 + k\tau_{CSTR}} \tag{3.167}$$

Substitute for X_{A1}, the conversion from the PFR from Equation 3.163:

$$X_{AE} = 1 - \frac{\exp(-k\tau_{PFR})}{(1 + k\tau_{CSTR})} \tag{3.168}$$

Equation 3.168 is the same as Equation 3.161, the solution for Case 1. Therefore, for a first-order reaction and isothermal conditions, the order of the reactors is not important. This is owing to the fact that *conversion* does not depend on initial concentration for a first-order reaction. For non-first-order reactions, the order of the reactors is important. For power law kinetics

$$(-r_A) = kC_A^n$$

the reactor sequence to give the highest conversion would be

$n = 1$ The order is not important for any PFR–CSTR combination.

$n > 1$ The PFR should be first, and then followed by the CSTR. If more than one CSTR is used, place them in order of increasing size.

$0 < n < 1$ The CSTR should be first, and then followed by the PFR. If more than one CSTR is used, place them in order of decreasing size.

3.8.6 Reactors with Side Streams

Sometimes the feed to a series of reactors is not all fed to the first reactor, but rather portions are added to different reactors. This procedure might be followed to control the product distribution, for example. When reactors in series are operated in such a manner, it is possible to analyze each reactor in the system by writing the mole balances, taking into account all of the feed streams to each reactor. This type of problem is illustrated in the following example.

Example 3.6

Consider a liquid-phase reaction occurring in two CSTR connected in series. The volume of each reactor is 100 L. The reaction is

$$A + B \xrightarrow{\ k\ } C, \quad \text{where} \left(-r_A\right) = kC_A C_B$$

with $k = 1$ L/mol s to give the rate in mol/L s.

The feed to the first reactor consists of 1 L/s of a solution of 1 mol/L of A and 0.2 mol/L of B. The feed to the second reactor is the effluent from the first reactor plus another feed stream consisting of 1 L/s of a solution containing 0.8 mol/L of B (no A in this stream). This arrangement is shown in Figure 3.14. Calculate the molar flow rate of C from the two reactors.

SOLUTION

The reaction is in the liquid phase so the volumetric flow rate is assumed constant. Note that the volumetric flow rate to reactor two is higher than that to reactor one because of the additional feed stream.

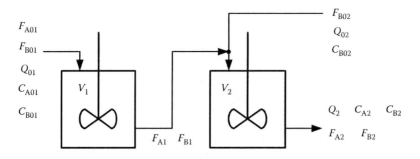

FIGURE 3.14
Reactor arrangement for Problem 3.6. Two CSTR in series with an additional feed stream to the second reactor.

Reactor 1

For the second-order reaction, the mole balance for A in reactor 1 is

$$F_{A0} - F_{A1} - kC_{A1}C_{B1}V_1 = 0 \tag{3.169}$$

In terms of volumetric flow rate

$$Q_0\left(C_{A0} - C_{A1}\right) - kC_{A1}C_{B1}V_1 = 0 \tag{3.170}$$

The concentration of A and B can be related using the stoichiometry, that is

$$C_{B1} = C_{B0} - \left(C_{A0} - C_{A1}\right) \tag{3.171}$$

Therefore, with Equation 3.171 and the following values:

$$Q_0 = 1\,\text{L/s}, \quad k = 1\,\text{L/mol} \cdot \text{s}, \quad C_{A0} = 1\,\text{mol/L}, \quad \text{and} \quad C_{B0} = 0.2\,\text{mol/L}$$

Equation 3.170 becomes

$$1 - C_{A1} - C_{A1}\left(0.2 - 1 + C_{A1}\right)100 = 0 \tag{3.172}$$

Equation 3.172 can be simplified to the following quadratic equation:

$$C_{A1}^2 - 0.79C_{A1} - 0.01 = 0 \tag{3.173}$$

Solving this quadratic equation gives a value of $C_{A1} = 0.8025\,\text{mol/L}$. As the flow rate is $Q = 1$ L/s, the molar flow rate of A in the effluent is

$$F_{A1} = C_{A1}Q = 0.8025\,\text{mol/s}$$

The number of moles of A that react in reactor 1 is

$$F_{A0} - F_{A1} = 1 - 0.8025 = 0.1975$$

Thus, the molar flow rate of C in the effluent is 0.1975 mol/s. The molar flow rate of B is

$$F_{B1} = F_{B0} - \left(F_{A0} - F_{A1}\right) = 0.2 - (1 - 0.8025) = 0.0025\,\text{mol/s}$$

Reactor 2

Now consider the second reactor. As there are two streams entering the reactor, we shall write the mole balances for A and B. For species A:

$$F_{A1} - F_{A2} - kC_{A2}C_{B2}V_2 = 0 \tag{3.174}$$

Note that the second feed stream to the second reactor will have a dilution effect on the inlet concentration of A. To avoid confusion, this problem is solved in terms of molar flow rates rather than concentrations. The mole balance for A is

$$F_{A1} - F_{A2} - V_2\frac{kF_{A2}F_{B2}}{\left(Q_{01} + Q_{02}\right)^2} = 0 \tag{3.175}$$

The mole balance for B is

$$F_{B1} + F_{B02} - F_{B2} - V_2 \frac{kF_{A2}F_{B2}}{\left(Q_{01} + Q_{02}\right)^2} = 0 \tag{3.176}$$

From the analysis of reactor 1, the inlet molar flow rates F_{A1} and F_{B1} are

$$F_{A1} = 0.8025 \text{ mol/s} \quad \text{and} \quad F_{B1} = 0.0025 \text{ mol/s}$$

The second feed stream to reactor 2 contains only B, and the molar flow rate is

$$F_{B02} = Q_{02}C_{B02} = 1 \text{ L/s} \times 0.8 \text{ mol/L} = 0.8 \text{ mol/s}$$

Now it is seen that the total molar flow rate of B to reactor 2, $F_{B1} + F_{B02}$, is 0.8025 mol/s, which is the same as the inlet molar flow rate of A. As 1 mol of A reacts for each mole of B that reacts, it follows that the concentration of A and B in reactor two must be equal. The mole balance for A is then written as

$$F_{A1} - F_{A2} - \frac{V_2 k F_{A2}^2}{\left(Q_{01} + Q_{02}\right)^2} = 0 \tag{3.177}$$

Substituting the numbers into this equation gives

$$0.8025 \text{ mol/s} - F_{A2} - \frac{100}{4}F_{A2}^2 = 0 \tag{3.178}$$

Solve Equation 3.178 to get $F_{A2} = 0.160$ mol/s. The other molar flow rates are

$$F_{B2} = F_{A2} = 0.160 \text{ mol/s}$$

$$F_{C2} = F_{A01} - F_{A2} = 1 - 0.160 = 0.84 \text{ mol/s}$$

3.9 Further Thoughts on Defining Conversion

In the analysis of systems of reactors in series illustrated in the preceding sections, the fractional conversion was defined in terms of the molar feed flow rate to the first reactor in the series. This convention is usually used when dealing with reactors in series. However, as conversion is a defined quantity, there are alternatives that may be encountered in the literature. Consider, for example, two CSTR in series, as shown in Figure 3.10. The mole balance equations for species A for reactors 1 and 2 are

$$F_{A0} - F_{A1} - (-r_A)V_1 = 0 \tag{3.179}$$

$$F_{A1} - F_{A2} - (-r_A)V_2 = 0 \tag{3.180}$$

These two equations are always valid. The fractional conversion from reactor 1 defined in terms of the feed flow rate is

$$X_{A1} = \frac{F_{A0} - F_{A1}}{F_{A0}} \quad \text{or} \quad F_{A1} = F_{A0}(1 - X_{A1})$$

Therefore, the mole balance equation written in terms of fractional conversion is

$$\frac{V_1}{F_{A0}} = \frac{X_{A1}}{(-r_A)}$$

For reactor 2, we have previously defined the outlet conversion in terms of the inlet flow rate to reactor 1

$$X_{A2} = \frac{F_{A0} - F_{A2}}{F_{A0}} \quad \text{or} \quad F_{A2} = F_{A0}(1 - X_{A2})$$

This definition gives the mole balance equation for reactor 2 in terms of conversion:

$$\frac{V_2}{F_{A0}} = \frac{X_{A2} - X_{A1}}{(-r_A)} \tag{3.181}$$

Equation 3.181 is the mole balance that we have used previously. However, it is also possible to define the fractional conversion in reactor 2 as the fractional conversion of the A *that enters reactor 2.*

$$X_{A2} = \frac{F_{A1} - F_{A2}}{F_{A1}} \quad \text{or} \quad F_{A2} = F_{A1}(1 - X_{A2})$$

The mole balance equation for reactor 2 is then written as

$$\frac{V_2}{F_{A1}} = \frac{X_{A2}}{(-r_A)} \tag{3.182}$$

Then, using the definition of conversion for reactor 1, we can write

$$\frac{V_2}{F_{A0}} = \frac{(1 - X_{A1})X_{A2}}{(-r_A)} \tag{3.183}$$

Obviously, Equation 3.183 is not the same as Equation 3.181. Note that X_{A2} will have a different numerical value depending on which definition is used. The molar flow rates are unchanged; this is simply an artifact of the differing definitions of X. This should illustrate the importance of clearly understanding the definitions used.

3.10 Transient Reactor Operation

Transient reactor operation is synonymous with unsteady-state operation. In a transient reactor, the conditions at a specific point in the reactor change with time. The batch reactor is always operated as a transient reactor. The PFR and CSTR operate at unsteady state during start-up and shut-down, or when changes in inlet conditions occur. In Chapter 4, where nonisothermal reactors are discussed, it will also be seen that transient operation can result when rates of heat exchange with the surroundings are altered. Sometimes reactors are deliberately run in the transient mode as a way of increasing conversion. On other occasions, the nature of the feed makes transient operation inevitable. For example, the reactors that are used on automobiles to destroy unwanted exhaust gas emissions are transient reactors, owing to the fact that the engine is operated at variable loads, resulting in a continuously changing exhaust gas composition. Transient reactor operation may also result from inadvertent changes in feed composition or temperature. The analysis of transient flow reactors is usually more complex than the analysis of steady-state reactors, and involves at the least the solution of a differential equation. The description in the following sections focuses primarily on the transient operation of CSTR.

3.10.1 Transient Plug Flow Reactor

The steady-state mole balance equation for a PFR was developed in Section 3.3. In transient operation, a term for the accumulation of a species in the reactor must be added, which gives the following mole balance equation for a PFR:

$$-\frac{\partial F_j}{\partial V} + r_j = \frac{\partial C_j}{\partial t} \tag{3.184}$$

Equation 3.184 is a partial differential equation because there are two independent variables, time and reactor volume. To solve this equation, both the initial conditions and the inlet conditions for the reactor must be known. The inlet conditions can vary with time (such as in the automobile exhaust gas converter previously mentioned), making the problem more difficult to solve. Partial differential equations must be solved numerically. Detailed analysis of a transient PFR is beyond the scope of this text.

3.10.2 Single Transient CSTR

Many variations are possible for transient CSTR operation. For example, a process may begin with an empty reactor, and then at time zero a feed stream is added to the reactor. Fluid accumulates in the vessel and at the same time reaction occurs. Once the reactor is full, an effluent stream may commence, or, alternatively, the feed stream may be discontinued. Another possibility is to have a reactor start-up with the reactor initially full of an inert solvent; then a feed stream containing reactants is added with an effluent stream of equal mass flow rate. A third possibility is that the reactor is running at steady state and a change in inlet concentration of reactant occurs. Some of these possibilities will be considered in the following.

The general mole balance for species j in a transient CSTR is developed using the principles described earlier in the chapter. The accumulation of moles over a time interval Δt is included in the mole balance equation to give

$$F_{j0} - F_{jE} + r_j V = \frac{\left(N_j\right)_{t+t\Delta t} - \left(N_j\right)_t}{\Delta t} \tag{3.185}$$

Taking the limit of an infinitesimally small time interval, the equation in differential form is

$$F_{j0} - F_{jE} + r_j V = \frac{dN_j}{dt} = \frac{d\left(C_{jE}V\right)}{dt} \tag{3.186}$$

In Equation 3.186, N_j is the number of moles of j in the reactor. As shown in the last term, the number of moles can be expressed in terms of the concentration and the reactor volume. Both F_{j0} and V can be functions of time, depending on the method of reactor operation. The concentration of j in the effluent stream (if present) is the same as the instantaneous concentration within the tank, but also changes with time.

3.10.2.1 Transient CSTR with Constant Volume

We consider first the case of a CSTR with the reactant in liquid solution (constant-density system) and the CSTR initially filled to the working volume, which will be held constant. At time zero, the reactant concentration in the feed stream is changed. The inlet and outlet mass flow rates are set equal, which means that the reactor volume is constant. Once the change to the inlet concentration of reactant A has been made, this concentration is held constant. This type of change is known as a step input change to the reactor. The inlet volumetric flow rate of A is also constant. Therefore, the mole balance equation is

$$F_{A0} - F_{AE} - (-r_A)V = V\frac{dC_{AE}}{dt} + C_{AE}\frac{dV}{dt} \tag{3.187}$$

The term $C_{AE}(dV/dt)$ is equal to zero because the reactor volume is constant for all time. Equation 3.187 can be rearranged to give

$$\frac{dC_{AE}}{dt} = \frac{F_{A0} - F_{AE}}{V} - (-r_A)_E \tag{3.188}$$

Equation 3.188 can be solved to give the reactor concentration as a function of time. For most forms of the rate expression, the solution has to be performed numerically.

Consider the case of a first-order irreversible reaction with a rate function

$$(-r_A)_E = kC_{AE} \tag{3.189}$$

For a constant-density fluid, we can write the mole balance in terms of concentration, noting that $F_A = QC_A$. For constant volume and volumetric flow rate, it is also convenient to write the mole balance using the space time, $\tau = (V/Q)$. The mole balance for A is then written as

$$\tau\frac{dC_{AE}}{dt} = C_{A0} - C_{AE} - \tau(-r_A) \tag{3.190}$$

For a first-order reaction, $(-r_A) = kC_{AE}$, we can substitute and rearrange the mole balance, Equation 3.190, to obtain

$$\frac{dC_{AE}}{dt} + \left(\frac{1 + \tau k}{\tau}\right)C_{AE} = \frac{C_{A0}}{\tau} \tag{3.191}$$

The solution to this equation is

$$C_{AE} = \frac{C_{A0}}{1 + \tau k} - \left(\frac{C_{A0}}{1 + \tau k} - C_{Ai}\right)\exp\left[-(1 + \tau k)\frac{t}{\tau}\right]$$

C_{Ai} is the initial concentration in the reactor at time $t = 0$. If the reactor is initially full of an inert solvent only, then $C_{Ai} = 0$. The initial concentration of A in the reactor would not be equal to zero if the reactor were initially running at some steady-state condition and the molar flow rate of A to the reactor was changed.

Consider the case where the initial concentration is zero, that is $C_{Ai} = 0$, and calculate the time required for the effluent concentration of A to reach 99% of the value computed from the steady-state mole balance equation. At steady state, the final effluent concentration is

$$C_{A,SS} = \frac{C_{A0}}{1 + \tau k} \tag{3.192}$$

Substitution into Equation 3.192 and setting $C_{Ai} = 0$ give

$$C_{AE} = C_{A,SS}\left(1 - \exp\left[-(1 + \tau k)\frac{t}{\tau}\right]\right) \tag{3.193}$$

Rearrange Equation 3.193 to an expression explicitly in time

$$t = \left(\frac{\tau}{1 + \tau k}\right)\ln\left(\frac{C_{A,SS}}{C_{A,SS} - C_{AE}}\right) \tag{3.194}$$

Substitute the desired value, $C_{AE} = 0.99C_{A,SS}$, into Equation 3.194 and denote t_S as the time required to reach this value. After simplification, this operation yields

$$t_S = \left(\frac{\tau}{1 + \tau k}\right)\ln\left(\frac{C_{A,SS}}{C_{A,SS} - 0.99C_{A,SS}}\right) = \left(\frac{\tau}{1 + \tau k}\right)\ln(100) = 4.6\left(\frac{\tau}{1 + \tau k}\right) \tag{3.195}$$

Two limiting cases can be identified. When $\tau k \ll 1$, it may be seen that $t_S \approx 4.6\tau$. When $\tau k \gg 1$, it may be seen that $t_S \approx (4.6/k)$. In other words, the time to reach 99% of the steady-state outlet concentration for a first-order reaction falls between the two limits:

$$\frac{4.6}{k} < t_S < 4.6\tau \tag{3.196}$$

Typical CSTR reach steady state after 3 to 4 τ.

3.10.2.2 Transient CSTR with Variable Volume: Semibatch Operation

A reactor in which there is an inlet feed stream and no outlet stream must have an increasing reaction volume with time, provided that the mass density remains constant. In a gas-phase system, the total volume of the reacting mixture could remain constant if the pressure in the reactor increased with time. In the following, we consider constant-density systems only. There are two scenarios under which such an operating condition might prevail:

1. A reactor may initially contain a solution of reactant A, and a feed stream containing reactant B is added. The reaction may be stopped when the reactor is full. This mode of operation is referred to as a semibatch operation. Alternatively, the feed stream may be turned off after a time, and the reactor contents might be left to react over time. Once the feed stream is discontinued, the reactor behaves as a batch reactor.

2. A CSTR may be started up from the empty state. In this situation, there is no effluent until the tank is full. Once the reactor operating volume is reached, the outlet mass flow rate is set equal to the inlet mass flow rate and the volume remains constant. The reactor will behave like the transient CSTR with constant volume, as described previously.

The reactor has no effluent stream; therefore, the mole balance can be written as

$$F_{A0} - (-r_A)V = \frac{dN_A}{dt} = \frac{d(VC_A)}{dt} \tag{3.197}$$

Note that the reactor volume, V, varies with time. In terms of inlet volumetric flow rate and concentration, the mole balance is

$$Q_0 C_{A0} - (-r_A)V = V\frac{dC_A}{dt} + C_A\frac{dV}{dt} \tag{3.198}$$

An expression for the change in reactor volume with time can be obtained from the overall mass balance equation. The overall mass balance is

$$\rho_0 Q_0 = \frac{d(\rho V)}{dt} \tag{3.199}$$

If the density of the system is constant, the density of the material inside the reactor is equal to the density of the entering feed stream. The change in volume with time in this case is given by

$$\frac{dV}{dt} = Q_0 \tag{3.200}$$

Equation 3.200 can be integrated to give

$$V = V_0 + Q_0 t \quad \text{or} \quad \tau = \tau_0 + t$$

Note that V_0 is the reactor volume at time zero, which might be equal to zero. Consider now a simple first-order reaction, where

$$(-r_A) = kC_A \tag{3.201}$$

Substituting the rate function into the mole balance equation gives

$$Q_0 C_{A0} - kC_A (V_0 + Q_0 t) = (V_0 + Q_0 t)\frac{dC_A}{dt} + C_A Q_0 \tag{3.202}$$

If the reactor is initially empty, $V_0 = 0$ and

$$Q_0 C_{A0} - kC_A Q_0 t = Q_0 t \frac{dC_A}{dt} + C_A Q_0 \tag{3.203}$$

Rearranging Equation 3.203 gives

$$C_{A0} - (1 + kt)C_A = t\frac{dC_A}{dt} \tag{3.204}$$

Equation 3.204 can be solved numerically to give the concentration of A as a function of time.

3.10.3 Transient CSTR in Series: Constant Volume

The transient analysis can easily be extended to a series of CSTR. The approach is once more to write down transient mole balances for each reactor, then to solve the resulting system of equations simultaneously. The methodology is illustrated for the case of two transient CSTR in series in which a reversible reaction occurs. The overall reaction is

$$A \underset{k_2}{\overset{k_1}{\rightleftharpoons}} B$$

The rate equation for the reversible reaction is

$$(-r_A) = k_1 C_A - k_2 C_B$$

At time zero, both reactors are full of inert solvent and do not contain any A or B. At $t = 0$, the feed of A is started to the first reactor. The reaction is conducted in the liquid phase and the density is constant. The reactor system is shown Figure 3.15. The component balances for species A and B may be written for each reactor in turn.

Reactor 1

The component balance for species A for reactor 1 is

$$\frac{d(V_1 C_{A1})}{dt} = F_{A0} - F_{A1} - (-r_{A1})V_1 \tag{3.205}$$

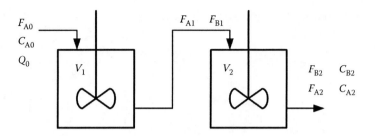

FIGURE 3.15
Two CSTR in series in which a reversible reaction is occurring.

The reactor volume, V_1, and volumetric flow rate, Q, are constant; therefore

$$F_{A0} = QC_{A0} \quad \text{and} \quad F_{A1} = QC_{A1} \tag{3.206}$$

Substitution of Equation 3.206 into Equation 3.205 gives

$$V_1 \frac{dC_{A1}}{dt} = Q(C_{A0} - C_{A1}) - (-r_{A1})V_1 \tag{3.207}$$

Equation 3.207 can be rearranged to the form

$$\frac{dC_{A1}}{dt} = \frac{C_{A0}}{\tau_1} - \frac{C_{A1}}{\tau_1} - k_1 C_{A1} + k_2 C_{B1} \tag{3.208}$$

Now, C_{B1} appears in Equation 3.208 because the reaction is reversible. We need a mole balance on B. Note that the mole balance based on simple stoichiometry

$$F_{B1} = F_{B0} + (F_{A0} - F_{A1}) \tag{3.209}$$

is not valid in the transient case, because A is accumulating in the reactor. Equation 3.209 would only be true at steady state. The mole balance for component B is developed in the same manner as for component A to give

$$\frac{dC_{B1}}{dt} = \frac{C_{B0}}{\tau_1} - \frac{C_{B1}}{\tau_1} + k_1 C_{A1} - k_2 C_{B1} \tag{3.210}$$

Note the change in sign on rate, because B is product

$$\frac{dC_{B1}}{dt} = -\frac{C_{B1}}{\tau_1} + k_1 C_{A1} - k_2 C_{B1} \tag{3.211}$$

Solve Equations 3.208 and 3.211 simultaneously to give C_{A1} and C_{B1} as a function of time.

Reactor 2

The component balance for species A for reactor 2 is

$$F_{A1} - F_{A2} - (-r_{A2})V_2 = V_2 \frac{dC_{A2}}{dt} \tag{3.212}$$

The component balance for species B for reactor 2 is

$$F_{B1} - F_{B2} + (-r_{A2})V_2 = V_2 \frac{dC_{B2}}{dt}$$

(3.213)

The space time for reactor 2 is given by

$$\tau_2 = \frac{V_2}{Q_0}$$

(3.214)

After simplification, Equations 3.212 and 3.213 can be written as

$$\frac{dC_{A2}}{dt} = \frac{C_{A1}}{\tau_2} - \left(\frac{1}{\tau_2} + k_1\right)C_{A2} + k_2 C_{B2}$$

(3.215)

$$\frac{dC_{B2}}{dt} = \frac{C_{B1}}{\tau_2} + k_1 C_{A2} - \left(\frac{1}{\tau_2} + k_2\right)C_{B2}$$

(3.216)

Now solve Equations 3.208, 3.211, 3.215, and 3.216 simultaneously to get C_{B1}, C_{A1}, C_{B2}, and C_{A2} as a function of time. Typically, a numerical solution method would be used to get the solution to these equations.

Case Study 3.2: Transient Operation of CSTR for the Hydrogenation of Oil

In this case study, the transient behavior of different CSTR configurations used for the catalytic hydrogenation of gas oil is examined. The reactions, catalyst, oil, operating conditions, and so on are the same as for Case Study 3.1, where it was seen that the rate of reaction of sulfur and nitrogen are given in terms of lumped kinetic models. For the rate of disappearance of sulfur

$$(-r_S) = A_S \exp\left(\frac{-E_S}{R_g T}\right)C_S^{1.5} P_{H_2}^{0.8} = k_1 C_S^{1.5} = 772.5 C_S^{1.5} \text{ mol of S/h gcat}$$

(3.217)

For the rate of disappearance of nitrogen compounds, the rate equation is

$$(-r_N) = A_N \exp\left(\frac{-E_N}{R_g T}\right)C_N P_{H_2}^{1.3} = k_2 C_N = 4.28 C_N \text{ mol of N/h gcat}$$

(3.218)

We will calculate the values of two ratios, Z_S and Z_N, defined as

$$Z_S = \frac{C_{SE}}{C_{S0}} \quad \text{and} \quad Z_N = \frac{C_{NE}}{C_{N0}}$$

(3.219)

These variables are the ratio of the concentration in the effluent to the concentration in the feed. The feed is assumed to have a constant density. The concentration of the sulfur in the gas oil feed stream is 0.00125 mol of S per cm³. The feed flow rate in all cases is 39.33 cm³/h.

SINGLE REACTOR: CONSTANT VOLUME

The first example is a single, constant-volume reactor that contains 10 g of catalyst. The volume is 100 cm³ and the reactor is initially filled with "clean" oil that does not contain any sulfur or nitrogen. At time $t = 0$, the gas oil feed is started and maintained at a flow rate of 39.33 cm³/h. The transient nitrogen mole balance for the constant-volume CSTR is

$$V \frac{dC_{NE}}{dt} = F_{N0} - F_{NE} - W\left(-r_N\right)_E \tag{3.220}$$

Substituting for the reaction rate of nitrogen from Equation 3.218 and converting the molar flow rate to a volumetric flow rate, and then rearranging give

$$\frac{dC_{NE}}{dt} = \frac{Q}{V}C_{N0} - \frac{Q}{V}C_{NE} - \frac{W}{V}k_2 C_{NE} \tag{3.221}$$

Substituting Z_N from Equation 3.219 gives

$$\frac{dZ_N}{dt} = \frac{Q}{V}\left(1 - Z_N\right) - \frac{W}{V}k_2 Z_N \tag{3.222}$$

Substitute the values for flow rate, volume, catalyst mass, and rate constant to give

$$\frac{dZ_N}{dt} = \frac{39.33}{100}\left(1 - Z_N\right) - \frac{10}{100}4.28 Z_N = 0.3933 - 0.8213 Z_N \tag{3.223}$$

The transient sulfur mole balance for the reaction of sulfur is given by

$$V \frac{dC_{SE}}{dt} = F_{S0} - F_{SE} - W\left(-r_S\right) \tag{3.224}$$

Substituting for the reaction rate of sulfur from Equation 3.217 and converting the molar flow rate to a volumetric flow rate, and then rearranging give

$$\frac{dC_{SE}}{dt} = \frac{Q}{V}C_{S0} - \frac{Q}{V}C_{SE} - \frac{W}{V}k_2 C_{SE}^{1.5} \tag{3.225}$$

Substituting Z_S from Equation 3.219, and noting that $C_{S0}^{0.5}Z_S^{1.5} = (C_{SE}^{1.5}/C_{S0})$ give

$$\frac{dZ_S}{dt} = \frac{Q}{V}\left(1 - Z_S\right) - \frac{W}{V}k_1 C_{S0}^{0.5} Z_S^{1.5} \tag{3.226}$$

Substitution of the values for flow rate, volume, catalyst mass, and rate constant gives

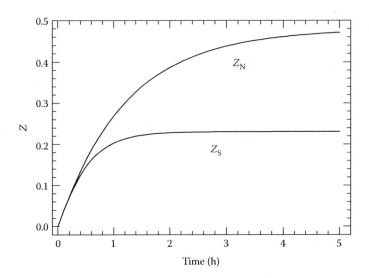

FIGURE 3.16
Dimensionless sulfur and nitrogen concentrations in reactor effluent for the single constant-volume transient CSTR. Results for Case Study 3.2.

$$\frac{dZ_S}{dt} = \frac{39.33}{100}(1 - Z_S) - \frac{10}{100}772.5 \times 0.00125^{0.5} Z_S^{1.5}$$

$$= 0.3933(1 - Z_S) - 2.731 Z_S^{1.5} \tag{3.227}$$

The two mole balance equations, Equations 3.223 and 3.227, can be integrated numerically over a 5 h period, with the initial condition that at $t = 0$, $Z_N = 0$ and $Z_S = 0$. The results of such an integration are shown in Figure 3.16. Note that the solution at large times must always approach the steady-state solution. Note that the exit sulfur concentration approaches its steady-state value more rapidly than does the nitrogen concentration.

TWO CSTR IN SERIES: CONSTANT VOLUME

In the next illustration, we use two CSTR in series, each with a volume of 50 cm³ and containing 5 g of catalyst. Both reactors are initially filled with clean oil that contains no sulfur or nitrogen. The mole balances for each reactor can then be written and solved simultaneously. For the first reactor, the mole balance equations are the same as for the single CSTR case. The subscripts 1 and 2 are used to denote the first and second reactor respectively, and the dimensionless concentration ratios are defined as

$$Z_{N1} = \frac{C_{N1}}{C_{N0}}, Z_{N2} = \frac{C_{N2}}{C_{N0}}, Z_{S1} = \frac{C_{S1}}{C_{S0}}, \text{ and } Z_{S2} = \frac{C_{S2}}{C_{S0}} \tag{3.228}$$

The nitrogen mole balance for the first reactor in terms of the dimensionless concentration, Z, is

$$\frac{dZ_{N1}}{dt} = \frac{Q}{V_1}(1 - Z_{N1}) - \frac{W_1}{V_1}k_2 Z_{N1} \tag{3.229}$$

Substitute the values for flow rate, volume, catalyst mass, and rate constant

$$\frac{dZ_{N1}}{dt} = \frac{39.33}{50}(1 - Z_{N1}) - \frac{5}{50}4.28Z_{N1} = 0.7866 - 1.2146Z_{N1} \tag{3.230}$$

The sulfur mole balance for the first reactor is

$$\frac{dZ_{S1}}{dt} = \frac{Q}{V_1}(1 - Z_{S1}) - \frac{W_1}{V_1}k_1C_{S0}^{0.5}Z_{S1}^{1.5} \tag{3.231}$$

Substitute values for flow rate, volume, catalyst mass, and rate constant, giving

$$\frac{dZ_{S1}}{dt} = \frac{39.33}{50}(1 - Z_{S1}) - \frac{5}{50}772.5 \times 0.00125^{0.5}Z_{S1}^{1.5}$$

$$= 0.7866(1 - Z_{S1}) - 2.7312Z_{S1}^{1.5} \tag{3.232}$$

The two mole balances can be solved independently of the mole balance equations for the second reactor. The mole balances for the second reactor depend on the effluent from the first reactor. The mole balance for nitrogen for the second reactor is

$$V_2\frac{dC_{N2}}{dt} = F_{N1} - F_{N2} - W_2(-r_N)_2 \tag{3.233}$$

Substituting for the reaction rate of nitrogen and the volumetric feed flow rate, and then rearranging give

$$\frac{dC_{N2}}{dt} = \frac{Q}{V_2}C_{N1} - \frac{Q}{V_2}C_{N2} - \frac{W_2}{V_2}k_2C_{N2} \tag{3.234}$$

In terms of the dimensionless concentrations, Z, we have

$$\frac{dZ_{N2}}{dt} = \frac{Q}{V_2}(Z_{N1} - Z_{N2}) - \frac{W_2}{V_2}k_2Z_{N2} \tag{3.235}$$

Substituting for the reaction rate of nitrogen and the volumetric feed flow rate, and then rearranging give

$$\frac{dZ_{N2}}{dt} = \frac{39.33}{50}(Z_{N1} - Z_{N2}) - \frac{5}{50}4.28Z_{N2}$$

$$= 0.7866Z_{N1} - 1.2146Z_{N2} \tag{3.236}$$

The transient sulfur mole balance for the second reactor is developed in a similar way. The inlet is the outlet from reactor 1:

$$V_2\frac{dC_{S2}}{dt} = F_{S1} - F_{S2} - W_2(-r_S)_2 \tag{3.237}$$

Substituting for the reaction rate of sulfur and the volumetric feed flow rate, and then rearranging give

$$\frac{dC_{S2}}{dt} = \frac{Q}{V_2} C_{S1} - \frac{Q}{V_2} C_{S2} - \frac{W_2}{V_2} k_2 C_{SE}^{1.5} \tag{3.238}$$

Noting that $C_{S0}^{0.5} Z_{S2}^{1.5} = \left(C_{S2}^{1.5} / C_{S0} \right)$, Equation 3.238 can be written in terms of Z_S as

$$\frac{dZ_{S2}}{dt} = \frac{Q}{V_2} \left(Z_{S1} - Z_{S2} \right) - \frac{W_2}{V_2} k_1 C_{S0}^{0.5} Z_{S2}^{1.5} \tag{3.239}$$

Substituting for the reaction rate of sulfur and the volumetric feed flow rate, and then rearranging give

$$\frac{dZ_{S2}}{dt} = \frac{39.33}{50} \left(Z_{S1} - Z_{S2} \right) - \frac{5}{50} 772.5 \times 0.00125^{0.5} Z_{S2}^{1.5} \tag{3.240}$$

Simplify Equation 3.240 to obtain the final mole balance:

$$\frac{dZ_{S2}}{dt} = 0.7866 \left(Z_{S1} - Z_{S2} \right) - 2.7312 Z_{S2}^{1.5} \tag{3.241}$$

There are four mole balance equations to solve, Equations 3.230, 3.232, 3.236, and 3.241. The equations are coupled, and must be solved simultaneously using an appropriate numerical method (e.g., Runga–Kutta method). The initial conditions, that is, the dimensionless concentrations at time zero, are

$$Z_{S1} = Z_{S2} = Z_{N1} = Z_{N2} = 0 \quad \text{at } t = 0 \tag{3.242}$$

The solutions of the four mole balance equations are presented in Figures 3.17 and 3.18 as plots of dimensionless concentration versus time.

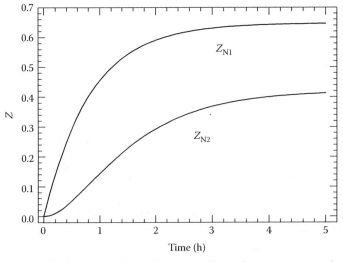

Figure 3.17   Dimensionless nitrogen concentrations in reactor effluent for two constant-volume transient CSTR in series. The conversion in the second reactor is higher than the first. Results for Case Study 3.2.

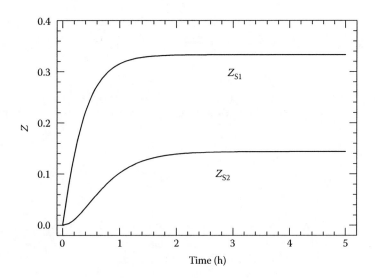

FIGURE 3.18
Dimensionless sulfur concentrations in reactor effluent for two constant-volume transient CSTR in series. The conversion in the second reactor is higher than the first. Results for Case Study 3.2.

SINGLE TRANSIENT REACTOR: VARIABLE VOLUME

In the third illustration, the reactor initially contains 50 cm^3 of clean oil and 10 g of catalyst. At time zero, the feed flow rate is set to 39.33 cm^3/h of gas oil. There is no effluent stream until the reactor volume is 100 cm^3, at which time the volume is held constant and the effluent flow rate equals the feed flow rate.

This problem must be solved in two stages because the mole balance equations for the filling stage are different from those when the reactor is full and the volume is held constant. The first step in the solution is to determine the length of time required to fill the reactor. It is necessary to add 50 cm^3 of oil to the reactor before there is an effluent stream. At an addition rate of 39.33 cm^3 of oil each hour, this operation will take $(50/39.33) = 1.271$ h. Therefore, the mole balance equation for the nitrogen for time less than 1.271 h is

$$\frac{d(VC_N)}{dt} = F_{N0} - W\left(-r_N\right) \tag{3.243}$$

Expanding the left-hand side using the chain rule, and substituting for the molar flow rate of nitrogen compounds give

$$C_N \frac{dV}{dt} + V \frac{dC_N}{dt} = F_{N0} - W\left(-r_N\right) \tag{3.244}$$

The change in volume with time depends on the volumetric flow rate, which is constant. The rate of volume change is given by

$$\frac{dV}{dt} = Q \tag{3.245}$$

The volume at any time is obtained by integration:

$$V = V_0 + Qt \tag{3.246}$$

In Equation 3.246, V_0 is the initial volume, 50 cm^3. Substituting the reaction rate of nitrogen, the volumetric feed flow rate, and the reactor volume into Equation 3.244, and then rearranging give

$$\frac{dC_N}{dt} = \left(\frac{Q}{V_0 + Qt}\right)C_{N0} - \left(\frac{Q}{V_0 + Qt}\right)C_N - \left(\frac{W}{V_0 + Qt}\right)k_2 C_N \tag{3.247}$$

Substituting the dimensionless nitrogen concentration, Z_N, gives

$$\frac{dZ_N}{dt} = \left(\frac{Q}{V_0 + Qt}\right) - \left(\frac{Q + W k_2}{V_0 + Qt}\right)Z_N \tag{3.248}$$

Substituting the numerical values for the reaction rate of nitrogen and the volumetric feed flow rate, and then rearranging give

$$\frac{dZ_N}{dt} = \left(\frac{39.33}{50 + 39.33t}\right) - \left(\frac{39.33 + 10 \times 4.28}{50 + 39.33t}\right)Z_N \tag{3.249}$$

Equation 3.249 can be simplified to the following result:

$$\frac{dZ_N}{dt} = \frac{39.33 - 82.13 \times Z_N}{50 + 39.33t} \tag{3.250}$$

Using a similar procedure, the transient balance for the reaction of sulfur is given by

$$C_S \frac{dV}{dt} + V \frac{dC_S}{dt} = F_{S0} - W(-r_S) \tag{3.251}$$

Substituting for the reaction rate of sulfur, the volumetric feed flow rate, and the change in volume with time, and then rearranging give

$$\frac{dC_S}{dt} = \left(\frac{Q}{V_0 + Qt}\right)C_{S0} - \left(\frac{Q}{V_0 + Qt}\right)C_S - \left(\frac{W}{V_0 + Qt}\right)k_1 C_S^{1.5} \tag{3.252}$$

Noting that $C_{S0}^{0.5} Z_S^{1.5} = \left(C_S^{1.5}/C_{S0}\right)$, the mole balance is written in terms of Z_S as

$$\frac{dZ_S}{dt} = \left(\frac{Q}{V_0 + Qt}\right) - \left(\frac{Q}{V_0 + Qt}\right)Z_S - \left(\frac{W}{V_0 + Qt}\right)C_{S0}^{0.5} k_1 Z_S^{1.5} \tag{3.253}$$

Substitution of the numbers gives

$$\frac{dZ_S}{dt} = \left(\frac{39.33}{50 + 39.33t}\right)(1 - Z_S) - \left(\frac{10}{50 + 39.33t}\right)0.00125^{0.5} \times 772.5 \times Z_S^{1.5}$$

Simplification gives the final mole balance for sulfur:

$$\frac{dZ_S}{dt} = \frac{39.33(1 - Z_S) - 273.12Z_S^{1.5}}{50 + 39.33t} \tag{3.254}$$

The transient mole balance for nitrogen and sulfur compounds must be numerically integrated over the time interval 0–1.271 h. Using the Runga–Kutta method, the solution of Equations 3.250 and 3.254 are obtained, and at the end of 1.271 h, the dimensionless sulfur and nitrogen concentrations are

$$Z_S = 0.2260 \quad \text{and} \quad Z_N = 0.3662$$

After 1.271 h has passed and the reactor is full, the operation is described by the mole balance equations for the constant-volume transient reactor with a volume of 100 cm³. This situation is the same as in the first part of the case study. The mole balance equations are

For sulfur

$$\frac{dZ_S}{dt} = 0.3933(1 - Z_S) - 2.731Z_S^{1.5} \tag{3.255}$$

For nitrogen

$$\frac{dZ_N}{dt} = 0.3933 - 0.8213Z_N \tag{3.256}$$

The two mole balance equations, Equations 3.255 and 3.256, are now integrated numerically over the time interval starting at 1.271 h, and finishing at the final time desired. The initial conditions are the dimensionless concentrations obtained from the solutions of Equations 3.250 and 3.254 at $t = 1.271$ h, that is

$$Z_S = 0.2260 \quad \text{and} \quad Z_N = 0.3662$$

As before, the Runga–Kutta method is used to perform the numerical solution of the mole balance equations. The transient operation of the reactor during both the filling stage and the constant-volume stage is shown in Figure 3.19 . The top graph represents the filling operation and the bottom graph the constant volume stage. Note that in the case of the sulfur, the conversion has almost reached the steady-state value at the end of the filling stage, while the nitrogen steady-state conversion takes much longer to achieve.

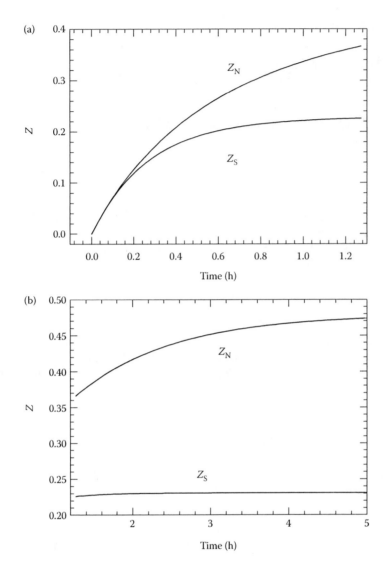

FIGURE 3.19
Dimensionless nitrogen and sulfur concentrations in the reactor for a variable volume transient CSTR. (a)
The graph shows the behavior during the first 1.271 h when the reactor is being filled. (b) The graph corresponds to constant volume operation. Note that for the sulfur conversion, most of the reaction occurs during
the filling stage. Results for Case Study 3.2.

3.11 Summary

In this chapter, the mole balances for the three ideal reactors, the perfectly mixed batch
reactor, the perfectly mixed continuous stirred tank reactor, and the PFR, were developed.
The mole balances are either differential or algebraic equations, depending on the reactor
type and the method of operation. In many cases, the mole balance equations are either
nonlinear or coupled systems of equations that require a numerical solution. Regardless of

the complexity of the reacting system, whether there are a multitude of reactors or a plethora of chemical reactions, the system can always be analyzed simply by developing and solving a system of mole balance equations. One of the most important points to remember when writing the mole balances is to consider density changes that may occur as a result of the chemical reaction. This factor is most important when gas-phase reactions occur and there is a change in the number of moles on reaction. The second point to remember, especially when there are density changes, is the effect of the presence of inert material. Although inert material does not participate in the reaction, it can influence the reaction rate and the equilibrium composition, if the reaction is reversible.

PROBLEMS

3.1 An irreversible gas-phase reaction is conducted in an isothermal batch reactor. The overall reaction is

$$A \rightarrow 2B$$

The reactor is filled with a mixture of 70 mol% A and 30 mol% inert gas. The initial pressure is 90 kPa and the temperature is 400 K. The rate expression is second order, that is $(-r_A) = kC_A^2$ and the rate constant has a value of $k = 0.75$ m^3/mol-h at 400 K. Assume ideal gases.

a. Calculate the time required to achieve 60% conversion of A in a constant-volume batch reactor.

b. Calculate the time required to achieve 60% conversion of A in a constant-pressure batch reactor.

3.2 An irreversible gas-phase reaction is conducted in an isothermal batch reactor operated at 400 K. The overall reaction may be represented by

$$A \rightarrow P + 2S$$

The reactor is initially filled with pure A and the initial pressure is 100.0 kPa. The rate expression for this reaction is second order in the concentration of A, that is, $(-r_A) = kC_A^2$, and the value of the rate constant at 400 K is $k = 0.75$ m^3/mol-h. Assume ideal gases.

a. Calculate the time required to achieve an 80% conversion of A if the reactor is a constant-volume batch reactor.

b. Calculate the time required to achieve an 80% conversion of A if the reactor is a variable-volume, constant-pressure batch reactor.

3.3 Take the same reaction as in Problem 3.1, but the reaction will now be carried out in an isothermal plug flow reactor (PFR). The feed is 70 mol% A and 30 mol% inert gas at a pressure of 90 kPa and a temperature of 400 K. The pressure drop in the reactor is negligible. The total feed flow rate to the reactor is 0.04 mol/min.

a. Calculate the reactor volume required to achieve 60% conversion of A.

b. Calculate the space time (based on feed inlet conditions) for the reactor in part (a).

c. Calculate the mean residence time for the reactor in part (a).

3.4   Consider the same reaction as in Problem 3.2, but the reaction is now carried out in an isothermal plug flow reactor (PFR). A feed stream of pure A, at 100 kPa and 400 K, is fed to the reactor at a rate of 0.040 mol/min. The reactor temperature is 400 K and the pressure drop is negligible.

    a. Calculate the reactor volume required to obtain an 80% exit fractional conversion of A.

    b. Calculate the space time (based on feed flow rate at reactor inlet conditions) for the operation in part (a).

    c. Calculate the average mean residence time for the operation in part (a).

3.5   Repeat Problem 3.3 for a CSTR. That is, the reactor feed composition and flow rate are the same as in problem 3.3 but the PFR is replaced by a CSTR. The temperature and the pressure are also the same.

3.6   Repeat Problem 3.4 for a CSTR. That is, the reactor feed composition and flow rate are the same as in problem 3.4 but the PFR is replaced by a CSTR. The temperature and the pressure are also the same.

3.7   Consider an ideal gas-phase reaction carried out in an isothermal batch reactor. The reaction is given by the following stoichiometry:

$$A + B \rightarrow C$$

The rate expression and rate constant are

$$(-r_A) = kC_A C_B, \quad k = 1 \times 10^{-5} \text{ m}^3/\text{mol} \cdot \text{s}$$

The reactor is initially filled with a mixture of 40 mol% A, 40 mol% B, and 20 mol% inert gas. The temperature is held constant at 100°C. The initial reactor pressure is 500 kPa. The reactor pressure is controlled during the course of the reaction such that the following relationship is valid:

$$P_T = P_{T0} - at, \quad a = 1 \text{ Pa/s}$$

where t is time in seconds and PT and PT0 are the total pressure and initial total pressure in Pa, respectively. Determine the time required to achieve 80% conversion of reactant A and the total reactor pressure at that time.

3.8   Consider an ideal gas reaction conducted in a plug flow reactor. The reaction stoichiometry is

$$A + B \xrightarrow{k} C + D$$

The rate equation and rate constant are

$$(-r_A) = kC_A C_B, \quad k = 5 \times 10^{-4} \text{ m}^3/\text{mol} \cdot \text{s}$$

The feed is 50 mol% A, 30 mol% B, and 20 mol% inert gas. The total inlet molar flow rate is 10 mol/s which, at the temperature and pressure of the reactor, gives a total volumetric flow rate of 0.1 m^3/s.

a. Determine the reactor volume required to achieve 50% conversion of A.

b. Determine the reactor volume required to achieve 80% conversion of A.

3.9 Consider the system of series/parallel reactions shown below:

$$A \xrightarrow{k_1} B \quad (-r_A) = k_1 C_A$$
$$B \xrightarrow{k_1} C \quad (r_C) = k_2 C_B$$
$$B \xrightarrow{k_1} R \quad (r_R) = k_3 C_B$$

The reactions are assumed to be irreversible. The reactions take place in two CSTR connected in series. The volume of each reactor is 300 L. The rate constants are

$$k_1 = 0.07\,\text{s}^{-1}, \quad k_2 = 0.09\,\text{s}^{-1}, \quad k_3 = 0.10\,\text{s}^{-1}$$

Pure A is fed to the first reactor at a concentration of 4 mol/L and a flow rate of 30 L/s. Calculate the molar flow rate of R in the effluent from the second reactor.

3.10 The following irreversible first-order reactions occur at constant density:

$$A \xrightarrow{k_1} R \xrightarrow{k_2} S$$

where $k_1 = 0.15$ min^{-1}; $k_2 = 0.05$ min^{-1}. This reaction system is to be analyzed in continuous flow reactors with a volumetric feed rate of 150 L/min and feed composition C_{A0}. Find the production rate of R (i.e., C_R/C_{A0}) for each of the following reactor systems:

a. A single CSTR with volume of 300 L

b. Two CSTRs in series, each with volume of 150 L

c. Two CSTRs in parallel, each with volume of 150 L and with the feed stream equally divided between them

3.11 The vapor-phase dehydration of ethanol over an ion exchange resin has been studied. The rate of ethanol dehydration to diethyl ether at temperatures of 110–135°C is given by the following rate function:

$$(-r_A) = \frac{k_s K_A^2 \left(P_A^2 - (1/K_1) P_W P_E \right)}{\left(1 + K_A P_A + K_W P_W \right)^2} \quad \text{mol/s} \cdot \text{gcat}$$

where $A = C_2H_5OH$, $W = H_2O$ and $E = (C_2H_5)_2O$. The constants are

$$k_s = 1.5 \times 10^8 \exp\left(\frac{-12{,}424}{T}\right) \text{mol/s} \cdot \text{gcat}, \quad K_A = 2.0 \times 10^{-7} \exp\left(\frac{4741}{T}\right) \text{kPa}^{-1},$$

$$K_1 = 0.25 \exp\left(\frac{1842}{T}\right), \quad K_W = 1.0 \times 10^{-9} \exp\left(\frac{7060}{T}\right) \text{kPa}^{-1}$$

with T in K and P_i in kPa. The overall reaction for ethanol dehydration to diethyl ether is

$$2A \rightleftharpoons W + E$$

a. Convert the above rate expression to concentration units, that is, replace the P_A, P_E, and P_W and C_A, C_E, and C_W. Use mol/m^3 for concentration units and assume ideal gas behavior.

b. Calculate the equilibrium fractional conversion of ethanol if pure ethanol is fed to a reactor at a total pressure of 300 kPa and 130°C.

c. The ethanol dehydration is to be carried out with a two-reactor system, specifically, a PFR followed by a CSTR. Pure ethanol vapor is fed to the PFR at a rate of 0.25 mol/h. Each reactor contains 20.0 g of catalyst (ion exchange resin) and both reactors are operated at 300 kPa and 130°C. Calculate the steady-state fractional conversion of ethanol at the exit of the PFR and at the exit of the CSTR.

d. Repeat part (c), but change the order of the two reactors.

3.12 Consider the system of reactions:

$$A + B \xrightarrow{k_1} C \quad \text{and} \quad A + C \xrightarrow{k_2} D$$

The reactions take place in liquid solution in two CSTR connected in series. Each reactor has a volume of 0.3 m^3. The kinetics of the reactions are

$$(-r_B) = k_1 C_A C_B \quad k_1 = 2 \times 10^{-5} \text{ m}^3/\text{mol} \cdot \text{s}$$

$$(+r_D) = k_2 C_A C_C \quad k_2 = 8 \times 10^{-6} \text{ m}^3/\text{mol} \cdot \text{s}$$

At time zero, both reactors are full of inert solvent. The feed to the first reactor consists of a flow of 0.03 m^3/s containing 4000 mol/m^3 of A and 3000 mol/m^3 of B in a solvent. Plot the concentrations of A, B, C, and D in the effluent streams from each reactor as a function of time, and determine the steady-state effluent compositions. How long does it take to achieve 99% of the steady-state composition?

3.13 A semibatch reactor has an inlet feed stream and no effluent stream. The following liquid reaction occurs in the reactor:

$$A \rightarrow B$$

The rate equation and rate constant are

$$(-r_A) = kC_A, \quad k = 1 \times 10^{-3}\,s^{-1}$$

The reactor initially contains 0.1 m^3 of a solution containing 1000 mol/m^3 of A. The feed stream flows at 0.001 m^3/s with a concentration of A of 1000 mol/m^3. Calculate the number of moles of A and B in the reactor after 900 s has passed.

3.14 Consider a plug flow reactor with a reversible isothermal gas-phase reaction:

$$A + B \rightleftharpoons C$$

The reactor feed consists of a mixture of 50% A and 50% B. The reaction rate is

$$(-r_A) = k_1 C_A C_B - k_2 C_C^2$$

The inlet temperature is 500 K and the total pressure is 200 kPa. The volumetric feed flow rate is 10 m^3/s. The rate constants have values of

$$k_1 = 0.1\,m^3/mol \cdot s \quad \text{and} \quad k_2 = 0.05\,m^3/mol \cdot s$$

a. Calculate the conversion of A in a reactor of volume 10 m^3.

b. Calculate the maximum conversion of A in an infinitely long reactor.

3.15 The reaction between ammonia and formaldehyde to give hexamine is

$$4NH_3 + 6HCHO \rightarrow (CH_2)_6 N_4 + 6H_2O$$

A 0.5 L CSTR is used for the reaction. Each reactant is fed to the reactor in a separate stream, at the rate of 1.5×10^{-3} L/s each. The ammonia concentration is 4.0 mol/L and the formaldehyde concentration is 6.4 mol/L. The reactor temperature is 36°C. Calculate the concentration of ammonia and formaldehyde in the effluent stream. The reaction rate equation and the rate constant are

$$(-r_A) = kC_A^2 C_B \text{ mol A/L} \cdot s, \quad k = 1.42 \times 10^3 \exp\left(\frac{-3090}{T}\right)$$

where A is ammonia and B is formaldehyde.

3.16 The following reaction occurs in aqueous solution:

$$NaOH + C_2H_5(CH_3CO) \rightarrow Na(CH_3CO) + C_2H_5OH$$
$$A + B \rightarrow C + D$$

This reaction is second order and irreversible, that is

$$(-r_A) = kC_A C_B$$

A well-mixed laboratory batch reactor is filled with a solution of 0.1 M NaOH and ethyl acetate (component B). After 15 min, the conversion of NaOH is observed to be 15%.

a. For a commercial-size reactor, how much time is required to achieve 30% conversion of a solution containing 0.2 M ethyl acetate and 0.2 M NaOH?

b. What reactor volume is required to produce 50 kg of sodium acetate (product C, molecular weight 66) at this 30% conversion (conditions of part (a))?

3.17 Consider the reaction of species A to form product B in a constant-density isothermal system. The rate of disappearance of A is given by a second-order rate, that is, $(-r_A) = kC_A^2$. A process plant carries out the reaction using a single PFR with 95% conversion of A. To increase production, a second reactor is to be added. The second reactor is to have the same volume as the first one, and the overall conversion from the system of two reactors is to remain at 95%. Four scenarios are imagined, as given in the following. For each case, determine by what factor the plant can increase production of B. Justify your answer using the appropriate mathematics or impeccable logic.

a. The second reactor is also a PFR and is operated in parallel.

b. The second reactor is a PFR and is operated in series.

c. The second reactor is a CSTR in series placed before the PFR.

d. The second reactor is a CSTR in series placed after the PFR.

4

Energy Balances in Ideal Reactors

The mole balance equation was developed in Chapter 3 and applied to a variety of ideal reactors. Energy changes were not considered, which led to some constraints on the systems that were analyzed. For the batch reactor case, the contents of the reactor had a uniform temperature, and this temperature did not change with time. For flow reactors, the temperature was constant with respect to both time and spatial position, and the inlet and outlet process stream temperatures were the same. With these imposed constraints, the mole balance equation, along with a rate expression, is sufficient to determine the conversion in a reactor. In the operation of most real reactors, regardless of whether they are laboratory units, pilot-scale, or full-size reactors, these constraints on the temperature do not apply. Thus, in a batch reactor, the temperature of the reaction mixture may change with time. In a flow reactor, the temperature may change with time and position, or the feed and effluent streams may have different temperatures. To account for these effects, reactor analysis must include the energy balance.

In this chapter, the energy balance equation is first developed in a general way—then it is applied to the three ideal reactor cases, namely the perfectly mixed batch reactor, the perfectly mixed CSTR, and the PFR. In the case of the CSTR, both transient and steady-state behavior is discussed. The reader may wish to study again the thermodynamics review presented in Chapter 2 prior to reading this chapter.

4.1 Influence of Temperature on Reactor Operation

The effects of temperature on the reaction rate were discussed in Chapters 1 and 2, and are briefly summarized here. Temperature changes influence the reaction rate, and hence the mode of reactor operation in two ways. The first is through the temperature dependence of the rate parameters; the second is through the temperature dependence of the equilibrium constant.

4.1.1 Temperature Dependence of Reaction Rate Constants

The temperature dependence of the reaction rate constants was mentioned in Chapter 1. The temperature dependence of each constant in the reaction rate expression has the general form (where $N = 0$ for the Arrhenius law)

$$k = AT^N \exp\left(\frac{-E}{R_g T}\right) \tag{4.1}$$

The exponential dependence of the rate constants on the temperature means that the reaction rate can change rapidly as the temperature changes. The larger the value of the activation energy, the greater will be the change in reaction rate for a given change in temperature. This effect can cause operating problems. For example, consider the case of an exothermic chemical reaction. As the reaction proceeds, thermal energy is released. If this energy is not removed from the reactor, the process stream will increase in temperature, causing the reaction rate to increase. Then, as the reaction rate increases, the rate of heat release also increases, which further increases the rate of reaction, and so on. In some cases, this effect may lead to reactor runaway and catastrophic failure, which should be avoided.

4.1.2 Effect of Temperature on Equilibrium Conversion

The equilibrium constant is also a function of temperature; hence, the equilibrium conversion will be affected by changes in temperature. The temperature dependence of the equilibrium constant is given by the van't Hoff equation (see Section 2.3.3):

$$\frac{\partial(\ln K)}{\partial T} = \frac{\Delta H_R^\circ}{R_g T^2} \tag{4.2}$$

For an exothermic reaction, an increase in temperature decreases the equilibrium yield, while for an endothermic reaction an increase in temperature increases the equilibrium yield, everything else being equal.

4.1.3 Energy Balance for Ideal Reactors

The mole and energy balance equations are a coupled set of equations that together describe the performance of an ideal reactor. Typically, the energy balance is expressed in terms of the reactor temperature, which may vary with space and time. The nature of the variations depends on the reactor type and mode of operation. In perfectly mixed batch reactors, the temperature is the same at all locations in the reactor, but changes with time as the reaction proceeds. Sometimes the reactor is operated with heat transfer through the reactor walls, or via heating or cooling coils inserted into the reactor. Adiabatic operation is also possible. In a plug flow reactor, the temperature varies with axial position (i.e., in the direction of flow) in the reactor. There may or may not be heat transfer through the reactor walls. In a CSTR, the temperature is uniform everywhere in the reactor and is equal to the outlet temperature. However, the inlet and outlet temperatures are different. There may or may not be heat transfer with the surroundings, either through the vessel walls or via heating or cooling coils within the reactor.

In the following sections, the general energy balance equation is introduced. This introduction is followed by a detailed development of the energy balance equation for the three ideal reactors (batch, PFR, and CSTR).

4.2 General Energy Balance

We start the derivation of the energy balance equation by considering the energy changes experienced by a general open system. The energy of the system may change with time as a result of heat exchange between the system and the surroundings, work done on the system by the surroundings, or by accumulation of mass in the system. Energy crosses the system boundary as mass flows into or out of the system. Energy changes also occur as chemical reactions proceed. The general energy balance equation can thus be written in words as

$$
\begin{bmatrix}
\text{Rate of} \\
\text{accumulation} \\
\text{of energy in} \\
\text{the system}
\end{bmatrix}
=
\begin{bmatrix}
\text{Heat} \\
\text{added to} \\
\text{system}
\end{bmatrix}
-
\begin{bmatrix}
\text{Work} \\
\text{done by} \\
\text{system}
\end{bmatrix}
+
\begin{bmatrix}
\text{Energy of} \\
\text{mass entering} \\
\text{the system}
\end{bmatrix}
-
\begin{bmatrix}
\text{Energy of} \\
\text{mass leaving} \\
\text{the system}
\end{bmatrix}
$$

The reader may recall that in books on heat transfer the general energy balance equation includes a term for thermal energy generation. Such a term does not appear in the above energy balance because this equation represents a *total* energy balance. Although a chemical reaction may generate or absorb thermal energy, this energy change is balanced by an equal change in energy owing to a compositional change. Strictly speaking, energy is never generated in any system unless there are nuclear reactions where matter is transformed into energy.

The energy of a system or a flowing stream consists of potential energy (which depends on the system height relative to a reference point, z), kinetic energy (which depends on the fluid velocity, v), and the internal energy, U (which is the sum of the molecular energies). The total energy per unit mass is the sum of these three components, and is written as

$$
E = U + \frac{v^2}{2} + gz^2
\tag{4.3}
$$

In most chemical reactors, changes in kinetic and potential energy are considered to be negligible and are ignored. This is usually a good approximation because the velocities of most process streams are relatively small, and the differences in elevation are minor. The largest energy effect is that owing to chemical reaction. The following approximation is thus made:

$$
E \approx U
\tag{4.4}
$$

The heat transfer between the system and the surroundings is denoted as q, and has the units of watts. It has a positive value if heat is transferred to the system from the surroundings. The work done by the system is given the symbol W and also has units of watts. It has a positive value if the work is done by the system on the surroundings. For a system with multiple species crossing the system boundary with molar flow rates of F_j, the energy balance may be written as

$$
\frac{dU}{dt} = q - W + \left[\sum_{j=1}^{n} F_j U_j \right]_{\text{in}} - \left[\sum_{j=1}^{n} F_j U_j \right]_{\text{out}}
\tag{4.5}
$$

Each of the terms in Equation 4.5 has units of power, joules/s (J/s) or watts (W). The work done on or by the system, W, may be divided into flow work, W_f, expansion work (also called pressure/volume work), W_E, and shaft work, W_S.

$$\frac{dU}{dt} = q - W_f - W_S - W_E + \left[\sum_{j=1}^{n} F_j U_j \right]_{in} - \left[\sum_{j=1}^{n} F_j U_j \right]_{out} \qquad (4.6)$$

Flow work is the work that a fluid stream does on the system or the surroundings by entering or leaving the system. The flow work depends on the volumetric flow rate, Q, of the stream crossing the system boundary and the pressure of the system, as follows:

$$W_f = -PQ \qquad (4.7)$$

The volumetric flow rate can be expressed in terms of the molar flow rate by introducing the molar concentration of each species, C_j:

$$W_f = -P \sum_{j=1}^{n} \frac{F_j}{C_j} = -\sum_{j=1}^{n} F_j P V_j \qquad (4.8)$$

The volume, V_j, in Equation 4.8 is the molar volume of component j in the process stream. The shaft work, denoted W_S, is the work done on or by the system by a mechanical device such as a mixer. The expansion work is the work done on or by the system as a result of volume changes of the system. For a constant-volume system, it is equal to zero: substitution of Equation 4.8 into Equation 4.5 and rearrangement thus give

$$\frac{dU}{dt} = q - W_S - W_E + \left[\sum_{j=1}^{n} F_j \left(U_j + P V_j \right) \right]_{in} - \left[\sum_{j=1}^{n} F_j \left(U_j + P V_j \right) \right]_{out} \qquad (4.9)$$

We now introduce the enthalpy, H. The enthalpy is a defined thermodynamic quantity and is related to the internal energy by the following equation:

$$H = U + PV \qquad (4.10)$$

Equation 4.10 can be substituted into Equation 4.9 to obtain

$$\frac{dA}{dV} = \frac{(\pi D)dz}{(\pi D^2/4)dz} = \frac{4}{D} \qquad (4.11)$$

Equation 4.11 is one form of the basic energy balance equation. We shall apply this energy balance to the batch reactor, the PFR, and the CSTR.

4.3 Batch Reactor

It was seen in Chapter 3 that the batch reactor can be operated in either constant- or variable-volume mode. The form of the energy balance equation for the batch reactor depends on which of these two modes is used. Most batch reactors are operated at constant volume, and this case will be treated first and in more detail, followed by the variable-volume case.

4.3.1 Constant-Volume Batch Reactor

The CVBR has no inlet or outlet streams; thus, it is a closed system. The energy terms in Equation 4.11 for flow streams are thus dropped. The expansion work in a constant-volume system is zero. The internal energy change of the system is the sum of the changes of all of the species in the system. For this analysis, the energy input from work done by mixing devices (shaft work) is ignored. The energy balance for a CVBR becomes

$$\frac{dU}{dt} = \frac{d\left(\sum_{i=1}^{n} N_j U_j\right)}{dt} = q \qquad (4.12)$$

The internal energy change of the reactor contents is thus equal to the rate of heat transfer with the surroundings. Equation 4.12 can be expanded using the chain rule:

$$\sum_{j=1}^{n} N_j \frac{dU_j}{dt} + \sum_{j=1}^{n} U_j \frac{dN_j}{dt} = q \qquad (4.13)$$

It is not convenient to work directly in terms of internal energy changes: temperature is a much more useful variable. The internal energy change of a species can be expressed in terms of the temperature and the constant-volume heat capacity. For any species j, this substitution gives

$$\frac{dU_j}{dt} = C_{Vj} \frac{dT}{dt} \qquad (4.14)$$

Therefore, the first term on the left-hand side of Equation 4.13 becomes

$$\sum_{j=1}^{n} N_j \frac{dU_j}{dt} = \left(\sum_{j=1}^{n} N_j C_{Vj}\right) \frac{dT}{dt} \qquad (4.15)$$

Consider a reaction given by the following overall expression:

$$aA + bB \rightarrow cC + dD \qquad (4.16)$$

The reaction can be expressed on the basis of 1 mol of A as

$$A + \frac{b}{a}B \rightarrow \frac{c}{a}C + \frac{d}{a}D \qquad (4.17)$$

Let I denote any inert substances present in the system. For the reaction of Equation 4.16, the right-hand side of Equation 4.15 is equal to

$$\left(\sum_{j=1}^{n} N_j C_{Vj}\right) \frac{dT}{dt} = \left(N_A C_{VA} + N_B C_{VB} + N_C C_{VC} + N_D C_{VD} + N_I C_{VI}\right) \frac{dT}{dt} \qquad (4.18)$$

The second term on the left-hand side of Equation 4.13, for the reaction of Equation 4.16, is

$$\sum_{j=1}^{n}\left(U_j \frac{dN_j}{dt}\right) = U_A \frac{dN_A}{dt} + U_B \frac{dN_B}{dt} + U_C \frac{dN_C}{dt} + U_D \frac{dN_D}{dt} \qquad (4.19)$$

Equation 4.19 contains terms for the reaction rates of each species in the reaction. Each rate can be expressed in terms of the rate of change of the number of moles of A. From the stoichiometry, the relationship between the rates of reaction of the different reactants and products is

$$q + F_{T0}\bar{C}_{P0}(T_0 - T_E) - (F_{A0} - F_{AE})\Delta H_{R,A} = 0$$

Substitution of these terms into Equation 4.19 and simplification gives

$$\sum_{j=1}^{n}\left(U_j \frac{dN_j}{dt}\right) = -\frac{dN_A}{dt}\left[\frac{d}{a}U_D + \frac{c}{a}U_C - U_A - \frac{b}{a}U_B\right] \qquad (4.20)$$

The difference between the internal energies of the products and the reactants, multiplied by the respective stoichiometric coefficients, equals the internal energy change of reaction, which is

$$\frac{d}{a}U_D + \frac{c}{a}U_C - U_A - \frac{b}{a}U_B = \Delta U_{R,A} \qquad (4.21)$$

In Equation 4.21, $\Delta U_{R,A}$ is the internal energy change owing to reaction per mole of A. The energy balance equation for a CVBR can therefore be written as

$$\left(\sum_{j=1}^{n} N_j C_{Vj}\right)\frac{dT}{dt} - \frac{dN_A}{dt}(\Delta U_{R,A}) = q \qquad (4.22)$$

In Equation 4.22, A is the reference reactant (or product), and the energy balance is expressed in terms of this reference component. Equation 4.22 contains two time derivative terms. The derivative of the moles of A with respect to time can be eliminated by incorporating the mole balance. The general mole balance for a batch reactor is

$$-\frac{1}{V}\frac{dN_A}{dt} = (-r_A) \qquad (4.23)$$

Substituting Equation 4.23 into Equation 4.22 gives the energy balance as

$$\left(\sum_{j=1}^{n} N_j C_{Vj}\right)\frac{dT}{dt} + (-r_A)V(\Delta U_{R,A}) = q \qquad (4.24)$$

Equation 4.24 can be rearranged to give an explicit expression in temperature; thus

$$\frac{dT}{dt} = \frac{q}{\left(\sum_{j=1}^{n} N_j C_{Vj}\right)} - \frac{(-r_A)V(\Delta U_{R,A})}{\left(\sum_{j=1}^{n} N_j C_{Vj}\right)} \tag{4.25}$$

The mole balance and energy balance equations between them describe the behavior of a nonisothermal CVBR. The two equations are coupled because the energy balance includes concentration terms and the mole balance includes the temperature. Owing to the exponential temperature dependence of the rate constants, a numerical solution of the two equations is required. In the general case, both C_V and $(\Delta U_{R,A})$ depend on the temperature, giving very nonlinear equations. This nonlinearity can lead to difficulties in the numerical solution.

4.3.2 Relationship between ΔU_R and ΔH_R

The energy balance in the CVBR includes the internal energy of reaction, ΔU_R. It is common to express the energy change on reaction in terms of the enthalpy of reaction, ΔH_R (see Chapter 2), and this latter quantity is the one usually encountered in reference texts. The relationship between ΔU_R and ΔH_R is given by the following equation:

$$\Delta U_{R,A} = \Delta H_{R,A} - \left[P + \left(\frac{\partial \Delta U_{R,A}}{\partial V}\right)_T\right]\left[\sum_{j=1}^{n} v_j V_j\right] \tag{4.26}$$

In Equation 4.26, v_j is the stoichiometric coefficient, which is negative for a reactant and positive for a product, and V_j is the molar volume of component j. For most liquid reactions, the following is usually a good approximation:

$$\sum_{j=1}^{n} v_j V_j \approx 0 \tag{4.27}$$

With this approximation, $\Delta H_R \approx \Delta U_R$ and $C_P \approx C_V$.

In a gas-phase reacting system, we consider two cases. In the first case, there is no change in moles on reaction, and it follows that

$$\sum_{j=1}^{n} v_j = 0 \quad \text{and} \quad \sum_{j=1}^{n} v_j V_j = 0$$

and therefore $\Delta H_R = \Delta U_R$. Furthermore, the change in heat capacity is the same for both constant-pressure and constant-volume heat capacity, that is

$$\Delta C_P = \Delta C_V \tag{4.28}$$

Therefore, either C_P or C_V values may be used in calculating energy changes. In the second case, there is a change in moles on reaction. Thus, $\Sigma v_j = \Delta N$ and therefore $\Delta H_R \neq \Delta U_R$.

To determine the difference between ΔH_R and ΔU_R, consider the case of an ideal gas mixture. For an ideal gas, the internal energy depends only on the temperature, and the molar volume is given by the ideal gas law; therefore

$$\left(\frac{\partial \Delta U_R}{\partial V}\right)_T = 0 \quad \text{and} \quad V_j = \frac{R_g T}{P}$$

It follows that

$$\sum_{j=1}^{n} v_j V_j = \frac{R_g T}{P} \sum_{j=1}^{n} v_j \tag{4.29}$$

The sum of the stoichiometric coefficients equals the change in moles on reaction, that is, $\Sigma v_j = \Delta N$, where ΔN is the change in moles on reaction when v_A is set equal to 1. Therefore

$$\Delta U_{R,A} = \Delta H_{R,A} - P\left(\Delta N \frac{R_g T}{P}\right) \tag{4.30}$$

Simplifying gives the result

$$\Delta U_{R,A} = \Delta H_{R,A} - \Delta N R_g T \tag{4.31}$$

In this scenario, the value of $\Delta U_{R,A}$ is different from the value of $\Delta H_{R,A}$, and error would result if they were assumed to be the same. The magnitude of the difference that can arise between the two is illustrated in Example 4.1.

Example 4.1

The gas-phase hydrogenation of benzene produces cyclohexane. The overall reaction is

$$C_6H_6 + 3H_2 \rightarrow C_6H_{12} \tag{4.32}$$

Compute the enthalpy and internal energy changes of this reaction over the temperature range 300–800 K. Assume that the species behave as ideal gases. The following data are available. The temperature dependence of the heat capacity is

$$C_P = a + bT + cT^2 + dT^3 \, \text{J/mol} \cdot \text{K} \tag{4.33}$$

The values of the constants in Equation 4.33 are

Compound	a	$b \times 10^2$	$c \times 10^5$	$d \times 10^9$
C_6H_6	−33.92	47.39	−30.17	71.30
H_2	27.14	0.9274	−1.381	7.645
C_6H_{12}	−54.54	61.13	−25.23	13.21

The enthalpies of formation of benzene and cyclohexane at 298.15 K are

$$\Delta H_{f,298,C6H6} = 8.298 \times 10^4 \, \text{J/mol} \tag{4.34}$$

and

$$\Delta H_{f,298,C6H12} = -1.232 \times 10^5 \, \text{J/mol} \tag{4.35}$$

SOLUTION

We start the solution by computing an expression for the enthalpy of reaction at 298.15 K. The enthalpy of reaction as a function of temperature can be computed from the enthalpies of formation and the heat capacity data. The standard enthalpy of formation of hydrogen is zero by definition, and therefore, the enthalpy change of reaction at 298.15 K is

$$\Delta H_{R,298,C6H6} = -1.232 \times 10^5 - 8.298 \times 10^4 = -2.062 \times 10^5 \, \text{J/mol} \tag{4.36}$$

Using this value for the enthalpy of reaction and the heat capacity data, a temperature-dependent expression is developed for the enthalpy of reaction (see Example 2.6 for the methodology):

$$\Delta H_{R,C6H6} = -1.813 \times 10^5 - 102.0T + 0.0548T^2 + 3.024 \times 10^{-5}T^3 - 2.026 \times 10^{-8}T^4 \, \text{J/mol} \tag{4.37}$$

The relationship between the internal energy change of reaction and the enthalpy change of reaction for an ideal gas is given by Equation 4.31:

$$\Delta U_{R,C6H6} = \Delta H_{R,C6H6} - \Delta N R_g T \tag{4.38}$$

For this hydrogenation reaction, there are 4 mol of reactants consumed for every mole of products that is produced. Therefore, it follows that

$$\Delta N = 1 - 4 = -3 \tag{4.39}$$

Equation 4.38 is written as

$$\Delta U_{R,C6H6} = \Delta H_{R,C6H6} + 3 \times 8.314 \times T \tag{4.40}$$

Combining Equations 4.37 and 4.40 gives

$$\Delta U_{R,C6H6} = -1.813 \times 10^5 - 77.058T + 0.0548T^2 + 3.024 \times 10^{-5}T^3 - 2.026 \times 10^{-8}T^4 \, \text{J/mol} \tag{4.41}$$

The following table gives some values for the two energy changes at various temperatures, computed from Equations 4.37 and 4.41:

T (K)	$\Delta H_{R,C_6H_6}$ (J/mol)	$\Delta U_{R,C_6H_6}$ (J/mol)
300	-2.063×10^5	-1.988×10^5
400	-2.119×10^5	-2.019×10^5
500	-2.161×10^5	-2.036×10^5
600	-2.189×10^5	-2.039×10^5
700	-2.203×10^5	-2.029×10^5
800	-2.206×10^5	-2.007×10^5

Note that the percent difference between the two energy values is about 3.8% at 300 K, and rises to about 9.9% at 800 K.

4.3.2.1 Assuming a Constant Value for Heat Capacity

It was seen in Chapter 2 that the heat capacity is usually a function of temperature. This temperature dependence must be included in the solution of the energy balance equation to obtain an exact solution. Furthermore, as the composition of the reaction mixture changes with time, the heat capacity of the mixture will usually change, even if the temperature is held constant. It is, however, not uncommon to assume that the heat capacity is either independent of the temperature or composition, or both. The energy balance is then expressed in terms of an average heat capacity value.

When the heat capacity is assumed to be independent of composition, the heat capacity is often based on the mass of the system. With this assumption, the temperature change term in Equation 4.24 becomes

$$\sum_{j=10}^{n} N_j \frac{dU_j}{dt} = m_t \bar{C}_V \frac{dT}{dt} \tag{4.42}$$

The units of $\bar{C}_V$ are J/kg $\cdot$ K. Now, substitute Equation 4.42 into Equation 4.24:

$$q = m_t \bar{C}_V \frac{dT}{dt} + (\Delta U_{R,A})(-r_A) V \tag{4.43}$$

Note that the use of an average heat capacity based on the mass of the system does not preclude the temperature dependence of the heat capacity from being used. Writing Equation 4.43 explicitly in terms of the temperature change of the reactor as a function of time gives

$$\frac{dT}{dt} = \frac{q}{m_t \bar{C}_V} - \frac{(\Delta U_{R,A})(-r_A) V}{m_t \bar{C}_V} \tag{4.44}$$

The mole balance equation for a CVBR may be written in terms of concentration:

$$\frac{dC_A}{dt} = -(-r_A) \tag{4.45}$$

Equations 4.44 and 4.45 are a system of two coupled ordinary differential equations and are initial value problems. The simultaneous numerical solution of these two equations yields the temperature and concentration in the reactor as a function of time.

4.3.3 External Heat Transfer in CVBR

The heat transfer term, q, may have various forms depending on the reactor and surroundings. The reactor may be heated by a heater that supplies a constant heat flux, in which case the value of q would be a constant. If there is no heat transfer with the surroundings, the reactor is adiabatic and q has a value of zero. Alternatively, q might be varied continuously to achieve a specified heating or cooling rate in the reactor.

A common method used to provide heat exchange with a reactor is to place heating or cooling coils inside the reactor, or to place a jacket containing a heat transfer fluid around the reactor surface. In this case, the rate of heat transfer is governed by the temperature difference between the reactor and the heat transfer fluid, and the value of the overall heat transfer coefficient, U. The heat transfer equation is written as

$$q = UA\left(T_\infty - T\right) \tag{4.46}$$

In Equation 4.46, A is the heat transfer area and T_∞ is the temperature of the heat transfer fluid. If the temperature of the heat transfer fluid is held constant during the course of reaction, and the reacting fluid temperature changes, then the rate of heat transfer is a function of time. Refer to Case Study 4.1 in Section 4.3 for an example of a batch reactor with external heat transfer.

4.3.4 Adiabatic Temperature Change for CVBR

When the reactor is adiabatic, Equation 4.22 can be written as

$$\left(\sum_{j=1}^{n} N_j C_{Vj}\right)\frac{dT}{dt} = \frac{dN_A}{dt}\left(\Delta U_{R,A}\right) \tag{4.47}$$

Eliminating the time derivative term in Equation 4.47 and rearranging give

$$\frac{dT}{dN_A} = \frac{\left(\Delta U_{R,A}\right)}{\left(\sum_{j=1}^{n} N_j C_{Vj}\right)} \tag{4.48}$$

Provided that ΔU_R and C_V are known as a function of T, Equation 4.48 can be integrated to give T as a function of N_A. If N_A is assumed to be constant, and furthermore we assume a constant average C_V, the integration is straightforward and we obtain the result

$$T - T_0 = \frac{\Delta U_{R,A}}{m_t \bar{C}_V}\left(N_A - N_{A0}\right) \tag{4.49}$$

Equation 4.49 can be written using fractional conversion, with an initial conversion of zero:

$$T - T_0 = \frac{-\Delta U_{R,A}}{m_t \bar{C}_V} N_{A0} X_A \tag{4.50}$$

Equation 4.50 can be used to check the maximum or minimum reactor temperature that can be attained with adiabatic operation by setting the fractional conversion equal to one.

Example 4.2

Consider the oxidation of carbon monoxide in an adiabatic constant-volume batch reactor. If the reaction is assumed to be essentially irreversible, it is described by the overall stoichiometry:

$$CO + \frac{1}{2}O_2 \rightarrow CO_2 \tag{4.51}$$

The rate of reaction in the presence of water is given by a power law rate expression:

$$(-r_{CO}) = 1.26 \times 10^{10} \exp\left(\frac{-20,131}{T}\right) C_{CO} C_{O_2}^{0.25} C_{H_2O}^{0.5} \; \text{mol/m}^3 \cdot \text{s} \tag{4.52}$$

A mixture consisting of 1% CO, 1% O_2, 1% H_2O, and 97% N_2 (mole percentages) is placed in a constant-volume batch reactor of 0.1 m^3 volume at an initial pressure of 1 bar and an initial temperature of 700 K. Calculate the time required to achieve 99% conversion of the CO, and calculate the pressure and temperature in the reactor at that time.

SOLUTION

The oxidation of CO is a highly exothermic chemical reaction, and, as the reactor is adiabatic, the energy balance will be quite important. It is thus necessary to solve both the mole and energy balance equations. We start the solution by considering the mole balance equation. The mole balance equation for CO can be written in terms of the concentration because the volume of the reactor is constant. The mole balance is

$$-\frac{dC_{CO}}{dt} = 1.26 \times 10^{10} \exp\left(\frac{-20,131}{T}\right) C_{CO} C_{O_2}^{0.25} C_{H_2O}^{0.5} (\text{mol/m}^3 \cdot \text{s}) \tag{4.53}$$

Equation 4.53 can be simplified. First note that the number of moles of water, and hence the concentration, is constant during the reaction, as water does not react and the reactor volume is constant. The water concentration is given by

$$C_{H_2O} = \left(\frac{P}{R_g T}\right)_0 (Y_{H_2O})_0 = \left(\frac{100,000}{8.314 \times 700}\right) \times 0.01 = 0.172 \; \text{mol/m}^3 \tag{4.54}$$

where Y denotes the mole fraction. The primary reactant concentrations can be written in terms of the fractional conversion of CO:

$$C_{CO} = (C_{CO})_0(1 - X_{CO}) \quad \text{and} \quad C_{O_2} = (C_{O_2})_0 - \frac{X_{CO}}{2}(C_{CO})_0$$

With these substitutions and resulting simplification, Equation 4.53 becomes

$$\frac{dX_{CO}}{dt} = 5.23 \times 10^9 \exp\left(\frac{-20,131}{T}\right)(1 - X_{CO})\left((C_{O_2})_0 - \frac{X_{CO}}{2}(C_{CO})_0\right)^{0.25} \qquad (4.55)$$

The initial concentrations of CO and oxygen can be easily calculated:

$$(C_{CO})_0 = \left(\frac{P}{R_g T}\right)_0 (Y_{CO})_0 = \left(\frac{100,000}{8.314 \times 700}\right) \times 0.01 = 0.172 \, \text{mol/m}^3 \qquad (4.56)$$

$$(C_{O_2})_0 = \left(\frac{P}{R_g T}\right)_0 (Y_{O_2})_0 = \left(\frac{100,000}{8.314 \times 700}\right) \times 0.01 = 0.172 \, \text{mol/m}^3 \qquad (4.57)$$

The mole balance equation in its final form is obtained by the substitution of Equations 4.56 and 4.57 into Equation 4.55:

$$\frac{dX_{CO}}{dt} = 3.37 \times 10^9 \exp\left(\frac{-20,131}{T}\right)(1 - X_{CO})\left(1 - \frac{X_{CO}}{2}\right)^{0.25} \qquad (4.58)$$

Having developed the mole balance, we now turn to the energy balance equation. The energy balance for an adiabatic constant-volume batch reactor is given by Equation 4.47, which is shown below in rearranged form, written in terms of fractional conversion:

$$\frac{dT}{dt} = \frac{dN_{CO}}{dt}\frac{(\Delta U_R)}{\left(\sum_{j=1}^{n} N_j C_{Vj}\right)} = -(C_{CO})_0 V \frac{dX_{CO}}{dt}\frac{(\Delta U_R)}{\left(\sum_{j=1}^{n} N_j C_{Vj}\right)} \qquad (4.59)$$

We start the solution of the energy balance by assembling some of the necessary data. The constant-pressure heat capacity for each component is a function of temperature, as follows:

$$C_P = a + bT + cT^2 + dT^3 \qquad (4.60)$$

The values of the constants in the heat capacity polynomial are obtained from Appendix 2:

Compound	a	$b \times 10^2$	$c \times 10^5$	$d \times 10^9$
O_2	25.44	1.518	−0.7144	1.310
N_2	28.85	−0.1569	0.8067	−2.868
H_2O	32.19	0.1920	1.054	−3.589
CO	28.11	0.1672	0.5363	−2.218
CO_2	22.22	5.9711	−3.495	7.457

The enthalpy of reaction as a function of temperature can be computed from the enthalpy of formation and the heat capacity data (see Chapter 2) to give

$$\Delta H_R = -2.796 \times 10^5 - 18.61T + 2.522 \times 10^{-2}T^2$$

$$-1.225 \times 10^{-5}T^3 + 2.255 \times 10^{-9}T^4 \, \text{J/mol} \tag{4.61}$$

Now, the energy balance for the constant-volume batch reactor requires the use of C_V and ΔU_R. Recall that the relationship between C_V and C_P is

$$C_V = C_P - R_g \tag{4.62}$$

The table of C_P data presented above can be changed to C_V data, where

$$C_V = a' + bT + cT^2 + dT^3 \tag{4.63}$$

Because $R_g = 8.314$, the following values may be used:

Compound	a'	$b \times 10^2$	$c \times 10^5$	$d \times 10^9$
O_2	17.13	1.518	−0.7144	1.310
N_2	20.54	−0.1569	0.8067	−2.868
H_2O	23.88	0.1920	1.054	−3.589
CO	19.80	0.1672	0.5363	−2.218
CO_2	13.91	5.9711	−3.495	7.457

The values of b, c, and d remain unchanged. Following the methodology of Example 4.1, the following equation for ΔU_R may be obtained:

$$\Delta U_R = \Delta H_R - \Delta N R_g T \tag{4.64}$$

For this reaction, there are 1.5 mol of reactants consumed for every mole of products that is produced. Therefore, it follows that

$$\Delta N = 1 - 1.5 = -0.5 \tag{4.65}$$

which leads to

$$\Delta U_R = \Delta H_R + 0.5 \times 8.314 \times T \tag{4.66}$$

The internal energy change of reaction is therefore given by

$$\Delta U_R = -2.796 \times 10^5 - 14.45T + 2.522 \times 10^{-2}T^2$$

$$-1.225 \times 10^{-5}T^3 + 2.255 \times 10^{-9}T^4 \, \text{J/mol} \tag{4.67}$$

The total heat capacity of the reactor contents changes with composition as well as temperature. Therefore, we write

$$\left(\sum_{j=1}^{n} N_j C_{Vj} \right) = V \left[C_{N2} C_{V,N2} + C_{CO} C_{V,CO} + C_{O2} C_{V,O2} \right.$$

$$\left. + C_{H2O} C_{V,H2O} + C_{CO2} C_{V,CO2} \right] \tag{4.68}$$

The concentrations of nitrogen ($16.68 \, mol/m^3$) and water ($0.172 \, mol/m^3$) remain unchanged during the reaction. Substituting in the known values and writing in terms of the fractional conversion of CO yields

$$\left(\sum_{j=1}^{n} N_j C_{Vj} \right) = 0.1 \left[16.68 C_{V,N2} + 0.172(1 - X_{CO}) C_{V,CO} \right.$$

$$\left. + 0.172(1 - 0.5 X_{CO}) C_{V,O2} + 0.172 C_{V,H2O} + 0.172 X_{CO} C_{V,CO2} \right] \tag{4.69}$$

In Equation 4.69, the values of C_V are given by Equation 4.63. Now substitute the known volume and initial concentration of CO into Equation 4.59 to obtain

$$\frac{dT}{dt} = -0.0172 \frac{dX_{CO}}{dt} \frac{(\Delta U_R)}{\left(\sum_{j=1}^{n} N_j C_j \right)} \tag{4.70}$$

Equations 4.67 and 4.69 can be substituted into Equation 4.70 to give the differential equation that relates the temperature to time. The resulting equation must be solved simultaneously with the mole balance, Equation 4.58, to give the fractional conversion and reactor temperature as a function of time. The solution of this system of coupled nonlinear ordinary differential equations is performed numerically, using the Runga–Kutta method.

A plot of the fractional conversion of CO as a function of time is shown in Figure 4.1a. The plot of the temperature as a function of time is shown in Figure 4.1b. The temperature increases as the conversion increases, which is the expected result, because the reaction is exothermic and the reactor is adiabatic. The fractional conversion reaches 99% after 318 s, at which point the temperature is 821 K.

The pressure inside the reactor at this point can be computed using the ideal gas law. The final composition is required for this calculation. Following the technique used in Chapter 3 for isothermal reactors, we first develop a stoichiometric table:

Compound	Initial Moles	Final Moles
CO	0.0172	$0.0172(1 - X_{CO})$
O_2	0.0172	$0.0172(1 - 0.5 X_{CO})$
CO_2	0	$0.0172 \, X_{CO}$
H_2O	0.0172	0.0172
N_2	1.668	1.668
Total	1.72	$1.72(1 - 0.005 X_{CO})$

Substituting for $X_{CO} = 0.99$, we calculate the total number of moles present to be 1.711 mol. The volume is unchanged; therefore, the pressure in the reactor is

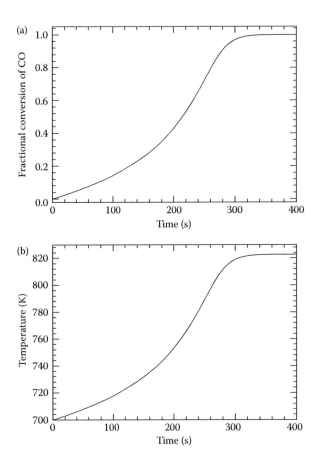

FIGURE 4.1
Time dependence of (a) conversion and (b) temperature for the oxidation of CO in a constant volume batch reactor. Because the reactor is adiabatic and the reaction is exothermic, the temperature rises as the reaction proceeds.

$$P = \frac{NR_gT}{V} = \frac{1.711 \times 8.314 \times 821}{0.1} = 116,823 \text{ Pa}$$

The reactor pressure is 1.17 bar, compared to the initial pressure of 1 bar.

Finally, a plot of temperature versus conversion is shown in Figure 4.2. Note that this line is almost straight, indicating that the temperature dependencies of the enthalpy of reaction and the heat capacity are not very significant over this temperature range.

4.3.5 Energy Balance for CVBR Containing a Catalyst

The CVBR energy balance equation can be used when the reactor contains a solid catalyst. In this case, it is necessary to consider the heat absorbed by the solid catalyst as well as that absorbed by the reacting fluid. The term described by Equation 4.15 must be modified to include the heat capacity of the catalyst, as follows:

$$\sum_{j=1}^{n}\left(N_j \frac{dU_j}{dt}\right) = \left(\sum_{j=1}^{n}N_jC_{Vj}\right)\frac{dT}{dt} + WC_{V\text{cat}}\frac{dT}{dt} \tag{4.71}$$

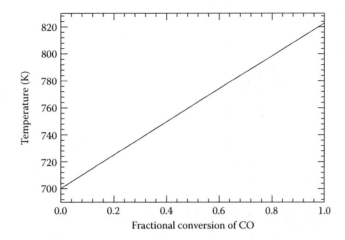

FIGURE 4.2
Conversion verses temperature for the oxidation of CO. The line is almost straight, which indicates that the temperature dependence of the heat capacity and internal energy of reaction is not significant.

The mass of catalyst is denoted W. Equation 4.71 is valid provided that the fluid and catalyst temperatures are the same. This assumption is usually valid in liquid reactions but may not be true for gas-phase reactions. The fluid heat capacity may be written using an average value, as in Equation 4.44. Including the energy change of the catalyst in Equation 4.44 would give

$$\frac{dT}{dt} = \frac{q}{m_{tf}\bar{C}_{Vf} + WC_{Vcat}} - \frac{(\Delta U_{R,A})(-r_A)V}{m_{tf}\bar{C}_{Vf} + WC_{Vcat}} \tag{4.72}$$

In Equation 4.72, the subscript f is used to denote the fluid phase. Note that the reaction rate in Equation 4.72 is expressed in terms of the reactor (fluid) volume. In catalytic reactors, the rate may be expressed in terms of catalyst mass, and in such a case, Equation 4.72 would be written as

$$\frac{dT}{dt} = \frac{q}{m_{tf}\bar{C}_{Vf} + WC_{Vcat}} - \frac{(\Delta U_{R,A})(-r_A)W}{m_{tf}\bar{C}_{Vf} + WC_{Vcat}} \tag{4.73}$$

Case Study 4.1 at the end of Section 4.3 illustrates a batch reactor with a catalyst.

4.3.6 Multiple Reactions in a Batch Reactor

The energy balance equation can be extended to include more than one reaction. With multiple reactions, the internal energy change of reaction for each reaction must be included in the energy balance equation. For example, consider a reactor in which the following two reactions occur:

$$A \rightarrow B$$
$$C \rightarrow D \tag{4.74}$$

The rates of disappearance of A and C are given by $(-r_A)$ and $(-r_C)$, respectively. Two mole balance equations would have to be solved, one each for A and C:

$$-\frac{1}{V}\frac{dN_A}{dt} = (-r_A) \quad \text{and} \quad -\frac{1}{V}\frac{dN_C}{dt} = (-r_C) \tag{4.75}$$

The energy balance equation for a homogeneous reactor would be

$$\frac{dT}{dt} = \frac{q}{\left(\sum_{j=1}^{n} N_j C_{Vj}\right)} - \frac{V}{\left(\sum_{j=1}^{n} N_j C_{Vj}\right)}\left[(-r_A)(\Delta U_{R,A}) + (-r_C)(\Delta U_{R,C})\right] \tag{4.76}$$

The general energy balance for a reactor containing n species (including inert material) in which m reactions occur is

$$\frac{dT}{dt} = \frac{q}{\left(\sum_{j=1}^{n} N_j C_{Vj}\right)} - \frac{V}{\left(\sum_{j=1}^{n} N_j C_{Vj}\right)}\sum_{k=1}^{m}(-r_k)(\Delta U_{R,k}) \tag{4.77}$$

Case Study 4.1 at the end of Section 4.3 illustrates a batch reactor with multiple reactions.

4.3.7 Variable-Volume Batch Reactor

The variable-volume batch reactor is similar in many respects to the constant-volume reactor. The difference is that the change in volume of the system owing to reaction results in work being done on or by the surroundings. If we denote the expansion work as W_E, then the batch reactor energy balance can be written as

$$\frac{dU}{dt} = \frac{d\left(\sum_{j=1}^{n} N_j U_j\right)}{dt} = q - W_E \tag{4.78}$$

The expansion work is related to changes in pressure and volume of the system

$$W_E = P\frac{dV}{dt} \tag{4.79}$$

The enthalpy $(H = U + PV)$ can be introduced, and Equation 4.78 would then be written as

$$\frac{dH}{dt} - \frac{d(PV)}{dt} = \frac{dH}{dt} - P\frac{dV}{dt} - V\frac{dP}{dt} = q - P\frac{dV}{dt} \tag{4.80}$$

Simplifying Equation 4.80 gives the result

$$\frac{dH}{dt} - V\frac{dP}{dt} = \frac{d\left(\sum_{j=1}^{n} N_j H_j\right)}{dt} - V\frac{dP}{dt} = q \tag{4.81}$$

If the volume change occurs at a constant pressure, the energy balance becomes

$$\frac{d\left(\sum_{j=1}^{n} N_j H_j\right)}{dt} = q \tag{4.82}$$

Expanding Equation 4.82 using the chain rule gives

$$\sum_{j=1}^{n} N_j \frac{dH_j}{dt} + \sum_{j=1}^{n} H_j \frac{dN_j}{dt} = q \tag{4.83}$$

The enthalpy change of a species can be expressed in terms of the temperature using the constant-pressure heat capacity:

$$\frac{dH_j}{dt} = C_{Pj} \frac{dT}{dt} \tag{4.84}$$

Therefore, write

$$\sum_{j=1}^{n} N_j \frac{dH_j}{dt} = \left(\sum_{j=1}^{n} N_j C_{Pj}\right) \frac{dT}{dt} \tag{4.85}$$

Consider, for illustration purposes, the reaction given by

$$A + \frac{b}{a} B \rightarrow \frac{c}{a} C + \frac{d}{a} D \tag{4.86}$$

Let the symbol I denote inert substances present in the system. The left-hand side of Equation 4.85 is then equal to

$$\left(\sum_{j=1}^{n} N_j C_{Pj}\right) \frac{dT}{dt} = \left(N_A C_{PA} + N_B C_{PB} + N_C C_{PC} + N_D C_{PD} + N_I C_{PI}\right) \frac{dT}{dt} \tag{4.87}$$

The enthalpy change owing to the change in moles is given by

$$\sum_{j=1}^{n} \left(H_j \frac{dN_j}{dt}\right) = H_A \frac{dN_A}{dt} + H_B \frac{dN_B}{dt} + H_C \frac{dN_C}{dt} + H_D \frac{dN_D}{dt} \tag{4.88}$$

Using the reaction stoichiometry, it is seen that the relationship between the rates of reaction of the different reactants and products can be expressed as

$$\frac{dN_B}{dt} = \frac{b}{a} \frac{dN_A}{dt}, \quad \frac{dN_C}{dt} = -\frac{c}{a} \frac{dN_A}{dt}, \quad \frac{dN_D}{dt} = -\frac{d}{a} \frac{dN_A}{dt}$$

Substitution of these terms into Equation 4.88 and simplification give

$$\sum_{j=1}^{n}\left(H_j\frac{dN_j}{dt}\right) = -\frac{dN_A}{dt}\left[\frac{d}{a}H_D + \frac{c}{a}H_C - H_A - \frac{b}{a}H_B\right] \qquad (4.89)$$

The difference between the enthalpies of the products and the reactants, multiplied by the respective stoichiometric coefficients, is the enthalpy change of reaction:

$$\frac{d}{a}H_D + \frac{c}{a}H_C - H_A - \frac{b}{a}H_B = \Delta H_{R,A} \qquad (4.90)$$

In Equation 4.90, $\Delta H_{R,A}$ is the enthalpy change due to reaction per mole of A. The complete energy balance equation can therefore be written as

$$\left(\sum_{j=1}^{n}N_jC_{Pj}\right)\frac{dT}{dt} - \frac{dN_A}{dt}\left(\Delta H_{R,A}\right) = q \qquad (4.91)$$

The mole balance is

$$-\frac{1}{V}\frac{dN_A}{dt} = \left(-r_A\right) \qquad (4.92)$$

Substituting Equation 4.92 into Equation 4.91 gives the energy balance as

$$\left(\sum_{j=1}^{n}N_jC_{Pj}\right)\frac{dT}{dt} + \left(-r_A\right)V\left(\Delta H_{R,A}\right) = q \qquad (4.93)$$

For an adiabatic reactor, the heat transfer term, q, is zero. The mole and energy balance equations describe the behavior of a nonisothermal variable-volume batch reactor. Note that Equation 4.93 is similar to Equation 4.24, except that $\Delta H_{R,A}$ is used rather than $\Delta U_{R,A}$, and C_P is used instead of C_V. Forms of the energy equation for the variable-volume batch reactor using an average C_P value can be developed in the same manner as for the constant-volume batch reactor.

Example 4.3

Consider the oxidation of carbon monoxide in an adiabatic constant-pressure batch reactor. The reaction is described by the overall stoichiometry

$$CO + \frac{1}{2}O_2 \rightarrow CO_2 \qquad (4.94)$$

We use the same kinetics and initial conditions as Example 4.2. The only difference will be that in this example the reactor will be maintained under constant pressure, whereas in Example 4.2, the reactor volume was held constant. The rate of reaction in the presence of water is given by the rate expression

$$(-r_{CO}) = 1.26 \times 10^{10} \exp\left(\frac{-20,131}{T}\right) C_{CO} C_{O_2}^{0.25} C_{H_2O}^{0.5} (mol/m^3 \cdot s) \qquad (4.95)$$

A mixture of 1% CO, 1% O_2, 1% H_2O, and 97% N_2 (mole percent) is placed in a batch reactor, with an initial volume of 0.1 m³, a pressure of 1 bar, and an initial temperature of 700 K. Calculate the time required to achieve 99% conversion of CO, and calculate the volume and temperature of the reactor at that time.

SOLUTION

As in Example 4.2, it is necessary to compute both the mole balance and the energy balance. The mole balance equation for CO should be written in terms of the number of moles because the reactor volume changes. The mole balance is

$$-\frac{1}{V}\frac{dN_{CO}}{dt} = 1.26 \times 10^{10} \exp\left(\frac{-20,131}{T}\right) \frac{N_{CO}}{V} \left(\frac{N_{O_2}}{V}\right)^{0.25} \left(\frac{N_{H_2O}}{V}\right)^{0.5} \qquad (4.96)$$

Equation 4.96 can be simplified. First note that the moles of water are constant during the reaction, as it is inert. The number of moles of water is given by

$$N_{H_2O} = \left(\frac{PV}{R_g T}\right)_0 Y_{H_2O} = \left(\frac{100,000 \times 0.1}{8.314 \times 700}\right)_0 0.01 = 0.0172 \text{ mol} \qquad (4.97)$$

where Y denotes the mole fraction. The moles of the reactants can be written in terms of the fractional conversion of CO:

$$N_{CO} = (N_{CO})_0 (1 - X_{CO}) \quad \text{and} \quad N_{O_2} = (N_{O_2})_0 - \frac{X_{CO}}{2}(N_{CO})_0$$

With these substitutions and resulting simplification, Equation 4.96 becomes

$$\frac{dX_{CO}}{dt} = 1.65 \times 10^9 \exp\left(\frac{-20,131}{T}\right)(1 - X_{CO})\left((N_{O_2})_0 - \frac{X_{CO}}{2}(N_{CO})_0\right)^{0.25} V^{-0.75}$$

The initial number of moles of CO and oxygen can be easily calculated:

$$(N_{CO})_0 = \left(\frac{PV}{R_g T}\right)_0 (Y_{CO})_0 = \left(\frac{100,000 \times 0.1}{8.314 \times 700}\right) \times 0.01 = 0.0172 \text{ mol} \qquad (4.98)$$

$$(N_{O_2})_0 = \left(\frac{PV}{R_g T}\right)_0 (Y_{O_2})_0 = \left(\frac{100,000 \times 0.1}{8.314 \times 700}\right) \times 0.01 = 0.0172 \text{ mol} \qquad (4.99)$$

The mole balance equation is simplified using Equations 4.98 and 4.99:

$$\frac{dX_{CO}}{dt} = 5.98 \times 10^8 \exp\left(\frac{-20,131}{T}\right)(1 - X_{CO})\left(1 - \frac{X_{CO}}{2}\right)^{0.25} V^{-0.75} \qquad (4.100)$$

It is now necessary to derive an expression for the reactor volume as a function of temperature and fractional conversion. We first develop a stoichiometric table:

Compound	Initial Moles	Final Moles
CO	0.0172	$0.0172(1 - X_{CO})$
O_2	0.0172	$0.0172(1 - 0.5X_{CO})$
CO_2	0	$0.0172\, X_{CO}$
H_2O	0.0172	0.0172
N_2	1.668	1.668
Total	1.72	$1.72\,(1 - 0.005X_{CO})$

The volume is related to the pressure and temperature using the ideal gas law:

$$\left(\frac{PV}{NT}\right)_0 = \left(\frac{PV}{NT}\right) \tag{4.101}$$

The pressure is constant at 1 bar, the initial temperature is 700 K, and the initial volume is 0.1 m^3. The volume of the reactor is therefore given by

$$V = V_0 \frac{T}{T_0} \frac{N}{N_0} = \frac{0.1T}{700} \frac{1.72(1 - 0.005X_{CO})}{1.72} = 1.429 \times 10^{-4} T(1 - 0.005X_{CO}) \tag{4.102}$$

Having developed the mole balance, we now turn to the energy balance equation. The energy balance for an adiabatic constant-pressure batch reactor is given by Equation 4.93 with the heat transfer term set equal to zero. Shown below in rearranged form, written in terms of fractional conversion, this equation becomes

$$\frac{dT}{dt} = \frac{dN_{CO}}{dt} \frac{(\Delta H_R)}{\left(\sum_{j=1}^{n} N_j C_{Pj}\right)} = -(N_{CO})_0 \frac{dX_{CO}}{dt} \frac{(\Delta H_R)}{\left(\sum_{j=1}^{n} N_j C_{Pj}\right)} \tag{4.103}$$

As with Example 4.2, we start the solution of the energy balance by assembling some of the necessary data. The constant-pressure heat capacity for each component is a function of temperature, as follows:

$$C_P = a + bT + cT^2 + dT^3 \tag{4.104}$$

The values of the constants are obtained from Appendix 2:

Compound	a	$b \times 10^2$	$c \times 10^5$	$d \times 10^9$
O_2	25.44	1.518	−0.7144	1.310
N_2	28.85	−0.1569	0.8067	−2.868
H_2O	32.19	0.1920	1.054	−3.589
CO	28.11	0.1672	0.5363	−2.218
CO_2	22.22	5.9711	−3.495	7.457

The enthalpy of reaction as a function of temperature can be computed from the enthalpy of formation and the heat capacity data (see Chapter 2) to give

$$\Delta H_R = -2.796 \times 10^5 - 18.61T + 2.522 \times 10^{-2}T^2$$

$$- 1.225 \times 10^{-5}T^3 + 2.255 \times 10^{-9}T^4 \, \text{J/mol} \tag{4.105}$$

The total heat capacity of the reactor contents change with composition as well as temperature. Therefore, we write

$$\left(\sum_{j=1}^{n} N_j C_{Pj}\right) = \left[N_{N_2}C_{P,N_2} + N_{CO}C_{P,CO} + N_{O_2}C_{P,O_2} + N_{H_2O}C_{P,H_2O} + N_{CO_2}C_{P,CO_2}\right] \tag{4.106}$$

The moles of nitrogen (0.859 mol) and water (0.172 mol) remain unchanged during the reaction. Substituting the known values and writing in terms of the fractional conversion of CO yields

$$\left(\sum_{j=1}^{n} N_j C_{Pj}\right) = \left[1.668C_{P,N_2} + 0.0172(1 - X_{CO})C_{P,CO}\right.$$

$$\left. + 0.0172(1 - 0.5X_{CO})C_{P,O_2} + 0.0172C_{P,H_2O} + 0.0172X_{CO}C_{P,CO_2}\right] \tag{4.107}$$

In Equation 4.107, the values of C_P are given by Equation 4.104. Returning to Equation 4.103, substitute the known initial number of moles of CO to obtain

$$\frac{dT}{dt} = -0.0172\frac{dX_{CO}}{dt}\frac{(\Delta H_R)}{\left(\sum_{j=1}^{n} N_j C_{Pj}\right)} \tag{4.108}$$

Equations 4.105 and 4.107 can be substituted into Equation 4.108 to give the differential equation that relates the temperature to time. The resulting equation must be solved simultaneously with the mole balance, which is obtained by substituting Equation 4.102 into Equation 4.100. The solution to the two equations must be performed numerically to give the fractional conversion and reactor temperature as a function of time. A plot of the fractional conversion of CO as a function of time is shown in Figure 4.3a and the temperature as a function of time is shown in Figure 4.3b. The fractional conversion reaches 99% after 534 s, at which point the temperature is 790 K. Note that the temperature is lower than in the constant-volume reactor at 99% conversion. The lower temperature of the gas, and hence lower energy content, reflects the energy that was expended in doing work on the surroundings.

The reactor volume at 99% conversion can be computed from Equation 4.102. Therefore, the volume is

$$V = 1.429 \times 10^{-4}T(1 - 0.005X_{CO}) = 1.429 \times 10^{-4} \times 790 \times (1 - 0.005 \times 0.99) = 0.112 \, \text{m}^3$$

The reactor volume has thus increased by about 12%.

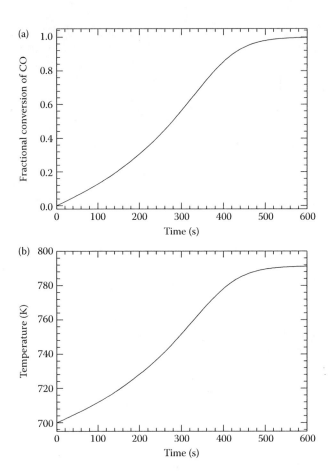

FIGURE 4.3

Time dependence of (a) conversion and (b) temperature for the oxidation of CO in a constant pressure batch reactor. Because the reactor is adiabatic and the reaction is exothermic, the temperature rises as the reaction proceeds.

Case Study 4.1:   Hydrogenation of Gas Oil in a Nonisothermal Batch Reactor

In this case study, the catalytic hydrogenation reactions occurring in a gas oil that contains sulfur and nitrogen compounds are considered. These reactions are similar to those considered in Case Studies 2.1 and 2.2, and the general characteristics of the oil, kinetic expressions, and so on are the same as those used in those case studies. The rate of reaction of molecules that contain sulfur, based on total sulfur molar concentration expressed as S, is

$$(-r_S) = A_S \exp\left(\frac{-E_S}{R_g T}\right) C_S^{1.5} P_{H_2}^{0.8} (\text{mol of S/h} \cdot \text{gcat}) \tag{4.109}$$

The rate of reaction for the nitrogen-containing compounds is given by

$$(-r_N) = A_N \exp\left(\frac{-E_N}{R_g T}\right) C_N P_{H_2}^{1.3} (\text{mol of N/h} \cdot \text{gcat}) \tag{4.110}$$

In addition to the reactions between hydrogen and the sulfur and nitrogen compounds, there are also reactions between hydrogen- and oxygen-containing compounds, hydrogenation of unsaturated hydrocarbons and also some hydrocracking. All of these latter reactions will be represented here by a single lumped expression that correlates the rate of disappearance of compounds with a boiling point higher than 343°C. The mass of these compounds in the reactor is denoted m_{343}. The rate expression for the disappearance of these compounds is

$$(-r_{343}) = A_{343} \exp\left(\frac{-E_{343}}{R_g T}\right)[m_{343}](\text{g of } 343^+/\text{h} \cdot \text{gcat}) \tag{4.111}$$

The concentrations in these three rate equations are

$C_S \equiv$ molar concentration of sulfur in liquid, mol/cm^3
\$C_N \equiv$ molar concentration of nitrogen in liquid, mol/cm^3
$[m_{343}] \equiv$ mass concentration of 343°C$^+$ material in liquid, g/cm^3
$m_{343} \equiv$ mass of 343°C$^+$ material in liquid, g

Note that the rate is based on the mass of the catalyst and not on the reactor volume. The unit gcat in the rate equations means grams of catalyst. The values of the kinetic parameters are

$$A_S = 5.22 \times 10^{11} \, \text{cm}^{4.5}/(\text{h} \cdot \text{gcat} \, \text{mol}^{0.5} \cdot \text{MPa}^{0.8}) \quad \frac{E_S}{R_g} = 15,100 \, \text{K}$$

$$A_N = 1.21 \times 10^7 \, \text{cm}^3/(\text{h} \cdot \text{gcat} \cdot \text{MPa}^{1.3}) \quad \frac{E_N}{R_g} = 12,300 \, \text{K}$$

$$A_{343} = 1.3 \times 10^{11} \, \text{cm}^3/\text{h} \cdot \text{gcat} \quad \frac{E_{343}}{R_g} = 16,300 \, \text{K}$$

The oil contains 4.3 wt.% sulfur (S), 0.30 wt.% nitrogen (N), and 74.2 wt.% 343°C$^+$ material. The density of the oil is a function of temperature:

$$\rho_L = \sqrt{1.286 - 0.0011T} \, \text{g/cm}^3 \tag{4.112}$$

The temperature in Equation 4.112 is in K. The reactor initially contains 94.7 cm^3 of oil at 20°C (293 K), which has a density of 0.982 g/cm^3, calculated from Equation 4.112. Therefore, 94.7 cm^3 of oil has a mass of 92.957 g. The reactor contains 10 g of catalyst. The hydrogen partial pressure is a constant 13.9 MPa during the reaction.

At time zero, the reactor is surrounded by an external heating jacket that is maintained at a constant temperature of 400°C (673 K). The heat transfer rate is expressed in terms of the temperature difference between the reactor and the jacket:

$$q = UA(T_\infty - T) \tag{4.113}$$

The product of the reactor heat transfer area, A, and the overall heat transfer coefficient, U, is given as $UA = 0.06$ W/K, or 216 J/h·K. Other physical parameters are as follows:

Heat capacity of the fluid:

$$C_{Pf} = -0.5167 + 9.964 \times 10^{-3}T - 8.686 \times 10^{-6}T^2 \text{ J/g} \cdot \text{K} \tag{4.114}$$

Heat capacity of the catalyst:

$$C_{P,\text{cat}} = 0.90 \text{ J/g} \cdot \text{K} \tag{4.115}$$

Enthalpy of reaction for the sulfur compounds:

$$(-\Delta H_{R,S}) = 1.0 \times 10^5 \text{ J/mol of S converted} \tag{4.116}$$

Enthalpy of reaction for the nitrogen compounds:

$$(-\Delta H_{R,N}) = 2.0 \times 10^5 \text{ J/mol of N converted} \tag{4.117}$$

Enthalpy of reaction for the 343°C material:

$$(-\Delta H_{R,343}) = 2.0 \times 10^2 \text{ J/g of 343}^\circ\text{C}^+ \text{ converted} \tag{4.118}$$

Note that all the reactions are exothermic. It will be assumed that the enthalpies of reaction are independent of temperature. The enthalpy change of reaction is assumed to be equal to the change in internal energy of reaction, and $C_P = C_V$. The temperature of the reactor and the conversions of S, N, and m_{343} will be calculated as a function of time.

This case study requires that the mole and energy balance equations for a batch reactor be solved. Note that the reactor volume (i.e., the liquid volume) changes during the course of the reaction. The concentrations of the species thus change as the volume changes. The reaction rates, however, are expressed in terms of the catalyst mass, which remains constant. The three mole balance equations are written by incorporating the rate equations into the general balance for a batch reactor. Because the liquid volume changes, the balances are written in terms of the number of moles (or mass). The equations are

$$-\frac{1}{W}\frac{dN_S}{dt} = A_S \exp\left(\frac{-E_S}{R_g T}\right)\left(\frac{N_S}{V}\right)^{1.5} P_{H_2}^{0.8} \text{ mol of S/h} \cdot \text{gcat} \tag{4.119}$$

$$-\frac{1}{W}\frac{dN_N}{dt} = A_N \exp\left(\frac{-E_N}{R_g T}\right)\left(\frac{N_N}{V}\right) P_{H_2}^{1.3} \text{ mol of N/h} \cdot \text{gcat} \tag{4.120}$$

$$-\frac{1}{W}\frac{dm_{343}}{dt} = A_{343} \exp\left(\frac{-E_{343}}{R_g T}\right)\frac{m_{343}}{V} \text{ g of 343}^+/\text{h} \cdot \text{gcat} \tag{4.121}$$

The reactor volume is the liquid volume, which depends on the density.

$$V = \frac{m_f}{\rho_L} = \frac{92.957}{\sqrt{1.286 - 0.0011T}} \tag{4.122}$$

where m_f is the total mass of oil, which was calculated previously. Substitution of the numerical values for the rate parameters, the catalyst mass, the hydrogen partial pressure, and the liquid density gives the following three equations:

$$\frac{dN_S}{dt} = -4.78 \times 10^{10} \exp\left(\frac{-15,100}{T}\right) N_S^{1.5} [1.286 - 0.0011T]^{0.75} \tag{4.123}$$

$$\frac{dN_N}{dt} = -4.0 \times 10^{7} \exp\left(\frac{-12,300}{T}\right) N_N \sqrt{1.286 - 0.0011T} \tag{4.124}$$

$$\frac{dm_{343}}{dt} = -1.4 \times 10^{10} \exp\left(\frac{-16,300}{T}\right) m_{343} \sqrt{1.286 - 0.0011T} \tag{4.125}$$

The initial composition of the oil is computed from the total mass and the initial concentrations.

$$\text{Sulfur} \Rightarrow 4.3 \times 10^{-2} \text{ g S/g oil} \times 92.957 \text{ g oil} = 4.0 \text{ g S} \equiv 0.125 \text{ mol}$$

$$\text{Nitrogen} \Rightarrow 0.3 \times 10^{-2} \text{ g N/g N} \times 92.957 \text{ g oil} = 0.279 \text{ g N} \equiv 0.020 \text{ mol}$$

$$m_{343} \Rightarrow 74.2 \times 10^{-2} \text{ g } m_{343}/\text{g oil} \times 92.957 \text{ g oil} = 68.974 \text{ g } m_{343}$$

The reactor is nonisothermal; therefore, it is necessary to perform an energy balance equation. The form of the energy balance equation must account for the energy changes of the liquid and the catalyst, and must include the effects of the three reactions. From Equations 4.72 and 4.77, it follows that the energy balance equation for this reactor has the form

$$\frac{dT}{dt} = \frac{q - W \sum_{k=1}^{m} (-r_k)(\Delta U_{R,k})}{\left(\sum_{j=1}^{n} N_j C_{Vj}\right) + W C_{Vcat}} \tag{4.126}$$

The heat capacity of the fluid is assumed to be independent of the composition, and is expressed as a polynomial in temperature in terms of the catalyst mass. The rate of heat transfer is given by Equation 4.113. Making these two substitutions, Equation 4.126 is written as

$$\frac{dT}{dt} = \frac{UA(T_\infty - T) - W \sum_{k=1}^{m} (-r_k)(\Delta U_{R,k})}{m_f C_{Pf} + W C_{Vcat}} \tag{4.127}$$

Finally, it is necessary to substitute for the reaction rates of the three pseudocomponents. From Equations 4.123, 4.124, and 4.125, noting that

$$-\frac{dN_S}{dt} = (-r_S)W; \quad -\frac{dN_N}{dt} = (-r_N)W; \quad -\frac{dm_{343}}{dt} = (-r_{343})W$$

and incorporating Equation 4.114, Equation 4.127 is written as

$$\frac{dT}{dt} = \frac{UA(T_\infty - T) + (dN_S/dt)(\Delta U_{R,S}) + (dN_N/dt)(\Delta U_{R,N}) + (dm_{343}/dt)(\Delta U_{R,343})}{m_f(-0.5167 + 9.964 \times 10^{-3}T - 8.686 \times 10^{-6}T^2) + WC_{Vcat}}$$

(4.128)

Substituting the numbers into Equation 4.128 gives the final energy balance

$$\frac{dT}{dt} = \frac{216(673.2 - T) + (dN_S/dt)(-10^5) + (dN_N/dt)(-2 \times 10^5) + (dm_{343}/dt)(-2 \times 10^2)}{92.957 \times (-0.5167 + 9.964 \times 10^{-3}T - 8.686 \times 10^{-6}T^2) + (10)(0.9)}$$

(4.129)

Although not written out here, the reader should note that the units in Equation 4.129 are consistent. Equations 4.123 through 4.125 and 4.129 represent a coupled system of four ordinary first-order differential equations. A numerical solution is required using, for example, the Runga–Kutta method. The result of such a solution from time zero to 300 min is shown in Figures 4.4 and 4.5. Figure 4.4 shows the fractional conversion of the three components as a function of time, while Figure 4.5 shows the reactor temperature as a function of time. Note from Figure 4.4 that the fractional conversion of the sulfur components is the highest. Note also from Figure 4.5 that the reactor

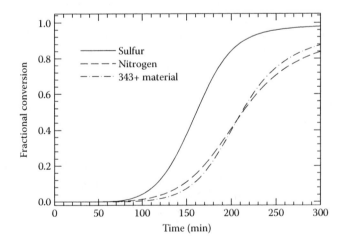

FIGURE 4.4
Fractional conversion as a function of time for the three pseudocomponents, sulfur, nitrogen, and 343+ material. The sulfur has the highest fractional conversion after 300 min.

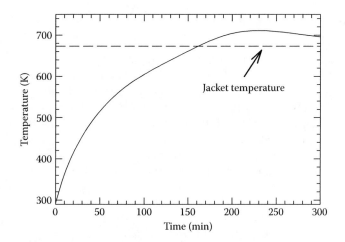

FIGURE 4.5
Temperature as a function of time for the batch reactor of Case Study 4.1. Note that the temperature initially increases owing to the exothermic chemical reaction and heat transfer from the surroundings. Later, the temperature starts to decrease as heat is transferred to the surroundings.

temperature initially rises and then starts to fall. The temperature rises for two reasons: the first reason is the exothermic reaction that releases heat, and the second is the heat transfer from the jacket. Heat is transferred into the reactor from the jacket as long as the reactor temperature is below the jacket temperature of 673.15 K (400°C). However, as the reactor temperature rises above this value, the direction of heat transfer is reversed, and in effect the reactor is cooled by the surrounding fluid. If the rate of heat transfer to the jacket is greater than the rate of heat release owing to chemical reaction, the reactor temperature falls. Note that once all the reactions are complete, the final reactor temperature must be 673.15 K (400°C).

4.4 Plug Flow Reactor

In this section, we consider the energy balance equation for a plug flow reactor. In Chapter 3, the mole balance for the PFR was developed, and it was seen that the concentration of the reacting species changes as the process fluid moves along the reactor, leading to axial gradients. Recall also that the plug flow assumptions state that radial mixing is complete; therefore, there are no radial temperature gradients. Because thermal energy is either released or absorbed during chemical reactions, the natural tendency is for the process stream either to increase or to decrease in temperature as it moves through the reactor. Isothermal operation in such a case can only be maintained if the rate of heat transfer with the surroundings exactly matches the rate of evolution of thermal energy in the reaction. In practice, this degree of control is rare, and most PFR are operated nonisothermally, either with or without heat transfer with the surroundings. It is therefore necessary to include the energy balance when analyzing such reactors. In the following sections, we develop the energy balance for the steady-state PFR. Both adiabatic and nonadiabatic cases are considered. We start by deriving the general energy balance equation, and then show how variations of the balance are used for the different modes of reactor operation.

4.4.1 Basic Energy Balance Equation

The energy balance equation for the PFR is simply a variation of the general balance that was introduced in Section 4.2. The energy balance can be derived by considering an incremental reactor volume, ΔV, as shown in Figure 4.6. The mechanisms of energy transport include the energy associated with the fluid that flows into and out of the volume element, the heat transfer with the surroundings and the flow work. There is seldom any mechanical device (such as a mixer) located in the reactor; hence the shaft work is usually equal to zero, and the volume element is fixed in size, so the expansion work is also zero. The accumulation of internal energy in the volume element with time equals the net rate of energy transfer, which gives the following energy balance equation:

$$\frac{\Delta U}{\Delta t} = -W_f + \left(\sum_{j=1}^{n} F_j U_j \right)_V - \left(\sum_{j=1}^{n} F_j U_j \right)_{V+\Delta V} + \Delta q \tag{4.130}$$

As seen in Section 4.2, the flow work is expressed in terms of the pressure and volume:

$$W_f = -P \sum_{j=1}^{n} \frac{F_j}{C_j} = -\sum_{j=1}^{n} F_j P V_j \tag{4.131}$$

Substitution for the expansion work and introduction of the enthalpy (see Section 4.2) give

$$\frac{\Delta U}{\Delta t} = \left(\sum_{j=1}^{n} F_j H_j \right)_V - \left(\sum_{j=1}^{n} F_j H_j \right)_{V+\Delta V} + \Delta q \tag{4.132}$$

In Equation 4.132, Δq is the incremental heat transfer to the surroundings in ΔV. At steady state, the accumulation term equals zero. If we take the limit as ΔV tends to the differential element, dV, the steady-state form of Equation 4.132 becomes a differential equation:

$$\frac{dq}{dV} = \frac{d\left(\sum_{j=1}^{n} F_j H_j \right)}{dV} \tag{4.133}$$

Equation 4.133 can be expanded using the chain rule to give

$$\frac{dq}{dV} = \sum_{j=1}^{n} F_j \frac{dH_j}{dV} + \sum_{j=1}^{n} H_j \frac{dF_j}{dV} \tag{4.134}$$

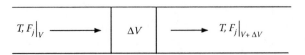

FIGURE 4.6
Incremental reactor volume used to derive the energy balance for a plug flow reactor.

The total enthalpy change therefore depends on the change in molar flow rate of each species, as well as the change in enthalpy in each species. The change in enthalpy of a component, dH_j, can be expressed in terms of C_P, the constant-pressure heat capacity.

$$\frac{dH_j}{dV} = C_{Pj}\frac{dT}{dV}$$ (4.135)

Equation 4.134 can thus be written as

$$\frac{dq}{dV} = \sum_{j=1}^{n} F_j C_{Pj}\frac{dT}{dV} + \sum_{j=1}^{n} H_j \frac{dF_j}{dV}$$ (4.136)

To develop the methodology for dealing with the change in molar flow rates of each species, we consider the following model reaction, written in terms of 1 mol of reactant A:

$$A + \frac{b}{a}B \rightarrow \frac{c}{a}C + \frac{d}{a}D$$ (4.137)

Let I represent the inert components in the process stream. The first term on the right-hand side of Equation 4.136 written for the model reaction is

$$\sum_{j=1}^{n} F_j C_{Pj}\frac{dT}{dV} = \left(F_A C_{PA} + F_B C_{PB} + F_C C_{PC} + F_D C_{PD} + F_I C_{PI}\right)\frac{dT}{dV}$$ (4.138)

Note that the molar flow rate of each species in Equation 4.138 changes owing to reaction as the process stream moves through the reactor. The second term on the right-hand side of Equation 4.136 written for the reaction of Equation 4.137 is

$$\sum_{j=1}^{n} H_j \frac{dF_j}{dV} = H_A \frac{dF_A}{dV} + H_B \frac{dF_B}{dV} + H_C \frac{dF_C}{dV} + H_D \frac{dF_D}{dV}$$ (4.139)

The relationship among the reaction rates of the various reactants and products can be obtained from the stoichiometry, as follows:

$$\frac{dF_B}{dV} = \frac{b}{a}\frac{dF_A}{dV}; \quad \frac{dF_C}{dV} = -\frac{c}{a}\frac{dF_A}{dV}; \quad \frac{dF_D}{dV} = -\frac{d}{a}\frac{dF_A}{dV}$$ (4.140)

Substituting these values into Equation 4.139 gives

$$\sum_{j=1}^{n} H_j \frac{dF_j}{dV} = -\frac{dF_A}{dV}\left[\frac{d}{a}H_D + \frac{c}{a}H_C - H_A - \frac{b}{a}H_B\right]$$ (4.141)

The difference in enthalpy between reactants and products is the enthalpy change of reaction

$$\Delta H_{R,A} = \frac{d}{a} H_D + \frac{c}{a} H_C - H_A - \frac{b}{a} H_B \tag{4.142}$$

The subscript A on the enthalpy change of reaction is used to emphasize that the value is based on 1 mol of A reacting. Therefore, Equation 4.141 becomes

$$\sum_{j=1}^{n} H_j \frac{dF_j}{dV} = \left(-\Delta H_{R,A}\right) \frac{dF_A}{dV} \tag{4.143}$$

Substituting Equation 4.143 into Equation 4.136 gives the energy balance equation for a steady-state PFR.

$$\frac{dq}{dV} = \left(\sum_{j=1}^{n} F_j C_{Pi}\right) \frac{dT}{dV} + \left(-\Delta H_{R,A}\right) \frac{dF_A}{dV} \tag{4.144}$$

The steady-state mole balance equation for species A for a PFR is

$$-\frac{dF_A}{dV} = \left(-r_A\right) \tag{4.145}$$

We can substitute Equation 4.145 into Equation 4.56 to give

$$\frac{dq}{dV} = \sum_{j=1}^{n} \left(F_j C_{Pj}\right) \frac{dT}{dV} - \left(-\Delta H_{R,A}\right) \left(-r_A\right) \tag{4.146}$$

Provided that the rate of heat transfer to the reactor is known, simultaneous solution of Equations 4.145 and either 4.144 or 4.146 gives the concentration and temperature profiles in the reactor as a function of V.

4.4.1.1 Use of Average Heat Capacity

Equation 4.146 contains the heat capacity of each species in the process stream. The chemical reaction changes the molar flow rate of each of the reactants and products; hence the heat capacity of the process stream has a compositional dependence. In addition, the heat capacities are generally functions of temperature. In some cases, approximate solutions are computed using an average heat capacity value. The heat capacity may be assumed to be independent of composition or independent of temperature, or both. As the mass flow rate is constant in a steady-state PFR, it is common to use the mass flow rate when assuming that the heat capacity does not change with composition. An approximate energy balance equation can be written as

$$\frac{dq}{dV} = \dot{m}\bar{C}_P \frac{dT}{dV} - \left(-\Delta H_{R,A}\right) \left(-r_A\right) \tag{4.147}$$

The average heat capacity is denoted $\bar{C}_P$. Even though the compositional dependence may be ignored, the temperature dependence may still be included. Alternatively, the heat capacity may also be assumed to be independent of temperature.

4.4.2 Temperature and Concentration Profiles as Function of Reactor Length

Many plug flow reactors consist of a circular vessel through which the process stream flows. It is common in such a case to write the mole and energy balance equations in terms of the reactor length, rather than the volume. The volume of the reactor is related to the axial length, z, and the vessel inside diameter, D:

$$V = \frac{\pi D^2}{4} z \quad \text{and} \quad dV = \frac{\pi D^2}{4} dz \tag{4.148}$$

Incorporating Equation 4.148 into the mole balance, Equation 4.145, gives

$$-\frac{dF_A}{dz} = \frac{\pi D^2}{4}(-r_A) \tag{4.149}$$

Incorporating Equation 4.148 into the energy balance, Equation 4.146, gives

$$\frac{dq}{dz} = \sum_{i=1}^{n}\left(F_j C_{Pj}\right)\frac{dT}{dz} - \frac{\pi D^2}{4}\left(-\Delta H_{R,A}\right)\left(-r_A\right) \tag{4.150}$$

The simultaneous solution of Equations 4.149 and 4.150 gives the temperature and concentration profiles along the axial length of the reactor, provided that the rate of external heat transfer is known.

4.4.3 Adiabatic Operation of PFR: Adiabatic Reaction Line

A PFR may be operated with or without heat transfer to the surroundings. If the reactor does not exchange heat with the surroundings, it is called adiabatic. In such a case, the energy change as a result of the chemical reaction must be balanced by a corresponding temperature change of the process stream. Writing Equation 4.144 for the adiabatic case gives

$$0 = \left(\sum_{j=1}^{n} F_j C_{Pj}\right)\frac{dT}{dV} + \left(-\Delta H_{R,A}\right)\frac{dF_A}{dV} \tag{4.151}$$

Elimination of the term dV in Equation 4.151 gives

$$\left(\sum_{j=1}^{n} F_j C_{Pj}\right)dT = \Delta H_{R,A} dF_A \tag{4.152}$$

Equation 4.152 gives a relationship between the extent of reaction and the temperature of the process stream. Note that this relationship is independent of the kinetics of the reaction. When C_P and $\Delta H_{R,A}$ are assumed to be independent of temperature and composition, Equation 4.152 can be easily integrated to give

$$\dot{m}\bar{C}_P(T - T_0) = \Delta H_{R,A}(F_A - F_{A0}) \tag{4.153}$$

Equation 4.153 can also be expressed in terms of the fractional conversion, X_A:

$$\dot{m}\bar{C}_P(T - T_0) = -\Delta H_{R,A}F_{A0}X_A \tag{4.154}$$

Equation 4.154 can be rearranged to give an explicit equation for reactor temperature

$$T = T_0 - \frac{\Delta H_{R,A}}{\dot{m}\bar{C}_P}F_{A0}X_A \tag{4.155}$$

Equation 4.155 is an alternative energy balance equation for an adiabatic reactor. It can be used to calculate a theoretical maximum or minimum temperature in an adiabatic reactor by setting X_A equal to one, to give

$$T = T_0 - \frac{\Delta H_{R,A}}{\dot{m}\bar{C}_P}F_{A0} \tag{4.156}$$

Equation 4.156 can be used to determine potential safety hazards owing to excessive temperature rises.

Example 4.4

Consider the oxidation of carbon monoxide in an adiabatic constant-pressure PFR. The reaction is described by the overall stoichiometry:

$$CO + \frac{1}{2}O_2 \rightarrow CO_2 \tag{4.157}$$

The rate of reaction in the presence of water is given by the rate expression

$$(-r_{CO}) = 1.26 \times 10^{10}\exp\left(\frac{-20,131}{T}\right)C_{CO}C_{O_2}^{0.25}C_{H_2O}^{0.5}\, mol/m^3 \cdot s \tag{4.158}$$

A mixture consisting of 1% CO, 1% O_2, 1% H_2O, and 97% N_2 (mole percentages) is fed to a PFR. The total volumetric flow rate of the feed is 1.00×10^{-4} m^3/s at 1 bar pressure and a temperature of 700 K. Calculate the volume required for 99% conversion and the corresponding outlet temperature.

SOLUTION

It is necessary to compute both the mole balance and the energy balance. The mole balance equation for CO should be written in terms of the molar flow rate because the reaction is a gas-phase reaction in which both the temperature and the number of moles change with reaction. The mole balance is

$$-\frac{dF_{CO}}{dV} = 1.26 \times 10^{10} \exp\left(\frac{-20,131}{T}\right) \frac{F_{CO}}{Q} \left(\frac{F_{O2}}{Q}\right)^{0.25} \left(\frac{F_{H2O}}{Q}\right)^{0.5} \tag{4.159}$$

Equation 4.159 can be simplified. First note that the moles of water are constant during the reaction, as it does not react. The molar flow rate of water in mol/s is

$$F_{H2O} = \left(\frac{PQ}{R_g T}\right)_0 Y_{H2O} = \left(\frac{100,000 \times 1.0 \times 10^{-4}}{8.314 \times 700}\right) \times 0.01 = 1.72 \times 10^{-5} \tag{4.160}$$

The molar flow rates of the reactants can be written in terms of the fractional conversion of CO:

$$F_{CO} = (F_{CO})_0 (1 - X_{CO}) \quad \text{and} \quad F_{O2} = (F_{O2})_0 - \frac{X_{CO}}{2}(F_{CO})_0$$

With these substitutions and resulting simplification, Equation 4.159 becomes

$$\frac{dX_{CO}}{dV} = 5.27 \times 10^7 \exp\left(\frac{-20,131}{T}\right)(1 - X_{CO})\left((F_{O2})_0 - \frac{X_{CO}}{2}(F_{CO})_0\right)^{0.25} Q^{-1.75}$$

The inlet molar flow rates of CO and oxygen in mol/s can be calculated as

$$(F_{CO})_0 = \left(\frac{PQ}{R_g T}\right)_0 (Y_{CO})_0 = \left(\frac{100,000 \times 1.0 \times 10^{-4}}{8.314 \times 700}\right) \times 0.01 = 1.72 \times 10^{-5} \tag{4.161}$$

$$(F_{O2})_0 = \left(\frac{PQ}{R_g T}\right)_0 (Y_{O2})_0 = \left(\frac{100,000 \times 1.0 \times 10^{-4}}{8.314 \times 700}\right) \times 0.01 = 1.72 \times 10^{-5} \tag{4.162}$$

The mole balance equation is simplified using Equations 4.161 and 4.162:

$$\frac{dX_{CO}}{dV} = 3.37 \times 10^6 \exp\left(\frac{-20,131}{T}\right)(1 - X_{CO})\left(1 - \frac{X_{CO}}{2}\right)^{0.25} Q^{-1.75} \tag{4.163}$$

The final step in developing the mole balance equation is to derive an expression for the volumetric flow rate as a function of temperature and fractional conversion. Following the technique used in Chapter 3 for isothermal reactors, we first develop a stoichiometric table:

Compound	Inlet Moles	Effluent Moles
CO	1.72×10^{-5}	$1.72 \times 10^{-5}(1 - X_{CO})$
O_2	1.72×10^{-5}	$1.72 \times 10^{-5}(1 - 0.5X_{CO})$
CO_2	0	$1.72 \times 10^{-5} X_{CO}$
H_2O	1.72×10^{-5}	1.72×10^{-5}
N_2	1.668×10^{-3}	1.668×10^{-3}
Total	1.72×10^{-3}	$1.72 \times 10{-3}\,(1 - 0.005X_{CO})$

The volumetric flow rate depends on the pressure and temperature:

$$\left(\frac{PQ}{F_T T}\right)_0 = \left(\frac{PQ}{F_T T}\right) \tag{4.164}$$

The pressure is constant at 1 bar, the initial temperature is 700 K, and the inlet volumetric flow rate is 1.0×10^{-4} m^3/s. The volumetric flow rate in the reactor therefore is

$$Q = Q_0 \frac{T}{T_0} \frac{F_T}{F_{T0}} = 1.0 \times 10^{-4} \frac{T}{700} \frac{1.72 \times 10^{-3}(1 - 0.005 X_{CO})}{1.72 \times 10^{-3}}$$

$$= 1.429 \times 10^{-7} T (1 - 0.005 X_{CO}) \tag{4.165}$$

Having developed the mole balance, we now turn to the energy balance equation. The energy balance for an adiabatic PFR is given by Equation 4.151. Written in terms of fractional conversion, this equation can be written as

$$\frac{dT}{dV} = \frac{dF_{CO}}{dV} \frac{(\Delta H_R)}{\left(\sum_{j=1}^{n} F_j C_{Pj}\right)} = -(F_{CO})_0 \frac{dX_{CO}}{dV} \frac{(\Delta H_R)}{\left(\sum_{j=1}^{n} F_j C_{Pj}\right)} \tag{4.166}$$

We start the solution of the energy balance by assembling the necessary data. The constant-pressure heat capacity for each component is a function of temperature, as follows:

$$C_P = a + bT + cT^2 + dT^3 \tag{4.167}$$

The values of the constants are obtained from Appendix 2:

Compound	a	$b \times 10^2$	$c \times 10^5$	$d \times 10^9$
O_2	25.44	1.518	−0.7144	1.310
N_2	28.85	−0.1569	0.8067	−2.868
H_2O	32.19	0.1920	1.054	−3.589
CO	28.11	0.1672	0.5363	−2.218
CO_2	22.22	5.9711	−3.495	7.457

The enthalpy of reaction as a function of temperature can be computed from the enthalpy of formation and the heat capacity data (see Chapter 2) to give

$$\Delta H_R = -2.796 \times 10^5 - 18.61T + 2.522 \times 10^{-2} T^2$$

$$- 1.225 \times 10^{-5} T^3 + 2.255 \times 10^{-9} T^4 \, \text{J/mol} \tag{4.168}$$

The total heat capacity of the process stream changes with composition as well as with temperature. Therefore, we write

$$\left(\sum_{j=1}^{n} F_j C_{Pj}\right) = \left[F_{N2} C_{P,N2} + F_{CO} C_{P,CO} + F_{O2} C_{P,O2} F_{H2O} C_{P,H2O} + F_{CO2} C_{P,O2}\right] \tag{4.169}$$

Substituting the known values of the molar flow rates and writing in terms of the fractional conversion of CO yields

$$\left(\sum_{j=1}^{n} F_j C_{Pj} \right) = \Big[1.668 \times 10^{-3} C_{P,N_2} + 1.72 \times 10^{-5} (1 - X_{CO}) C_{P,CO}$$

$$+ 1.72 \times 10^{-5} (1 - 0.5 X_{CO}) C_{P,O_2} + 1.72 \times 10^{-5} C_{P,H_2O}$$

$$+ 1.72 \times 10^{-5} X_{CO} C_{P,CO_2} \Big] \tag{4.170}$$

In Equation 4.170, the values of C_P are given by Equation 4.167. Returning to Equation 4.166, substitute the known initial number of moles of CO to obtain

$$\frac{dT}{dV} = -1.72 \times 10^{-5} \frac{dX_{CO}}{dV} \frac{(\Delta H_R)}{\left(\sum_{j=1}^{n} F_j C_{Pj} \right)} \tag{4.171}$$

Equations 4.168 and 4.170 can be substituted into Equation 4.171 to give the differential equation that relates the temperature to reactor volume. The resulting equation must be solved simultaneously with the mole balance, which is obtained by substituting Equation 4.165 into Equation 4.163. The solution to the two equations must be performed numerically to give the fractional conversion and reactor temperature. Solution of these equations for 99% conversion of the CO gives a required reactor volume of 0.0567 m³. The corresponding outlet temperature is 790 K, which is the same as observed in the constant-pressure batch reactor.

4.4.4 External Heat Transfer to PFR

Example 4.4 illustrated the operation of an adiabatic PFR. Adiabatic operation is common in industrial reactors and, indeed, has many advantages over the nonadiabatic case. Generally, a reactor designed for adiabatic operation is much simpler and cheaper to build and to operate because it requires less equipment. However, many reactions cannot be carried out successfully using an adiabatic mode of operation. Many reactions have very large heat effects associated with them. For these types of reactions, adiabatic operation would lead to temperature extremes in the reactor. In the former case, unwanted side reactions may occur or, in severe cases, reactor failure. In the latter case, the decrease in temperature results in a low reaction rate, which can lead to the requirement of very large reactors. In such cases, the reactor may be operated with heat transfer to the surroundings to control the temperature within a specified range.

A nonadiabatic PFR exchanges heat with the surroundings. Several different designs are used industrially. One possibility would be to impose a constant heat flux at the wall of the reactor, which could be implemented using electrical heating. It is more common, however, to surround the reactor with a heat transfer fluid: the rate of heat transfer then depends on the local temperature difference between the fluid in the reactor and the heat transfer fluid. Such a reactor may be constructed in different ways but the simplest scenario is illustrated in Figure 4.7. This design is very similar to a double pipe heat exchanger, consisting of two concentric tubes with the central tube acting as the reactor and the surrounding jacket containing the heat transfer fluid.

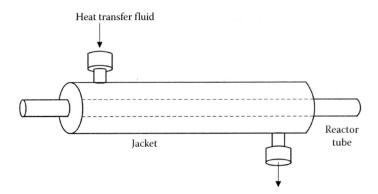

FIGURE 4.7
Typical configuration for an externally heated or cooled PFR. The central tube serves as the reactor and the jacket contains an appropriate heat transfer fluid. Many other designs are possible.

When reactors are operated in the nonadiabatic mode, the diameter of the reactor is usually relatively small so that an effective rate of heat transfer can be achieved. Use of a small reactor diameter can lead to a requirement of very long reactor length if a large volume is required. Rather than have a single very long tube, a common design uses multiple reactor tubes surrounded by a large jacket containing the heat transfer fluid, resembling a shell and tube heat exchanger. This type of arrangement is shown in Figure 4.8.

For the situation illustrated in either of Figures 4.7 or 4.8, the local rate of heat transfer is expressed in terms of the local temperature difference between the reactor temperature and the jacket temperature, as follows:

$$\frac{dq}{dV} = U\left(T_\infty - T\right)\frac{dA}{dV} \tag{4.172}$$

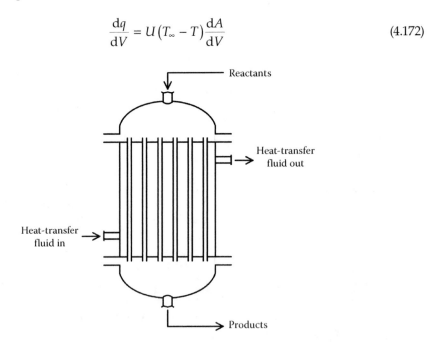

FIGURE 4.8
Multitubular reactor used for nonadiabatic operation. The reacting mixture flows through the tubes while the heat transfer fluid flows over the outside. The heat transfer fluid frequently undergoes a phase change.

In Equation 4.172, the symbols have the following meanings:

$U \equiv$ overall heat transfer coefficient

$T_\infty \equiv$ temperature of the surrounding heat transfer fluid

$T \equiv$ temperature inside of the reactor

$A \equiv$ heat transfer area of the reactor tube upon which U is based

Typically, a PFR has a circular cross section. For a reactor of length L, the total reactor volume and the volume of the differential element can be expressed as

$$V = \frac{\pi D^2}{4} L \quad \text{and} \quad dV = \frac{\pi D^2}{4} dz \tag{4.173}$$

The total heat transfer area is the inside surface area of the reactor tube. The total area and the differential area can be written as

$$A = \pi D L \quad \text{and} \quad dA = \pi D \, dz \tag{4.174}$$

Therefore, the change in heat transfer area for a change in reactor volume is

$$\frac{dA}{dV} = \frac{(\pi D) dz}{(\pi D^2 / 4) dz} = \frac{4}{D} \tag{4.175}$$

Substitution of Equation 4.175 into Equation 4.172 gives

$$\frac{dq}{dV} = \frac{4U}{D} (T_\infty - T) \tag{4.176}$$

Equation 4.176 is the rate of heat transfer per unit volume. On substitution of Equation 4.176 into the PFR energy balance, Equation 4.146 gives

$$\frac{4U}{D} (T_\infty - T) = \sum_{j=1}^{n} (F_j C_{Pj}) \frac{dT}{dV} - (-\Delta H_{R,A})(-r_A) \tag{4.177}$$

Equation 4.177 is one form of the energy balance for a PFR with external heat exchange. It may be written explicitly in terms of the temperature gradient by rearranging to give

$$\frac{T}{dV} = \frac{4U}{D \sum_{j=1}^{n} (F_j C_{Pj})} (T_\infty - T) + \frac{(-\Delta H_{R,A})(-r_A)}{\sum_{j=1}^{n} (F_j C_{Pj})} \tag{4.178}$$

Writing Equation 4.178 in terms of the reactor length by substituting Equation 4.173

$$\frac{dT}{dz} = \frac{\pi D U}{\sum_{j=1}^{n} (F_j C_{Pj})} (T_\infty - T) + \frac{\pi D^2}{4 \sum_{j=1}^{n} (F_j C_{Pj})} (-\Delta H_{R,A})(-r_A) \tag{4.179}$$

Use of an average heat capacity value (composition independent) gives the following equation:

$$\frac{dT}{dz} = \frac{\pi D U}{\dot{m}\overline{C}_P}\left(T_\infty - T\right) + \frac{\pi D^2}{4\dot{m}\overline{C}_P}\left(-\Delta H_{R,A}\right)\left(-r_A\right) \tag{4.180}$$

Equation 4.179 or 4.180 is solved simultaneously with the mole balance equation, Equation 4.149, to obtain the temperature and concentration profiles along the reactor.

Equations 4.179 and 4.180 include the temperature of the heat transfer fluid in the reactor jacket. One common method of reactor operation is to have a phase change occurring in the heat transfer fluid, that is, a boiling or condensing fluid is used. In such an operating mode, T_∞ is constant, and the solution of the energy balance equation is made using this constant value. Alternatively, the coolant could be a single-phase fluid and undergo a temperature change as it flows through the jacket.

If the temperature of the heat transfer fluid has a changing temperature, an additional energy balance equation must be solved for the heat transfer fluid. This energy balance is developed in the following, where the subscript HTF is used to denote the heat transfer fluid. Two modes of operation are possible for the reactor arrangement illustrated in Figure 4.6. In the first mode of operation, the heat transfer fluid enters the jacket at the same end of the unit as the reactant stream enters the reactor, at coordinate $z = 0$. Both fluids then flow in the same direction down the unit and leave at the position $z = L$, where L is the length of the reactor. This mode of operation is called cocurrent flow. The heat transferred to or from the heat transfer fluid over an incremental heat transfer area, dA, is equal to the rate of change of enthalpy of the heat transfer fluid. This equality is given by

$$-\left(\dot{m}C_P\right)_{HTF}\frac{dT_\infty}{dz} = U\frac{dA}{dz}\left(T_\infty - T\right) \tag{4.181}$$

From Equation 4.181, it may be seen that if T_∞ is greater than T, the right-hand side is a positive number, implying that heat is transferred to the reactor. Therefore, the left-hand side must be positive, which implies that dT_∞/dz is negative; in other words, the temperature of the heat transfer fluid falls as it flows through the jacket, which is consistent with heat transfer to the reactor.

The second mode of reactor operation is to have the heat transfer fluid enter the jacket at the opposite end of the unit as the reactant stream, that is at $z = L$. The heat transfer fluid therefore flows in the opposite direction to the process stream. This mode of operation is called countercurrent flow and, in this case, the energy balance equation for the heat transfer fluid is

$$\left(\dot{m}C_P\right)_{HTF}\frac{dT_\infty}{dz} = U\frac{dA}{dz}\left(T_\infty - T\right) \tag{4.182}$$

For a reactor of circular cross section, $dA = \pi D dz$; thus, Equations 4.181 and 4.182 may respectively be written as

$$\text{Cocurrent flow}\ \frac{dT_\infty}{dz} = -\frac{U\pi D\left(T_\infty - T\right)}{\left(\dot{m}C_P\right)_{HTF}} \tag{4.183}$$

$$\text{Countercurrent flow } \frac{dT_\infty}{dz} = +\frac{U\pi D(T_\infty - T)}{(\dot{m}C_P)_{\text{HTF}}} \tag{4.184}$$

Note that D is the inside tube diameter. These equations are first-order ordinary differential equations and require a boundary condition. For cocurrent flow, both the reactor fluid temperature and the heat transfer fluid temperature are known at position $z = 0$, that is,

$$T_\infty = (T_\infty)_0 \quad \text{and} \quad T = T_0 \quad \text{at } z = 0$$

For countercurrent flow, the fluid inlets are at opposite ends of the reactor; therefore, the following conditions are known:

$$T_\infty = (T_\infty)_0 \quad \text{at } z = L \quad \text{and} \quad T = T_0 \quad \text{at } z = 0$$

The countercurrent flow case is more complex from a mathematical point of view because it is a boundary value problem, whereas the cocurrent flow case is an initial value problem. An externally cooled PFR is illustrated in Example 4.5.

Example 4.5

Consider the oxidation of carbon monoxide in a PFR. The stoichiometry of the reaction is

$$CO + \frac{1}{2}O_2 \rightarrow CO_2 \tag{4.185}$$

The rate of reaction in the presence of water is given by the rate expression

$$(-r_{CO}) = 1.26 \times 10^{10} \exp\left(\frac{-20{,}131}{T}\right) C_{CO} C_{O_2}^{0.25} C_{H_2O}^{0.5} \text{ mol/m}^3 \cdot \text{s} \tag{4.186}$$

A mixture consisting of 1% CO, 1% O_2, 1% H_2O, and 97% N_2 (mole percentages) is fed to a PFR. The volumetric feed flow rate is 1.00×10^{-4} m³/s at a pressure of 1 bar and a temperature of 700 K. The reactor has an internal diameter of 0.25 m and a length of 1.0 m. Perform the following calculations:

a. Plot the axial temperature and concentration profiles for adiabatic operation.
b. Plot the axial temperature and concentration profiles when the reactor is externally cooled by a constant-temperature fluid at 600 K. Compare the outlet temperature and convert to that achieved for adiabatic operation. The heat transfer coefficient is

$$U = 0.005 \text{ W/m}^2 \cdot \text{K}$$

c. Plot the axial temperature and concentration profiles for the cocurrent flow case where

$$U = 0.005 \text{ W/m}^2 \cdot \text{K} \quad \text{and} \quad \left(\dot{m}C_P\right)_{\text{HTF}} = 0.025 \text{ W/K}$$

and the inlet temperature of the heat transfer fluid is 600 K.

d. Plot the axial temperature and concentration profiles for the countercurrent flow case where the inlet temperature of the heat transfer fluid is 600 K. The other conditions are as in part (c).

SOLUTION

It is necessary to express both the mole balance and the energy balance in terms of the reactor length. The chemical reaction and the inlet conditions to the reactor are the same as those for Example 4.4; therefore, we take the following information from the solution of that example. The mole balance equation in terms of reactor volume for the given reactor conditions is

$$\frac{dX_{CO}}{dV} = 3.37 \times 10^6 \exp\left(\frac{-20,131}{T}\right)(1 - X_{CO})\left(1 - \frac{X_{CO}}{2}\right)^{0.25} Q^{-1.75} \tag{4.187}$$

Equation 4.187 can be written in terms of the axial position: Equation 4.173 gives the relationship between volume and length for a circular reactor:

$$dV = \frac{\pi D^2}{4} dz = \frac{\pi 0.25^2}{4} dz = 4.91 \times 10^{-2} dz \tag{4.188}$$

Substituting Equation 4.188 into Equation 4.187 and rearranging give

$$\frac{dX_{CO}}{dz} = 1.65 \times 10^5 \exp\left(\frac{-20,131}{T}\right)(1 - X_{CO})\left(1 - \frac{X_{CO}}{2}\right)^{0.25} Q^{-1.75} \tag{4.189}$$

The volumetric flow rate inside the reactor for this inlet was calculated in Example 4.4 to be

$$Q = 1.429 \times 10^{-7} T(1 - 0.005 X_{CO}) \tag{4.190}$$

The constant-pressure heat capacity for each component is a function of temperature, as follows:

$$C_P = a + bT + cT^2 + dT^3 \tag{4.191}$$

The values of the constants are obtained from Appendix 2:

Compound	a	$b \times 10^2$	$c \times 10^5$	$d \times 10^9$
O_2	25.44	1.518	−0.7144	1.310
N_2	28.85	−0.1569	0.8067	−2.868
H_2O	32.19	0.1920	1.054	−3.589
CO	28.11	0.1672	0.5363	−2.218
CO_2	22.22	5.9711	−3.495	7.457

From Example 4.4, the enthalpy of reaction as a function of temperature is

$$\Delta H_R = -2.796 \times 10^5 - 18.61T + 2.522 \times 10^{-2}T^2$$

$$- 1.225 \times 10^{-5}T^3 + 2.255 \times 10^{-9}T^4 \, \text{J/mol} \tag{4.192}$$

The total heat capacity of the reactor contents changes with composition as well as with temperature. From Example 4.4, we have

$$\left(\sum_{j=1}^{n} F_j C_{Pj} \right) = \left[1.668 \times 10^{-3} C_{P,N_2} + 1.72 \times 10^{-5} (1 - X_{CO}) C_{P,CO} \right.$$

$$+ 1.72 \times 10^{-5} (1 - 0.5X_{CO}) C_{P,O_2} + 1.72 \times 10^{-5} C_{P,H_2O}$$

$$\left. + 1.72 \times 10^{-5} X_{CO} C_{P,CO_2} \right] \tag{4.193}$$

In Equation 4.193, the values of C_P are given by Equation 4.191. We now develop the energy balance equation for each part of the problem.

Part (a)

From Example 4.4, the energy balance for the adiabatic reactor with the given inlet flow rate is

$$\frac{dT}{dV} = -1.72 \times 10^{-5} \frac{dX_{CO}}{dV} \frac{(\Delta H_R)}{\sum_{j=1}^{n} (F_j C_{Pj})} \tag{4.194}$$

Equation 4.194 written in terms of the axial reactor coordinate is

$$\frac{dT}{dz} = -1.72 \times 10^{-5} \frac{dX_{CO}}{dz} \frac{(\Delta H_R)}{\sum_{j=1}^{n} (F_j C_{Pj})} \tag{4.195}$$

The temperature and fractional conversion as a function of the axial coordinate can now be determined by the simultaneous numerical solution of Equations 4.189 and 4.195, using the auxiliary Equations 4.190 through 4.193. Figure 4.9 shows the plot of the fractional conversion and temperature thus obtained, as a function of the axial coordinate. From the solution, the outlet fractional conversion of CO is 0.954 and the outlet temperature is 787 K.

Part (b)

Part (b) requires the solution of a PFR energy balance with heat exchange to a fluid that is maintained at a constant temperature. The energy balance for a cooled tubular PFR with a circular cross section has the form given by Equation 4.179:

$$\frac{dT}{dz} = \frac{\pi D U}{\sum_{j=1}^{n} (F_j C_{Pj})} (T_\infty - T) + \frac{\pi D^2}{4 \sum_{j=1}^{n} (F_j C_{Pj})} (-\Delta H_{R,A})(-r_A) \tag{4.196}$$

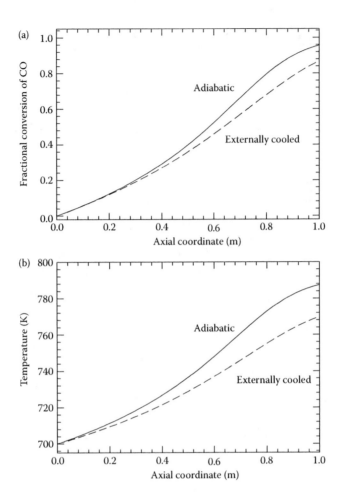

FIGURE 4.9
Axial conversion (a) and temperature (b) profiles for the PFR of Example 4.5. The solid line represents the adiabatic reactor case and the dashed line shows an externally cooled reactor with a constant-temperature heat transfer fluid.

Equation 4.149 can be substituted into Equation 4.196 to obtain

$$\frac{dT}{dz} = \frac{\pi D U}{\sum_{j=1}^{n}(F_j C_{Pj})}(T_\infty - T) - (F_{CO})_0 \frac{dX_{CO}}{dz} \frac{(\Delta H_R)}{\sum_{j=1}^{n}(F_j C_{Pj})} \tag{4.197}$$

Substitute the inlet molar flow rate of CO, reactor diameter, heat transfer coefficient, and temperature of the heat transfer fluid. The final energy balance is

$$\frac{dT}{dz} = \frac{3.93 \times 10^{-3}}{\sum_{j=1}^{n}(F_j C_{Pj})}(600 - T) - 1.72 \times 10^{-5} \frac{dX_{CO}}{dz} \frac{(\Delta H_R)}{\sum_{j=1}^{n}(F_j C_{Pj})} \tag{4.198}$$

The numerical solution of the mole and energy balance equations is performed as in Part (a), except that the energy balance is now given by Equation 4.198. The axial conversion and temperature profiles are shown in Figure 4.9 for direct comparison to the adiabatic case. For Part (b), the outlet fractional conversion is calculated to be 0.868 and the outlet temperature is 770 K.

In this case of the cooled reactor, the outlet temperature and conversion are both lower than for the adiabatic case, which is what might be expected.

Part (c)

In Part (c), the heat transfer fluid undergoes a temperature change as it flows through the jacket. It is therefore necessary to add an energy balance equation for the heat transfer fluid. The energy balance for the reactor remains unchanged, and is given by

$$\frac{dT}{dz} = \frac{3.93 \times 10^{-3}}{\sum_{j=1}^{n}\left(F_j C_{Pj}\right)}\left(T_\infty - T\right) - 1.72 \times 10^{-5}\frac{dX_{CO}}{dz}\frac{\left(\Delta H_R\right)}{\sum_{j=1}^{n}\left(F_j C_{Pj}\right)} \tag{4.199}$$

For cocurrent flow, the energy balance for the heat transfer fluid is given by Equation 4.183

$$\frac{dT_\infty}{dz} = -\frac{U\pi D\left(T_\infty - T\right)}{\left(\dot{m}C_P\right)_{HTF}} \tag{4.200}$$

Substitution of the numerical values gives

$$\frac{dT_\infty}{dz} = -0.157\left(T_\infty - T\right) \tag{4.201}$$

The initial condition for Equation 4.201 is

$$T_\infty = 600 \quad \text{at } z = 0$$

The simultaneous solution of the three differential equations, Equations 4.189, 4.199, and 4.201, gives the temperature and conversion profiles along the reactor. These plots are illustrated in Figure 4.10. The outlet fractional conversion of CO is calculated to be 0.873 and the outlet temperature is 771 K. The outlet temperature of the heat transfer fluid is 619 K.

The reactor temperature exhibits a similar type of profile as in Part (b). Note that the coolant temperature increases in a fairly regular manner from an inlet temperature of 600 K to the outlet temperature of 619 K. The outlet conversion is higher than for Part (b) because the reactor outlet temperature is higher. This is a consequence of the increasing temperature of the heat transfer fluid.

Part (d)

The solution to Part (d) is the same as the solution to Part (c) except that the energy balance equation for the heat transfer fluid is

$$\frac{dT_\infty}{dz} = +0.157\left(T_\infty - T\right) \tag{4.202}$$

The boundary condition for Equation 4.202 is

$$T_\infty = 600 \quad \text{at } z = L$$

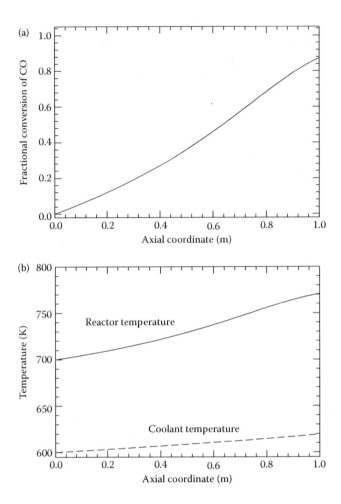

FIGURE 4.10
Axial conversion (a) and temperature (b) profiles for the PFR of Example 4.5, Part (c). Both the reactor temperature and the coolant temperature exhibit a steady rise. The reactor fluid and the coolant flow in the same direction.

The simultaneous solution of Equations 4.189, 4.199, and 4.202 gives the axial temperature and conversion profiles. The solution is illustrated in Figure 4.11. The outlet fractional conversion of CO is calculated to be 0.881 and the outlet temperature is 772 K. The outlet temperature of the heat transfer fluid is 619 K, which is essentially the same value observed for the cocurrent flow case.

It is interesting to note that the temperature profile of the process stream for the countercurrent flow case is similar to the cocurrent flow case. This result is by no means universal, and, in other situations, dramatically different results might be obtained by changing the direction of flow of the heat transfer fluid. It is usually necessary to solve the mole and energy balance equations for the different scenarios to determine which flow pattern gives the optimal solution.

4.4.5 Multiple Reactions in PFR

The energy balance equation for the PFR is easily extended to account for the presence of more than one reaction. In the multiple reaction case, the enthalpy change of reaction for

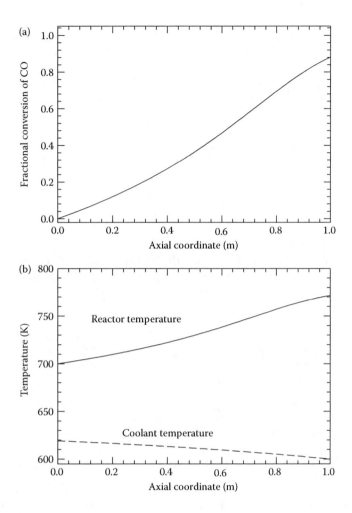

FIGURE 4.11
Axial conversion (a) and temperature (b) profiles for the PFR of Example 4.5, Part (d). Both the reactor temperature and the coolant temperature exhibit a steady increase. Note that the coolant flows in the direction opposite to the fluid in the reactor.

each reaction must be included in the energy balance equation. A mole or mass balance equation must be written for each independent reaction. For example, consider a reactor in which the following two reactions occur:

$$A \rightarrow B$$
$$C \rightarrow D \tag{4.203}$$

The rates of disappearance of A and C are given by $(-r_A)$ and $(-r_C)$, respectively. Two mole balance equations must be solved, one each for A and C:

$$-\frac{dF_A}{dV} = (-r_A) \quad \text{and} \quad -\frac{dF_C}{dV} = (-r_C) \tag{4.204}$$

The energy balance equation for a PFR must include the enthalpy changes owing to both reactions. For example, taking the general PFR energy balance, Equation 4.146, we obtain

$$\frac{dq}{dV} = \sum_{j=1}^{n}\left(F_j C_{Pj}\right)\frac{dT}{dV} - \left[\left(-\Delta H_{R,A}\right)\left(-r_A\right) + \left(-\Delta H_{R,C}\right)\left(-r_C\right)\right] \tag{4.205}$$

The general energy balance for a reactor containing n species (including inert material) in which m reactions occur is

$$\frac{dq}{dV} = \sum_{j=1}^{n}\left(F_j C_{Pj}\right)\frac{dT}{dV} - \sum_{k=1}^{m}\left(-\Delta H_{R,k}\right)\left(-r_k\right) \tag{4.206}$$

For a tubular PFR of circular cross section that is surrounded by a heat transfer fluid, the energy balance for the multiple reaction case may be written in terms of the axial coordinate as

$$\frac{dT}{dz} = \frac{\pi D U}{\displaystyle\sum_{j=1}^{n}\left(F_j C_{Pj}\right)}\left(T_\infty - T\right) + \frac{\pi D^2}{4\displaystyle\sum_{j=1}^{n}\left(F_j C_{Pj}\right)}\sum_{k=1}^{m}\left(-\Delta H_{R,k}\right)\left(-r_k\right) \tag{4.207}$$

The energy balance equation is coupled to each of the mole balance equations because each mole balance equation includes the temperature. See Case Study 4.2 for an illustration of an externally cooled PFR in which multiple reactions occur.

Case Study 4.2:   Hydrogenation of Gas Oil in a Nonisothermal PFR

In this case study, we return to the catalytic desulfurization (HDS) and denitrogenation (HDN) reactions in an oil. We consider a nonisothermal PFR with external heat transfer. The properties of the oil, kinetic expressions, and so on are the same as those used in Case Study 4.1. The rate of reaction of sulfur molecules, based on total sulfur concentration expressed as S, is

$$(-r_S) = A_S \exp\left(\frac{-E_S}{R_g T}\right)C_S^{1.5}P_{H_2}^{0.8}\ \text{mol of S/h}\cdot\text{gcat} \tag{4.208}$$

The rate expression for the nitrogen-containing compounds is

$$(-r_N) = A_N \exp\left(\frac{-E_N}{R_g T}\right)C_N P_{H_2}^{1.3}\ \text{mol of N/h}\cdot\text{gcat} \tag{4.209}$$

As in Case Study 4.1, the other reactions (hydrogenation, deoxygenation) that occur are represented by a single lumped expression that is expressed as the rate of disappearance of compounds with a boiling point higher than 343°C. The mass flow rate of these compounds is denoted $\dot{m}_{343}$ and the rate expression for their disappearance is

$$(-r_{343}) = A_{343} \exp\left(\frac{-E_{343}}{R_g T}\right)[m_{343}] \text{ g of } 343^+/\text{h} \cdot \text{gcat} \tag{4.210}$$

The concentrations in these three rate equations are

$C_S \equiv$ molar concentration of sulfur in liquid
$C_N \equiv$ molar concentration of nitrogen in liquid
$[m_{343}] \equiv$ mass concentration of 343°C$^+$ material in liquid
$\dot{m}_{343} \equiv$ mass flow rate of 343°C$^+$ material in liquid

Note that the rate of reaction for each of the three species is based on the mass of the catalyst, and not on the reactor volume. The unit gcat in the rate equations means grams of catalyst. The values of the kinetic parameters are

$$A_S = 5.22 \times 10^{11} \text{ cm}^{4.5}/(\text{h} \cdot \text{gcat mol}^{0.5} \cdot \text{MPa}^{0.8}), \quad \frac{E_S}{R_g} = 15{,}100 \text{ K}$$

$$A_N = 1.21 \times 10^7 \text{ cm}^3/(\text{h} \cdot \text{gcat} \cdot \text{MPa}^{1.3}), \quad \frac{E_N}{R_g} = 12{,}300 \text{ K}$$

$$A_{343} = 1.3 \times 10^{11} \text{ cm}^3/\text{h} \cdot \text{gcat}, \quad \frac{E_{343}}{R_g} = 16{,}300 \text{ K}$$

The reactor feed contains 4.3 wt.% sulfur (S), 0.30 wt.% nitrogen (N), and 74.2 wt.% 343°C$^+$ material. The oil density depends on temperature according to the relation

$$\rho_L = \sqrt{1.286 - 0.0011T} \text{ g/cm}^3 \tag{4.211}$$

The temperature in Equation 4.211 is expressed in kelvin.

The oil is treated in a tubular plug flow reactor having an internal diameter of 2.50 cm and a length of 17.0 cm. The reactor contains 100 g of catalyst. The oil feed stream has a temperature 400°C (673.15 K) and a flow rate of 15.0 g/min. The hydrogen partial pressure is constant at 13.9 MPa throughout the reactor.

The reactor is surrounded by a heat transfer fluid that is maintained at a constant temperature of 400°C (673.15 K). The overall heat transfer coefficient based on the inside reactor area is 5×10^{-3} W/cm^2 · K. The following information is available:

Heat capacity of the fluid:

$$C_{Pf} = -0.5167 + 9.964 \times 10^{-3}T - 8.686 \times 10^{-6}T^2 \text{ J/g} \cdot \text{K} \tag{4.212}$$

Enthalpy of reaction for the sulfur compounds:

$$(-\Delta H_{R,S}) = 1.0 \times 10^5 \text{ J/mol of S converted} \tag{4.213}$$

Enthalpy of reaction for the nitrogen compounds:

$$(-\Delta H_{R,N}) = 2.0 \times 10^5 \text{ J/mol of N converted} \tag{4.214}$$

Enthalpy of reaction for the 343°C$^+$ material:

$$(-\Delta H_{R,343}) = 2.0 \times 10^2 \text{ J/g of } 343°C^+ \text{ converted} \tag{4.215}$$

Note that all of the reactions are exothermic. It shall be assumed that all of the enthalpies of reaction are independent of temperature.

This case study requires that we solve the mole and energy balance equations for a PFR. The liquid density changes during the reaction, giving a change in the volumetric flow rate. The concentrations of the species thus change as the fluid density changes. The reaction rates, however, are expressed in terms of the catalyst mass, which remains constant. The three species balance equations are written by incorporating the rate equations into the general balance for a PFR. Because the liquid density changes, the balances are written in terms of the molar (or mass) flow rates. It is also necessary to convert the reaction rate per unit mass to a rate per volume. The starting point is the generic mole balance, written using the reactor length, Equation 4.149:

$$-\frac{dF_A}{dz} = \frac{\pi D^2}{4}(-r_A) \tag{4.216}$$

Equation 4.216 is valid when the reaction rate is expressed in terms of the total reactor volume. When the reaction rate is expressed in terms of the catalyst mass, it is necessary to add the density of the catalyst bed, ρ_{bed}, to the equation; thus

$$-\frac{dF_A}{dz} = \frac{\pi D^2}{4}\rho_{bed}(-r_A) = \frac{\pi D^2}{4}\frac{W}{V}(-r_A) \tag{4.217}$$

In Equation 4.217, W is the mass of catalyst in the reactor and V is the total reactor volume. Following the generic form represented by Equation 4.217, the mole/mass balance equations for this reactor may be written as

$$\frac{dF_S}{dz} = -\frac{\pi D^2}{4}\frac{W}{V}A_S \exp\left(\frac{-E_S}{R_g T}\right)\left(\frac{F_S}{Q}\right)^{1.5} P_{H_2}^{0.8} \tag{4.218}$$

$$\frac{dF_N}{dz} = \frac{\pi D^2}{4}\frac{W}{V}A_N \exp\left(\frac{-E_N}{R_g T}\right)\frac{F_N}{Q} P_{H_2}^{1.3} \tag{4.219}$$

$$\frac{d\dot{m}_{343}}{dz} = \frac{\pi D^2}{4}\frac{W}{V}A_{343} \exp\left(\frac{-E_{343}}{R_g T}\right)\frac{m_{343}}{Q} \tag{4.220}$$

The density of the fluid, and hence the volumetric flow rate, is a function of the temperature. The volumetric flow rate at any point, Q, can be related to the volumetric flow rate at the inlet, Q_0, using the density, that is

$$Q = Q_0 \frac{\rho_0}{\rho} \tag{4.221}$$

The mass flow rate in the reactor, which is equal to the product of the density and the volumetric flow rate, is a constant 15 g/min. It therefore follows that the volumetric flow rate at any point in the reactor is

$$Q = \frac{15}{\rho_L} \frac{\text{cm}^3}{\text{min}} \equiv \frac{900}{\sqrt{1.286 - 0.0011T}} \frac{\text{cm}^3}{\text{h}} \tag{4.222}$$

The volume of the 17.0 cm long reactor is computed from

$$V = \frac{\pi D^2}{4} L = \frac{\pi 2.5^2}{4} 17.0 = 83.45 \, \text{cm}^3 \tag{4.223}$$

Substitution of the numerical values of the rate parameters, the catalyst mass, reactor volume, the hydrogen partial pressure, the liquid density, and so on gives the following three equations (ensuring that the units are consistent):

$$\frac{dF_S}{dz} = -9.35 \times 10^8 \exp\left(\frac{-15,100}{T}\right) F_S^{1.5} [1.286 - 0.0011T]^{0.75} \, \text{mol/h} \cdot \text{cm} \tag{4.224}$$

$$\frac{dF_N}{dz} = -2.42 \times 10^6 \exp\left(\frac{-12,300}{T}\right) F_N \sqrt{1.286 - 0.0011T} \, \text{mol/h} \cdot \text{cm} \tag{4.225}$$

$$\frac{d\dot{m}_{343}}{dz} = -8.51 \times 10^8 \exp\left(\frac{-16,300}{T}\right) \dot{m}_{343} \sqrt{1.286 - 0.0011T} \, \text{g/h} \cdot \text{cm} \tag{4.226}$$

Note that the molar flow rates of S and N have units of mol/h, and the mass flow rate of the 343°C+ material has units of g/h. The reactor length is expressed in cm. To solve these three equations, it is necessary to know the flow rates of the three components at the reactor inlet. This calculation uses the inlet mass flow rate, the composition, and the molar mass of the components:

$$\text{Sulfur} \Rightarrow 15\,\text{g/min} \times 4.3 \times 10^{-2}\,\text{g S/g oil} \times \frac{1}{32}\,\text{mol/g} = 0.0202\,\text{mol S/min}$$

$$\equiv 1.21\,\text{mol S/h}$$

$$\text{Nitrogen} \Rightarrow 15\,\text{g/min} \times 0.3 \times 10^{-2}\,\text{g N/g oil} \times \frac{1}{14}\,\text{mol/g} = 0.00321\,\text{mol N/min}$$

$$\equiv 0.193\,\text{mol N/h}$$

$$\dot{m}_{343} \Rightarrow 15\,\text{g/min} \times 74.2 \times 10^{-2}\,\text{g }343^+/\text{g oil} = 11.13\,\text{g }343^+/\text{min} \equiv 667.8\,\text{g }343^+/\text{h}$$

Because the reactor is nonisothermal, it is necessary to solve an energy balance equation. Equation 4.207 describes the temperature change as a function of distance for a circular externally cooled reactor in which multiple reactions occur:

$$\frac{dT}{dz} = \frac{\pi D U}{\sum_{j=1}^{n}\left(F_j C_{Pj}\right)}(T_{\infty} - T) + \frac{\pi D^2}{4\sum_{j=1}^{n}\left(F_j C_{Pj}\right)}\sum_{k=1}^{m}\left(-\Delta H_{R,k}\right)\left(-r_k\right) \tag{4.227}$$

The heat capacity of the fluid is assumed to be independent of the composition, and is expressed as a polynomial in temperature. Furthermore, the reaction rate is given in terms of the catalyst mass which must be included. With these two adjustments, Equation 4.227 is written in terms of the total mass flow rate:

$$\frac{dT}{dz} = \frac{\pi D U}{\dot{m} C_{Pf}}(T_{\infty} - T) + \frac{\pi D^2}{4\dot{m} C_{Pf}}\frac{W}{V}\sum_{k=1}^{m}\left(-\Delta H_{R,k}\right)\left(-r_k\right) \tag{4.228}$$

Finally, it is necessary to substitute for the reaction rates of the three components. The reaction rates of the species are

$$-\frac{dF_S}{dz} = \frac{\pi D^2}{4}\frac{W}{V}(-r_S); \quad -\frac{dF_N}{dz} = \frac{\pi D^2}{4}\frac{W}{V}(-r_N); \quad -\frac{d\dot{m}_{343}}{dz} = \frac{\pi D^2}{4}\frac{W}{V}(-r_{343})$$

Substitution of these quantities into Equation 4.228 gives

$$\frac{dT}{dz} = \frac{\pi D U}{\dot{m} C_{Pf}}(T_{\infty} - T)$$

$$-\frac{1}{\dot{m} C_{Pf}}\left[\frac{dF_S}{dz}\left(-\Delta H_{R,S}\right) + \frac{dF_N}{dz}\left(-\Delta H_{R,N}\right) + \frac{d\dot{m}_{343}}{dz}\left(-\Delta H_{R,343}\right)\right] \tag{4.229}$$

Substitution of the known numerical values, ensuring that the units are consistent, gives the final energy balance equation

$$\frac{dT}{dz} = \frac{0.1571}{C_{Pf}}(673.15 - T)$$

$$-\frac{1}{C_{Pf}}\left[\frac{dF_S}{dz}(111.11) + \frac{dF_N}{dz}(222.22) + \frac{d\dot{m}_{343}}{dz}(0.222)\right] \tag{4.230}$$

The units of Equation 4.230 are K/cm. Simultaneous numerical solution of Equations 4.224 through 4.226 and 4.230 gives the temperature and concentration profiles as a function of length. Figure 4.12a shows the fractional conversion of the three components, while Figure 4.12b shows the temperature profile. Sulfur is the most reactive component which is consistent with previous case studies on these reactions. The reactor temperature rises initially as the rate of heat release is greater than the rate of

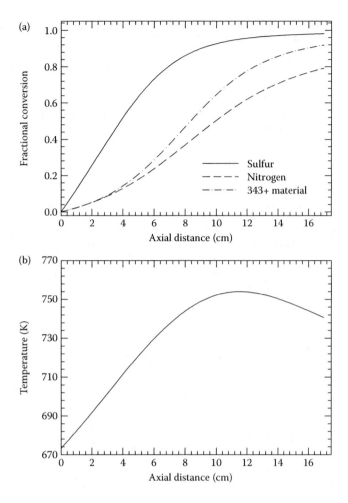

FIGURE 4.12
Axial conversion (a) and temperature (b) profiles for the PFR of Case Study 4.2. The reactor temperature initially rises and then falls, typical of cooled PFR with an exothermic reaction.

heat transfer to the coolant. As the concentration of the reactants falls, the rate slows and the rate of heat transfer exceeds the rate of heat release. The reactor temperature then starts to fall.

4.4.6 Parametric Sensitivity of PFR

The parametric sensitivity of a reactor refers to its response to changes in process conditions, for example, a change in inlet feed conditions (flow rate, temperature, reactant concentration, etc.) or a change in the rate of external heat transfer. The parametric sensitivity of a reactor is an important part of safety analysis. In general, a small change in a process variable should not lead to large changes in temperature at any point in the reactor. Detailed analysis of parametric sensitivity is beyond the scope of this book; however, a typical problem is illustrated in Case Study 4.3. See Lee (1985) or Westerterp (1984) for additional details.

Case Study 4.3:   Parametric Sensitivity of a Cooled Plug Flow Reactor

This case study examines the sensitivity of the plug flow reactor used in Case Study 4.2. Two of the process variables will be changed to determine their effect on reactor operation. In Part (a), the effect of changing the value of the heat transfer coefficient is investigated. In Part (b), the effect of a changing enthalpy of reaction for the 343°C⁺ material is investigated. Note that in both cases the mole or mass balance equations for the reactor remain unchanged from Case Study 4.2, and are given by Equations 4.224 through 4.226. The energy balance will, however, change as we alter the values of the overall heat transfer coefficient, U, or the enthalpy of reaction of the 343°C⁺ material $(-\Delta H_{R,343})$.

PART (A): SENSITIVITY OF THE REACTOR TO CHANGES IN THE HEAT TRANSFER COEFFICIENT

In this part of the case study, we examine the effect of changing the overall heat transfer coefficient, U. All other conditions are the same as those used in Case Study 4.2. The energy balance equation for the conditions of Case Study 4.2 is given by Equation 4.230. Rearranging this equation slightly so that the value of U is left as a variable gives

$$\frac{dT}{dz} = \frac{31.42U}{C_{Pf}}\left(673.15 - T\right)$$

$$-\frac{1}{C_{Pf}}\left[\frac{dF_S}{dz}(111.11) + \frac{dF_N}{dz}(222.22) + \frac{dm_{343}}{dz}(0.222)\right] \qquad (4.231)$$

Note that the units on U in Equation 4.231 are W/cm²K. The axial temperature profile can be obtained by solving the four differential equations, as done in Case Study 4.1. This profile is obtained for the following values of U:

$$0.0 \text{ (adiabatic reactor)}, 2.5 \times 10^{-3}, 5 \times 10^{-3} \text{ (Case Study 4.2)}, 1 \times 10^{-2}, 5 \times 10^{-2}$$

The axial temperature profiles resulting from the different values of U are illustrated in Figure 4.13. It can be seen that the axial temperature profile is not extremely sensitive to changes in the value of the heat transfer coefficient. The worst-case scenario, from a plant operating standpoint, would be a failure of the cooling system, which would result in the reactor operation becoming adiabatic. The maximum reactor temperature can thus be obtained from the profile where U is equal to zero. Note that as the heat transfer coefficient increases, the reactor temperature profile becomes flatter. In the limit as the heat transfer coefficient tends to infinity, the reactor temperature becomes constant at 673.15 K. In other words, the reactor behaves like an isothermal reactor. This result occurs because the inlet reactor temperature is the same as the coolant temperature. If the reactor inlet temperature is not equal to the coolant temperature, the reactor will always exhibit a temperature change regardless of the value of U.

PART (B): SENSITIVITY OF THE REACTOR TO CHANGES IN ENTHALPY OF REACTION

In Part (b), we examine the scenario of an increase in the enthalpy of reaction of the 343°C⁺ material, $(-\Delta H_{R,343})$. This increase requires that the feed composition change. We shall assume, however, that all of the other properties remain the same as in case

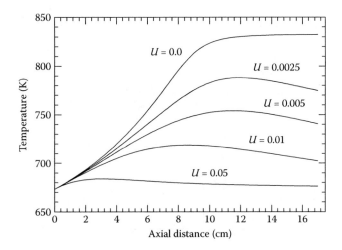

Axial temperature profile at different values of the heat transfer coefficient. The maximum reactor temperature occurs when the heat transfer coefficient is zero, which corresponds to adiabatic operation.

study 4.2. The energy balance equation for the conditions of Case Study 4.2 is given by Equation 4.230. Rearranging this equation slightly so that the value of $(-\Delta H_{R,343})$ is left as a variable gives

$$\frac{dT}{dz} = \frac{0.1571}{C_{Pf}}(673.15 - T)$$

$$- \frac{1}{C_{Pf}}\left[\frac{dF_S}{dz}(111.11) + \frac{dF_N}{dz}(222.22) + \frac{dm_{343}}{dz}\frac{(-\Delta H_{R,343})}{900}\right] \quad (4.232)$$

The axial temperature profile can be obtained by solving the four differential equations (three mole balances and the energy balance), as done in Case Study 4.2. This profile is shown in Figure 4.14 for the following values of $(-\Delta H_{R,343})$:

$$200 \text{ (Case Study 4.2), 300, 500, and 600 J/g} \cdot \text{K}$$

The temperature profile shown in Figure 4.14 exhibits a large peak as the value of $(-\Delta H_{R,343})$ increases. At a value of $(-\Delta H_{R,343}) = 600$ J/g $\cdot$ K, a large spike is observed. This type of rapid temperature rise is usually considered undesirable, and attempts should be made to design the reactor system to avoid such a phenomenon. It is also instructive to examine the fractional conversion of the reactants as a function of axial distance for the case of $(-\Delta H_{R,343}) = 600$ J/g $\cdot$ K. This plot is shown in Figure 4.15. It is seen that the spike in the temperature profile corresponds to a very rapid increase in the fractional conversion: in fact the temperature only starts to decrease after all of the reactants have been consumed. Although 100% conversion of the reactants is desirable, this rapid acceleration of reaction rate may cause excessive reactor temperatures, leading to harmful side effects.

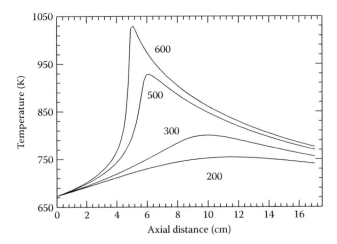

FIGURE 4.14

Axial temperature profile at different values of the enthalpy of reaction. The numbers next to each curve are the values of $(-\Delta H_{R,343})$. Note the large spike in the temperature profile that occurs when the enthalpy of reaction is large. In this situation, the temperature in the reactor is limited by conversion of the reactants.

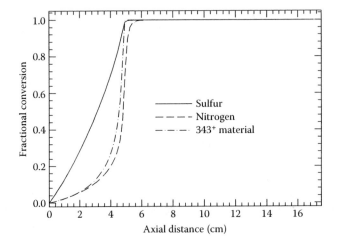

FIGURE 4.15

Axial conversion profile when $(-\Delta H_{R,343}) = 600$ J/g. The conversion of all three components rapidly increases to a value of one as the temperature rises. This type of reactor operation is usually considered to be undesirable.

4.4.7 Equilibrium Effects in PFR

It was observed at the beginning of this chapter that the composition of a reacting mixture at equilibrium is affected by the temperature. For reactions in which the conversion is limited by equilibrium, it is necessary to pay close attention to the coupling between the heat that is released or absorbed by the chemical reaction, the reaction rate, and the equilibrium composition. We recall here that increasing the temperature generally increases the reaction rate, owing to the exponential temperature dependence of the rate constants. As a higher reaction rate usually leads to a smaller reactor volume, it is usually desirable to operate at as high a temperature as possible, provided that unwanted side reactions do not occur or the reactor does not fail. For equilibrium limited reactions, however, it is also necessary to consider the effect of temperature on the equilibrium constant. We consider two cases, endothermic and exothermic reactions.

4.4.7.1 Equilibrium Effects in Endothermic Reactions

The temperature dependence of the equilibrium constant is given by the van't Hoff equation:

$$\frac{\partial (\ln K)}{\partial T} = \frac{\Delta H_R^\circ}{R_g T^2} \tag{4.233}$$

The value of ΔH_R° is positive for an endothermic reaction, and therefore an increase in the temperature leads to an increase in the value of K. Higher values of K in turn imply that the equilibrium yield is higher, and therefore an endothermic reaction should be carried out at as high a temperature as possible to maximize the yield. As a high temperature also increases the reaction rate, it is evident that high temperatures are desirable both to maximize rate and conversion in endothermic reactions, and higher temperatures will lead to smaller reactors.

4.4.7.2 Equilibrium Effects in Exothermic Reactions

In an exothermic reaction, the value of ΔH_R° is negative. Therefore, an increase in the temperature decreases the value of K, with a concomitant decrease in the equilibrium yield. A conflict arises between the need for a high temperature to maximize the reaction rate, and a low temperature to maximize the yield.

Figure 4.16 shows a typical equilibrium composition line that might be obtained for an exothermic reaction. The line has a fractional conversion near one at low temperature but

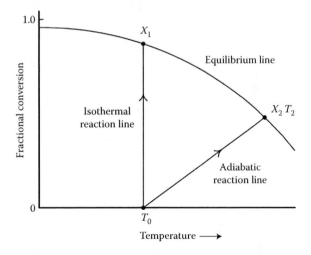

FIGURE 4.16

The equilibrium line shows the relationship between the temperature and fractional conversion at equilibrium for a specified starting composition. The adiabatic reaction line shows the maximum conversion that can be obtained for adiabatic operation for a given inlet temperature. The isothermal reaction line shows the best conversion for isothermal operation.

the conversion decreases as the temperature increases. Now consider the performance of two types of reactor, the adiabatic reactor and the isothermal reactor. For a given set of operating conditions, the adiabatic reaction line governs the relationship between the temperature and the fractional conversion. Assuming constant ΔH_R° and C_P, Equation 4.155 gives

$$T = T_0 - \frac{\Delta H_{R,A}}{\dot{m}\bar{C}_P} F_{A0}X_A \qquad (4.234)$$

A typical adiabatic reaction line is shown in Figure 4.16 for an arbitrary value of T_0. In the reactor, both the temperature and the fractional conversion will increase along this line until the equilibrium line is encountered at X_2 and T_2. At this point, the rate is zero, and the temperature and composition of the process stream remain constant for any additional reactor volume.

If the same reactor was operated isothermally, the vertical operating line shown in Figure 4.16 would be obtained. As the temperature is constant, the fractional conversion finally attained, equal to X_1, is higher than that achieved in the adiabatic reactor.

Isothermal operation may not be practical, and does not result in a minimum reactor volume. It is possible to determine an optimal temperature progression along the reactor so as to maximize the conversion for the minimum possible reactor volume. Consider the diagram shown in Figure 4.17. This diagram shows lines of constant reaction rate corresponding to set values of temperature and conversion. Each solid line represents a locus of the set of temperature and fractional conversion that give a specified value of the rate. The top line corresponds to a rate of zero; in other words, the equilibrium line. We have already seen that the equilibrium conversion drops as the temperature rises. As we move from the top to the bottom of the figure, the value of the reaction rate increases. Note, however, that as the value of the reaction rate increases, the maximum conversion achievable at that rate decreases. This observation suggests a route for minimizing the reactor volume. One could start the reaction at a high temperature to take advantage of the high rate at that temperature,

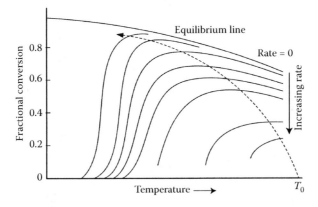

FIGURE 4.17
Each solid line in this diagram represents a constant value of reaction rate. The top line is the equilibrium line, where the rate is equal to zero. At higher temperatures, the maximum fractional conversion that can be achieved at any rate drops. The dashed line shows a method of reactor operation that would achieve a minimum total reactor volume.

and then progressively lower the temperature to increase the equilibrium yield. Although the details are beyond the scope of this text, it is possible to determine an optimal temperature profile that will achieve a desired conversion for a minimum reactor volume. The dashed line in Figure 4.17 illustrates the type of temperature and conversion pathway that must be followed to achieve this optimal design. This reactor design would require a complex cooling pattern along the reactor length, which might not be practical.

4.4.7.3 Multibed Adiabatic Reactors for Equilibrium Limited Reactions

As it may be either difficult or complicated to achieve an optimal temperature profile within a reactor, other designs are often used in practice. For relatively fast reactions that are equilibrium limited, the adiabatic reactor is commonly used in industry. As seen above, the adiabatic reactor experiences a temperature rise that limits the conversion. Therefore, in applications where very high conversions are desired, it is common to use multiple reactors in series, with cooling between each reactor.

Consider, for example, an exothermic reaction that occurs in a series of adiabatic fixed bed catalytic reactors. The operating path would be similar to the one shown in Figure 4.18. This plot shows the equilibrium line as a function of temperature and conversion. For adiabatic operation, the temperature and conversion in the first reactor follow the adiabatic reaction line denoted Bed 1. When the fraction conversion reaches a value of X_1, which is approaching the equilibrium line, the process stream is cooled by passing it through a heat exchanger. As this is a catalytic reaction, no further conversion occurs in this stage, as shown by the flat line. When the process stream is cooled by a predetermined amount, the stream enters reactor number two, in which it undergoes a further adiabatic reaction until the conversion is X_2. The stream is cooled as before, then fed into the last bed, in which the conversion is brought to X_3. Using this strategy, a high level of conversion can be achieved using adiabatic reactors. A diagram of the reactor that corresponds to Figure 4.18 is given

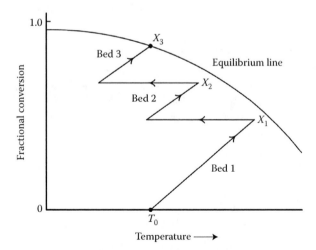

FIGURE 4.18
Multiple adiabatic reactors in series can be used to achieve a high level of conversion without cooling the reactor. Each bed is run adiabatically until the equilibrium line is approached. At this point, the process stream is cooled prior to entry to the next bed. The number of beds and the extent of reaction in each bed depend on the reaction and a variety of cost factors.

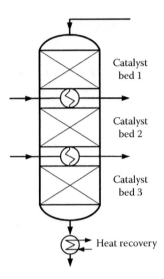

FIGURE 4.19
Multiple bed adiabatic catalytic reactor with interstage cooling. This type of design is often used for achieving high conversions in exothermic reactions where the equilibrium tends to limit the reaction yield. A classic industrial application of this design is the oxidation of sulfur dioxide to sulfur trioxide.

in Figure 4.19. The exact number of beds and the degree of cooling between the beds depends on the reaction involved and operating and capital costs. The design of such a system is a complex optimization problem that is beyond the scope of this text.

The most classic application of the multibed adiabatic fixed bed reactor is the oxidation of sulfur dioxide, an equilibrium limited reversible reaction

$$SO_2 + \frac{1}{2}O_2 \rightleftharpoons SO_3 \tag{4.235}$$

This reaction is used to make sulfur trioxide, which, when added to water, makes sulfuric acid. Very high yields are desired (in excess of 99%), but the reaction is highly exothermic and equilibrium limited. As a result the multibed approach has been adopted, with typical industrial units having three or four beds. A detailed treatment of the optimization of a sulfur dioxide reactor is given in Froment et al. (2011), and is also discussed in Thomas and Thomas (1967) and Lee (1985), among others.

4.5 Continuous Stirred Tank Reactor

We now consider the energy balance for a CSTR. In contrast to the PFR, where a spatial variation in temperature is observed, in a CSTR, the perfect mixing assumption means that the temperature is uniform at every point inside the reactor. Therefore, just as the concentration in the effluent stream is equal to the concentration in the reactor, the temperature of the fluid in the reactor is equal to the temperature of the effluent stream. The energy balance is again derived from the general energy balance, Equation 4.11. If we ignore the shaft work done by the mixing impeller, but retain the expansion work owing to a volume change, the general energy balance equation, Equation 4.11, written for a CSTR is

$$\frac{dU}{dt} = q - W_E + \left[\sum_{j=1}^{n} F_j H_j \right]_{in} - \left[\sum_{j=1}^{n} F_j H_j \right]_{out} \tag{4.236}$$

We shall now use Equation 4.236 to examine various operating modes for the CSTR, starting with the steady-state case, then moving to the transient case.

4.5.1 Steady-State CSTR: Basic Energy Balance

We start by considering the steady-state reactor because it is the easiest system to analyze. When the reactor is operating at steady state, the transient term in Equation 4.236 is equal to zero, and the energy balance equation is written as

$$q + \left[\sum_{j=1}^{n} F_j H_j \right]_{in} - \left[\sum_{j=1}^{n} F_j H_j \right]_{out} = 0 \tag{4.237}$$

Equation 4.237 states that the enthalpy difference between the inlet and outlet streams depends on the change in molar flow rate of each species in the stream and the change in enthalpy of each species. To quantify these changes, we consider the same generic reaction that was used for the development of the energy balance equation for the batch reactor and the PFR, that is

$$A + \frac{b}{a}B \rightarrow \frac{c}{a}C + \frac{d}{a}D \tag{4.238}$$

The energy balance is performed with respect to all of the species in the reactor including inert species (such as nonreacting solvents). The inert material is denoted with the symbol I. Consider the change in enthalpy between the inlet and outlet streams. This change may be written in terms of the enthalpies of each of the components present in the stream, as follows:

$$\left[\sum_{j=1}^{n} F_j H_j \right]_{in} - \left[\sum_{j=1}^{n} F_j H_j \right]_{out} = F_{A0}H_{A0} + F_{B0}H_{B0} + F_{C0}H_{C0} + F_{D0}H_{D0} + F_{I0}H_{I0}$$

$$- F_{AE}H_{AE} - F_{BE}H_{BE} - F_{CE}H_{CE} - F_{DE}H_{DE} - F_{IE}H_{IE} \tag{4.239}$$

The molar flow rate of each species in the effluent stream can be related to the molar flow rate of that species in the inlet stream, and the stoichiometry of the reaction. The effluent molar flow rates of each species are thus given by

$$F_{AE} = F_{A0} - (F_{A0} - F_{AE}); \qquad F_{BE} = F_{B0} - \frac{b}{a}(F_{A0} - F_{AE});$$

$$F_{CE} = F_{C0} + \frac{c}{a}(F_{A0} - F_{AE}); \qquad F_{DE} = F_{D0} + \frac{d}{a}(F_{A0} - F_{AE}); \tag{4.240}$$

$$F_{I0} = F_{IE}$$

Substituting these relations into Equation 4.239 gives

$$\left[\sum_{j=1}^{n} F_j H_j\right]_{in} - \left[\sum_{j=1}^{n} F_j H_j\right]_{out} = F_{A0}H_{A0} + F_{B0}H_{B0} + F_{C0}H_{C0} + F_{D0}H_{D0} + F_{I0}H_{I0}$$

$$- \left[F_{A0} - (F_{A0} - F_{AE})\right]H_{AE} - \left[F_{B0} - \frac{b}{a}(F_{A0} - F_{AE})\right]H_{BE}$$

$$- \left[F_{C0} + \frac{c}{a}(F_{A0} - F_{AE})\right]H_{CE} - \left[F_{D0} + \frac{d}{a}(F_{A0} - F_{AE})\right]H_{DE} - F_{IE}H_{IE} \tag{4.241}$$

Rearranging Equation 4.241 by grouping common terms gives the result

$$\left[\sum_{j=1}^{n} F_j H_j\right]_{in} - \left[\sum_{j=1}^{n} F_j H_j\right]_{out} = F_{A0}(H_{A0} - H_{AE}) + F_{B0}(H_{B0} - H_{BE})$$

$$+ F_{C0}(H_{C0} - H_{CE}) + F_{D0}(H_{D0} - H_{DE}) + F_{I0}(H_{I0} - H_{IE})$$

$$- (F_{A0} - F_{AE})\left[\frac{c}{a}H_{CE} + \frac{d}{a}H_{DE} - H_{AE} - \frac{b}{a}H_{BE}\right] \tag{4.242}$$

The enthalpy of reaction per mole of A, $\Delta H_{R,A}$, is the difference in enthalpy between the reactants and the products:

$$\frac{c}{a}H_{CE} + \frac{d}{a}H_{DE} - H_{AE} - \frac{b}{a}H_{BE} = \Delta H_{R,A} \tag{4.243}$$

Equation 4.242 can be simplified using Equation 4.243 to give

$$\left[\sum_{j=1}^{n} F_j H_j\right]_{in} - \left[\sum_{j=1}^{n} F_j H_j\right]_{out} = \sum_{j=1}^{n} F_{j0}(H_{j0} - H_{jE}) - (F_{A0} - F_{AE})\Delta H_{R,A} \tag{4.244}$$

The enthalpy change of a component is related to its temperature change and the constant-pressure heat capacity, as follows:

$$\left(H_{j0} - H_{jE}\right) = \int_{T_E}^{T_0} C_{Pj} \, dT \tag{4.245}$$

Combining Equations 4.237, 4.244, and 4.245 gives the energy balance equation for a steady-state CSTR:

$$q + \sum_{j=1}^{n}\left[F_{j0}\int_{T_E}^{T_0} C_{Pj} \, dT\right] - \left(F_{A0} - F_{AE}\right)\Delta H_{R,A} = 0 \tag{4.246}$$

Note that in Equation 4.246, the term containing the heat capacity only involves the composition of the feed stream to the reactor. Therefore, for the CSTR, it is not necessary to consider the effect of the reaction (compositional effects) on the heat capacity of the process stream. Hence, Equation 4.246 can be expressed in terms of an average heat capacity for the feed stream, as follows:

$$q + F_{T0}\int_{T_E}^{T_0} \bar{C}_{P0} \, dT - \left(F_{A0} - F_{AE}\right)\Delta H_{R,A} = 0 \tag{4.247}$$

In Equation 4.247, $\bar{C}_{P0}$ is used to denote the average heat capacity of the inlet stream, expressed here on a molar basis. If the heat capacity is assumed to be independent of temperature, it is possible to write Equation 4.247 as

$$q + F_{T0}\bar{C}_{P0}\left(T_0 - T_E\right) - \left(F_{A0} - F_{AE}\right)\Delta H_{R,A} = 0 \tag{4.248}$$

In Equation 4.248, assuming that $\bar{C}_{P0}$ is not a function of temperature may or may not be a good approximation. If the value of $\bar{C}_{P0}$ does change with temperature, but it is still desirable to use a constant value for computational convenience, then use the value at the average of the reactor inlet and outlet temperatures, that is, use

$$\bar{C}_{P0} = \frac{\left(C_{P0}\right)_{T_0} + \left(C_{P0}\right)_{T_E}}{2} \tag{4.249}$$

If we introduce the fractional conversion of A into Equation 4.248, we obtain

$$q + F_{T0}\bar{C}_{P0}\left(T_0 - T_E\right) - \left(F_{A0}X_{AE}\right)\Delta H_{R,A} = 0 \tag{4.250}$$

The mole balance equation for a CSTR can also be substituted into Equation 4.250. Recall that the mole balance for a CSTR, in terms of fractional conversion, is

$$(-r_A)V = F_{A0}X_{AE} \tag{4.251}$$

Substituting Equation 4.251 into Equation 4.250 gives

$$q + F_{T0}\bar{C}_{P0}(T_0 - T_E) - (-r_A)V\Delta H_{R,A} = 0 \tag{4.252}$$

Equations 4.247, 4.248, 4.250, and 4.251 are all different forms of the energy balance equation for a CSTR. Note that the last three versions assume a temperature-independent value for the heat capacity.

4.5.2 Adiabatic Operation of CSTR: Adiabatic Reaction Line

The CSTR may be operated without external heat transfer; such a reactor is said to be adiabatic. The general enthalpy balance for an adiabatic reactor is

$$\sum_{j=1}^{n}\left[F_{j0}\int_{T_E}^{T_0} C_{Pj}\,dT\right] - (F_{A0} - F_{AE})\Delta H_{R,A} = 0 \tag{4.253}$$

In terms of an average inlet heat capacity and the fractional conversion, the enthalpy balance is

$$F_{T0}\bar{C}_{P0}(T_0 - T_E) - (F_{A0}X_{AE})\Delta H_{R,A} = 0 \tag{4.254}$$

Equation 4.254 can be rearranged to give

$$T_E = T_0 - \frac{\Delta H_{R,A}}{F_{T0}\bar{C}_{P0}} F_{A0}X_{AE} \tag{4.255}$$

Note that Equation 4.255 is essentially the same relationship between temperature and conversion that was derived for the plug flow reactor case. This relationship is basically valid for an adiabatic reactor and any reaction kinetics.

4.5.3 External Heat Exchange in CSTR

The heat transfer term, q, may have various forms depending on the reactor and surroundings. The reactor may be heated by a heater that supplies a constant heat flux, in which case the value of q would be a constant. If there is no heat transfer with the surroundings, the reactor is adiabatic, and q has a value of zero. Alternatively, q might be varied continuously to achieve a specified heating or cooling rate to the reactor.

Common methods of providing heat exchange include placing heating or cooling coils inside the reactor, or placing a jacket containing a heat transfer fluid around the reactor surface. In these cases, the rate of heat transfer depends on the reactor temperature and the temperature of the heat transfer fluid:

$$q = UA(T_\infty - T_E) \tag{4.256}$$

In Equation 4.256, U is the overall heat transfer coefficient, A is the area for heat transfer, and T_∞ is the temperature of the heat transfer fluid.

Example 4.6

Consider the oxidation of carbon monoxide in an adiabatic constant-pressure CSTR. The reaction is described by the overall stoichiometry:

$$CO + \frac{1}{2}O_2 \rightarrow CO_2 \tag{4.257}$$

The rate of reaction in the presence of water is given by the rate expression

$$(-r_{CO}) = 1.26 \times 10^{10} \exp\left(\frac{-20,131}{T}\right) C_{CO} C_{O_2}^{0.25} C_{H_2O}^{0.5} \; mol/m^3 \cdot s \tag{4.258}$$

A mixture consisting of 1% CO, 1% O_2, 1% H_2O, and 97% N_2 (mole percentages) is fed to a CSTR. The total volumetric flow rate of the feed is 1.00×10^{-4} m³/s at a pressure of 1 bar and a temperature of 700 K. Calculate the volume required for 99% conversion and the corresponding outlet temperature.

SOLUTION

It is necessary to compute both the mole balance and the energy balance. The mole balance equation for CO should be written in terms of the molar flow rate because the reaction is a gas-phase reaction in which both the temperature and the number of moles change with reaction. The mole balance for a CSTR is

$$\frac{(F_{CO})_0 - F_{CO}}{V} = 1.26 \times 10^{10} \exp\left(\frac{-20,131}{T_E}\right) \frac{F_{CO}}{Q} \left(\frac{F_{O_2}}{Q}\right)^{0.25} \left(\frac{F_{H_2O}}{Q}\right)^{0.5} \tag{4.259}$$

This equation can be simplified. First note that the moles of water are constant during the reaction, as it does not react. The molar flow rate of water is

$$F_{H_2O} = \left(\frac{PQ}{R_g T}\right)_0 Y_{H_2O} = \left(\frac{100,000 \times 1.0 \times 10^{-4}}{8.314 \times 700}\right) \times 0.01 = 1.72 \times 10^{-5} \tag{4.260}$$

The molar flow rates of the reactants can be written in terms of the fractional conversion of CO:

$$F_{CO} = (F_{CO})_0 (1 - X_{CO}) \quad \text{and} \quad F_{O_2} = (F_{O_2})_0 - \frac{X_{CO}}{2}(F_{CO})_0$$

With these substitutions and resulting simplification, Equation 4.259 becomes

$$\frac{X_{CO}}{V} = 5.23 \times 10^7 \exp\left(\frac{-20,131}{T_E}\right)(1 - X_{CO})\left((F_{O_2})_0 - \frac{X_{CO}}{2}(F_{CO})_0\right)^{0.25} Q^{-1.75}$$

The inlet molar flow rates of CO and oxygen in mol/s can be calculated as

$$(F_{CO})_0 = \left(\frac{PQ}{R_g T}\right)_0 (Y_{CO})_0 = \left(\frac{100,000 \times 1.0 \times 10^{-4}}{8.314 \times 700}\right) \times 0.01 = 1.72 \times 10^{-5} \tag{4.261}$$

$$\left(F_{O_2}\right)_0 = \left(\frac{PQ}{R_g T}\right)_0 \left(Y_{O_2}\right)_0 = \left(\frac{100,000 \times 1.0 \times 10^{-4}}{8.314 \times 700}\right) \times 0.01 = 1.72 \times 10^{-5} \qquad (4.262)$$

The mole balance equation is simplified using Equations 4.261 and 4.262:

$$\frac{X_{CO}}{V} = 3.37 \times 10^6 \exp\left(\frac{-20,131}{T_E}\right)(1 - X_{CO})\left(1 - \frac{X_{CO}}{2}\right)^{0.25} Q^{-1.75} \qquad (4.263)$$

The next step in developing the mole balance equation is to derive an expression for the volumetric flow rate as a function of temperature and fractional conversion. Following the technique used in Chapter 3 for isothermal reactors, we first develop a stoichiometric table:

Compound	Inlet Moles	Effluent Moles
CO	1.72×10^{-5}	$1.72 \times 10^{-5} (1 - X_{CO})$
O_2	1.72×10^{-5}	$1.72 \times 10^{-5} (1 - 0.5 X_{CO})$
CO_2	0	$1.72 \times 10^{-5} X_{CO}$
H_2O	1.72×10^{-5}	1.72×10^{-5}
N_2	1.668×10^{-3}	1.668×10^{-3}
Total	1.72×10^{-3}	$1.72 \times 10^{-3} (1 - 0.005 X_{CO})$

The volumetric flow rate depends on the pressure and temperature.

$$\left(\frac{PQ}{F_T T}\right)_0 = \left(\frac{PQ}{F_T T}\right) \qquad (4.264)$$

The pressure is a constant 1 bar, the initial temperature is 700 K, and the inlet volumetric flow rate is 1.00×10^{-4} m^3/s. The volumetric flow rate in the reactor is

$$Q = Q_0 \frac{T}{T_0} \frac{F_T}{F_{T0}} = 1.0 \times 10^{-4} \frac{T_E}{700} \frac{1.72 \times 10^{-3}(1 - 0.005 X_{CO})}{1.72 \times 10^{-3}}$$

$$= 1.429 \times 10^{-7} T_E (1 - 0.005 X_{CO}) \qquad (4.265)$$

Finally, substitute for the final conversion of 0.99 and simplify to obtain the mole balance with two unknowns, the effluent temperature and the reactor volume:

$$\frac{1}{V} = 2.754 \times 10^{16} \exp\left(\frac{-20,131}{T_E}\right) T_E^{-1.75} \qquad (4.266)$$

Having developed the mole balance, we now turn to the energy balance equation. The energy balance for an adiabatic CSTR is given by Equation 4.253. In terms of fractional conversion, this equation can be written as

$$\sum_{j=1}^{n}\left[F_{j0} \int_{T_E}^{T_0} C_{Pj} \, dT\right] - X_{CO}\left(F_{CO}\right)_0 \Delta H_R = 0 \qquad (4.267)$$

We start the solution of the energy balance by assembling some of the necessary data. The constant pressure heat capacity for each component is a function of temperature, as follows:

$$C_P = a + bT + cT^2 + dT^3 \tag{4.268}$$

The values of the constants are obtained from Appendix 1:

Compound	a	$b \times 10^2$	$c \times 10^5$	$d \times 10^9$
O_2	25.44	1.518	−0.7144	1.310
N_2	28.85	−0.1569	0.8067	−2.868
H_2O	32.19	0.1920	1.054	−3.589
CO	28.11	0.1672	0.5363	−2.218
CO_2	22.22	5.9711	−3.495	7.457

The enthalpy of reaction as a function of temperature was computed in Example 4.4. At the reactor effluent temperature, the value is

$$\Delta H_R = -2.796 \times 10^5 - 18.61 T_E + 2.522 \times 10^{-2} T_E^2$$

$$- 1.225 \times 10^{-5} T_E^3 + 2.255 \times 10^{-9} T_E^4 \text{ J/mol} \tag{4.269}$$

The heat capacity of the reactor feed changes with temperature. Therefore, write

$$\sum_{j=1}^{n} \left[F_{j0} \int_{T_E}^{T_0} C_{Pj} \, dT \right] = (F_{CO})_0 \int_{T_E}^{T_0} C_{P,CO} \, dT + (F_{O2})_0 \int_{T_E}^{T_0} C_{P,O_2} \, dT$$

$$+ (F_{N2})_0 \int_{T_E}^{T_0} C_{P,N2} \, dT + (F_{H2O})_0 \int_{T_E}^{T_0} C_{P,H2O} \, dT \tag{4.270}$$

The integral of the heat capacity can be written in the generic form

$$\int_{T_E}^{T_0} C_P \, dT = \left[aT + \frac{b}{2}T^2 + \frac{c}{3}T^3 + \frac{d}{4}T^4 \right]_{T_E}^{T_0} \tag{4.271}$$

Substituting the known values of the molar flow rates and the coefficients in the heat capacity equations into Equation 4.270, integrating and simplifying give

$$\sum_{j=1}^{n} \left[F_{j0} \int_{T_E}^{T_0} C_{Pj} \, dT \right] = 35.42 - 4.960 \times 10^{-2} T_E + 1.147 \times 10^{-6} T_E^2$$

$$- 4.537 \times 10^{-9} T_E^3 + 1.216 \times 10^{-12} T_E^4 \tag{4.272}$$

The final energy balance equation is obtained by substituting Equations 4.272 and 4.269 into Equation 4.267, and also substituting the inlet molar flow rate of CO and the final fractional conversion. The resulting energy balance equation is a simple polynomial with the effluent temperature as the single unknown:

$$40.17 - 4.917 \times 10^{-2} T_E + 4.126 \times 10^{-7} T_E^2 - 4.011 \times 10^{-9} T_E^3 + 1.066 \times 10^{-12} T_E^4 = 0 \qquad (4.273)$$

Solution of Equation 4.273 gives an outlet reactor temperature of 790 K. This temperature can be substituted into Equation 4.266 to obtain a reactor volume of 0.5 m³.

It is worthwhile to make two observations at this point. The first is that the CSTR volume for 99% conversion is about 10 times larger than that required for the PFR for the same inlet conditions. This size difference occurs because in the CSTR all of the reaction occurs at a CO concentration corresponding to 99% conversion, whereas in the PFR the reaction occurs over a continuously varying CO concentration. In the CSTR, all of the reaction occurs at the exit temperature, whereas in the PFR it occurs at a continuously varying temperature. The fact that all of the reaction occurs in the CSTR at the highest value of the rate constant (highest temperature) is an advantage that this reactor has over the PFR for adiabatic operation with exothermic reactions; however, in this example, the increase in the value of the rate constant is not sufficient to offset the decrease in rate caused by the low concentration.

The second observation is that the reactor outlet temperature is the same as for the PFR with identical inlet conditions (Example 4.4). Close inspection of the equations for the adiabatic reaction for a PFR and a CSTR shows that these two equations are in fact the same, and therefore, for the same inlet conditions and outlet fractional conversion, the outlet temperature must be the same (provided that the pressure is assumed to be constant in both reactors.)

4.5.4 Multiple Reactions in CSTR

In a CSTR with multiple reactions, the enthalpy change for each reaction must be included in the energy balance equation. A mole balance equation must be written for each independent reaction. For example, consider a reactor in which the following two reactions occur:

$$A \rightarrow B$$
$$C \rightarrow D \qquad (4.274)$$

The rates of disappearance of A and C are given by $(-r_A)$ and $(-r_C)$, respectively. Two mole balance equations must be solved, one each for A and C:

$$F_{A0} - F_{AE} = (-r_A)V \quad \text{and} \quad F_{C0} - F_{CE} = (-r_C)V \qquad (4.275)$$

The energy balance equation for a CSTR must include the enthalpy changes owing to both reactions. For example, taking the general CSTR energy balance, Equation 4.246, we obtain

$$q + \sum_{j=1}^{n} \left[F_{j0} \int_{T_E}^{T_0} C_{Pj} \, dT \right] - (F_{A0} - F_{AE}) \Delta H_{R,A} - (F_{C0} - F_{CE}) \Delta H_{R,C} = 0 \qquad (4.276)$$

The general energy balance for a reactor containing n species (including inert material) in which m reactions occur is

$$q + \sum_{j=1}^{n} \left[F_{j0} \int_{T_E}^{T_0} C_{Pj} \, dT \right] - \sum_{k=1}^{m} (F_{k0} - F_{kE}) \Delta H_{R,k} = 0 \tag{4.277}$$

4.5.5 Multiple Steady States in a CSTR

One of the interesting operating features of a nonisothermal CSTR is the existence of multiple steady states under certain operating conditions. In a reactor with multiple steady states, there is more than one set of steady-state operating conditions possible. From a mathematical perspective, the existence of multiple steady states implies that there is more than one solution (temperature and fractional conversion) that satisfies the mole and energy balance equations. In a reactor in which multiple steady states occur, it is not possible to deduce from a steady-state analysis which steady state will be achieved in practice. In the following, we use some graphical analysis to illustrate how multiple steady states arise and the implications for reactor operation. A simple first-order reaction with a constant-density fluid is chosen for simplicity, but the principles may be extended to more complex cases, with a concomitant increase in mathematical complexity.

For illustration purposes, consider a first-order reaction occurring in a nonisothermal CSTR. The mole balance equation is

$$\frac{V}{QC_{A0}} = \frac{X_{AE}}{kC_{A0}(1 - X_{AE})} \tag{4.278}$$

Equation 4.278 can be rearranged to give an expression for the fractional conversion of A:

$$X_{AE} = \frac{\tau k}{1 + \tau k} \tag{4.279}$$

The steady-state energy balance written in terms of the fractional conversion and constant heat capacity is given by Equation 4.250. If the reactor exchanges heat with its surroundings according to Equation 4.256, the energy balance is

$$UA(T_\infty - T_E) + F_{T0}\bar{C}_{P0}(T_0 - T_E) - (F_{A0}X_{AE})\Delta H_{R,A} = 0 \tag{4.280}$$

Note that q is positive if heat is added to the reactor, that is, when $(T_\infty > T_E)$. Equation 4.280 can be rearranged as follows:

$$(-\Delta H_{R,A})F_{A0}X_{AE} = (F_{T0}\bar{C}_{P0} + UA)T_E - (F_{T0}\bar{C}_{P0}T_0 + UAT_\infty) \tag{4.281}$$

Substituting for conversion from Equation 4.279 gives

$$(-\Delta H_{R,A})F_{A0}\left(\frac{\tau k}{1+\tau k}\right) = \left(F_{T0}\bar{C}_{P0} + UA\right)T_E - \left(F_{T0}\bar{C}_{P0}T_0 + UAT_\infty\right) \qquad (4.282)$$

Equation 4.282 is simply a combination of the mole and energy balance equations for a CSTR with a first-order reaction, constant heat capacity, and constant density. The rate constant is an exponential function of temperature. If it follows the Arrhenius law, k is given by

$$k = A\exp\left(\frac{-E}{R_g T_E}\right) \qquad (4.283)$$

Equation 4.282 is a nonlinear equation. For a given set of operating conditions, the reactor temperature is the only unknown value in the equation. However, owing to the nature of the equation, it is possible that under some sets of operating conditions there will be more than one solution to Equation 4.282. When two or more values of T_E will satisfy Equation 4.282, the reactor is operating in a region of multiple steady states. It is instructive to examine the nature of Equation 4.282 graphically. We define two functions corresponding to the left-hand side and right-hand side of Equation 4.282, respectively. The first function is

$$G(T) = \left(-\Delta H_{R,A}\right)F_{A0}\left(\frac{\tau k}{1+\tau k}\right) \qquad (4.284)$$

The function $G(T)$ represents the rate of heat generation in the reactor. The second function is

$$R(T) = \left(F_{T0}\bar{C}_{P0} + UA\right)T_E - \left(F_{T0}\bar{C}_{P0}T_0 + UAT_\infty\right) \qquad (4.285)$$

The function $R(T)$ represents the rate of heat removal by heat transfer with the surroundings and the rate of fluid removal from the reactor. Note that $R(T)$ is a linear function of temperature in this case because C_P and ΔH_R are assumed to be constant. The intercept of the straight line depends on, for example, the value of the reactor inlet temperature. A typical behavior of the function $R(T)$ at different inlet temperatures is shown in Figure 4.20. An increase in the value of the reactor inlet temperature simply shifts the line to the right, while maintaining the value of the slope. The slope of the curve can be adjusted by changing the value of $(F_{T0}\bar{C}_{P0} + UA)$, and a typical behavior is shown in Figure 4.21. Note that some of the operating parameters are usually fixed, so it is rare that complete freedom to adjust all of the parameters will exist. If it was desired to process a given molar flow rate of feed, then to change the slope would require a change in the value of the heat transfer coefficient or the heat transfer area.

A plot of the generation function, $G(T)$, verses T gives a sigmoidal curve because of the exponential temperature dependence of the rate constant. The shape of the curve depends in large part on the value of the activation energy. A typical plot is shown in Figure 4.22.

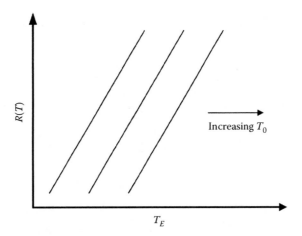

FIGURE 4.20
Behavior of the heat removal function at different feed inlet temperatures, with all other operating conditions being the same. An increase in the inlet temperature simply shifts the line to the right.

All possible reactor operating temperatures must satisfy both functions, $G(T)$ and $R(T)$, such that the two functions have equal values. If both of the functions are plotted on the same graph, the points of intersection of the two lines represent permissible solutions to the reactor energy and mole balance equations. Figure 4.23 illustrates such a graph. In this figure, a single $G(T)$ curve is shown, which would correspond to specific values of the kinetic parameters and enthalpy of reaction. Multiple $R(T)$ curves are shown, each of which corresponds to a different reactor inlet temperature, with all other parameters the same.

Consider first the $R(T)$ line that corresponds to the lowest value of the inlet temperature. This line is at the extreme left of Figure 4.23. This line intersects the $G(T)$ at a single point, labeled 1 in Figure 4.23. This point of intersection corresponds to the only solution to the mole and energy balance equations; thus this point represents the reactor operating tem-

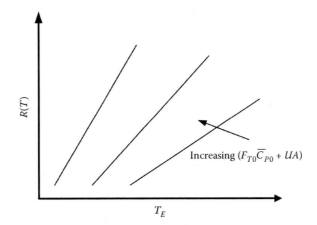

FIGURE 4.21
Behavior of the heat removal function at different values of $(F_{T_0}\overline{C}_{P_0} + UA)$ with all other operating conditions the same. The slope of the line changes.

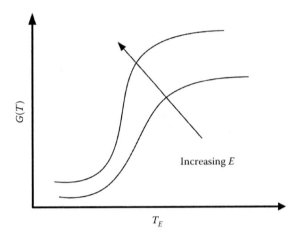

FIGURE 4.22
Shape of the heat generation function at different reactor temperatures. The curve shape depends on the activation energy.

perature. If the reactor inlet temperature is increased, with all other conditions held constant, the $R(T)$ line moves to the right. The sigmoidal $G(T)$ curve is not affected by the change in the inlet temperature because T_0 does not appear in the expression for $G(T)$. The second $R(T)$ curve from the left intersects the $G(T)$ curve at two points, marked with points 2 and 6, respectively. Both these points represent possible operating temperatures for the reactor, the low- and high-temperature stable operating conditions for this reactor. The next $R(T)$ line in Figure 4.23 intersects the $G(T)$ curve at three points, labeled 3, 5, and 7 respectively. Points 3 and 7 represent stable-steady states. Point 5 is an unstable steady state which cannot be achieved in practice. Continuing to increase the inlet temperature, the next $R(T)$ curve illustrated has two stable steady states at 4 and 8, the final curve has a single steady state at point 9. The steady state that is realized in the reactor depends on either the initial conditions or

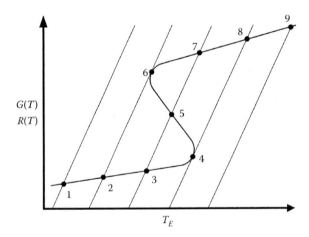

FIGURE 4.23
Plot of a single heat generation curve and multiple heat removal curves. Each heat removal curve corresponds to a different reactor inlet temperature.

the history of the reactor. The best method to determine the final steady state is to use a transient analysis; this method will be illustrated shortly in an example. However, we can visualize some behavior using a quasi-steady-state analysis.

In a quasi-steady-state analysis, we analyze the reactor using a steady-state approach, that is, using the steady-state mole and energy balance equations. It is assumed that small changes are made to the operating conditions, and that the reactor is allowed to come to a new steady state before further changes are made.

We consider a CSTR that is initially running at steady state with a low feed temperature. The feed temperature is then slowly increased through the region of multiple steady states (see Figure 4.23) with steady state prevailing at each point. The feed temperature is then slowly decreased back to the original value. We plot the outlet reactor temperature as a function of the feed temperature. The resulting plot is shown in Figure 4.24. The numbered points in Figure 4.24 correspond to the same numbers in Figure 4.23.

At the inlet temperature corresponding to point 1, there is only one possible operating point for the reactor. As the feed temperature increases, the reactor temperature also increases until point 2 is reached. At this inlet temperature, there are two possible steady-state operating points for the reactor; however, the reactor will operate at point 2. The reactor temperature continues to increase with the feed temperature, passing through point 3, until it reaches point 4. At this point the reactor leaves the region of multiple steady states and any slight increase in feed temperature will send the reactor temperature rapidly to point 8 on the plot. Further increases in the feed temperature will simply increase the reactor temperature along the curve, in the direction of point 9.

If the reactor feed temperature is now slowly decreased, the reactor temperature will also decline. It will move down along the upper line, passing through points 8 and 7 until it reaches point 6. At this point, any further reduction in the feed temperature will send the reactor temperature down to point 2 on the curve, and then in the direction of point 1. It is thus evident that the heating and cooling behavior of the reactor are not the same, and hysteresis is observed. A curve of the type shown in Figure 4.23 is often referred to as an ignition extinction curve.

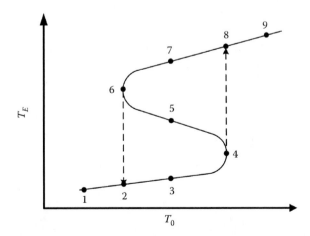

FIGURE 4.24
Ignition extinction curve for a CSTR operated at pseudo-steady state. The reactor temperature follows the lower branch of the curve as the feed temperature increases, and then falls along the upper branch as the feed temperature is reduced.

The existence of multiple steady states has implications for the control of the reactor. Suppose, for example, that the reactor was operating at steady state at point 7 in Figure 4.23. If there was a change in the process variables and the reactor feed temperature fell for a period of time, the reactor operating temperature might fall to the lower operating line (point 1 or 2 for example). It would then be necessary to increase the feed temperature past point 4 to get the reactor operating temperature back to the upper line and point 7.

Note that the exact form of the ignition extinction curve depends on the operating conditions and the reaction kinetics. Complex kinetic expressions can lead to very complex multiple steady-state performance. It is also important to emphasize that when a steady-state analysis predicts the existence of multiple steady states, it is not possible to know which steady state will be achieved in practice. The final operating point depends on the history of the reactor, which can only be determined using transient analysis. Transient analysis is discussed after Example 4.7, which illustrates a CSTR problem with multiple steady states.

Example 4.7

A first-order liquid-phase homogeneous reaction is carried out in an adiabatic CSTR. The reactor volume is $1.8 \times 10^{-2} m^3$, the feed flow rate is $60 \times 10^{-6} m^3/s$, the feed density is 1000 kg/m^3, the heat capacity is a constant 4182 J/kg · K, and the enthalpy change of reaction is $\Delta H_R = -2.09 \times 10^5$ J/mol. The inlet concentration of reactant A is 3000 mol/m^3. The rate expression is

$$(-r_A) = kC_A = 4.48 \times 10^6 \exp\left(\frac{-7554}{T}\right) C_A \text{ mol/m}^3 \cdot \text{s} \qquad (4.286)$$

where the units of k are 1/s and the concentration of A is in mol/m^3.

    a. Calculate the steady-state outlet temperature for an inlet temperature of 298 K.
    b. Plot the pseudo-steady-state ignition extinction curve.

SOLUTION

(a) The solution to this problem requires both the mole and energy balance equations. The steady-state mole balance equation is

$$F_{A0} - F_{AE} - (-r_A)V = 0 \qquad (4.287)$$

The fluid density is constant, so the mole balance can be written in terms of concentration as

$$C_{A0} - C_{AE} - (-r_A)\frac{V}{Q} = 0 \qquad (4.288)$$

Equation 4.288 can be written in terms of the fractional conversion and the space time. Also substituting the rate function, Equation 4.286, gives

$$C_{A0} - (C_{A0} - C_{A0}X_{AE}) - kC_{A0}(1 - X_{AE})\tau = 0 \qquad (4.289)$$

Simplify Equation 4.289 by eliminating the inlet concentration

$$X_{AE} - k(1 - X_{AE})\tau = 0 \tag{4.290}$$

The space time can be readily calculated:

$$\tau = \frac{V}{Q} = \frac{1.8 \times 10^{-2}\, m^3}{60 \times 10^{-6}\, m^3/s} = 300\, s \tag{4.291}$$

Substitute the value of the space time and the rate constants into Equation 4.290:

$$X_{AE} - (300)(4.48 \times 10^6)(1 - X_{AE})\exp\left(\frac{-7554}{T_E}\right) = 0 \tag{4.292}$$

Equation 4.292 can be simplified and rearranged to give an expression in terms of conversion:

$$X_{AE} = \frac{1.344 \times 10^9 \exp(-7554/T_E)}{1 + 1.344 \times 10^9 \exp(-7554/T_E)} \tag{4.293}$$

The steady-state energy balance for an adiabatic reactor, assuming constant C_P, can be written in terms of the total mass flow rate of the inlet stream, as follows:

$$\dot{m}_T C_P (T_0 - T_E) - F_{A0} X_{AE} (\Delta H_{R,A}) = 0 \tag{4.294}$$

Equation 4.294 gives a relationship between X_{AE} and T_E. The mass flow rate is the product of the volumetric flow rate and the density, or

$$\dot{m}_T = Q\rho = 60 \times 10^{-6}\, m^3/s \times 1000\, kg/m^3 = 60 \times 10^{-3}\, kg/s \tag{4.295}$$

The inlet molar flow rate of A is the product of the concentration and the volumetric flow rate:

$$F_{A0} = QC_{A0} = 60 \times 10^{-6}\, m^3/s \times 3000\, mol/m^3 = 0.18\, mol/s \tag{4.296}$$

Substituting the numerical values into Equation 4.294 gives

$$60 \times 10^{-3}\, kg/s \times 4182\, J/kg \cdot K\, (298 - T_E) - 0.18\, mol/s \times (-2.09 \times 10^5)\, J/mol X_{AE} = 0$$

After simplification, the energy balance is

$$(298 - T_E) = -150 X_{AE} \tag{4.297}$$

Equation 4.297 can be rearranged to give an expression for the outlet temperature as a function of conversion. Note that this relation is nothing more than the adiabatic reaction line.

$$T_E = 150X_{AE} + 298 \qquad\qquad (4.298)$$

Equation 4.298 can be substituted into the mole balance, Equation 4.293:

$$X_{AE} = \frac{1.344 \times 10^9 \exp\left(-7554/150X_{AE} + 298\right)}{1 + 1.344 \times 10^9 \exp\left(\dfrac{-7554}{150X_{AE} + 298}\right)} \qquad\qquad (4.299)$$

The solution of this nonlinear equation gives the outlet conversion, X_{AE}. Solving Equation 4.299 numerically yields the following three values for X_{AE}, and corresponding values for T_E:

$$X_{AE} = 0.0158 \quad T_E = 300.38 \text{ K}$$
$$X_{AE} = 0.333 \quad T_E = 347.8 \text{ K}$$
$$X_{AE} = 0.9831 \quad T_E = 445.5 \text{ K}$$

It is seen that for an inlet temperature of 298 K there are three solutions possible. Only the upper and lower values of X_{AE} represent stable operating points that could be achieved in practice. The central point is an unstable state which would not exist in the reactor. To determine which of the stable operating points are achieved in actual operation from the given operating conditions, it is necessary to perform a transient analysis. Transient reactor analysis is discussed immediately following this example.

(b) To develop the pseudo-steady-state ignition/extinction curve, it is necessary to solve the mole and energy balance equations at a variety of inlet temperatures. Therefore, we express Equation 4.299 in the more general form:

$$X_{AE} = \frac{1.344 \times 10^9 \exp\left(-7554/150X_{AE} + T_0\right)}{1 + 1.344 \times 10^9 \exp\left(-7554/150X_{AE} + T_0\right)} \qquad\qquad (4.300)$$

Solution of Equation 4.300 over a range of inlet temperatures gives the values shown in the table below. Note that either one or three solutions are possible, depending on the value of T_0. For two values of T_0, two of the solutions are the same: these points are the "turning points," and identify the extremities of the multiple steady-state region.

T_0 (K)		T_E (K)		
260.0	260.1			
264.0	264.1			
264.1	264.2			
264.2	264.3	390.0	390.0	Turning point
265.0	265.1	383.8	396.9	
305.0	310.0	340.0	456.0	
310.0	322.0	330.0	458.0	
310.2	328.9	328.9	458.6	Turning point
310.5			459.0	
311.0			460.0	
315.0			463.0	

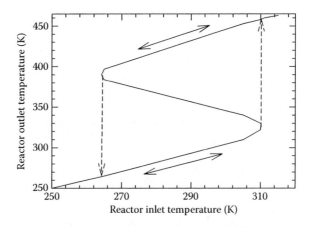

FIGURE 4.25
Ignition extinction curve for Example 4.7. The curve shows the possible outlet reactor temperatures for the given inlet temperature. The upper and lower curves represent stable operating points, while the central line represents unstable points.

The possible outlet temperatures are plotted as a function of the reactor inlet temperature in Figure 4.25. This figure can be used to determine the performance of a reactor operating at pseudostate with a slowly increasing inlet reactor temperature. An increase in inlet temperature from 250 K would see the outlet temperature increase along the bottom curve until an inlet temperature of 310.20 K was reached (reactor temperature of 328.85 K). At this point, the reactor would experience a rapid temperature increase to 458.63 K. If the inlet temperature was then slowly lowered, the reactor temperature would decrease along the upper curve until the inlet temperature fell to 264.18 K (reactor temperature of about 390 K). The reactor would then experience a rapid temperature change to 264.26 K. Provided that steady state was reached, the reactor inlet temperature would have to be raised to 310.2 K to attain the upper operating line.

4.5.6 Transient CSTR with Energy Effects: Constant-Volume Case

The transient operation of a CSTR was considered in Chapter 3, where the transient mole balance was derived. In this section, the transient energy balance for a CSTR is developed, starting with the general energy balance, Equation 4.236. The right-hand side of Equation 4.236 is developed in the same manner as for the steady-state case. For constant-volume operation, the expansion work is zero and the transient energy balance is

$$q + \sum_{j=1}^{n}\left[F_{j0}\int_{T_E}^{T_0} C_{Pj}\,\mathrm{d}T \right] - \left(F_{A0} - F_{AE}\right)\Delta H_{R,A} = \frac{\mathrm{d}U}{\mathrm{d}t} \tag{4.301}$$

The right-hand side of Equation 4.301 is the change in internal energy with time and is the sum of the internal energy change of every species in the reactor. If there are m species in the reactor, the change in internal energy is given by

$$\frac{\mathrm{d}U}{\mathrm{d}t} = \frac{\mathrm{d}\left(\sum_{j=1}^{m} N_j U_j \right)}{\mathrm{d}t} \tag{4.302}$$

The right-hand side of Equation 4.302 can be expanded using the chain rule

$$\frac{dU}{dt} = \sum_{j=1}^{m} N_j \frac{dU_j}{dt} + \sum_{j=1}^{m} U_j \frac{dN_j}{dt} \tag{4.303}$$

Equation 4.303 contains terms that reflect the internal energy change owing to reaction and temperature. For a constant-volume transient CSTR, it can be simplified following an analagous procedure to that used in developing the energy balance equation for the constant-volume batch reactor. The final result is

$$\frac{dU}{dt} = \left(\sum_{j=1}^{m} N_j C_{Vj} \right) \frac{dT}{dt} - \frac{dN_A}{dt} (\Delta U_{R,A}) \tag{4.304}$$

Combining Equations 4.301 and 4.304 gives the energy balance for the constant-volume transient CSTR:

$$q + \sum_{j=1}^{n} \left[F_{j0} \int_{T_E}^{T_0} C_{Pj}\, dT \right] - (F_{A0} - F_{AE}) \Delta H_{R,A} = \left(\sum_{j=1}^{m} N_j C_{Vj} \right) \frac{dT_E}{dt} - \frac{dN_A}{dt} (\Delta U_{R,A}) \tag{4.305}$$

Note that Equation 4.58 contains both heat capacities, C_V and C_P, and both energy changes of reaction, ΔH_R and ΔU_R. For liquid-phase reactions, the two heat capacities are equal, while for ideal gases, we recall from Equation 2.34

$$C_P = C_V + R_g \tag{4.306}$$

Refer to Section 4.1, where the difference between ΔH_R and ΔU_R was discussed. For liquid-phase reactions, the two are approximately equal.

If the reactor is operated at *both* constant pressure and constant volume, the enthalpy change and the internal energy changes are the same, and the energy balance can be written as

$$q + \sum_{j=1}^{n} \left[F_{j0} \int_{T_E}^{T_0} C_{Pj}\, dT \right] - (F_{A0} - F_{AE}) \Delta H_{R,A} = \left(\sum_{j=1}^{m} N_j C_{Pj} \right) \frac{dT_E}{dt} - \frac{dN_A}{dt} (\Delta H_{R,A}) \tag{4.307}$$

The transient energy balance is illustrated in the following example. If the fluid density is constant, Equation 4.307 can be written in terms of concentration of A and inlet volumetric flow rate:

$$q + \sum_{j=1}^{n} \left[F_{j0} \int_{T_E}^{T_0} C_{Pj}\, dT \right] - Q(C_{A0} - C_{AE}) \Delta H_{R,A} = \left(\sum_{j=1}^{m} N_j C_{Pj} \right) \frac{dT_E}{dt} - V \frac{dC_A}{dt} (\Delta H_{R,A}) \tag{4.308}$$

Example 4.8

In this example, we will apply transient analysis to the adiabatic CSTR of Example 4.7. The reactor volume is constant at $1.8 \times 10^{-2} \text{m}^3$, that is, the reactor is initially full. The initial reactant concentration in the reactor is the same as the concentration of A in the feed, 3000 mol/m³, and the feed temperature is 298 K. The feed flow rate is 60×10^{-6} m³/s, the feed density is 1000 kg/m³, the heat capacity is a constant 4182 J/kg · K, and the enthalpy change of reaction is $\Delta H_R = -2.09 \times 10^5$ J/mol. The rate expression is given by

$$(-r_A) = kC_A = 4.48 \times 10^6 \exp\left(\frac{-7554}{T}\right) C_A \,\text{mol/m}^3 \cdot \text{s} \tag{4.309}$$

where the units of k are 1/s and the concentration of A is in mol/m³. It was seen in Example 4.7 that this set of operating conditions corresponds to the region in which multiple steady-state solutions exist.

Determine the outlet temperature at various initial reactor temperatures and determine the critical initial temperature. The critical initial temperature is defined as the temperature which is required for the reactor to attain the high-temperature steady-state solution in the region of multiple steady states.

SOLUTION

This problem involves a transient reactor with constant volume V (because the reactor is initially full) and constant Q (because the density is constant). The transient mole balance equation was developed in Chapter 3, and is given by

$$F_{A0} - F_{AE} - (-r_A)V = \frac{dN_A}{dt} \tag{4.310}$$

Equation 4.310 can be expressed in terms of concentration. As the reactor volume and volumetric flow rate are constant, we can write

$$C_{A0} - C_{AE} - (-r_A)\frac{V}{Q} = \frac{V}{Q}\frac{dC_A}{dt} \tag{4.311}$$

Equation 4.311 is simplified by introducing the space time, the rate expression, and by dividing through by the inlet concentration:

$$\frac{1}{\tau}\left(1 - \frac{C_{AE}}{C_{A0}}\right) - k\frac{C_{AE}}{C_{A0}} = \frac{d\left(C_{AE}/C_{A0}\right)}{dt} \tag{4.312}$$

We now define a dimensionless concentration, Z:

$$Z = \frac{C_{AE}}{C_{A0}} \tag{4.313}$$

Substituting Equation 4.313 into Equation 4.312, and also substituting the value of the space time, gives the final transient mole balance equation

$$\frac{1}{300}(1 - Z) - 4.48 \times 10^6 \exp\left(\frac{-7554}{T}\right)Z = \frac{dZ}{dt} \tag{4.314}$$

Equation 4.314 is a first-order ordinary differential equation, and requires an initial condition. The initial concentration of A in the reactor is the same as in the feed stream, and therefore the initial condition is $Z_i = 1$.

The energy balance is developed from the generic Equation 4.305. The reactor is adiabatic, so q is equal to zero. As the fluid is a liquid, it is assumed that C_V is equal to C_P, and $\Delta H_{R,A} \approx \Delta U_{R,A}$. Furthermore, the heat capacity based on the mass of the fluid is assumed to be independent of both temperature and composition. Thus, Equation 4.85 can be written as

$$\dot{m}_T C_P \left(T_0 - T_E\right) - \left(F_{A0} - F_{AE}\right)\Delta H_{R,A} = m_t C_P \frac{\mathrm{d}T_E}{\mathrm{d}t} - \frac{\mathrm{d}N_A}{\mathrm{d}t}\left(\Delta H_{R,A}\right) \tag{4.315}$$

In Equation 4.315, $\dot{m}_T$ is the total mass flow rate of the feed and m_t is the mass of the fluid in the reactor (which is a constant). Introducing the concentration and the constant reactor volume gives the equation

$$\dot{m}_T C_P \left(T_0 - T_E\right) - Q\left(C_{A0} - C_{AE}\right)\Delta H_{R,A} = m_t C_P \frac{\mathrm{d}T_E}{\mathrm{d}t} - V\frac{\mathrm{d}C_{AE}}{\mathrm{d}t}\left(\Delta H_{R,A}\right) \tag{4.316}$$

Writing Equation 4.316 using the dimensionless concentration ratio, Z, gives

$$\frac{\dot{m}_T C_P}{C_{A0}}\left(T_0 - T_E\right) - Q\left(1 - Z\right)\Delta H_{R,A} = \frac{m_t C_P}{C_{A0}}\frac{\mathrm{d}T_E}{\mathrm{d}t} - V\frac{\mathrm{d}Z}{\mathrm{d}t}\left(\Delta H_{R,A}\right) \tag{4.317}$$

Equation 4.317 can be rearranged to give

$$\frac{\mathrm{d}T_E}{\mathrm{d}t} = \frac{\dot{m}_T}{m_t}\left(T_0 - T_E\right) - \frac{QC_{A0}}{m_t C_P}\left(1 - Z\right)\Delta H_{R,A} + \frac{VC_{A0}}{m_t C_P}\frac{\mathrm{d}Z}{\mathrm{d}t}\left(\Delta H_{R,A}\right) \tag{4.318}$$

Substituting the values gives the final transient energy equation

$$\frac{\mathrm{d}T_E}{\mathrm{d}t} = 3.33 \times 10^{-3}\left(298 - T_E\right) + 0.5\left(1 - Z\right) - 150\frac{\mathrm{d}Z}{\mathrm{d}t} \tag{4.319}$$

Equation 4.319 and the appropriate initial condition must be solved simultaneously with the transient mole balance equation, Equation 4.314, to obtain the temperature and concentration as a function of time. At a sufficiently long time, the steady-state solution is reached.

The steady-state outlet temperature that is computed by the numerical solution of the transient mole and energy balance equations must be the same as one of the two stable steady states computed in Example 4.7. The low steady-state temperature is 300.38 K and the high steady-state temperature is 445.5 K. By using a numerical solution tool, and changing the value of the initial reactor temperature, it is found that for temperatures less than or equal to 341.104 K, the steady-state solution corresponds to 300.38 K, while for temperatures greater than this, the steady-state solution corresponds to a reactor temperature of 445.5 K.

In Figure 4.26, the reactor temperature as a function of time is shown for these two cases, that is, for initial temperatures of 341.104 and 341.105 K. In both cases, the reactor temperature initially rises as the reaction proceeds. Once the operation starts there are then two competing forces in the energy balance. One force is the exothermic chemical reaction that acts to increase the reactor temperature, while the second force is the incoming cold feed that acts to reduce the reactor temperature. The critical temperature

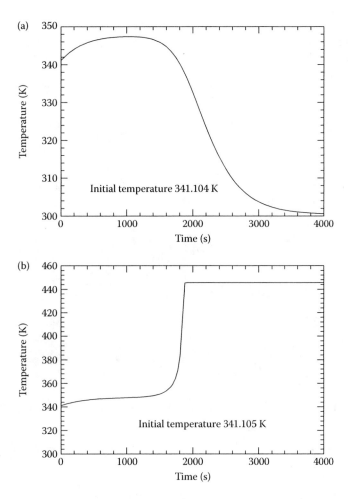

FIGURE 4.26
Reactor temperature as a function of time for inlet temperatures of (a) 341.104 K and (b) 341.105 K. At an initial temperature of 341.105 K, the reactor temperature crosses the critical threshold, and the final reactor temperature lies on the upper operating line.

that must be reached in the reactor corresponds to the central unstable steady-state solution. This latter value is 347.8 K, which was calculated in Example 4.7. Although it may not be obvious from Figure 4.26, when the initial temperature in the reactor is 341.104 K, the maximum temperature that is achieved is 347.4 K. When the initial reactor temperature is 341.105 K, however, the temperature exceeds 347.8 K, and the reaction rapidly proceeds, pushing the steady-state temperature to 455.5 K. Note that in practice it is unlikely that temperature control would be as precise as that used here: the use of six significant figures is simply for illustration purposes.

4.5.7 Semibatch Operation

We now consider a perfectly mixed semibatch reactor. In this reactor, there is an inlet stream but no outlet stream, and the reactor contents at any time are uniform in temperature and concentration. This operational mode might prevail when a CSTR is being filled during startup. Because the volume of the reactor changes with time, expansion work is done by the system on the surroundings and an expansion work term must be included in the energy balance. We start the derivation by writing the general energy balance for a system of n components:

$$\frac{d\left(\sum_{j=1}^{n} N_j U_j\right)}{dt} = q + \left[\sum_{j=1}^{n} F_j H_j\right]_{in} - W_E \tag{4.320}$$

In a procedure analogous to that used for the variable-volume batch reactor, the expansion work is related to the change in the product of the pressure and volume

$$W_E = P \frac{dV}{dt} \tag{4.321}$$

Equations 4.320 and 4.321 can be combined to give:

$$\frac{d\left(\sum_{j=1}^{n} N_j U_j\right)}{dt} = q + \left[\sum_{j=1}^{n} F_j H_j\right]_{in} - P\frac{dV}{dt} \tag{4.322}$$

The expansion work term can be incorporated most readily by using the enthalpy. We recall that

$$H = U + PV \tag{4.323}$$

Taking the derivative and using the chain rule give

$$dH = dU + PdV + VdP \tag{4.324}$$

For a constant-pressure process, the last term is zero. If we limit the filling process to a constant-pressure operation, we can combine Equations 4.322 and 4.324:

$$\frac{d\left(\sum_{j=1}^{n} N_j U_j\right)}{dt} = q + \left[\sum_{j=1}^{n} F_{j0} H_{j0}\right] \tag{4.325}$$

The subscript 0 denotes inlet conditions. The right-hand side of Equation 4.325 is expanded using the chain rule to give

$$\frac{d\left(\sum_{j=1}^{n} N_j H_j\right)}{dt} = \sum_{j=1}^{n} N_j \frac{dH_j}{dt} + \sum_{j=1}^{n} H_j \frac{dN_j}{dt} \tag{4.326}$$

The enthalpy change of a species can be expressed in terms of the temperature change and the constant-pressure heat capacity

$$\frac{dH_j}{dt} = C_{Pj}\frac{dT}{dt} \tag{4.327}$$

Therefore, it is possible to write

$$\sum_{j=1}^{n} N_j \frac{dH_j}{dt} = \left(\sum_{j=1}^{n} N_j C_{Pj}\right)\frac{dT}{dt} \tag{4.328}$$

Consider, for illustration purposes, the reaction given by

$$A + \frac{b}{a}B \rightarrow \frac{c}{a}C + \frac{d}{a}D \tag{4.329}$$

The enthalpy change owing to the change in moles is given by

$$\sum_{i=1}^{n}\left(H_j\frac{dN_j}{dt}\right) = H_A\frac{dN_A}{dt} + H_B\frac{dN_B}{dt} + H_C\frac{dN_C}{dt} + H_D\frac{dN_D}{dt} \tag{4.330}$$

The change in moles with time for each species can be expressed in terms of the inlet flow rate and the rate of reaction:

$$\frac{dN_A}{dt} = F_{A0} - (-r_A)V \quad \frac{dN_B}{dt} = F_{B0} - \frac{b}{a}(-r_A)V$$

$$\frac{dN_C}{dt} = F_{C0} + \frac{c}{a}(-r_A)V : \frac{dN_D}{dt} = F_{D0} + \frac{d}{a}(-r_A)V \tag{4.331}$$

Substitution of these terms into Equation 4.330 and simplification give

$$\sum_{i=1}^{n}\left(H_j\frac{dN_j}{dt}\right) = (H_A F_{A0} + H_B F_{B0} + H_C F_{C0} + H_D F_{D0})$$

$$+ (-r_A)V\left[\frac{d}{a}H_D + \frac{c}{a}H_C - H_A - \frac{b}{a}H_B\right] \tag{4.332}$$

The difference between the enthalpies of the products and the reactants, multiplied by the stoichiometric coefficients, is the enthalpy change of reaction

$$\frac{d}{a}H_D + \frac{c}{a}H_C - H_A - \frac{b}{a}H_B = \Delta H_{R,A} \tag{4.333}$$

The energy balance is therefore written by combining Equations 4.328, 4.332, and 4.333

$$\left(\sum_{j=1}^{n} N_j C_{Pj}\right)\frac{dT}{dt} + \left[\sum_{j=1}^{n} F_{j0}H_j\right] + (-r_A)V\Delta H_{R,A} = q + \left[\sum_{j=1}^{n} F_{j0}H_{j0}\right] \tag{4.334}$$

Rearranging Equation 4.334 gives an explicit expression in temperature change

$$\left(\sum_{j=1}^{n} N_j C_{Pj}\right)\frac{dT}{dt} = q + \left[\sum_{j=1}^{n} F_{j0}\left(H_{j0} - H_j\right)\right] - (-r_A)V\Delta H_{R,A} \tag{4.335}$$

The change in enthalpy of the inlet components as their temperature increases from the inlet temperature to the temperature of the reactor contents can be expressed as an integral in terms of the heat capacity:

$$\left(\sum_{j=1}^{n} N_j C_{Pj}\right)\frac{dT}{dt} = q + \left[\sum_{j=1}^{n} F_{j0}\int_{T}^{T_0} C_{Pj}\,dT\right] - (-r_A)V\Delta H_{R,A} \tag{4.336}$$

The heat capacity integral can be solved provided that the temperature dependence of the heat capacities is known. The energy balance equation must be solved simultaneously with the mole balance equation, given by Equation 4.331. Note that the volume, V, will be a function of time.

The solution of Equation 4.336 requires an initial temperature, T_i, of the reactor contents. In the event that the reactor is empty at time zero, the initial temperature is equal to the inlet feed temperature, T_0. As in previous energy balance equations, the heat capacity may be taken as independent of composition or temperature or both. In the special case where the reactants are present in a dilute solution comprised mainly of solvent, the heat capacity is often assumed to be independent of composition. In terms of mass, the energy balance in such a reactor can be written as

$$m_t\left(C_P\right)_{sol}\frac{dT}{dt} = q + \dot{m}\left(C_P\right)_{sol}\left(T_0 - T\right) - (-r_A)V\Delta H_{R,A} \tag{4.337}$$

where m_t is the total mass of the reactor contents and $\dot{m}$ is the mass flow rate of the feed. The mass of the reactor contents and the volume are related with the density. If the heat capacity is given on a volumetric basis, that is with units of, for example, $J/m^3 K$, the energy balance is

$$V\left(C_P\right)_{sol}\frac{dT}{dt} = q + Q\left(C_P\right)_{sol}\left(T_0 - T\right) - (-r_A)V\Delta H_{R,A} \tag{4.338}$$

The solution of a semibatch reactor problem is illustrated in Example 4.9.

Example 4.9

Consider an elementary second-order reaction conducted in the liquid phase:

$$A + B \rightarrow C \quad \text{where } (-r_A) = kC_A C_B$$

The reaction is carried out in a semibatch adiabatic reactor. At time zero, the reactor volume is 5 m^3 and contains 500 mol of A in a 0.1 molar solution at a temperature of 25°C. Species B is then fed to the reactor at 50°C (323 K) at a molar flow rate of 10 mol/min and a concentration of 0.1 mol/L. The feed is stopped when 500 mol of B has been added to the reactor.

Enthalpies of formation:

$$A: -80 \text{ kJ/mol}, \quad B: -60 \text{ kJ/mol}, \quad C: -160 \text{ kJ/mol}$$

Heat capacities:

$$A: 60 \text{ J/mol} \cdot K, B: 60 \text{ J/mol} \cdot K, C: 120 \text{ J/mol} \cdot K$$

The rate constant, k, has a value of 0.01 L/mol · s at 300 K and the activation energy is 50,000 J/mol. The heat capacity of the liquid solution may be taken to be a constant $(C_P)_{\text{sol}} = 4000$ J/L · K. Determine the number of moles of each species present, and the reactor temperature, as a function of time. Plot the results for the first 2 h of operation.

SOLUTION

As a first step in the solution, identify the initial volume and the volumetric feed flow rate. If the reactor contains 500 mol of A in a solution containing 0.1 mol/L, it follows that the initial volume, V_0, is

$$V_0 = \frac{N_{A,i}}{C_{A,i}} = \frac{500 \text{ mol}}{0.1 \text{ mol/L}} = 5000 \text{ L} \tag{4.339}$$

where the subscript i denotes initial conditions. The volumetric feed flow rate depends on the addition rate of B and its concentration; thus

$$Q = \frac{F_{B0}}{C_{B0}} = \frac{10 \text{ mol/min}}{0.1 \text{ mol/L}} = 100 \text{ L/min} \equiv 1.667 \text{ L/s} \tag{4.340}$$

The feed is stopped after 500 mol of B is added. At an addition rate of 10 mol/min, it takes 50 min to add 500 mol. Therefore, the feed flow is stopped after 50 min, and thereafter the reactor acts as a well-mixed batch reactor. The analysis of the 2 h period is therefore performed in two stages: the first 50 min when the reactor operates in semibatch mode, and the remaining 70 min of pure batch operation.

Step 1: First 50 min of Operation

The transient mole balances for the three species, A, B, and C, can be written in the general form for a semibatch reactor as

$$\frac{dN_A}{dt} = F_{A0} - (-r_A)V \tag{4.341}$$

$$\frac{dN_B}{dt} = F_{B0} - (-r_B)V \tag{4.342}$$

$$\frac{dN_C}{dt} = F_{C0} + (+r_C)V \tag{4.343}$$

The inlet molar flow rates of both A and C are zero, and the reaction rates are related through the stoichiometry as

$$(-r_A) = (-r_B) = (+r_C) = kC_AC_B \tag{4.344}$$

The reaction rates should be expressed in terms of the number of moles present because the volume of the solution changes with time. With these substitutions, the three mole balance equations are

$$\frac{dN_A}{dt} = -k\frac{N_A}{V}\frac{N_B}{V}V \tag{4.345}$$

$$\frac{dN_B}{dt} = F_{B0} - k\frac{N_A}{V}\frac{N_B}{V}V \tag{4.346}$$

$$\frac{dN_C}{dt} = k\frac{N_A}{V}\frac{N_B}{V}V \tag{4.347}$$

The volume changes linearly with the added fluid; thus

$$V = V_0 + Qt = 5000 + 1.667t \tag{4.348}$$

The time has units of seconds and the volume is in liters. Substitution of Equation 4.348 into the mole balances gives

$$\frac{dN_A}{dt} = -k\left(\frac{N_AN_B}{V_0 + Qt}\right) \tag{4.349}$$

$$\frac{dN_B}{dt} = F_{B0} - k\left(\frac{N_AN_B}{V_0 + Qt}\right) \tag{4.350}$$

$$\frac{dN_C}{dt} = k\left(\frac{N_AN_B}{V_0 + Qt}\right) = -\frac{dN_A}{dt} \tag{4.351}$$

The rate constant depends on the temperature. The value of k is known at 300 K and the activation energy is also known. From these numbers, the preexponential factor can be calculated using the Arrhenius law

$$A = k\exp\left(\frac{+E}{R_gT}\right) = 0.01\exp\left(\frac{+50,000}{8.314 \times 300}\right) = 5.08 \times 10^6 \text{ L/mol} \cdot \text{s} \tag{4.352}$$

The numerical values of the constants are substituted into the mole balance equations to obtain

$$\frac{dN_A}{dt} = -5.08 \times 10^6 \exp\left(\frac{-6014}{T}\right)\left(\frac{N_A N_B}{5000 + 1.667t}\right) \tag{4.353}$$

$$\frac{dN_B}{dt} = \frac{10}{60} - 5.08 \times 10^6 \exp\left(\frac{-6014}{T}\right)\left(\frac{N_A N_B}{5000 + 1.667t}\right) \tag{4.354}$$

$$\frac{dN_C}{dt} = -\frac{dN_A}{dt} = 5.08 \times 10^6 \exp\left(\frac{-6014}{T}\right)\left(\frac{N_A N_B}{5000 + 1.667t}\right) \tag{4.355}$$

The three mole balance equations are coupled, and furthermore contain the temperature. The energy balance equation is required to obtain the temperature. The reactor is adiabatic and the heat capacity is a constant for the solution. Furthermore, the heat capacity of the solution is expressed on a volumetric basis. The energy balance is thus Equation 4.338 with the value of q set equal to zero:

$$V(C_P)_{sol}\frac{dT}{dt} = Q(C_P)_{sol}(T_0 - T) - (-r_A)V\Delta H_{R,A} \tag{4.356}$$

Rearranging Equation 4.356 and substituting the rate function give

$$\frac{dT}{dt} = \frac{Q(C_P)_{sol}(T_0 - T) - k(N_A N_B/V_0 + Qt)\Delta H_{R,A}}{V(C_P)_{sol}} \tag{4.357}$$

The enthalpy of reaction is the enthalpy of formation of the products minus the enthalpy of formation of the reactants:

$$\Delta H_{R,A} = \Delta H_{f,C} - \Delta H_{f,A} - \Delta H_{f,B}$$

$$= (-160,000) - (-80,000) - (-60,000) = -20,000 \text{ J/mol} \tag{4.358}$$

Note that $\Delta H_{R,A}$ does not change with temperature because $\Delta C_P = 0$. Substitute the numerical values of the variables and simplify to give the final energy balance

$$\frac{dT}{dt} = \frac{(6668)(323 - T) + 1.016 \times 10^{11}\exp(-6014/T)(N_A N_B/5000 + 1.667t)}{(5000 + 1.667t)(4000)} \tag{4.359}$$

Equations 4.353 through 4.355 and 4.359 comprise a system of four coupled nonlinear differential equations which must be solved numerically. The initial conditions are

$$T_i = 298 \text{ K}; \ N_{A,i} = 500; \ N_{B,i} = 0; \ N_{C,i} = 0$$

The four equations are integrated between time zero and 50 min (3000 s) using the Runga–Kutta method. The solution at 3000 s is

$$T_{3000} = 310.6 \text{ K}; \ N_{A,3000} = 225.44; \ N_{B,3000} = 225.44; \ N_{C,3000} = 274.56$$

Note that at 3000 s, the number of moles of A and B are equal, which follows from the stoichiometry. That is, 500 mol of A was initially present, 500 mol of B was added, the reaction stoichiometry is 1 mol of A reacting for each mole of B, and therefore the number of moles remaining in the reactor must be the same.

Step 2: Remaining 70 min of Operation

After 500 mol of B has been added to the reactor (which originally contained 500 mol of A), the concentrations (and number of moles) of A and B must be the same at all time because no further additions of B are made. The volume of the reactor remains constant at 10,000 L. The transient mole balance for A can thus be written as

$$-\frac{dN_A}{dt} = (-r_A)V = k\frac{N_A}{V}\frac{N_B}{V}V = k\frac{N_A^2}{V} \tag{4.360}$$

Substitute the numbers to give

$$\frac{dN_A}{dt} = -5.08 \times 10^6 \exp\left(\frac{-6014}{T}\right)\left(\frac{N_A^2}{10,000}\right) \tag{4.361}$$

The number of moles of B and C can be obtained from the stoichiometry as

$$N_B = N_A; \quad N_C = 500 - N_A \tag{4.362}$$

The energy balance for this period of operation is the energy balance for an adiabatic batch reactor, which is written as

$$V(C_P)_{sol}\frac{dT}{dt} = -(-r_A)V\Delta H_{R,A} = -k\frac{N_A^2}{V}\Delta H_{R,A} \tag{4.363}$$

The energy balance uses the enthalpy rather than the internal energy because the reaction is carried out in the liquid phase at constant volume. Equation 4.363 can be rearranged, and after substitution of the numbers becomes

$$\frac{dT}{dt} = \frac{1.016 \times 10^{11} \exp(-6014/T)\left(N_A^2/10,000\right)}{(10,000)(4000)} \tag{4.364}$$

Equations 4.361 and 4.364 are a system of two coupled nonlinear ordinary differential equations. The solution of these equations requires the initial conditions. These conditions are simply the solution at the end of the first 50 min of semibatch operation, that is

$$T_{3000} = 310.6 \text{ K}; \quad N_{A,3000} = 225.44; \quad N_{B,3000} = 225.44; \quad N_{C,3000} = 274.56$$

A numerical solution of Equations 4.361 and 4.364 is performed using the Runga–Kutta method, along with the auxiliary Equation 4.362, from a starting time of 3000 s to the final time of 7200 s. The final number of moles of each component and the temperature in the reactor is thereby determined to be

$$T_{7200} = 310.7 \text{ K}; \quad N_{A,7200} = 78.12; \quad N_{B,7200} = 78.12; \quad N_{C,7200} = 421.88$$

A plot of the concentration of the three components and the reactor temperature over the entire 2 h is shown in Figure 4.27. Several observations can be made from this figure. From the plot of the moles of species as a function of time (plot a), it is seen that the number of moles of A falls continuously with time. This observation is logical because all of the A is initially present in the reactor. The number of moles of B in the reactor rises for the first 3000 s, because the rate of addition is greater than the rate of disappearance by reaction. At 3000 s, the moles of A and B in the reactor are the same, and thereafter decline at the same rate because of the reaction stoichiometry. The number of moles of product C rises continuously over the 7200 s time period.

The reactor temperature (plot b) rises during the first 3000 s by just over 12°C, and rises only a fraction of a degree over the remaining time. This behavior results because of the relatively large heat capacity of the reactor contents compared to the heat released on reaction. Most of the temperature rise during the first 3000 s of reactor operation results from the addition of feed which is at a higher temperature than the reactor contents. This type of behavior is not unusual when reactions are conducted in the liquid phase with relatively large quantities of inert solvent.

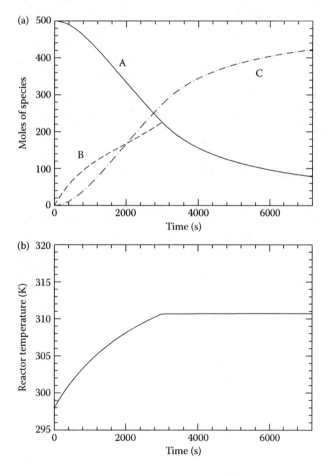

FIGURE 4.27
Time dependence of (a) the moles of the reacting species and (b) the temperature in a semibatch reactor. The feed stream is switched off at 3000 s. Results from Example 4.9.

Semibatch reactors can be operated under a variety of different scenarios. The operating conditions are often dictated by the kinetics of the reaction. For example, when several reactions occur in the vessel the rate of feed addition may be controlled to achieve a preferred product distribution by controlling the relative concentrations of reactants. Alternatively, the rate of heat release may be the deciding factor. Regardless of the complexity, the analysis can be performed using the material described in this section as a starting basis.

4.6 Summary

This chapter has shown how the energy balance equation is used to determine the temperature in an ideal chemical reactor. The energy balance equation is always coupled to the mole balance equation through the temperature dependence of the reaction rate constants and the temperature dependence of the physical properties of the reacting fluid. The energy balance equation is almost invariably a nonlinear equation, and with the coupling to the mole balance requires a numerical solution. When performing an energy balance, it is important to remember that both internal energies and enthalpies of reaction are used. In many cases, the two are effectively interchangeable; however, in other cases, the differences are quite significant.

The introduction of the energy balance into the reactor analysis problem can lead to some interesting complexities, especially in the CSTR. In that reactor, even simple kinetic expressions can result in the existence of multiple solutions for the mole and energy balance. In such a case, the reactor operation can exhibit unstable operation if the inlet conditions fluctuate. In this case, it is necessary to perform a transient analysis to determine the operating point of the reactor.

The combination of the mole and energy balance equations can be used to solve for the concentration and temperatures in any ideal reactor system, regardless of the number of reactions and reactors, or the complexity of the reaction kinetics. If one starts from the basic balance equations, it is possible to derive the appropriate system of equations and then to solve them. The most important factor to remember is that the correct solution depends on making the appropriate choices when deriving the equations.

PROBLEMS

4.1 An elementary second-order liquid-phase reaction occurs in a CSTR. The reactor has a volume of 1200 L and is heated by a steam jacket.

$$A + B \xrightarrow{k} C \quad (-r_A) = kC_A C_B$$

The feed temperature is 27°C, and the flow rate is 30 L/min. The inlet concentrations of A and B are 2 mol/L. The desired conversion of A is 60%.

a. Determine the reactor temperature required to achieve the desired conversion.

b. Determine the temperature of the steam in the jacket required to operate the reactor at the temperature determined in part (a).

Activation energy	10,000 J/mol
Rate constant at 27°C	0.01725 L/mol·min
ΔH_R (assume constant)	41,840 J/mol A
Heat capacity of mixture	4184 J/L·K
Reactor heat transfer area	6 m^2
Heat transfer coefficient	70 W/m^{2}K

4.2 A tubular flow reactor is to be designed to produce butadiene from butene by dehydrogenation (gas-phase reaction). The first-order reaction rate expression is written in terms of the partial pressure of butene as

$$C_4H_8 \rightarrow C_4H_6 + H_2 \quad (-r_{C_4H_8}) = kP_{C_4H_8} \text{ mol/L} \cdot h$$

The constant k has units of $\text{mol} \cdot (h \cdot L \cdot atm)^{-1}$. The reaction is endothermic, the reactor is often operated adiabatically; therefore, steam is often added to the feed to provide thermal energy for the reaction. For this reactor, the feed is a mixture of 10 mol of steam for each mole of butene. The reactor pressure is 2 atm and the feed temperature is 650°C. Values of the rate constant at different temperatures are given in the following table:

T(K)	922	900	877	855	832
k	11.0	4.90	2.04	0.85	0.32

Assume that the heat of reaction is a constant 1.1×10^5 J/mol butene. The heat capacity of the feed stream may be considered constant at 2.1 kJ/kg · K.

a. Calculate the reactor volume for a 20% conversion of butene if the reactor is operated isothermally at 650°C with a total inlet molar flow rate of 11,000 mol/h.

b. Determine the reactor volume for 20% conversion of butene for a total inlet molar feed rate of 11,000 mol/h in an adiabatic reactor.

4.3 A chemical plant is planning to install a reactor to produce ethanoic acid by the hydrolysis of methyl acetate. The reaction is

$$CH_3COOCH_3 + H_2O \rightarrow CH_3CO_2H + CH_3OH$$

The reaction is pseudo-first-order overall because of the excess of water

$$-r_{CH_3COOCH_3} = k[CH_3COOCH_3]$$

At 25°C, the rate constant has a value $k = 3.51 \times 10^{-3}$ min^{-1}, and at 35°C, $k = 8.84 \times 10^{-3}$ min^{-1}. A 1 mol/L aqueous solution of methyl acetate resides in a storage tank at 0°C and will be pumped to the reactor at a rate of 30,000 L/h. The production rate of ethanoic acid is to be 8.64 tonnes/day, based on 24 h a day operation. The reactor is to be an adiabatic CSTR and the feed is to be preheated in an electrical heater. The operating cost of the reactor (including capital cost allowance, stirring cost, pumps, etc.) is $1 a minute for each liter of capacity (e.g., a 4 L reactor costs $4 a minute, a 6 L reactor $6 a minute, etc.). Electricity costs $0.10 a kilowatt hour and the preheater is 90% efficient. Determine the reactor volume which will minimize the production cost of ethanoic acid. Assume that the specific heat of the mixture is a constant equal to 4180 J/kg · K. The heat of formation data, $\Delta H^\circ_{f,298}$, are given below:

Methanol	−238.7 kJ/mol	Ethanoic acid	−484.4 kJ/mol
Water	−286.0 kJ/mol	Methyl acetate	−447.8 kJ/mol

4.4 The hydrogenation of aromatic compounds in an oil is carried out in an internal recycle reactor at a constant hydrogen pressure. The recycle rate is sufficiently high so that the reactor behaves like a CSTR. The reactor contains 20 g of catalysts and is operated adiabatically. The rate function for the hydrogenation of the aromatic compounds is

$$r_{HYD} = -r_{aro} = A \exp\left(\frac{-E}{R_g T}\right)[C_{aro}] \text{ moles of aromatics/h} \cdot \text{gcat}$$

The oil has the following properties:

$x_{f,aro}$	Mass fraction of feed that is aromatic	0.30
M_{wt}	Average molecular mass of feed	210
C_P	Average heat capacity of oil	1.75 J/(g · K)
$\Delta H_{R,aro}$	Heat of reaction of aromatics	-2.0×10^5 J/mol

The values of the kinetic parameters are:

$$A = 1.3 \times 10^{11} \text{ cm}^3/\text{h} \cdot \text{gcat} \quad \frac{E}{R_g} = 16{,}300 \text{ K}$$

Oil is fed to the reactor at a constant rate of 0.5 cm³/min. The temperature of the feed stream is increased very slowly from 250°C to 375°C and then decreased very slowly from 375°C to 250°C. Assume that the density of the oil is constant at 0.85 g/cm³ and that the hydrogenation of the aromatics is the only reaction. The pseudo-steady-state approximation can be made for the slow changes in feed temperature, that is, it can be assumed that steady state is achieved at all feed temperatures because the changes in feed temperature occur very slowly. Prepare a plot of fractional conversion of the aromatic content as a function of feed temperature for the above changes in feed temperature.

4.5 The following reaction is carried out in an adiabatic plug flow reactor:

$$A + B \rightleftharpoons C + D$$

Calculate the maximum conversion that may be achieved if the feed enters at 27°C. Only A and B are in the feed. The following data are available:

$$\Delta H_R = -120,000 \text{ J/mol}$$

$$C_{P,A} = C_{P,B} = C_{P,C} = C_{P,D} = 100 \text{ J/mol K}$$

$$K_e(\text{equilibrium constant}) = 500,000 \text{ at } 50°C$$

$$F_{A0} = F_{B0} = 10 \text{ mol/min}$$

4.6 A first-order endothermic reaction is carried out in a constant-volume batch reactor.

$$A \rightarrow B \quad \text{the rate constant is: } k = 0.20 \exp\left(\frac{-10,000}{R_g T}\right) \text{min}^{-1}$$

The enthalpy of reaction is 50,000 J/mol. The reactor volume is 1 m³ and contains 1000 kg of mixture. There are initially 10,000 mol of reactant A in the reactor. The reactor is heated to 400°C, during which time 10% of the initial A reacts. When the temperature reaches 400°C, the reactor operates adiabatically. The heat capacity of the mixture is a constant 2000 J/kg · K.

a. After the heating stops, how much time is required to achieve a final conversion of 70% of the A originally present before the heating started?

b. What is the final temperature in the reactor?

4.7 A first-order liquid-phase exothermic reaction

$$A \rightarrow B$$

is carried out in a batch reactor. There is no heat transfer through the reactor wall. The reaction mixture is to be held at a constant 40°C by the addition of an inert coolant at a temperature of 25°C. The flow rate of coolant varies with time so as to maintain the reactor temperature at 40°C. The following data are available:

Heat capacity of all components	$C_P = 2000$ J/kg K (assume constant)
Density of all components	1000 kg/m³
Heat of reaction	$\Delta H_R^\circ = -200,000$ J/mol
Value of rate constant at 40°C	1.0×10^{-4} s^{-1}

At time $t = 0$, only compound A is present in the tank at a concentration of 8000 mol/m³. The initial volume is 1.5 m³.

a. Calculate the conversion of A 2 h after the start of the reaction.

b. Calculate the rate of coolant addition 2 h after the start of the reaction.

c. Calculate the total amount of coolant added after 2 h.

4.8 Consider an adiabatic constant pressure batch reactor in which a second-order gas-phase reaction occurs

$$A + B \rightarrow C \quad (-r_A) = kC_A C_B$$

The rate constant is independent of temperature and equals $0.01 \text{m}^3/\text{mol} \cdot \text{s}$. The initial reactor temperature is 450 K and the pressure is 100 kPa. The initial volume of the reactor is 1.0 m^3. The reactor initially contains 30% by volume A, 30% B, and 40% inerts. The heat capacity of the mixture is a constant $C_V = 1000$ J/kg K. The average molecular mass of the initial material is 30 g/mol. The internal energy change on reaction is a constant $\Delta U_R = -60$ kJ/mol. Calculate

a. The time required to reach 80% conversion

b. The reactor volume at 80% conversion

c. The reactor temperature at 80% conversion

4.9 Consider a constant-volume batch reactor in which a liquid-phase reaction occurs. The reaction is a catalytic reaction (solid catalyst is used), and is represented by the following overall reaction, with second-order kinetics:

$$A + B \rightarrow C + D \quad (-r_A) = kC_A C_B \text{ mol of A/s} \cdot \text{kgcat}$$

The kgcat term means kg of catalyst and concentrations are in mol/m^3. The reaction rate constant is given by the following expression:

$$k = 1 \times 10^{-5} \exp\left(\frac{-2500}{T}\right) \text{m}^6/\text{mol} \cdot \text{s} \cdot \text{kgcat}$$

One cubic meter (1000 kg) of a fluid solution containing 2000 mol of A and 2000 mol of B are placed in the reactor. At time zero, 10 kg of catalyst is added to the reactor and the reaction starts.

a. If the reactor is operated isothermally at a temperature of 100°C, calculate how much time is required to reach 80% conversion.

b. If the reactor is operated adiabatically, and the initial temperature of the mixture is 27°C, calculate the temperature of the reaction mixture when the conversion is 80%.

$$\Delta H_{R,A} = -50,000 \text{ J/mol}$$

Heat capacity of fluid = 4000 J/kg · K

Heat capacity of catalyst = 10,000 J/kg · K

4.10 Consider an adiabatic plug flow reactor in which a second-order ideal gas-phase reaction occurs

$$A + B \rightarrow C$$

The reactor feed is composed of 40 mol% A, 50 mol% B, and 10 mol% inert material. The total inlet molar flow rate is 10 mol/s. The reactor inlet temperature is 300 K and the pressure is 101.325 kPa. The reaction rate is

$$(-r_A) = k C_A C_B \text{ mol/s} \cdot \text{m}^3 \quad \text{where } k = 1 \times 10^{13} \exp\left(\frac{-12,500}{T}\right) \text{m}^3/\text{mol} \cdot \text{s}$$

The concentrations are in mol/m³. The enthalpy of reaction at 298 K is equal to −50,000 J/mol. The heat capacities are

A: 10 J/mol·K, B: 5 J/mol·K, C: 15 J/mol·K, and inert: 15 J/mol·K

The desired fractional conversion of A is 80%. Calculate the reactor volume required.

4.11 Consider the reaction

$$A + B \rightarrow C + D$$

The reaction is zero order. That is, the rate equation is

$$(-r_A) = A \exp\left(\frac{-E}{R_g T}\right) \quad \text{where } A = 5 \times 10^9 \text{ mol/m}^3\text{s} \quad \text{and} \quad \frac{E}{R_g} = 7000 \text{ K}$$

The reaction is carried out in a CSTR of volume 0.02 m³ that is operated adiabatically. The inlet molar flow rates of A and B are 0.20 and 0.25 mol/s, respectively. The total mass flow rate is 0.40 kg/s. The constant enthalpy of reaction is −600 kJ/mol (of A) and the heat capacity of the feed stream is 4200 J/mol·K (assume constant).

a. Calculate the outlet conversion if the reactor is operated at 340 K.

b. Derive a general steady-state mole balance for A. The result should be an equation which gives the outlet conversion of A as a function of outlet temperature.

c. For adiabatic operation with an inlet feed temperature of 300 K, develop the steady-state energy balance. The result should be an equation that gives outlet conversion as a function of outlet temperature.

d. Plot the mole and energy balances and identify the steady-state solution(s).

e. Comment on the stability of the steady state(s) obtained in part (d).

References

Froment, G.B, K.B. Bischoff, and J. De Wilde, 2011, *Chemical Reactor Analysis and Design*, 3rd Ed., Wiley, New York.

Lee, H.H., 1985, *Heterogeneous Reactor Design*, Butterworth, London.

Thomas, J.M. and W.J. Thomas, 1967, *Introduction to the Principles of Heterogeneous Catalysis*, Academic Press, London.

Westerterp, K.R., W.P.M. van Swaaij, and A.A.C.M. Beenackers, 1984, *Chemical Reactor Design and Operation*, John Wiley & Sons, New York.

5

Chemical Kinetics for Homogeneous Reactions

No study of chemical reaction engineering can be complete without an examination of chemical kinetics. Kinetic analysis deals with the molecular processes involved in a chemical reaction that lead to a formulation of the rate equation. In Chapter 3, the rate of reaction was seen to be an integral part of the mole balance equation for a chemical reactor. In that chapter, the rate equation was not considered in much detail and this chapter supplies some of the necessary details. This chapter starts with a discussion of the general nature of rate equations, and continues with a brief look at reaction rate theories. Rate equations for both simple reactions and those that proceed via complex multistep mechanisms are considered. Some of the common methods for deriving rate expressions from postulated mechanisms are then given.

Reaction rate theories can provide some insight into the general form of a rate equation, but cannot be used to obtain quantitative values for the constants in them. These constants must be determined experimentally, and the latter part of this chapter is thus devoted to an introduction to experimental techniques, including data analysis and modeling.

5.1 General Nature of Rate Functions

In Chapter 1, the concept of the rate equation or rate function was introduced. The rate equation endeavors to quantify the rate of a chemical reaction. The rate of a chemical reaction depends on the temperature and the concentrations of the reactants, products, and, sometimes, on the presence of other materials. The reaction rate function (or rate equation) is the mathematical relationship between the rate of a chemical reaction, and the temperature and concentration. It is expressed either as a rate of disappearance of a reactant or as the rate of formation of a product. For example, as seen in Chapter 3, the rate of disappearance of a reactant A can be written in the generic form

$$(-r_A) = f(T, C_1, C_2, C_3, \ldots, C_j) \tag{5.1}$$

where $(-r_A)$ is the rate of disappearance of reactant A. C_j represents the concentration of any species, j, present in the reacting mixture that influences the rate, and T is the temperature. The concentrations and temperatures are local values at a given reactor position. As seen in Chapter 3, in a plug flow reactor, the concentration varies with axial position and thus so does the value of the reaction rate. In perfectly mixed reactors, either batch or flow, the reaction rate is uniform over the reactor volume. There are a number of general observations that can be made about the nature of rate equations.

It is often the case that the rate of reaction decreases as the concentration of reactant A decreases (i.e., as the conversion of A increases), although this is not always true. Reactions that exhibit a decrease in reaction rate with a decrease in reactant concentration over the entire range of reactant concentration are often referred to as having "normal" kinetics. Recall that the concept of "normal kinetics" was introduced in Chapter 3 when comparing reactor sizes for PFR and CSTR. However, we emphasize that this type of kinetic behavior is not always observed.

Often, the concentration and temperature dependence can be separated into two separate and independent functions. Rate equations of this type can be represented in the form

$$(-r_A) = f_1(T) \times f_2(C_1, C_2, \ldots, C_j) \tag{5.2}$$

The temperature dependence of the reaction rate may vary, but it usually contains an exponential term. The various types of temperature dependence are discussed in more detail shortly, but it is often expressed as the Arrhenius law:

$$f_1(T) = A \exp\left(\frac{-E}{R_g T}\right) \tag{5.3}$$

The origin of this temperature dependence is discussed below. In Equation 5.3, the constant A is known as the preexponential factor, E is the activation energy, R_g is the universal gas constant, and T is the absolute temperature (i.e., the temperature in kelvin). The units of E/R_g are kelvin and $E/R_g T$ is dimensionless.

The concentration dependence in rate equations exhibits many forms. One of the most commonly encountered forms of concentration dependence is the power law-type expression. For example, if the rate of reaction depended on the concentrations of three species (denoted with subscripts 1, 2, and 3), the power law form of the concentration dependence might be

$$f_2(C) = C_1^{\alpha_1} C_2^{\alpha_2} C_3^{\alpha_3} \tag{5.4}$$

In Equation 5.4, α_j is called the order of reaction with respect to component j. The overall order of the reaction is given by the sum of all of the α_j values. Note that the value of α_j may or may not be an integer, and that there is in general no relationship between the α_j's and the stoichiometry of the reaction. Rate equations that have a concentration dependence of the power law type offer the advantage of computational simplicity in many calculations. Such expressions may be based on theoretical foundations or simply represent empirical equations that correlated experimental data. In fact, a large number of chemical reactions, even those that involve complex multistep processes, can be described by power law models, especially over fairly narrow ranges of operating conditions.

When the reverse reaction is significant, the reaction rate equation may be written as a difference between forward and reverse reactions. Consider, for example, the reversible reaction

$$A + B \rightleftharpoons C + D \tag{5.5}$$

If the rate equation exhibits a separable concentration and temperature dependence, the rate equation would typically have the form

$$(-r_A) = f_1(T) f_2(C_A, C_B) - f_3(T) f_4(C_C, C_D) \tag{5.6}$$

There are many rate expressions in which the dependence of the rate on temperature and concentration cannot be separated into two independent functions. For example, consider the simple irreversible reaction

$$A \rightarrow B \qquad (5.7)$$

An example of a rate expression with nonseparable temperature and concentration dependence for this reaction is given below.

$$(-r_A) = \frac{k_1 C_A}{(1 + k_2 C_A)} \qquad (5.8)$$

Both rate parameters, k_1 and k_2, are functions of temperature, and thus the rate cannot be expressed neatly as the product of two functions, one each for temperature and concentration dependence. As will be seen later, kinetic expressions that cannot be neatly divided into separate functions of temperature and concentration occur often in catalytic reactions.

The variety of rate expressions that are encountered in practice is large: brevity necessitates that only a few of these will be discussed in this book.

5.2 Reaction Mechanism

One of the most important factors in understanding reaction rates is the reaction mechanism. Broadly speaking, the reaction mechanism is a description of the molecular-scale process by which reactants are turned into products. The reaction mechanism consists of a sequence of elementary steps, each of which involves a collision between moieties, which might be atoms, molecules, or free radicals. Each reactive collision produces a product or a reaction intermediate. A reaction intermediate is an identifiable species that is not considered a final product. Intermediates are often short-lived species such as free radicals. A comprehensive understanding of a reaction mechanism can be used to develop theoretically based rate expressions, which in turn can be used to design reactors to work more efficiently and effectively.

Reaction mechanisms vary greatly in complexity. Simple reactions may proceed via a single reactive collision, while in complex systems, multiple-step mechanisms may involve literally hundreds of reactive intermediates.

5.2.1 Rate Expressions for Elementary Reactions

If the reaction mechanism consists of a single collision (in which reactants collide and form products in a single step), the reaction is said to be elementary. The rate equations for elementary reactions are straightforward to write. The rate is proportional to the concentration of each reactant raised to the power of its stoichiometric coefficient. For example, consider the irreversible reaction

$$A + B \rightarrow C + D \qquad (5.9)$$

If this reaction were elementary, then the rate equation for the forward reaction would be deduced directly from the stoichiometry and written as

$$(-r_A) = kC_A C_B \tag{5.10}$$

Note that only one form of the overall reaction can represent an elementary reaction. Thus, the same reaction written as

$$2A + 2B \rightarrow 2C + 2D \tag{5.11}$$

would not be elementary. There are many examples of elementary reactions; however, it is rarely obvious whether or not a reaction is elementary from the stoichiometry. There are a number of pointers, however, to when a reaction is not elementary. These are summarized in the following.

If the stoichiometric coefficients in the reaction are not integers, the reaction cannot be elementary. For example, a reaction might be represented by the following overall stoichiometry:

$$A \rightarrow 1.62R + 0.75S \tag{5.12}$$

This reaction is not elementary as written because the stoichiometric coefficients in the reaction are not integer values.

Not only must all of the stoichiometric coefficients be integers, but the sum of the stoichiometric coefficients for the reactants should be three or less. That is, only two or three molecules should be involved in the collision that produces the reaction. Although elementary reactions could theoretically involve more than three molecules, the number of such collisions that would occur in practice would give very low reaction rates. Thus, if a reaction proceeds at an observable rate, the probability that a four body or higher collision is required is remote. For example, we can write the reaction between nitrogen and hydrogen to produce ammonia as

$$N_2 + 3H_2 \rightarrow 2NH_3 \tag{5.13}$$

All of the coefficients in this reaction are integers. However, the sum of the stoichiometric coefficients for the reactants is four, which implies that an elementary reaction would require the simultaneous collision of one nitrogen molecule with three hydrogen molecules. Statistically speaking, the number of four body collisions is relatively low, which implies that the reaction rate from such a mechanism would be very low. Furthermore, both forward and reverse reactions must be elementary for the reaction to be considered an elementary reaction. For example, the decomposition of ammonia into hydrogen and nitrogen could be written as

$$2NH_3 \rightarrow N_2 + 3H_2 \tag{5.14}$$

The sum of the coefficients for this forward reaction is two, which is within the bounds of acceptability; however, the reverse reaction, involving a four body collision is unlikely. Therefore the forward reaction shown here is also unlikely to be elementary. In other words, both the forward and reverse reactions must be elementary or neither will be. This concept is known as the principle of microscopic reversibility.

Experimental data may also be used to test a hypothesis that a reaction is elementary. For example, experimental data may not be well correlated by the rate equation based on an elementary reaction. Consider the reaction

$$A + B \rightarrow C + D \tag{5.15}$$

If the reaction were elementary, the rate equation for the forward reaction would be

$$(-r_A) = kC_A C_B \tag{5.16}$$

If this type of rate expression does not correlate experimental data well, then the reaction is also not elementary. It should be noted, however, that even if the experimental data do obey the postulated rate equation, this does not necessarily imply that the reaction is elementary.

5.2.2 Rate Expressions for Nonelementary Reactions

For nonelementary reactions, the reaction mechanism is composed of a sequence of nonelementary steps. If the reaction mechanism can be deduced, it may be possible to derive a rate expression from it. Some of the methods for doing this are explained later in this chapter. However, it is not always possible to identify the mechanism of a reaction and, even if identified, it may not be possible to derive a rate equation from it. As a result, many rate expressions used in reactor design are purely empirical equations. These equations are generally referred to as models. It is important to appreciate that reaction models without any theoretical underpinning are widely used, and can be sufficient for design purposes. Indeed, few of the very successful processes in use today would exist if the developers had waited until all of the theoretical underpinnings were known.

5.3 Theoretical Analysis of Reaction Rate

Considerable effort has been made to develop fundamental theories for chemical reactions. Although much progress has been made in the understanding of reaction rates, reaction rate theory cannot as yet be used to predict the value of rate constants. However, it is useful to discuss briefly two reaction rate theories, as there are some useful concepts to be learned from them, such as the origin of the temperature dependence of the reaction rate. Two elementary reaction rate theories are *collision theory* and *absolute reaction rate theory*. Note that these rate theories deal with elementary steps.

5.3.1 Collision Theory

The collision theory was developed in an attempt to produce a working theory of reaction rates and thereby to calculate the value of a rate constant. Collision theory is based on a simplified version of the kinetic theory of gases, in which molecules are modeled as hard spheres. The hard sphere assumption is only approximately valid for very simple reactions, such as those between simple atoms. The underlying idea is, however, correct in that reactions occur as a result of collisions between molecules. For dilute gases, the kinetic theory of gases predicts that the frequency of collisions between molecules of A and molecules of B, Z_{AB}, is

$$Z_{AB} = \left[8\pi R_g T \frac{(M_A + M_B)}{M_A M_B} \right]^{0.5} \sigma_{AB}^2 C_A C_B N_\circ^2 \tag{5.17}$$

The units of Z_{AB} are collisions/volume-time. M_A and M_B are the molecular masses of A and B, respectively, and σ_{AB} is the collision diameter, defined as $\sigma_A + \sigma_B/2$. The diameter of the molecule, σ, is given in many references. $N_\circ$ is Avogadro's number, and C_A and C_B are the concentrations of A and B in units of mol/vol.

At normal temperature and pressure, the number of collisions predicted by Equation 5.17 is of the order of 10^{28} (collisions/cm$^3 \cdot$s), which is a very large number. This value is much higher than the observed reaction rate, and to account for the difference it is necessary to introduce the idea of an "effective" collision. An effective collision is one that results in a reaction, and, because the rate of most chemical reactions is substantially lower than the number of collisions between reactants, it necessarily follows that not all collisions result in a reaction. There are two reasons why not all collisions are effective. The first reason is that the orientation of the molecules when they collide may not be correct for reaction to occur. To account for this fact a steric factor, s, is introduced, which is defined as the fraction of collisions that occur in which the molecules have the correct orientation for reaction. Depending on the shape of the molecules involved in the reaction, this factor can be a fairly low number. The second reason that a collision may not result in a reaction is that a minimum energy, E, is required for the collision to result in reaction. Typically, the energy distribution among the molecules in a system follows a Maxwell distribution, and from this distribution, the fraction of molecules having or exceeding the minimum energy required for reaction is given by an exponential expression

$$\exp\left(\frac{-E}{R_g T} \right)$$

Adding the steric factor and the energy distribution function to Equation 5.17 gives the number of collisions resulting in reaction as

$$(-r_A) = s Z_{AB} \exp\left(\frac{-E}{R_g T} \right) C_A C_B = k C_A C_B \tag{5.18}$$

From Equation 5.17, it is seen that Z_{AB} is proportional to the square root of T; therefore, the temperature dependence of the rate constant, k, is predicted by collision theory to be of the form

$$k = k_0 T^{0.5} \exp\left(\frac{-E}{R_g T} \right) \tag{5.19}$$

Thus, the rate constant is predicted to depend on the temperature according to a square root and exponential dependence. This type of dependence is now compared to that deduced from the second theory, absolute reaction rate theory.

5.3.2 Absolute Reaction Rate Theory

Absolute reaction rate theory has its origins in statistical thermodynamics. The fundamental basis is beyond the scope of this book; the general idea and the result will be presented briefly. The theory states that for a chemical reaction to occur the reactants must first form an activated complex. This complex is regarded as being located at the top of an energy barrier lying between the initial and final states, and the rate of reaction depends on the rate at which the activated complex can travel over this barrier.

Consider some reversible reaction represented by the following overall reaction:

$$A + B \underset{k_2}{\overset{k_1}{\rightleftharpoons}} C + D \tag{5.20}$$

Assume that the route from reactants to products includes the formation of an activated complex as an intermediate, which is denoted AB^*. If the intermediate is included in the reaction scheme, and furthermore it is assumed that all steps are reversible, the reaction can be represented as

$$A + B \rightleftharpoons AB^* \rightleftharpoons C + D \tag{5.21}$$

The activated complex can be viewed as lying at the top of an energy barrier between reactants and products, as illustrated in Figure 5.1. Essentially, molecules of A and B must be involved in a sufficiently energetic collision to form the activated complex. Once formed, the activated complex may either decay into products or revert to the original reactants. The height of the energy barrier for the forward reaction (E_1) is the activation energy for the forward reaction. Similarly, the height of the barrier for the reverse reaction (E_2) is the activation energy for the reverse reaction. The difference between the forward and reverse activation

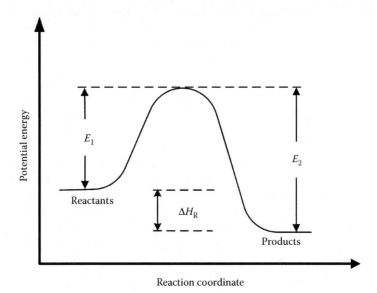

FIGURE 5.1
Potential energy plot of a chemical reaction. The reactants must pass over the energy barrier to become products. The activated complex sits at the top of the energy barrier.

energies is the enthalpy change on reaction, ΔH_R, as shown in Figure 5.1. The detailed derivation of the rate equation from the absolute reaction rate theory is beyond the scope of this text. After some theoretical analysis, the following expression for the reaction rate is obtained:

$$(-r_A) = k_0 T \exp\left(\frac{-E}{R_g T}\right) C_A C_B \tag{5.22}$$

It can be seen that the basic form of the rate equation is different from that produced by collision theory, in that the temperature dependence of the rate constant is different (see Equation 5.19).

5.3.3 Temperature Dependence of Rate Constant

Collision theory and absolute reaction rate theory as discussed in Sections 5.3.1 and 5.3.2 predict different temperature dependencies of the rate constant. These dependencies can be summarized in the following two equations:

$$k = k_0 T^{0.5} \exp\left(\frac{-E}{R_g T}\right) \quad \text{(collision theory)} \tag{5.23}$$

$$k = k_0 T \exp\left(\frac{-E}{R_g T}\right) \quad \text{(absolute reaction rate theory)} \tag{5.24}$$

Other kinetic theories, or different variations of these two theories, predict concentration dependence similar to collision theory and absolute reaction rate theory, but with different temperature dependence of k. A common feature of all of the theories is an exponential dependence of the rate constant on temperature. An additional power law term differs in form from theory to theory. A general equation for the temperature dependence of the rate constant, which incorporates the temperature dependence predicted by the different rate theories, is

$$k = k_0 T^N \exp\left(\frac{-E}{R_g T}\right) \tag{5.25}$$

In Equation 5.25, the power N can have values in the range from -2 to $+2$, depending on the assumptions that are made in the rate theory upon which the derivation of k is based. However, if the ln of the rate constant is plotted against the inverse temperature, a straight line plot frequently results, as illustrated in Figure 5.2. This linear behavior often results because the term T^N will tend to change much more slowly than the exponential term, especially if a small temperature range is considered. As a result, it is very common to write the temperature dependence of the rate constant using only the exponential portion, that is, using the Arrhenius law:

$$k = A \exp\left(\frac{-E}{R_g T}\right) \tag{5.26}$$

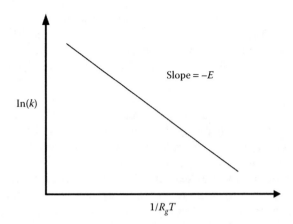

FIGURE 5.2
Typical temperature dependence of the reaction rate constant observed in practice. This plot is often called an Arrhenius plot.

Note that constants in more complex rate equations may also follow the Arrhenius temperature dependence, for example, the rate equation might have the form

$$(-r_A) = \frac{k_1 C_A}{1 + k_2 C_A} \tag{5.27}$$

The rate constants in Equation 5.27 might be given by

$$k_1 = A_1 \exp\left(\frac{-E_1}{R_g T}\right) \quad \text{and} \quad k_2 = A_2 \exp\left(\frac{-E_2}{R_g T}\right) \tag{5.28}$$

One important observation that should be made at this point is that the exponential dependence of the rate constants on temperature can lead to very large changes in the value of the rate constant for comparatively small changes in temperature.

5.4 Rate Equations for Nonelementary Reactions

Many reactions have been found to proceed via a sequence of steps, known as a reaction mechanism, rather than as the result of a simple molecular rearrangement that occurs with a single collision. Many types of rate expressions have been experimentally observed for such reactions, including first order, fractional order, and complex expressions. In some cases, it is possible to postulate a reaction mechanism and derive the form of the rate expression from this mechanism. Some of the techniques used to accomplish this derivation are explained in this section. Note that in the next few sections, to simplify the nomenclature, the concentration of a species will be denoted using the square bracket nomenclature, for example, the concentration of A is denoted [A].

5.4.1 Unimolecular Reactions

We begin the discussion of nonelementary reactions by considering the case of a unimolecular reaction. A unimolecular reaction is one that involves only a single molecule that can either undergo an isomerization reaction or decompose into two or more parts. In a unimolecular reaction, the activated complex must consist of only a single molecule of reactant that has attained sufficient energy to climb the energy barrier to reaction. The reactant molecules must obtain this energy from collisions with other molecules. It would appear that this requirement would imply a second-order reaction; however, the classic work of Lindeman (described below) demonstrated how first-order rate expressions could be achieved for unimolecular reactions.

5.4.1.1 Lindeman Mechanism

The Lindeman mechanism was developed to explain how a first-order rate expression could be achieved for what was in effect a second-order reaction. We consider here only the basics of the method, although the original theory has been modified to account for some inconsistencies in the original derivation. Consider the unimolecular reaction given by the overall expression

$$A \rightarrow B \tag{5.29}$$

In some cases, this type of reaction has been observed to follow a first-order rate expression, that is

$$(-r_A) = k[A] \tag{5.30}$$

Lindeman proposed that the reaction followed the two-step mechanism involving the collision of two molecules of A to form a high-energy molecule of A, which could then react to form a molecule of B. This proposed scheme can be represented by the following two-step mechanism:

$$A + A \underset{k_{-1}}{\overset{k_1}{\rightleftharpoons}} A^* + A \tag{5.31}$$

$$A^* \xrightarrow{k_2} B \tag{5.32}$$

In this scheme, the first step is a collision between two molecules of A, represented by Equation 5.31. In this collision, there is some energy transfer resulting in the formation of A*, which is an "activated" A molecule. This energized A can then react to form the product B, as described in Equation 5.32. If each of these reaction steps is treated as an elementary reaction, the following two reaction rates can be written. The rate of disappearance of nonactivated A is

$$(-r_A) = k_1[A]^2 - k_{-1}[A][A^*] \tag{5.33}$$

The net rate of formation of "energized" A molecules is

$$\left(+r_{A^*}\right) = k_1[A]^2 - k_{-1}[A][A^*] - k_2[A^*] \tag{5.34}$$

This mechanism must be analyzed to produce a rate equation consistent with observed behavior. It is desirable to eliminate the concentration of energized molecules, [A*], because it is not usually an easily measurable quantity. This elimination can be done by using the steady-state hypothesis.

5.4.1.2 Steady-State Hypothesis for Lindeman Mechanism

The steady-state hypothesis is a powerful tool for developing rate equations from mechanisms that include reactive intermediates. The hypothesis assumes that the net rate of formation of an intermediate is negligible and may be approximated by zero, often a reasonable assumption for highly reactive intermediates such as free radicals. Apply this hypothesis to the Lindeman mechanism by assuming that the net rate of formation of A* is zero. From Equation 5.34

$$\left(+r_{A^*}\right) = 0 = k_1[A]^2 - \left\{k_{-1}[A] + k_2\right\}[A^*] \tag{5.35}$$

Equation 5.35 can be rearranged to give an explicit equation in [A*]

$$[A^*] = \frac{k_1[A]^2}{k_2 + k_{-1}[A]} \tag{5.36}$$

Back substitute Equation 5.36 into Equation 5.33 to get

$$\left(-r_A\right) = k_1[A]^2 - k_{-1}[A]\left\{\frac{k_1[A]^2}{k_2 + k_{-1}[A]}\right\} \tag{5.37}$$

Equation 5.37 is expanded by expressing both terms on the right-hand side in terms of the lowest common denominator and combining

$$\left(-r_A\right) = \frac{k_1k_2[A]^2 + k_1k_{-1}[A]^3 - k_1k_{-1}[A]^3}{k_2 + k_{-1}[A]} \tag{5.38}$$

Simplification of Equation 5.38 gives the final rate equation

$$\left(-r_A\right) = \frac{k_1k_2[A]^2}{k_2 + k_{-1}[A]} \tag{5.39}$$

When the concentration of A is relatively large, the term k_{-1} [A] will be much greater than k_2, and will dominate the denominator. In this case, the rate will simplify to the expression

$$\left(-r_A\right) \approx \left(\frac{k_1k_2}{k_{-1}}\right)[A] = k[A] \tag{5.40}$$

Equation 5.40 is apparently first order. At low concentrations of A, the value of k_2 will dominate in the denominator, and thus the rate reduces to an apparently second-order equation

$$\left(-r_A\right) \approx k_1[A]^2$$

This type of behavior, where the reaction rate is first order at high reactant concentration, but switches to second order at low reactant concentration, has been observed for many reactions.

5.4.2 Rice–Herzfeld Mechanisms

Reactions involving organic substances are of major importance in the chemical industry and consequently have been the subject of much study. Many organic gas-phase reactions follow power law type rate equations of the following form:

$$\left(-r_A\right) = kC_A^n \tag{5.41}$$

In Equation 5.41, n can have simple values such as 0.5, 1, or 1.5. In spite of the comparatively simple rate equations, which in some cases might suggest that a single elementary reaction is responsible for the reaction, the mechanism of such reactions have been shown to proceed via complex free radical mechanisms comprised of the following overall steps:

Initiation: Formation of free radicals by the parent reactants

Propagation: Reaction of free radicals with reactants to produce products and more free radicals

Termination: Destruction of free radicals by reaction between two of them

The following examples will be used to illustrate the method proposed by Rice and Herzfeld to develop rate expressions based on proposed free radical mechanisms. The reader should note that the mechanisms illustrated in the following sections do not necessarily represent the "correct" mechanism. These types of reactions are quite complex, and there is often disagreement in the literature as to the exact steps occurring. Furthermore, the reaction mechanism may change under different operating conditions. The examples given are intended solely to illustrate the methodology of rate expression derivation using the steady-state hypothesis.

Case 1. Three-half-order kinetics: The decomposition of acetaldehyde

The thermal decomposition of acetaldehyde to form methane and carbon monoxide is represented by the following overall reaction:

$$CH_3CHO \rightarrow CH_4 + CO \tag{5.42}$$

Traces of C_2H_6 are also formed in the decomposition reaction. The reaction rate can be expressed in terms of the rate of disappearance of CH_3CHO or the rate of formation of

CH_4 (or CO). In this reaction, these rates are not equal because some of the CH_3CHO forms C_2H_6. The rate of formation of CH_4 has been observed to follow a rate expression of the form

$$r_{CH_4} = k[CH_3CHO]^{1.5} \tag{5.43}$$

A reaction order of 1.5 cannot result from an elementary reaction, and the following mechanism for this reaction has been suggested:

$$CH_3CHO \xrightarrow{k_1} CH_3^{\bullet} + CHO^{\bullet} \tag{5.44}$$

$$CH_3^{\bullet} + CH_3CHO \xrightarrow{k_2} CH_4 + CH_3CO^{\bullet} \tag{5.45}$$

$$CH_3CO^{\bullet} \xrightarrow{k_3} CH_3^{\bullet} + CO \tag{5.46}$$

$$2CH_3^{\bullet} \xrightarrow{k_4} C_2H_6 \tag{5.47}$$

$$CHO^{\bullet} \xrightarrow{k_5} CO + H^{\bullet} \tag{5.48}$$

$$H^{\bullet} + CH_3CHO \xrightarrow{k_6} H_2 + CH_3CO^{\bullet} \tag{5.49}$$

We want to use the steady-state hypothesis to derive a rate expression based on this proposed mechanism. To start the derivation, apply the steady-state hypothesis to the $CH_3^{\bullet}$ radicals by setting their net rate of formation equal to zero. $CH_3^{\bullet}$ radicals are formed in reactions (5.44) and (5.46), and consumed in reactions (5.45) and (5.47). Therefore, the net rate of formation, which is equal to zero, is given by the following equation:

$$k_1[CH_3CHO] - k_2[CH_3^{\bullet}][CH_3CHO] + k_3[CH_3CO^{\bullet}] - k_4[CH_3^{\bullet}]^2 = 0 \tag{5.50}$$

Using a similar approach, application of the steady-state assumption to the net rate of formation of the $CH_3CO^{\bullet}$ radicals gives the equation

$$k_2[CH_3^{\bullet}][CH_3CHO] - k_3[CH_3CO^{\bullet}] + k_6[H^{\bullet}][CH_3CHO] = 0 \tag{5.51}$$

Adding Equations 5.50 and 5.51, and rearranging gives

$$[CH_3^{\bullet}] = \left(\frac{k_1 + k_6[H^{\bullet}]}{k_4} \right)^{0.5} [CH_3CHO]^{0.5} \tag{5.52}$$

The hydrogen radicals must be eliminated next. The net rate of formation of $H^{\bullet}$ is obtained from Equations 5.48 and 5.49, as follows:

$$k_5[CHO^{\bullet}] - k_6[H^{\bullet}][CH_3CHO] = 0 \tag{5.53}$$

Rearranging this expression gives the following expression:

$$[H^\bullet] = \frac{k_5}{k_6} \frac{[CHO^\bullet]}{[CH_3CHO]} \tag{5.54}$$

This expression for $H^\bullet$ still contains a free radical concentration, that of $CHO^\bullet$. This species is formed in Step 1 and destroyed in Step 5, so at steady state, its net rate of formation is

$$k_1[CH_3CHO] - k_5[CHO^\bullet] = 0 \tag{5.55}$$

Combining Equations 5.54 and 5.55, then gives

$$[H^\bullet] = \frac{k_1}{k_6} \frac{[CH_3CHO]}{[CH_3CHO]} = \frac{k_1}{k_6} \tag{5.56}$$

When Equation 5.56 is combined with Equation 5.52, we obtain

$$[CH_3^\bullet] = \left(\frac{2k_1}{k_4}\right)^{0.5} [CH_3CHO]^{0.5} \tag{5.57}$$

CH_4 is involved only in the reaction given by Equation 5.45

$$r_{CH_4} = k_2[CH_3^\bullet][CH_3CHO] \tag{5.58}$$

Substituting the methyl radical concentration from Equation 5.57 into Equation 5.58 gives the rate of formation of methane:

$$r_{CH_4} = k_2\left(\frac{2k_1}{k_4}\right)^{0.5} [CH_3CHO]^{1.5} \tag{5.59}$$

The apparent rate constant is a function of the rate constants of three steps in the mechanism. Each of these rate constants is an exponential function of temperature, and each step has its own activation energy. The apparent activation energy of the reaction would thus be a combination of the three activation energies of the elementary steps, as follows:

$$E = E_2 + \frac{1}{2}(E_1 - E_4) \tag{5.60}$$

Case 2. First-order kinetics: The decomposition of ethane

The decomposition of ethane (ethylene) into ethene and hydrogen is an important reaction for the production of ethene, which is used in the production of many useful products. One possible mechanism that has been proposed is composed of the following steps:

$$C_2H_6 \xrightarrow{k_1} 2CH_3^\bullet \tag{5.61}$$

$$CH_3^{\bullet} + C_2H_6 \xrightarrow{k_2} CH_4 + C_2H_5^{\bullet} \tag{5.62}$$

$$C_2H_5^{\bullet} \xrightarrow{k_3} C_2H_4 + H^{\bullet} \tag{5.63}$$

$$H^{\bullet} + C_2H_6 \xrightarrow{k_4} H_2 + C_2H_5^{\bullet} \tag{5.64}$$

$$H^{\bullet} + C_2H_5^{\bullet} \xrightarrow{k_5} C_2H_6 \tag{5.65}$$

Once again the steady-state hypothesis can be applied to each of the free radicals. The $CH_3^{\bullet}$ radicals are formed in reaction (5.61) and are destroyed in reaction (5.62). The net rate of formation of methyl radicals is therefore

$$k_1[C_2H_6] - k_2[CH_3^{\bullet}][C_2H_6] = 0 \tag{5.66}$$

The $C_2H_5^{\bullet}$ radicals are formed in reactions (5.62) and (5.64), and destroyed in reactions (5.63) and (5.65). Therefore, the net rate of formation is

$$k_2[CH_3^{\bullet}][C_2H_6] - k_3[C_2H_5^{\bullet}] - k_5[C_2H_5^{\bullet}][H^{\bullet}] + k_4[H^{\bullet}][C_2H_6] = 0 \tag{5.67}$$

The $H^{\bullet}$ radicals are formed in reaction (5.63), and destroyed in reactions (5.64) and (5.65), therefore

$$k_3[C_2H_5^{\bullet}] - k_4[H^{\bullet}][C_2H_6] - k_5[C_2H_5^{\bullet}][H^{\bullet}] = 0 \tag{5.68}$$

If we add Equations 5.66 through 5.68, a number of terms cancel and we obtain, after rearrangement

$$[H^{\bullet}] = \frac{k_1}{2k_5} \frac{[C_2H_6]}{[C_2H_5^{\bullet}]} \tag{5.69}$$

Equation 5.69 is substituted into Equation 5.68 to give a quadratic equation

$$2k_5k_3[C_2H_5^{\bullet}]^2 - k_4k_1[C_2H_6]^2 - k_5k_1[C_2H_6][C_2H_5^{\bullet}] = 0 \tag{5.70}$$

Equation 5.70 is solved using the quadratic formula to give the solution in terms of $[C_2H_5^{\bullet}]$:

$$[C_2H_5^{\bullet}] = \left\{ \frac{k_1}{4k_3} + \left[\left(\frac{k_1}{4k_3} \right)^2 + \frac{k_1k_4}{2k_5k_3} \right]^{0.5} \right\} [C_2H_6] = k'[C_2H_6] \tag{5.71}$$

The rate constant k' is a lumped parameter comprised of the combination of constants. The rate of formation of C_2H_4 is given by reaction (5.63):

$$+r_{C_2H_4} = k_3[C_2H_5^{\bullet}] \tag{5.72}$$

Substituting Equation 5.71 into Equation 5.72 gives the rate of formation of ethene:

$$+r_{C_2H_4} = k''[C_2H_6] \tag{5.73}$$

Here, k'' is another lumped kinetic parameter. The overall reaction is first order in ethane concentration. An alternative reaction mechanism is used in a problem at the end of the chapter. The final form of the rate equation depends on the number of radicals that participate in the initiation and termination steps, and a variety of possible combinations have been observed.

The examples considered in Section 5.4.2, while complex, are still relatively simple. Although it is dangerous to generalize, it is usually observed that the reaction mechanism increases in complexity as the size of the molecules increases. For example, combustion reactions of hydrocarbons of the gasoline (petrol) molecular mass range involve hundreds of steps and an equal number of reaction intermediates. In such cases, it can be very difficult to identify the reaction mechanism and, if it is identified, to derive a simple rate equation from it.

5.4.3 Complex Gas-Phase Reactions

Complicated mechanisms can also give rise to rate expressions that are not of the power law form. A classic example is the formation of HBr from H_2 and Br_2, which can be represented by the following overall reaction:

$$H_2 + Br_2 \rightarrow 2HBr$$

The proposed mechanism is summarized below:

Initiation:

$$Br_2 \xrightarrow{k_1} 2Br^{\bullet} \tag{5.74}$$

Propagation:

$$Br^{\bullet} + H_2 \xrightarrow{k_2} H^{\bullet} + HBr \tag{5.75}$$

$$H^{\bullet} + Br_2 \xrightarrow{k_3} Br^{\bullet} + HBr \tag{5.76}$$

$$H^{\bullet} + HBr \xrightarrow{k_4} H_2 + Br^{\bullet} \tag{5.77}$$

Termination:

$$2Br^{\bullet} \xrightarrow{k_5} Br_2 \tag{5.78}$$

The steady-state hypothesis can be used to develop a rate expression for this mechanism by applying it to the reactive intermediates. We start by the process by writing the overall reaction rate for hydrogen bromide. HBr is formed in reactions (5.75) and (5.76), and destroyed in reaction (5.77). The net rate of formation is therefore given by

$$r_{HBr} = k_2[Br^\bullet][H_2] + k_3[H^\bullet][Br_2] - k_4[HBr][H^\bullet] \tag{5.79}$$

The steady-state hypothesis can be applied to the net rate of formation of both $Br^\bullet$ and $H^\bullet$ radicals to give the following equations. The $H^\bullet$ radicals are involved in the propagation steps, that is, reactions (5.75) through (5.77). Thus, we obtain an expression for the net reaction rate:

$$r_{H^\bullet} = 0 = k_2[Br^\bullet][H_2] - k_3[H^\bullet][Br_2] - k_4[H^\bullet][HBr] \tag{5.80}$$

The $Br^\bullet$ radicals are involved in every step, that is, in reactions (5.74) through (5.78). The net rate of formation is thus given by

$$r_{Br^\bullet} = 0 = 2k_1[Br_2] - k_2[Br^\bullet][H_2] + k_3[H^\bullet][Br_2] + k_4[H^\bullet][HBr_2] - 2k_5[Br^\bullet]^2 \tag{5.81}$$

Combining Equations 5.80 and 5.81 followed by simplification gives an expression for the concentration of $Br^\bullet$:

$$[Br^\bullet] = \left(\frac{k_1}{k_5}\right)^{0.5} [Br_2]^{0.5} \tag{5.82}$$

Equation 5.82 can then be substituted into Equation 5.80 and the result rearranged to give an equation for the concentration of $H^\bullet$:

$$[H^\bullet] = \frac{k_2 \left(k_1/k_5\right)^{0.5} [Br_2]^{0.5}[H_2]}{k_3[Br_2] + k_4[HBr]} \tag{5.83}$$

Equation 5.80 can be subtracted from Equation 5.79 to simplify the expression for the rate of formation of HBr:

$$r_{HBr} = 2k_3[H^\bullet][Br_2] \tag{5.84}$$

Finally, substitute Equation 5.83 into Equation 5.84 to give the rate expression for the formation of HBr:

$$r_{HBR} = \frac{2k_3k_2 \left(k_1/k_5\right)^{0.5} [H_2][Br_2]^{1.5}}{k_3[Br_2] + k_4[HBr]} \tag{5.85}$$

Equation 5.85 is obviously not of the power law form seen in previous cases. The temperature and concentration dependence of the reaction rate cannot be separated.

5.4.4 Rate-Determining Step Method for Deriving Rate Expressions

The pseudo-steady-state hypothesis is not the only technique that has been used to derive rate equations from proposed mechanisms. Although it works well in many cases, and leads to rate expressions that successfully describe experimentally observed reaction rates, it is by no means universally applicable. Another common approach used to derive rate expressions from proposed mechanisms is called the rate-determining step method. In this technique, it is assumed that one step in the mechanism is intrinsically slow compared to the others, and thus controls the rate. This step in the mechanism is essentially the "bottleneck" in the mechanism. The other steps are intrinsically very fast compared to the rate-determining step, and as a result may be assumed to be in equilibrium, that is, their net rate of reaction is negligibly small. This method is widely used in developing rate equations for catalytic reactions, as shown in Chapter 8. The method is illustrated here for a homogeneous reaction.

The formation of phosgene from chlorine and carbon monoxide can be described by the following overall reversible reaction:

$$CO + Cl_2 \underset{k_r}{\overset{k_f}{\rightleftharpoons}} COCl_2 \tag{5.86}$$

The rate of formation of phosgene has been observed to obey the following rate expression:

$$r_{COCl_2} = k_f [CO][Cl_2]^{1.5} - k_r [COCl_2][Cl_2]^{0.5} \tag{5.87}$$

Such a complex expression cannot result from an elementary reaction. We now examine two proposed mechanisms, and use the rate-determining step method to derive rate expressions based on each one. The first proposed reaction mechanism is composed of three elementary steps:

$$Cl_2 \underset{k_{-1}}{\overset{k_1}{\rightleftharpoons}} 2Cl^{\bullet} \tag{5.88}$$

$$CO + Cl^{\bullet} \underset{k_{-2}}{\overset{k_2}{\rightleftharpoons}} COCl^{\bullet} \tag{5.89}$$

$$COCl^{\bullet} + Cl_2 \underset{k_{-3}}{\overset{k_3}{\rightleftharpoons}} COCl_2 + Cl^{\bullet} \tag{5.90}$$

In the proposed mechanism, the rates of the reactions represented by Equations 5.88 and 5.89 are taken to be very fast and are assumed to proceed to equilibrium. The overall rate of the reaction is controlled by Equation 5.90, which is known as the rate-determining step. The rate of reaction is written in terms of the rate-determining step, that is

$$r_{COCl_2} = k_3[COCl^{\bullet}][Cl_2] - k_{-3}[COCl_2][Cl^{\bullet}] \tag{5.91}$$

It is necessary to eliminate the free radical concentrations from Equation 5.91. Reaction (5.88) is assumed to be in equilibrium and thus the net rate of reaction is equal to zero. Therefore

$$k_1[Cl_2] = k_{-1}[Cl^{\bullet}]^2 \tag{5.92}$$

The concentration of Cl$^\bullet$ can be expressed explicitly by rearrangement to give

$$[Cl^\bullet] = \left(\frac{k_1}{k_{-1}} [Cl_2] \right)^{0.5} \tag{5.93}$$

Reaction (5.89) is also assumed to be in equilibrium. Setting the net reaction rate to zero gives

$$k_2[Cl^\bullet][CO] = k_{-2}[COCl^\bullet] \tag{5.94}$$

Equation 5.94 is also rearranged to give

$$[COCl^\bullet] = \frac{k_2}{k_{-2}} [Cl^\bullet][CO] \tag{5.95}$$

Substituting Equation 5.93 into Equation 5.95 gives

$$[COCl^\bullet] = \frac{k_2}{k_{-2}} \left(\frac{k_1}{k_{-1}} \right)^{0.5} [CO][Cl_2]^{0.5} \tag{5.96}$$

Using Equations 5.93 and 5.95 to eliminate [COCl$^\bullet$] and [Cl$^\bullet$] in Equation 5.91 gives

$$r_{COCl_2} = k_3 \frac{k_2}{k_{-2}} \left(\frac{k_1}{k_{-1}} \right)^{0.5} [CO][Cl_2]^{1.5} - k_{-3} \left(\frac{k_1}{k_{-1}} \right)^{0.5} [COCl_2][Cl_2]^{0.5} \tag{5.97}$$

This rate equation is the same as the one observed to correlate the experimental data (Equation 5.87), provided that

$$k_f = k_3 \frac{k_2}{k_{-2}} \left(\frac{k_1}{k_{-1}} \right)^{0.5}$$

and

$$k_r = k_{-3} \left(\frac{k_1}{k_{-1}} \right)^{0.5}$$

The steady-state/rate-determining step approach involves different assumptions than the steady-state hypothesis. Depending on the assumptions that are made in proposing the mechanism, either method may provide a rate equation that is consistent with the experimental data. The rate-determining step approach is extensively used in developing rate expressions for catalytic reactions. Rate equations for catalytic reactions are discussed in Chapter 8.

It is seldom the case that a single reaction mechanism is considered when investigating the kinetics of a chemical reaction. The mechanism of any given reaction is often the subject of much controversy and debate in the literature. For example, another proposed mechanism for the formation and decomposition of phosgene is composed of the following sequence of steps:

$$Cl_2 \underset{k_{-1}}{\overset{k_1}{\rightleftharpoons}} 2Cl^\bullet \tag{5.98}$$

$$Cl^\bullet + Cl_2 \underset{k_{-2}}{\overset{k_2}{\rightleftharpoons}} Cl_3^\bullet \tag{5.99}$$

$$Cl_3^\bullet + CO \underset{k_{-3}}{\overset{k_3}{\rightleftharpoons}} COCl_2 + Cl^\bullet \tag{5.100}$$

Reaction (5.100) is the rate-determining step; thus, the reaction rate is given by

$$r_{COCl_2} = k_3[CO][Cl_3^\bullet] - k_{-3}[COCl_2][Cl^\bullet] \tag{5.101}$$

Using the same approach as for the first proposed mechanism, reaction (5.98) is assumed to be in equilibrium; therefore

$$[Cl^\bullet] = \left\{ \frac{k_1}{k_{-1}}[Cl_2] \right\}^{0.5} \tag{5.102}$$

Reaction (5.99) is in equilibrium and, therefore

$$[Cl_3^\bullet] = \frac{k_2}{k_{-2}}[Cl^\bullet][Cl_2] \tag{5.103}$$

Substituting for $[Cl^\bullet]$ from Equation 5.102 gives

$$[Cl_3^\bullet] = \frac{k_2}{k_{-2}} \left(\frac{k_1}{k_{-1}} \right)^{0.5} [Cl_2]^{1.5} \tag{5.104}$$

Substituting for $[Cl_3^\bullet]$ and $[Cl^\bullet]$ from Equations 5.104 and 5.102 into Equation 5.101 gives the rate expression

$$r_{COCl_2} = k_3 \frac{k_2}{k_{-2}} \left(\frac{k_1}{k_{-1}} \right)^{0.5} [CO][Cl_2]^{1.5} - k_{-3} \left(\frac{k_1}{k_{-1}} \right)^{0.5} [COCl_2][Cl_2]^{0.5} \tag{5.105}$$

This rate expression is identical to that derived from the other proposed mechanism. In this case

$$k_f = k_3 \frac{k_2}{k_{-2}} \left(\frac{k_1}{k_{-1}} \right)^{0.5}$$

and

$$k_r = k_{-3} \left(\frac{k_1}{k_{-1}} \right)^{0.5}$$

Either of these two mechanisms could explain the observed kinetic behavior, which leads to an important observation. Simply because a proposed mechanism is capable of generating a rate expression that is consistent with the observed experimental behavior, the mechanism is not necessarily valid. Additional information is often required to narrow the choice of a mechanism. For example, in the example given here, it is generally accepted that the first mechanism is more consistent with other experimental observations. It is very important to appreciate the difference between a reaction mechanism and a mathematical model for the reaction. From an engineering point of view, it may be more important to have a reliable mathematical model, and less important to understand the detailed mechanism.

5.5 Mechanisms and Models

In the previous sections of this chapter, we explained how rate equations could be derived from proposed mechanisms. It is worthwhile to summarize the key points of these previous sections prior to a description of some of the methods used to estimate reaction rate parameters. The main points from the previous sections are:

1. All reactions proceed via a sequence of steps at the molecular level, which, taken together, comprise the reaction mechanism.
2. The mechanism may comprise a single step (elementary reaction) or a sequence of steps.
3. The reaction mechanism, if known, can be used to develop a rate function, with the reaction rate expressed as a function of reactant and product concentration, and temperature. Usually, the development of the rate equation from the mechanism involves making some assumptions.
4. The rate function, whether for an elementary reaction or one based on a complex reaction mechanism, contains one or more temperature-dependent parameters. The numerical value of these parameters cannot be calculated from first principles and must be determined by experimental measurement.

The significance of these points should not be underestimated, especially point 4. The fact that rate functions contain experimentally determined parameters, obtained by statistical analysis of rate data, means that a certain degree of caution should be exercised when using them. This is especially true when it comes to making statements about the validity of the assumptions made in deriving the rate function. We have seen earlier that it is possible for more than one mechanism to give the same rate function. It is also possible for two different rate functions, derived from different mechanisms, to give an acceptable correlation of the experimental data. It is often difficult to determine exactly the reaction mechanism.

Furthermore, even if the reaction mechanism is reasonably well established, it may be so complex that its use in a practical reactor model is too unwieldy. In such a case, simplifying assumptions or "lumped" kinetics are often used. Although this latter point is of less concern as the power of computers continues to grow, it is still valid. The net result is that the rate function used to design or analyze a reactor should be regarded more as a model than a mechanistic description of the reaction process.

A model is developed using experimental data collected over a limited range of operating conditions. The model should not be extrapolated to new conditions outside of this range: even a model that is an exact description of the mechanism may not be correct if extrapolated to new operating conditions because the reaction mechanism may change. For example, a rate model that assumes that a certain step in the reaction mechanism is the rate-determining step would not be appropriate under a new set of operating conditions where the rate-determining step changed.

5.6 Experimental Methods in Rate Data Collection and Analysis

It follows from Section 5.5 that the essence of parameter estimation involves curve fitting experimental data to rate equations. Therefore, from an experimental perspective, the task is to collect data that are suitable for this purpose, either directly or after some degree of manipulation. Many methods are used by experimentalists in kinetic investigations, so this section is meant only to serve as a brief introduction to this area. It is important to note that there is no general "recipe" that can be used for all kinetic studies. The nature of the reaction is an important consideration; some experimental methods are better suited to gas-phase reactions, while others may be appropriate if a liquid-phase reaction is being studied. The presence of a heterogeneous catalyst can also be a significant factor when selecting an experimental technique, a theme developed further in Chapter 11. In addition, the heat effects of the reaction are important. The availability of equipment and computational tools may also influence the choice of experimental method.

The development of a rate function from experimental methods can be divided into a series of steps:

1. A set of experiments is conducted in a small experimental reactor. Experimental reactors used for obtaining rate data come in a variety of forms but usually they are small bench-scale units. Typically, the reaction conditions such as flow rate, inlet and outlet concentrations and temperatures, time, and so on are measured and recorded.

2. The raw data are manipulated in some fashion to produce a set of data that are in the correct form for the calculation of the parameters in the rate function. The nature of this manipulation depends on the choices made in steps 1 and 3.

3. The reduced data are used to determine the functional dependence of reaction rate on concentration and temperature. This dependence is determined using a variety of statistical and numerical methods. Depending on the results, new experiments may be planned.

The following sections describe some experimental reactor types and data analysis techniques used to determine parameters in rate equations. As noted, there is not a single best method for conducting an experimental investigation and each method offers advantages and disadvantages, which will have differing degrees of importance depending on the reaction being studied.

5.6.1 Rate Data in Constant-Volume Batch Reactor

The CVBR is often used in the initial stages of a kinetic investigation, especially for homogeneous liquid-phase reactions. The advantages of the batch reactor are that a feed system is not required and only a comparatively small amount of reactants and products are needed. A disadvantage is that the initial state is often not well defined because a finite and significant length of time may be required to reach the desired initial conditions: during this time, some reaction may occur. The usual experimental procedure is to use an isothermal CVBR, although sometimes adiabatic operation is used. Isothermal operation reduces the number of parameters that must be simultaneously fitted, so it is considered more desirable than adiabatic operation. Isothermal operation may, however, be difficult to achieve if there are strong heat effects associated with the reaction.

The usual practice is to monitor the concentration of a specific species or the total pressure as a function of time, that is, as the reaction proceeds. From the concentration (or pressure) versus time data, it is necessary to obtain rate of reaction versus concentration results, which are subsequently used to determine the rate parameters. There are two main methods of analyzing CVBR data, the differential method and the integral method. Each method has advantages and disadvantages, and neither is obviously the best for all cases. These two methods are briefly presented in the following sections, with some examples.

5.6.2 Differential Method of Analysis of CVBR Data

The constant-volume batch reactor produces a set of data: typically, these data are concentration of reactant in the reactor as a function of time. In the differential method of analysis, the desired quantity is the rate of reaction at a given value of the concentration, and therefore the experimental data must be manipulated to obtain these values. Recall from Chapter 3 the batch reactor mole balance equation for a constant-volume reactor:

$$(-r_A) = -\frac{1}{V}\frac{dN_A}{dt} = -\frac{dC_A}{dt} \tag{5.106}$$

It is seen from Equation 5.106 that the reaction rate is the rate of change of concentration of reactant A with time. Therefore, if we have collected data of concentration as a function of time, it is necessary to differentiate the data to obtain the reaction rate. There are a number of ways of differentiating experimental data; we shall consider two of them. A typical plot of C_A versus time that might be obtained in a CVBR is illustrated in Figure 5.3: the slope of the curve at any point is dC_A/dt. The slope (and hence the reaction rate) at any value of C_A can be obtained by choosing the desired concentration on the vertical axis of Figure 5.3, and then finding the slope that corresponds to this value. The problem is to obtain a reliable value of the slope from the experimental data, which may contain some experimental error. Two methods for obtaining the slope are to apply finite difference formulae to the data, or to fit a curve to the data points and then differentiate the curve to obtain the slope.

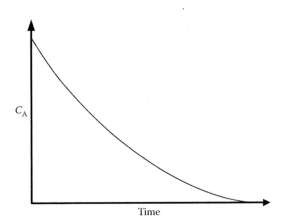

FIGURE 5.3
Typical concentration versus time curve that may be obtained in a constant-volume batch reactor. In this case, A is a reactant.

5.6.2.1 Finite Difference Formulae

The slope can be obtained in a discrete manner by applying the finite difference formula to the data points. Assume that we have N data points measured at various times. Thus, let t_0 represent time 0, at which time the concentration of reactant A is given by C_{A0}. The concentration at data point j, $C_{A,j}$ corresponds to time t_j, and the last data point is taken at time t_N and corresponds to concentration C_{AN}. The slope of the line at time zero is expressed in terms of the concentrations at data points 0, 1, and 2, that is

$$\frac{dC_A}{dt}(t_0) = \frac{-3C_{A0} + 4C_{A1} - C_{A2}}{2\Delta t} \tag{5.107}$$

The slope at the data points 1 to $N-1$ is written in terms of the concentrations at the preceding and following points:

$$\frac{dC_A}{dt}(t_j) = \frac{\left(C_{A,j+1} - C_{A,j-1}\right)}{2\Delta t} \tag{5.108}$$

Finally, the slope of the curve at the last data point, N, is given by

$$\frac{dC_A}{dt}(t_N) = \frac{\left(C_{A,N-2} - 4C_{A,N-1} + 3C_{A,N}\right)}{2\Delta t} \tag{5.109}$$

These finite difference formulae are valid when the time step size is the same between all data. The implementation of this method is straightforward.

5.6.2.2 Fitting a Curve to the Data

An alternative to the finite difference method is to fit a curve to the data points. This curve fitting may be done using a variety of numerical techniques and there are many programs available for this purpose. After the equation for the concentration versus time curve has been generated it can be differentiated, usually analytically, to determine the slope and hence the reaction rate at any desired concentration. It is not always easy to determine what type of equation should be used to fit the concentration–time data. A polynomial is often used for this purpose because it is conceptually straightforward to fit and differentiate. It is usually preferable to use as few adjustable parameters as possible in this equation. For example, a high-order polynomial that is curve fit to experimental data that contain a significant degree of error may lead to oscillations, and physically meaningless behavior of the reaction rate.

Once the derivative of the C_A versus time curve is determined, a reduced data set consisting of a table of reaction rates and concentrations can be generated. These data are used to determine the values of the constants in the rate model. The techniques for performing this operation are described in the next section.

5.6.3 Determining Form of Rate Expression

This section describes some of the methods that are used in evaluating rate constants from reaction rate data. Note that the methods described in this section are not limited to batch reactor data, but rather can be applied to rate data obtained in any reactor type.

5.6.3.1 Power Law Models

The first step in developing rate constants values is to assume some form for the rate expression. This expression may be an empirical model or a model that has been derived from an assumed mechanism using the methods outlined earlier in this chapter. For example, for a unimolecular reaction, one might start by assuming that the reaction rate obeys a simple power law model

$$(-r_A) = kC_A^n \tag{5.110}$$

It is necessary to determine if there are values of k and n that will correlate the experimental values. Equation 5.110 can be linearized by taking the logarithm of both sides, to obtain the following equation:

$$\ln(-r_A) = \ln(k) + n\ln(C_A) \tag{5.111}$$

The constants k and n can be determined by regression analysis. The data may be plotted as shown in Figure 5.4 to test for systematic deviation. In Figure 5.4, it is seen that the data points seem to follow a straight line with a random deviation of the points from the line. If a graph like Figure 5.4 is observed, the proposed model may correlate the data well over the experimental range of concentrations. The goodness of fit should be tested using the appropriate statistical parameters, and by comparing the model prediction to the observations.

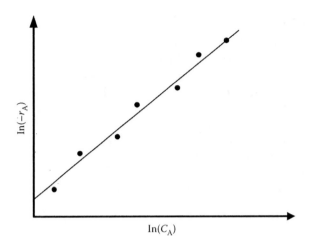

FIGURE 5.4
A plot of the ln of the rate as a function of the ln of the concentration of reactant A. The random nature of the deviation of the experimental points from the line indicate that the model may be appropriate.

The data points may not follow a straight line; therefore, the proposed model would be invalid. Alternatively, the data points may appear to be linear but show a systematic deviation about the line. For example, the graph may give a result like the one shown in Figure 5.5. In this figure, the data points are close to the line, but show a systematic deviation, indicating that another model should probably be investigated to achieve a better representation of the data.

For reactions that depend on more than one reactant concentration, such as bimolecular reactions, a power law model may be proposed. For example, an irreversible reaction between molecules of A and B might be assumed to have a rate equation of the form

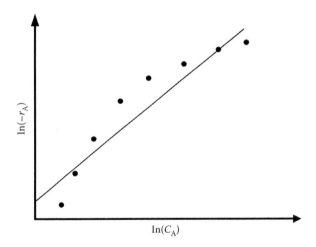

FIGURE 5.5
The plot ln of the rate vs. ln of the concentration shows a systematic deviation about the best fit line, indicating that a power law equation may not be the best model for this reaction.

$$(-r_A) = k C_A^\alpha C_B^\beta \tag{5.112}$$

Equation 5.112 can be linearized by taking the ln of both sides, which gives

$$\ln(-r_A) = \ln k + \alpha \ln C_A + \beta \ln C_B \tag{5.113}$$

To find the parameters in Equation 5.113, it is necessary to have values of $(-r_A)$ at different C_A and C_B. The presence of B in the rate equation means that a larger number of experiments must be performed to eliminate the possibility of cross correlation. Consider the reaction

$$A + B \rightarrow C \tag{5.114}$$

If experiments are conducted where the initial concentrations of A and B were the same, then at any given time the concentrations of A and B would be equal because of the stoichiometry of the reaction. Equation 5.113 would therefore simplify to

$$\ln(-r_A) = \ln k + (\alpha + \beta) \ln C_A \tag{5.115}$$

It would thus not be possible to determine α and β separately. Therefore, it is necessary to do a series of experiments with different ratios of C_{A0}/C_{B0}.

The use of the differential method of analysis is illustrated in Example 5.1.

Example 5.1

Consider the essentially irreversible liquid-phase reaction represented by

$$A \rightarrow C \tag{5.116}$$

The concentration of A as a function of time was measured in an isothermal constant-volume batch reactor.

Time (min)	C_A(mol/L)
0	2.50
10	2.10
20	1.78
30	1.50
40	1.30
50	1.15
60	1.01
70	0.90
80	0.81
90	0.73
100	0.66

Determine if a power law-type rate expression is able to correlate these data and, if so, determine the values of the kinetic parameters.

SOLUTION

The first step in the analysis is to convert the data to reaction rates at known values of concentrations. As discussed in the previous section, there are a number of options available to perform this calculation. In this example, the finite difference formulae, Equations 5.107, 5.108, and 5.109 will be used. For example, at time zero, Equation 5.107 is used:

$$\frac{dC_A}{dt}(t_0) = \frac{-3C_{A0} + 4C_{A1} - C_{A2}}{2\Delta t} = \frac{-3(2.50) + 4(2.10) - (1.78)}{2(10)}$$

$$= -0.044 \text{ mol/L} \cdot \text{min}$$

(5.117)

This result is a negative number, indicating that the concentration is decreasing with time. The reaction rate is the minus of this number, or 0.044. For the subsequent points, except for the last one, Equation 5.108 is used. As an example, for an elapsed time of 10 min, the rate is

$$\frac{dC_A}{dt}(10) = \frac{\left(C_{A,j+1} - C_{A,j-1}\right)}{2\Delta t} = \frac{(1.78 - 2.50)}{2(10)} = -0.036$$

(5.118)

Finally for 100 min, Equation 5.109 is used. The rate data thus computed are summarized in the following table:

Time (min)	C_A (mol/L)	$(-r_A)$ (mol/L · min)
0	2.50	0.044
10	2.10	0.036
20	1.78	0.030
30	1.50	0.024
40	1.30	0.0175
50	1.15	0.0145
60	1.01	0.0125
70	0.90	0.010
80	0.81	0.0085
90	0.73	0.0075
100	0.66	0.0065

Note that the reaction rate falls with concentration. Now that the reaction rates have been estimated, a plot can be made of reaction rate versus concentration. If this plot is prepared using log scales, a straight line will indicate that a power law rate model correlates the data. Such a plot is shown in Figure 5.6. From this figure, it is seen that a straight line does indeed appear to correlate the data. From the least squares regression analysis, the equation of the line is given as

$$(-r_A) = 0.012 C_A^{1.49} \text{ mol/L} \cdot \text{min}$$

(5.119)

The fact that the power is not an integer is an indication of a more complex mechanism than a simple collision of two molecules of A.

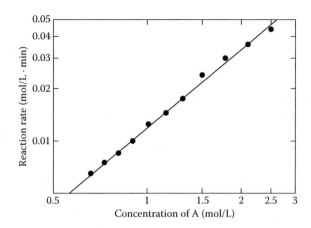

FIGURE 5.6
Reaction rate plotted as a function of reactant concentration on a log–log scale. The line is the least squares fit of the data points. It appears that these data follow a power law kinetic model.

5.6.3.2 Complex Rate Expressions

A large number of chemical reactions obey the relatively simple power law-type rate models. However, the reaction rate for a large number of reactions does not obey these types of models, and consequently more complex expressions are required. For example, free radical reactions may have rate models of high degrees of complexity. As rate expressions increase in complexity, the methods used to determine the rate parameters also have a concomitant increase in complexity. Consider, for example, the rate expression

$$(-r_A) = \frac{kC_A}{1 + K_1 C_A} \tag{5.120}$$

The rate constants in an equation like Equation 5.120 can be determined in two general ways. The first is to use nonlinear regression analysis. Historically, many workers in kinetics have avoided nonlinear regression analysis because of the computational complexity involved. However, at present there are many readily available software packages that make this technique relatively simple to use. The other method is to linearize the rate expression and then to use linear regression analysis. In the simpler cases, linearization of the equation allows graphical analysis to be used, which has the advantage of allowing a visual interpretation of the results. It is often possible to linearize the rate equation in a number of different ways. For example, Equation 5.120 can be linearized in three possible ways, to give the following equations:

$$\frac{C_A}{(-r_A)} = \frac{1}{k} + \frac{K_1}{k} C_A \tag{5.121}$$

$$\frac{1}{(-r_A)} = \frac{1}{kC_A} + \frac{K_1}{k} \tag{5.122}$$

$$\frac{(-r_A)}{C_A} = k - K_1(-r_A) \tag{5.123}$$

It is important to note that that a linear regression analysis performed on each of these three equations will yield different parameter values. The differences may be minor, and each set of parameters may predict the reaction rate to a similar degree of accuracy. There are some cases, however, where an inappropriate selection of a linearization may give a set of parameters that give a poor prediction of the reaction rate. For this reason, it may be preferable to use nonlinear regression analysis. The reader should also note that some choices may be less desirable from a statistical point of view. For example, use of Equation 5.121 or 5.123 involves plotting an experimental variable against itself (see Example 5.2), a practice which should usually be avoided. Example 5.2 illustrates the linearization of a rate equation.

Example 5.2

Consider the catalytic reaction represented by the following overall reaction:

$$A + \frac{1}{2}B \rightarrow C + D \tag{5.124}$$

In a large excess of B, the reaction rate follows the expression given by Equation 5.120, that is

$$(-r_A) = \frac{kC_A}{1 + K_1C_A} \quad \text{mol/min} \cdot \text{gcat} \tag{5.125}$$

Values for the rate of disappearance of A as a function of concentration are given in the following table:

C_A (mol/m³)	$(-r_A)$ (mol/min · gcat)
0.049	0.038
0.105	0.068
0.196	0.135
0.298	0.176
0.402	0.218
0.605	0.238
0.792	0.290
1.012	0.312
1.193	0.352

Use these data to determine the values of the reaction parameters in the rate equation. Compare the results obtained by using different linearized versions of the rate equation to that obtained using nonlinear regression analysis.

SOLUTION

Three possible ways to linearize the rate equation were given as Equations 5.121 through 5.123. Each of these forms requires that different variables be plotted to produce a straight line. The first form is

$$\frac{C_A}{(-r_A)} = \frac{1}{k} + \frac{K_1}{k} C_A$$

(5.126)

In this case, a plot of C_A against $C_A/(-r_A)$ should give a straight line of slope K_1/k and intercept of $1/k$, from which K_1 and k can be computed. The second linearized form is

$$\frac{1}{(-r_A)} = \frac{1}{kC_A} + \frac{K_1}{k}$$

(5.127)

In this case, a plot of $1/C_A$ against $1/(-r_A)$ should give a straight line of slope $1/k$ and intercept of K_1/k, from which K_1 and k can be computed. The third form is

$$\frac{(-r_A)}{C_A} = k - K_1(-r_A)$$

(5.128)

A plot of $(-r_A)$ against $(-r_A)/C_A$ should give a straight line of slope $-K_1$ and intercept of k. The reduced data required to generate the plots is given below:

C_A	$(-r_A)$	$C_A/(-r_A)$	$1/C_A$	$1/(-r_A)$	$(-r_A)/C_A$
0.049	0.038	1.289	20.408	26.316	0.776
0.105	0.068	1.544	9.524	14.706	0.648
0.196	0.135	1.452	5.102	7.407	0.689
0.298	0.176	1.693	3.356	5.682	0.591
0.402	0.218	1.844	2.488	4.587	0.542
0.605	0.238	2.542	1.653	4.202	0.393
0.792	0.290	2.731	1.263	3.448	0.366
1.012	0.312	3.244	0.988	3.205	0.308
1.193	0.352	3.389	0.838	2.841	0.295

The plots of the three linearized equations are shown in Figure 5.7. In each case, the symbols represent the reduced data and the line is the best-fit straight line that correlates the data. Linear least squares analysis on Equations 5.126, 5.127, and 5.128 give the following values for the reaction rate parameters:

Equation Number	Slope	Intercept	k	K_1
(5.126)	1.931	1.194	0.8377	1.618
(5.127)	1.215	1.888	0.8232	1.554
(5.128)	−1.563	0.8292	0.8292	1.563

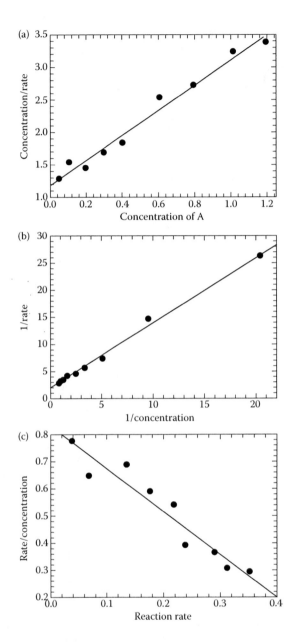

FIGURE 5.7
These plots are the linearized version of the reaction rate expression for Example 5.2. Graph (a) corresponds to Equation 5.126, graph (b) corresponds to Equation 5.127, and graph (c) corresponds to Equation 5.128. The straight lines are the regression lines obtained by linear least squares analysis.

A nonlinear regression analysis on Equation 5.125 gives the following values:

$$k = 0.8692 \quad \text{and} \quad K_1 = 1.723$$

The difference in values given by the different regression schemes is typical when linearization is performed. The predictive ability of the four sets of constants can be

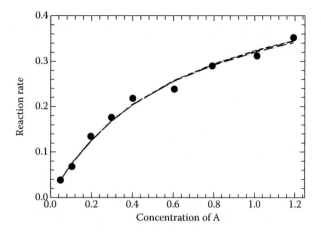

FIGURE 5.8
Comparison of the predicted reaction rate at various concentration of A from the four different sets of reaction rate parameters. The four lines are virtually indistinguishable, indicating that all four parameter sets give the same level of accuracy.

compared by plotting the predicted reaction rate at each concentration, as shown in Figure 5.8. The four curves are almost the same, indicating that each set of parameters has similar prediction accuracy.

5.6.4 Integral Method of Analysis of CVBR Data

The integral method of analysis does not require the generation of rate of reaction data; the experimental concentration versus time curves are used directly. The need to differentiate the experimental data is avoided, which is the primary advantage of the integral method. In the integral method, the form of the rate expression is first assumed. This rate expression is then substituted into the mole balance equation and the result is integrated to give a predictive equation for the concentration as a function of time. The values of the rate constants are adjusted until the best match between the observed and predicted concentration versus time curves is obtained. If a satisfactory match cannot be obtained, other forms for the rate equation are proposed.

5.6.4.1 Power Law Kinetics

Consider, for example, a unimolecular reaction with the proposed rate function:

$$(-r_A) = kC_A^n \tag{5.129}$$

Recall that for the differential method we would find $(-r_A)$ at various C_A, and then plot $\ln (-r_A)$ versus $\ln (C_A)$ to get k and n. In the integral method, Equation 5.129 is substituted directly into the CVBR mole balance equation to give

$$\int_{C_{A0}}^{C_A} \frac{-dC_A}{C_A^n} = kt \tag{5.130}$$

The objective is to find a value of n that will correlate the data, which in this case means that a plot of the left side of Equation 5.130 versus time will give a straight line, as illustrated in Figure 5.9. For any value of n except 1, Equation 5.131 can be written as

$$\frac{-1}{1-n}\left[(C_A)^{1-n} - (C_{A0})^{1-n}\right] = kt \tag{5.131}$$

A plot of $(-1/1-n)\left[(C_A)^{1-n} - (C_{A0})^{1-n}\right]$ versus kt should therefore give a straight line of slope equal to k passing through the origin. It is necessary to find the value of n to do this. One approach would be to assume values of n and iterate until the value is found that gives the best straight line. For this example of a unimolecular reaction, it is possible to be more systematic using what is known as the method of half-lives. The half-life of the reaction, $t_{0.5}$, is defined as the time required for the concentration to drop to half of its initial value. Substitution for $C_A = 0.5C_{A0}$ into Equation 5.131 gives

$$kt_{0.5} = \frac{1}{n-1}\left[\left(\frac{C_{A0}}{2}\right)^{1-n} - C_{A0}^{1-n}\right] \tag{5.132}$$

Equation 5.132 can be rearranged to give

$$(n-1)\,kt_{0.5} = (C_{A0})^{1-n}\left(2^{n-1} - 1\right) \tag{5.133}$$

Further rearrangement of Equation 5.133 gives

$$t_{0.5} = \left[\frac{2^{n-1}-1}{k(n-1)}\right](C_{A0})^{1-n} \tag{5.134}$$

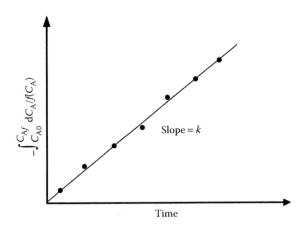

FIGURE 5.9
Integral method plot for a uni-molecular reaction with a power law rate equation. The value of n must give a straight line of slope equal to the value of the rate constant.

If we take the ln of both sides of Equation 5.134, we obtain

$$\ln(t_{0.5}) = (1 - n)\ln(C_{A0}) + \ln\left[\frac{2^{n-1} - 1}{k(n - 1)}\right] \tag{5.135}$$

From Equation 5.135, it can be seen that a plot of $\ln(t_{0.5})$ versus $\ln(C_{A0})$ should give a straight line of slope equal to $(1 - n)$. For an irreversible reaction, the data required for this plot can be obtained from a single CVBR experiment (see Example 5.3).

5.6.4.2 Complex Kinetics

When the rate equation is more complicated than a simple power law expression, the integral method becomes more complicated, although it is still possible to use it. Consider the rate equation for a unimolecular reaction

$$(-r_A) = \frac{kC_A}{1 + K_1C_A} \tag{5.136}$$

Substituting Equation 5.136 into the integral form of the CVBR mole balance equation gives

$$-\int_{C_{A0}}^{C_A} \left(\frac{1}{C_A} + K_1\right) dC_A = kt \tag{5.137}$$

Integration of Equation 5.138 between the limits gives

$$\ln\left(\frac{C_{A0}}{C_A}\right) + K_1(C_{A0} - C_A) = kt \tag{5.138}$$

It remains to find values of k and K_1 that give the best match between predicted and experimental concentration values. The procedure would be

1. Assume starting values for k and K_1.
2. Compute a set of values of C_A at values of time over which the experiment was performed.
3. Compute the sum of the square of the difference between the experimental and observed values.
4. Adjust the values of k and K_1 until the minimum value for the sum of the square of the difference (computed in step 3) is obtained.

The algorithm outlined in steps 1 through 4 is an optimization problem, and numerous methods exist for solutions of problems of this type.

Actually, the method of half-lives could also be applied to Equation 5.138. Substitution for $C_A = 0.5C_{A0}$ into Equation 5.138 gives

$$\frac{\ln(2)}{k} + \frac{K_1}{2k}C_{A0} = t_{0.5} \qquad (5.139)$$

A plot of the half-life versus initial concentration should thus give a straight line with the slope equal to $K_1/2k$ and the intercept equal to $\ln(2)/k$. This illustration shows that it may not always be obvious as to which technique to use.

Example 5.3

Consider the essentially irreversible, liquid-phase reaction that was examined in Example 5.2:

$$A \rightarrow C \qquad (5.140)$$

The concentration of A as a function of time was measured in an isothermal constant-volume batch reactor to yield the following data:

Time (min)	C_A (mol/L)
0	2.50
10	2.10
20	1.78
30	1.50
40	1.30
50	1.15
60	1.01
70	0.90
80	0.81
90	0.73
100	0.66

Assuming that a power law-type rate expression is able to correlate these data, use the integral method to determine the values of the kinetic parameters.

SOLUTION

This problem will be solved using the method of half-lives. We begin by estimating the half-life at as many of the concentration values as possible. At time zero, the concentration of A is 2.50. Half this value is 1.25, which falls between 40 and 50 min. Using linear interpolation between these two times, we obtain a time of 43.33 min corresponding to a concentration of 1.25. Similar analysis for the other data gives the following values:

Time (min)	C_{A0} (mol/L)	$C_{A0}/2$ (mol/L)	Half-life (min)
0	2.50	1.25	43.3
10	2.10	1.05	47.1
20	1.78	0.89	51.1
30	1.50	0.75	57.5
40	1.30	0.65	61.4
50	1.15	0.575	n/a

Notice that from the data supplied it is only possible to use the first five data points. Now, recall Equation 5.135:

$$\ln(t_{0.5}) = (1 - n)\ln(C_{A0}) + \ln\left[\frac{2^{n-1} - 1}{k(n - 1)}\right] \qquad (5.141)$$

A plot of ln $(t_{0.5})$ versus ln(C_{A0}) can now be made, where the slope should be equal to $(1 - n)$. This plot is shown in Figure 5.10. The slope of the line and the intercept are determined by least squares analysis. These values give

$$(1 - n) = -0.546 \quad \text{and} \quad \left[\frac{2^{n-1} - 1}{k(n - 1)}\right] = 70.96 \qquad (5.142)$$

Solving for n and k gives the following rate expression:

$$(-r_A) = 0.012 C_A^{1.54} \text{ mol/L} \cdot \text{min} \qquad (5.143)$$

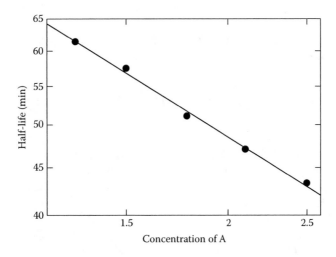

FIGURE 5.10
Half-life plotted as a function of reactant concentration on a log–log scale. The line is the least squares fit of the data points. The slope of the line is equal to –0.5462.

Note that the rate constant has the same value as that computed in Example 5.1 using the differential method, but the reaction order is higher. This fact can be attributed to the use of fewer data points as well as the different method of analysis.

Note: In this example of the integral method, the data were manipulated so that use could be made of linear regression analysis. However, it is possible to use the data directly if nonlinear regression analysis can be used. The integral form of the batch reactor equation for the power law model is

$$\frac{-1}{1-n}\left[(C_A)^{1-n} - (C_{A0})^{1-n}\right] = kt \tag{5.144}$$

Using a nonlinear solver to optimize the values of n and k gives the following rate expression:

$$(-r_A) = 0.012C_A^{1.52} \, \text{mol/L} \cdot \text{min} \tag{5.145}$$

This solution is close to that generated by the previous two methods. In practice, the value of the power in the rate expression would probably be rounded to 1.5, and thus all three methods give essentially the same results.

5.6.5 Temperature Dependence of Rate Constants

It was mentioned in Section 5.6.1 that rate data are frequently taken under isothermal reactor conditions. Subsequent analysis, either using the differential or integral methods, yields values of the rate parameters at the reaction temperature. It is usually desired to find the temperature dependence of the rate constants, and, therefore, the experiments are usually performed at a variety of reactor temperatures. The range of temperatures used in the experimental study should cover the range of operation of the full-size reactor that will be designed. It is standard practice to assume an Arrhenius type of temperature dependence for the rate constants, that is

$$k = A \, \exp\left(\frac{-E}{R_g T}\right) \tag{5.146}$$

The usual procedure is to plot $\ln k$ versus $1/R_g T$ to determine the values of A and E. The data points on this graph should be close to a straight line if the reaction rate constants follow an Arrhenius type temperature dependence. The slope of the line is equal to $-E$, as shown in Figure 5.2.

5.6.6 Methods of Isolation and Excess

The method of isolation and the method of excess are similar experimental techniques that are sometimes used to reduce the number of parameters that must be determined from any given set of data. The methods are especially useful when reversible reactions are being studied.

5.6.6.1 *Method of Isolation*

The method of isolation might be used to analyze a reversible reaction. In essence, the method consists of the analysis of rate data in the absence of products. For example, consider the following chemical reaction:

$$A \rightleftharpoons R + S$$

The proposed rate equation is

$$(-r_A) = \frac{k_1 C_A - k_2 C_R C_S}{(1 + k_3 C_S)} \tag{5.147}$$

Rate data could be collected in a batch reactor where the only reactant or product present at time zero was reactant A. Rate could be measured for a period of time where the concentrations of R and S remain relatively low, and thus the reverse reaction is not significant. During this time, the reaction rate can be approximated by

$$(-r_A) \approx k_1 C_A \tag{5.148}$$

These data could be used to determine the value of k_1 and an experiment could be performed with only R and S present at time zero, with data collected while the concentration of A is small.

$$+r_A = \frac{k_2 C_R C_S}{1 + k_3 C_S} \tag{5.149}$$

When fitting data to Equations 5.148 and 5.149, the number of parameters is lower than when fitting data to Equation 5.147. It is usually preferable to reduce the number of parameters to be determined from any given set of data because this increases the confidence in the result.

5.6.6.2 *Method of Excess*

Another method that is often used to reduce the number of parameters to be fit from a set of data is the method of excess. As the name implies, one of the reactants is present in a large excess, so that its concentration is essentially unchanged during the reaction. For example, consider the reaction between A and B that is governed by the following rate function:

$$(-r_A) = k C_A^\alpha C_B^\beta \tag{5.150}$$

If the initial concentration of B is very much larger than the initial concentration of A, the concentration of B will remain essentially constant, and therefore the rate equation can be approximated as

$$(-r_A) \approx \left(k C_{B0}^\beta \right) C_A^\alpha = k_1 C_A^\alpha \tag{5.151}$$

Data from this experiment could be used to compute the power α and the lumped reaction rate constant k_1, from which the value of k can be calculated. The experiment could be repeated for an excess of A to determine the value of β.

Example 5.4

Crystal violet dye (denoted CV) forms an intensely colored solution when mixed with water. Addition of a base causes a chemical reaction between the crystal violet and hydroxide ions with a subsequent loss of color. The rate of reaction is first order in both CV and OH, as follows:

$$(-r_{CV}) = kC_{CV}C_{OH} \tag{5.152}$$

A series of batch reactor experiments was carried out in which the concentration of sodium hydroxide was varied, and the concentration of crystal violet dye in the reactor was measured as a function of time. For each sodium hydroxide concentration, the time required for the crystal violet concentration to fall to one-half of its original value (the half-life) was recorded. The averaged results of several experiments are given in the table below.

Concentration NaOH (mol/L)	Half-life of CV (s)
0.10	69
0.15	46
0.20	35

Determine the value of the rate constant, k, from these data.

SOLUTION

The reaction rate data were recorded in a large excess of NaOH. Therefore, it is safe to assume that the concentration of NaOH remains unchanged during the course of the reaction. The reaction is thus a pseudo-first-order reaction, with an apparent rate constant that includes the value of the NaOH concentration, as follows:

$$(-r_{CV}) = -\frac{dC_{CV}}{dt} = k_{app}C_{CV} \quad \text{where } k_{app} = kC_{OH} \tag{5.153}$$

Equation 5.153 can be integrated analytically to give

$$\ln\left(\frac{C_{CV}}{C_{CV0}}\right) = -k_{app}t \tag{5.154}$$

Substituting for one-half the initial concentration gives the formula

$$\ln(2) = k_{app}t_{0.5} \tag{5.155}$$

We can substitute the measured half-life values from the data table into Equation 5.155 to determine values of the apparent rate constant for the three NaOH concentrations. The intrinsic rate constant, k, is calculated by dividing the apparent rate constant by the NaOH concentration. These results are shown in the table below.

C_{OH}	k_{app}	$k = k_{app}/C_{OH}$
0.10	0.010	0.10
0.15	0.015	0.10
0.20	0.020	0.10

We see from the table that the results consistently predict a value of $k = 0.10$. By using an excess of NaOH, the analysis is simplified, because the calculations are based on a first-order reaction.

5.6.7 Using a CSTR to Collect Rate Data

Reaction rate data may be obtained using other types of reactors, for example, flow reactors. The continuous stirred tank reactor is frequently chosen, especially for liquid-phase reactions. The reactor is usually run at steady state. The steady-state mole balance equation for the CSTR is

$$F_{A0} - F_{AE} - (-r_A)V = 0 \tag{5.156}$$

The inlet and outlet volumetric flow rates and concentrations are measured and the reactor volume is known. It is therefore easy to collect rate of reaction data directly from a CSTR without using the differential analysis necessary for the batch reactor. In this respect, the CSTR is better than the batch reactor. Once the rate data have been collected, the analysis proceeds in the same manner as described in Section 5.6.3. Furthermore, because the temperature in the CSTR is uniform and known, it is less important that the reactor be operated isothermally. It should be noted that each CSTR experiment will yield one data point, whereas a batch reactor experiment can be used to generate multiple data points. The CSTR therefore requires that more experiments be performed, which is the price that is paid for the ease of data analysis.

5.6.8 Using PFR to Collect Rate Data

A plug flow reactor can also be used to generate data for determining rate expressions. The data analysis is similar in many respects to the batch reactor. The PFR is less commonly used for kinetic experiments, and is often used when other reactors are not appropriate. The use of a tubular reactor to collect data for reaction rate analysis is discussed in more detail in Chapter 11, where its use in catalytic systems is discussed.

5.7 Summary

In this chapter, a somewhat detailed explanation of chemical kinetics has been presented. Both homogeneous reactions and an introduction to heterogeneous reactions were discussed. The area of chemical kinetics is a very broad one, and it should be emphasized that only the general outlines have been presented in this chapter, although all of the important kinetic concepts have been discussed. Several of the important points are summarized here.

The rate of chemical reaction depends on the temperature and the concentrations of the species present in the mixture. Fundamental rate theories can be used to glean some basic information about the temperature and concentration dependence of elementary reactions. The reaction may proceed via a simple elementary reaction involving only a single collision between two molecules, or it may involve a complex mechanism consisting of many steps. The reaction mechanism cannot in general be deduced from first principles, but rather is determined from experimental observations. It is important to distinguish between the mechanism of the reaction and a model for the reaction that is used in reactor analysis. The mechanism is the sequence of actual molecular steps in the reaction while a model is usually an approximation of the real situation. Models are used when the mechanism is not available, cannot be determined, or is too complex for practical use. Many industrial-scale reactors can be modeled with a sufficient degree of accuracy using empirical models for the rate. When using such a model for design, it is important to realize the danger of extrapolating beyond the range of experimental conditions upon which the original model was based.

Rate expressions may be derived from mechanisms, usually with some simplifying assumptions. However, a rate expression is developed; it contains constants that must be determined experimentally, by obtaining rate data in a laboratory reactor. There are many ways of obtaining and correlating rate data, and the chosen method usually depends on the nature of the reaction that is being studied. As mentioned above, because even mechanistically based rate equations contain constants that are determined from experimental data, that are in turn measured over a limited range of experimental conditions, the extrapolation of rate equations to other conditions should be avoided if possible. Conversely, when designing an experimental program to determine reaction rate parameters, it is important to cover the range of conditions at which the final design will operate.

PROBLEMS

5.1 The decomposition of ethane by a free radical mechanism was shown in Section 5.4.2. Other mechanisms have been proposed for this reaction, and one such possibility is represented by the following five-step free radical mechanism:

$$C_2H_6 \xrightarrow{k_1} 2CH_3^\bullet \text{ (second-order reaction)}$$
$$CH_3^\bullet + C_2H_6 \xrightarrow{k_2} CH_4 + C_2H_5^\bullet$$
$$C_2H_5^\bullet \xrightarrow{k_3} C_2H_4 + H^\bullet$$
$$H^\bullet + C_2H_6 \xrightarrow{k_4} H_2 + C_2H_5^\bullet$$
$$2C_2H_5^\bullet \xrightarrow{k_5} C_4H_{10}$$

Note that in contrast to the mechanisms examined in Chapter 5, the first step is assumed to be second order, that is, it involves a collision between two ethane molecules. Derive a rate expression for the formation of C_2H_4 using the steady-state hypothesis.

5.2 An important class of reactions involve enzymes, where the enzyme is used to promote the rate of reaction in a manner similar to a catalyst. Indeed, this area is sometimes referred to as enzyme catalysis. The enzyme reacts with the reactant molecule (often called the substrate), which forms an intermediate that decays to products, regenerating the enzyme. Consider the following enzyme reaction:

$$S + E \underset{k_2}{\overset{k_1}{\rightleftharpoons}} ES \underset{k_4}{\overset{k_3}{\rightleftharpoons}} P + E$$

where E represents an enzyme, and ES is an intermediate. We define the terms: $E_o = E + ES$ (the total enzyme concentration, including that contained in the intermediate), $k_m = (k_2 + k_3)/k_1$, and $K = k_1 k_3 / k_2 k_4$. Derive the following rate equation using the steady-state hypothesis:

$$(-r_S) = \frac{k_3 [E_o]\{[S] - ([P]/k)\}}{k_m + [S] + (k_4/k_1)[P]}$$

5.3 Consider the catalytic decomposition of ozone by chlorine. The proposed mechanism is

$$\text{Initiation} \quad Cl_2 + O_3 \xrightarrow{k_1} ClO^\bullet + ClO_2^\bullet \tag{1}$$

$$\text{Propagation} \quad ClO_2^\bullet + O_3 \xrightarrow{k_2} ClO_3^\bullet + O_2 \tag{2}$$

$$\text{Propagation} \quad ClO_3^\bullet + O_3 \xrightarrow{k_3} ClO_2^\bullet + 2O_2 \tag{3}$$

$$\text{Termination} \quad ClO_3^\bullet + ClO_3^\bullet \xrightarrow{k_4} Cl_2 + 3O_2 \tag{4}$$

Find the rate of disappearance of O_3.

5.4 The oxidation of phosgene is given by the following overall reaction:

$$2COCl_2 + O_2 \rightarrow 2CO_2 + 2Cl_2$$

The reaction can be activated by high-intensity light in a photochemical reaction. A proposed mechanism is

$$COCl_2 + hv \xrightarrow{k_1} COCl^\bullet + Cl^\bullet \tag{1}$$

$$COCl^\bullet + O_2 \xrightarrow{k_2} CO_2 + ClO^\bullet \tag{2}$$

$$COCl_2 + ClO^\bullet \xrightarrow{k_3} CO_2 + Cl_2 + Cl^\bullet \tag{3}$$

$$COCl^\bullet + Cl_2 \xrightarrow{k_4} COCl_2 + Cl^\bullet \tag{4}$$

$$Cl^\bullet + Cl^\bullet + M \xrightarrow{k_5} Cl_2 + M \tag{5}$$

where M is any molecule and $h\upsilon$ is the radiation. Apply the PSSH to $ClO^\bullet$ and $COCl^\bullet$ to develop a rate equation.

5.5 The original Rice–Herzfeld mechanism for the decomposition of acetone (CH_3COCH_3) is

$$CH_3COCH_3 \rightarrow CH_3^\bullet + COCH_3^\bullet$$
$$CH_3^\bullet + CH_3COCH_3 \rightarrow CH_4 + {}^\bullet CH_2COCH_3$$
$${}^\bullet CH_2COCH_3 \rightarrow CH_3^\bullet + CH_2CO$$
$$CH_3^\bullet + {}^\bullet CH_2COCH_3 \rightarrow C_2H_5COCH_3$$

Develop a rate equation for the decomposition of acetone based on this mechanism.

5.6 The pyrolysis of ethyl nitrate has the following proposed mechanism:

$$C_2H_5ONO_2 \xrightarrow{k_1} C_2H_5O^\bullet + NO_2 \tag{1}$$

$$C_2H_5O^\bullet \xrightarrow{k_2} CH_3^\bullet + CH_2O \tag{2}$$

$$CH_3^\bullet + C_2H_5ONO_2 \xrightarrow{k_3} CH_3NO_2 + C_2H_5O^\bullet \tag{3}$$

$$2C_2H_5O^\bullet \xrightarrow{k_4} CH_3CHO + C_2H_5OH \tag{4}$$

The reaction rates at different concentrations of ethyl nitrate are

$C_{C_2H_5ONO_2}$ (mol/m^3)	0.0975	0.0759	0.0713	0.2714	0.2436
$(-r_{C_2H_5ONO_2})$ (mol/m^{3}s)	0.0134	0.0122	0.0121	0.023	0.0209

Use the PSSH to deduce a rate equation and test it for consistency with the experimental data.

5.7 Consider the decomposition of acetaldehyde, given by the reaction

$$CH_3CHO \rightarrow CH_4 + CO$$

One proposed mechanism was illustrated in Section 5.4.2. Consider an alternative

$$CH_3CHO \xrightarrow{k_1} CH_3^\bullet + CHO$$

$$CH_3^\bullet + CH_3CHO \xrightarrow{k_2} CH_4 + CH_3CO^\bullet$$

$$CH_3CO^\bullet \xrightarrow{k_3} CH_3^\bullet + CO$$

$$2CH_3CO^\bullet \xrightarrow{k_4} CH_3COCOCH_3$$

Note the difference in the termination step for this mechanism, compared to the one in the text. Use the PSSH to derive a rate equation for this mechanism.

5.8 Consider the following overall reaction:

$$N_2O_5 \rightarrow 2NO_2 + \frac{1}{2}O_2$$

The observed rate expression for the decomposition of N_2O_5 is first order:

$$(-r_{N_2O_5}) = k[N_2O_5]$$

A proposed mechanism is

$$N_2O_5 \underset{k_2}{\overset{k_1}{\rightleftharpoons}} NO_2 + NO_3^*$$

$$NO_2 + NO_3^* \overset{k_3}{\longrightarrow} NO^* + O_2 + NO_2$$

$$NO^* + NO_3^* \overset{k_4}{\longrightarrow} 2NO_2$$

Show that this mechanism is consistent with the observed rate equation. Note that NO_2 does not react in step 2, but it affects the rate of decomposition of NO_3^*.

5.9 The gas-phase reaction $2A \rightarrow B + C + D$ is carried out isothermally in a constant volume batch reactor. Initially there is pure A at 1 atm pressure. The following pressures are recorded during the reaction:

Time (min)	0	1.2	1.95	2.90	4.14	5.7	8.1
Pressure	1	1.1	1.15	1.20	1.25	1.30	1.35

For a rate expression $(-r_A) = kC_A^n$, determine values of k and n.

5.10 An aqueous solution of ethyl acetate is reacted with sodium hydroxide in a batch reactor. The initial concentration of ethyl acetate is 5.0 g/L and that of caustic soda is 0.1 molar. The reaction is second order and irreversible. Values of the rate constant in L/mol·min are

$$k = 23.5 \text{ at } 0°C \quad \text{and} \quad 92.4 \text{ at } 20°C.$$

Calculate the time required to convert 95% of the ethyl acetate at 40°C. The formula for ethyl acetate is $C_2H_5(COOCH_3)$ and the molecular mass is 88.

5.11 Determine the reactor volume required to produce 50 kg a day of product R. The reactor is a stirred tank batch reactor which can be run for 8 h each day. The stoichiometry of the reaction is known to be 1 mol of reactant A consumed for each mole of R produced. The molecular weight of R is 50. The feedstock consists of an aqueous solution of A, concentration 1 mol/L. The kinetics of the

reaction are unknown, but the following kinetic data were obtained in a laboratory batch stirred tank reactor. The data are concentration–time data.

Time (h)	0	1	2	3	4	5	6	7	8	9	10
C_A (mol/L)	2	1.6	1.33	1.14	1.0	0.89	0.80	0.73	0.67	0.62	0.57

The data were obtained at the temperature at which the commercial reactor will
be run.

5.12 Consider the solid-catalyzed isomerization reaction

$$A \rightarrow B$$

The reaction takes place at 25°C and 100 kPa. The rate function and rate constant
are

$$(-r_A) = kC_A(\text{mol/kg cat} \cdot \text{s}) \quad \text{where } k = 2 \times 10^{-3}\, \text{m}^3/\text{kg cat} \cdot \text{s}$$

The feed to a recycle reactor consists of pure A at 25°C and 100 kPa, with a flow
rate of 1.2 m³/s. The reactor contains 500 kg catalyst. Calculate the fractional
conversion of A at recycle ratios of 0, 1.0, 5.0, 10.0, and ∞.

5.13 The half-life of a reaction in a batch reactor is defined as the time it takes for the
concentration of reactant to fall to one-half of its initial value. For a certain first-
order reaction, it is observed that at 25°C the half-life is 60 s and at 35°C it is 30 s.
Calculate the activation energy for the reaction.

5.14 A common guideline is that a reaction rate doubles for an increase in temperature of 10°C. Consider a second-order reaction, with rate expression given by
$(-r_A) = kC_A^2$.

   a. Calculate the activation energy required for the rate to double with an
      increase in temperature from 25°C to 35°C.

   b. Calculate the ratio of reaction rates at 100°C and 110°C for the case in part (a).

6

Nonideal Reactor Analysis

In Chapters 3 and 4, we developed the conservation equations used to calculate the concentrations and temperatures in reactors that obeyed the ideal reactor assumptions. Recall that in an ideal batch reactor, the contents were assumed to be perfectly mixed, and in flow reactors, the two extremes of plug flow and perfect mixing were used. These assumptions lead to simplicity in the mole and energy balances, and are valid in many industrial reactors, such that it is sufficient to use an ideal reactor model to analyze and predict the reactor performance to a level acceptable for design calculations. In other cases, however, the deviation from ideal reactor behavior may be sufficiently large that the use of an ideal reactor model will give an unacceptably large error. Deviations from ideal reactor performance are caused by many factors, and depend on the type of reactor and its mode of operation. The purpose of this chapter is to introduce the methods of analyzing nonideal systems, which can be used either to model a real reactor or to investigate the degree of nonideality of an existing system, with a view to improving its performance. The causes of deviation from the ideal case are first summarized, followed by a treatment of some methods of analyzing nonideal reactors. Finally, the topic of mixing is introduced. Note that the modeling of nonideal reactors is a complex subject and far from an exact science. Many techniques have been used to model nonideal reactors and what follows is an introduction to some possible methodology. Further information on the modeling of nonideal systems can be found in Levenspiel (1999) and Froment et al. (2011). A more advanced presentation of mixing in flow systems is given by Nauman and Buffham (1983), while those looking for an in-depth treatment of mixing phenomena can refer to Ottino (1989) or Baldyga and Bourne (1999).

6.1 Causes of Nonideal Reactor Behavior

Prior to discussing methods for accounting for deviations from ideal reactor behavior, some of the causes of such deviations are briefly discussed. Recall that ideal reactor models all involve an assumption about the mixing behavior inside the reactor. Mixing is defined and discussed in more detail later in this chapter: suffice to say at this point that the degree of mixing relates to the presence or absence of concentration and temperature gradients in the reactor vessel. Thus, ideal tank reactors (either batch or CSTR) are assumed to have a completely uniform composition and temperature, while the PFR has no radial gradients and no mixing in the axial direction. However, such assumptions may not be valid, especially in badly designed systems. The following sections discuss very briefly some of the causes of nonideal behavior.

6.1.1 Nonideal Behavior in CSTR

A typical CSTR consists of a tank, the contents of which are stirred by an impeller. It is assumed that the circulation of the fluid caused by the impeller rotation is sufficient to achieve perfect mixing. In practice, the fluid motion is not the same in all parts of the vessel; fluid velocities tend to be larger near the impeller and decrease as the distance from the impeller increases. Baffles are usually used as an aid in improving the circulation of fluid. Depending on the kinetics of the reaction and the circulation pattern in the vessel, it is possible for relatively stagnant zones of fluid to develop. These zones are sometimes referred to as "dead zones," and are illustrated in Figure 6.1. Stagnant zones can result from incorrectly sized impellers, inappropriate baffles, or incorrect tank geometry. Stagnant zones tend to lower the effective reactor volume. Imperfect mixing may also result from feed bypassing. The perfectly mixed assumption requires that fresh feed is instantly and completely mixed with the entire tank contents. However, if the inlet and outlet pipes are not placed correctly some of the feed may flow directly to the outlet without being mixed into the bulk fluid in the reactor: see Figure 6.1. The usual "rule of thumb" used when designing a CSTR is that a straight line drawn between the inlet and outlet stream locations should pass through the center of the impeller.

Good mixing also becomes more difficult to achieve as the viscosity of the fluid increases. In addition, the presence of multiple phases can also lead to nonideal behavior.

6.1.2 Deviations from Plug Flow Performance

The assumptions involved in the plug flow model were given in Chapter 3. The PFR is assumed to have zero axial mixing and perfect radial mixing as well as a flat velocity profile. All of these plug flow criteria are violated to a certain extent in real reactors. Even in highly turbulent flow, a radial velocity profile still exists in a tubular reactor. This velocity profile in turn results in concentration and temperature gradients. Furthermore, if a tubular reactor is operated with exchange of heat through the reactor wall, it is inevitable that a radial temperature gradient will develop in the reactor, which leads to a radial concentration gradient. In addition, there is always a degree of axial mixing in a tubular flow reactor, although the amount may be small. These factors cause a deviation from plug flow. If the deviation from plug flow is significant, the PFR model will lead to an unacceptable error, and a reactor model that more accurately reflects the true situation will be required. In packed beds, channeling of the fluid along the wall or flow maldistribution owing to poor packing can lead to deviations from plug flow.

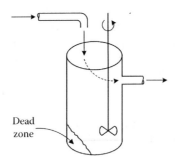

FIGURE 6.1
Typical CSTR illustrating possible causes of nonideal performance. Badly placed inlet and outlet lines can cause feed bypassing, and poorly designed impellers and baffles can result in the formation of dead zones.

6.1.3 Deviations from Perfect Mixing in Batch and Fed Batch Reactors

The deviations from perfect mixing in batch reactors have similar causes to the deviations in CSTR. Poorly designed impellers or baffles can lead to imperfect flow patterns. Incorrectly placed injection ports in semibatch reactors can also cause significant deviations from ideal performance, especially for relatively fast reactions.

6.2 Residence Time and Mixing

The relationship among the thermodynamics, transport phenomena, and kinetics governs the behavior of any chemical reactor. If the flow behavior does not follow one of the ideal models, a more complex methodology than that hitherto employed must be used for the reactor analysis. Theoretically, any reactor can be modeled by solving the appropriate equations of transport that describe the flow in the reactor. However, this can be a formidable task and, for turbulent flow especially, in many cases is still beyond the power of current analysis techniques. Furthermore, exact mass and energy transport models have not been formulated for all real reactor situations.

There are many factors that govern the performance of the reactor and these can be grouped into two broad classifications. The first is the amount of time that a reactant molecule spends in the reactor. The longer that reactants can spend in the reactor, the more chance there is for them to react. The time that molecules spend in the reactor is called the residence time, and the distribution of residence times for feed molecules is an important factor in determining the extent of reaction. The residence time is a key parameter in flow reactor performance.

The second major factor that determines reactor performance is the extent of mixing within the reactor. For all reactions involving more than one molecule, it is necessary that molecular collisions occur as a first step; therefore, the concentration at the molecular level is important. The degree of mixing depends not only on the reactor configuration and operating conditions, but also on the fluid and reactant properties. Mixing is a complex phenomenon that is analyzed at several scales, but ultimately it is the mixing at the molecular level that determines the extent of reaction.

In the following sections, the concept of the RTD and its use in evaluating and modeling reactor performance is presented. This explanation is followed by an introduction to mixing and its relationship to the RTD. At the end of this chapter, the reader should have some useful tools for the analysis of nonideal reactors.

6.3 RTD Function

This section introduces basic residence time theory in the context of a flowing system. The description is based on the RTD of nonreacting species in an arbitrary flow system with a constant-density fluid. The relationship to a chemically reacting species is explained at the end of this section.

Residence time theory deals with the age of particles within a flow system. For a chemical reactor, the system is simply the volume of the reactor; particles enter and leave the system in the process streams. The particles may be atoms, molecules, or fluid elements that are conserved as they flow through the system. The age, or residence time, of a particle in the system is the elapsed time between the time at which the particle enters the system and the time at which it leaves. If a particle reenters the system after it leaves, the ageing of that particle resumes from the value it had at its previous exit. It should also be noted that no particle is allowed to remain in the system forever; all particles must have an original entrance and a final departure, although, as noted above, a particle may leave and subsequently reenter the system. The age of the particle when it finally leaves the system is called the residence time.

6.3.1 Measurement of RTD

The RTD is usually measured using an inert tracer. A tracer is a material that is nonreactive, nonadsorbing, and has properties that closely resemble the species for which it is desired to have the RTD. The tracer must have a distinguishing feature that makes it possible to measure its concentration easily. A tracer might be a colored dye whose concentration can be measured using a spectrometer, or a radioactive species that is monitored using scintillation counting. Neutrally buoyant particles (particles whose density is the same as that of the fluid) are also used.

Consider a simple system of a fixed volume that has one entrance and one exit. Suppose that a finite quantity of tracer is injected into the inlet of the system. The tracer molecules flow through the system and leave over some period of time, until finally all of the tracer molecules leave the system. Depending on the flow pattern inside the system, the time at which the tracer molecules exit the system may vary, and thus the residence time of the tracer molecules has a distribution, which depends on the flow pattern in the system. By measuring the concentration of the tracer in the effluent stream, the RTD can be determined.

6.3.2 RTD Function

The distribution of the residence times of the tracer molecules can be represented by a density function, denoted $f(t)$. This function, which is also called the RTD function, is a measurement of the fraction of molecules that have the residence time t. A possible plot of the density function versus the residence time is shown in Figure 6.2.

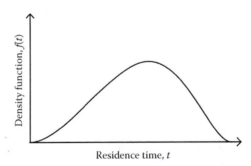

FIGURE 6.2
Typical residence time distribution function for a flow system. The density function can be used to determine deviations from ideal reactor performance.

The density function can be described in terms of probabilities. The probability that the residence time of a tracer molecule lies between time t and $t + dt$ is

$$f(t)\, dt \tag{6.1}$$

Since, all the molecules must enter and eventually leave the system, it follows from the probability addition rule that

$$\int_0^\infty f(t)\, dt = 1 \tag{6.2}$$

The RTD function may be used directly in the analysis of reactors, or related functions may be used depending on the application.

6.3.3 Cumulative Distribution and Washout Functions

Two useful functions that are related to the density function are the cumulative distribution function and the washout function. The cumulative distribution function gives the fraction of molecules with residence times from time 0 to time t, and is obtained by integrating the density function between 0 and t.

$$F(t) = \int_0^t f(t)\, dt \tag{6.3}$$

The washout function gives the fraction of molecules with residence times in the range t to ∞. It is also calculated from the density function:

$$W(t) = \int_t^\infty f(t)\, dt \tag{6.4}$$

It is clear that the addition of Equations 6.3 and 6.4 gives Equation 6.2; thus

$$F(t) + W(t) = 1 \tag{6.5}$$

Furthermore, the relationship between the functions in derivative form is

$$\frac{dF(t)}{dt} = -\frac{dW(t)}{dt} = f(t) \tag{6.6}$$

A typical cumulative distribution function is illustrated in Figure 6.3, and the corresponding washout function is shown in Figure 6.4.

6.3.4 Determination of $f(t)$ from Pulse Input

The direct experimental determination of $f(t)$ requires a pulse of tracer to be injected into the system. Essentially, this requires an amount of tracer to be injected into the inlet of the

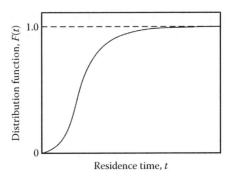

FIGURE 6.3
Typical cumulative distribution function for a flow system. The cumulative distribution function gives the fraction of particles with a residence time less than t. The cumulative distribution function can be obtained by integrating the density function.

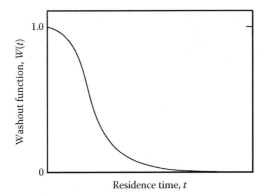

FIGURE 6.4
Typical washout function for a flow system. The washout function gives the fraction of particles with a residence time greater than t. It can be obtained from the residence time density function.

vessel at a single instant of time. The ideal shape of the injected pulse is mathematically described by a Dirac delta function. In practice, the tracer should be injected in as short an elapsed time as is experimentally possible to achieve. The concentration of the tracer in the effluent is then measured as a function of time, and the density function can be evaluated using the following analysis. The fraction of tracer molecules that leaves the system over some time interval dt is

$$\frac{dN}{N_0} = f(t)\,dt \tag{6.7}$$

In Equation 6.7, dN represents the number of tracer molecules that leave the system over the time interval dt, and N_0 is the total number of tracer molecules injected. The number of molecules can be expressed in terms of the concentration and the volumetric flow rate, Q:

$$dN = QC(t)\,dt \tag{6.8}$$

$C(t)$ is the concentration of tracer in the effluent stream as a function of time. The total number of molecules initially injected can also be calculated from the effluent profile, by integration over a long period of time:

$$N_0 = \int_0^\infty QC(t)\, dt \qquad (6.9)$$

Substitute Equations 6.8 and 6.9 into Equation 6.7 and rearrange to obtain the RTD function for a constant-density (constant volumetric flow rate) system:

$$f(t) = \frac{C(t)}{\displaystyle\int_0^\infty C(t)\, dt} \qquad (6.10)$$

Therefore, the RTD function may be obtained by measuring the effluent tracer concentration as a function of time with a pulse injection into the system inlet. The cumulative distribution and washout functions may be calculated from $f(t)$ by integration, using Equations 6.3 and 6.4. Note that it is not necessary to know how many molecules were injected into the system, although if the injected amount is independently known, the value can be used to check the mass balance by comparison with the result of Equation 6.9.

6.3.5 Determination of $f(t)$ from Step Change

Rather than inject a pulse of tracer into the inlet, another possibility is to increase or decrease the tracer concentration in a stepwise manner. Typically, this involves either a step increase or a step decrease in tracer concentration. In the case of a step increase from an initial value, $C(0)$, to some final value, $C(\infty)$, the outlet concentration as a function of time gives the cumulative distribution function, $F(t)$, directly from the following equation:

$$F(t) = \frac{C(t) - C(0)}{C(\infty) - C(0)} \qquad (6.11)$$

If the inlet tracer concentration has a step decrease from an initial value, $C(0)$, to a final value, $C(\infty)$, then the effluent concentration can be used to compute the washout function, $W(t)$:

$$W(t) = \frac{C(t) - C(\infty)}{C(0) - C(\infty)} \qquad (6.12)$$

The washout function is considered by many designers to be the best function to use for defining the moments of the distribution function.

6.3.6 Means and Moments

In addition to characterizing the RTD using the three functions $f(t)$, $F(t)$, and $W(t)$, the residence time can also be characterized using the moments of the RTD function. These moments are used in some reactor models for the RTD, as will be seen in the following sections. Several types of moments are used. The first type is the moment around the origin, which is defined in terms of the density function as

$$\mu_n = \int_0^\infty t^n f(t)\,dt \tag{6.13}$$

The value of n determines which moment is calculated. For example, when $n = 0$, the zeroth moment is obtained, which is equal to one (see Equation 6.2).

$$\mu_0 = \int_0^\infty f(t)\,dt = 1 \tag{6.14}$$

The first moment about the origin is the *mean residence time*, t_m

$$\mu_1 = t_m = \int_0^\infty t f(t)\,dt \tag{6.15}$$

The mean residence time is the average time that molecules spend in the system. For a constant-density system, the mean residence time can also be calculated from the system volume, V, and the volumetric flow rate, Q:

$$t_m = \frac{V}{Q} \tag{6.16}$$

Recall that for a constant-density system the mean residence time is equal to the space time. Calculating the mean residence time in the two different ways can be used as a check on the accuracy of the experimental data used to calculate the RTD. The moments can also be determined from the washout function, as follows:

$$\mu_n = n\int_0^\infty t^{n-1} W(t)\,dt \tag{6.17}$$

In terms of the washout function, the mean residence time is therefore equal to

$$t_m = \int_0^\infty W(t)\,dt \tag{6.18}$$

Rather than compute the moments about the origin, it is common for the moments of the RTD function to be computed about the mean residence time. These moments are called central moments, and are defined as

$$\mu_n' = \int_0^\infty (t - t_m)^n f(t)\, dt \tag{6.19}$$

The variance, σ^2, of the RTD is the second central moment:

$$\sigma^2 = \mu_2' = \int_0^\infty (t - t_m)^2 f(t)\, dt \tag{6.20}$$

The variance measures the spread of the distribution about the mean. The third central moment, the skewness, measures the symmetry of the distribution about the mean:

$$s^3 = \mu_3' = \int_0^\infty (t - t_m)^3 f(t)\, dt \tag{6.21}$$

The central moments are related to the moments about the origin by

$$\mu_0' = \mu_0 = 1 \tag{6.22}$$

$$\mu_1' = 0 \tag{6.23}$$

$$\mu_2' = \mu_2 - t_m \tag{6.24}$$

and so on. The use of the moments in RTD models will be shown shortly.

6.3.7 Normalized RTD Functions

It is common when using residence time theory to work with a normalized RTD, defined as the residence time divided by the mean residence time. This dimensionless time is given by

$$\theta = \frac{t}{t_m} \tag{6.25}$$

The washout function expressed in terms of θ follows the relationship

$$\int_0^\infty W(\theta)\, d\theta = 1 \tag{6.26}$$

The density functions for the two cases are related by

$$f(\theta)\, d\theta = f(t)\, dt \tag{6.27}$$

Because $t_m\,d\theta = dt$, it follows that

$$f(\theta) = t_m f(t) \tag{6.28}$$

The other relationships do not change. The normalized cumulative distribution function is

$$F(\theta) = \int_0^\theta f(\theta)\,d\theta \tag{6.29}$$

The normalized washout function is

$$W(\theta) = \int_\theta^\infty f(\theta)\,d\theta \tag{6.30}$$

The moments about the origin and the central moments for the normalized RTD function are

$$v_n = \int_0^\infty \theta^n f(\theta)\,d\theta \tag{6.31}$$

$$v_n' = \int_0^\infty (\theta - 1)^n f(\theta)\,d\theta = \frac{\mu_n'}{t_m^n} \tag{6.32}$$

The normalized RTD curves are useful for direct comparison of different-size reactors, for example, because the mean residence time is eliminated as a variable. As shown in the next section, the normalized RTD is the same for every perfectly mixed stirred tank reactor.

Example 6.1

Consider a vessel through which a fluid is flowing. To determine the RTD function, a pulse injection of the tracer is done and the outlet concentration is monitored as a function of time. The output concentration in mol/L measured as a function of time in seconds is

t	0	60	120	180	240	300	360	420	480	540	600	720	840
C	0	1	5	8	10	8	6	4	3	2.2	1.5	0.6	0

Plot the RTD, the cumulative distribution, and the washout functions.

SOLUTION

We first plot the data (C vs. t) to obtain the concentration time curve, which is shown in Figure 6.5. The points in the plot have simply been connected by straight lines. To obtain

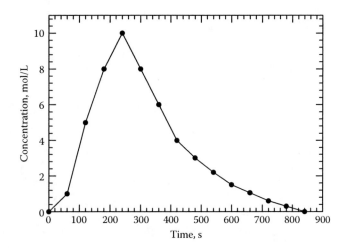

FIGURE 6.5
Concentration versus time plot of the experimental data for Example 6.1. The experimental points have simply been connected by straight lines.

the $f(t)$ curve, we need to integrate the concentration curve, that is, we must evaluate the integral

$$\int_0^\infty C(t)dt$$

Obviously it is not possible to integrate over infinite time. The practical limits on the integral are the time of first appearance of the tracer in the effluent to the time when the effluent concentration falls essentially to zero, which in this case is 0–840 s. The value of this integral is the area under the curve, which can be found using a numerical integration formula (e.g., Simpson's rule or the trapezoidal rule). If Simpson's rule is to be used, it is necessary to estimate values at 660 and 780 s to produce even intervals. These numbers were obtained by linear interpolation. Then, using numerical integration, we find the following value for the integral over the time interval of interest:

$$\int_0^{840} C(t)dt = 3008 \text{ mol} \cdot \text{s/L}$$

Once the cumulative concentration has been evaluated, the RTD function is evaluated directly from the concentration data:

$$f(t) = \frac{C(t)}{3008} \frac{\text{mol/L}}{\text{mol} \cdot \text{s/L}}$$

The curve has the same shape as the $C(t)$ curve but the scale on the vertical axis is different. The RTD function plot is shown in Figure 6.6. The cumulative distribution function, $F(t)$, can be obtained by integrating the $f(t)$ curve, again using numerical integration. The value of $F(t)$ at any time is obtained by integrating from time zero to each time value that corresponds to a data point, according to the formula

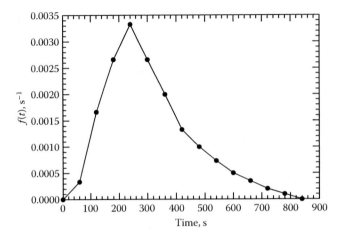

FIGURE 6.6
Residence time distribution function obtained by integrating the experimental concentration–time data for
Example 6.1. As with Figure 6.5, the data points have been connected by straight lines.

$$F(t) = \int_0^t f(t)\,dt$$

By employing numerical integration to the experimental $f(t)$ curve, the $F(t)$ data can be
calculated at each time point, and then the total curve generated. This curve is shown in
Figure 6.7. In a similar manner, the washout function may be generated, by integrating
from each time point to the upper practical limit of 840 s using the equation

$$W(t) = \int_t^{840} f(t)\,dt$$

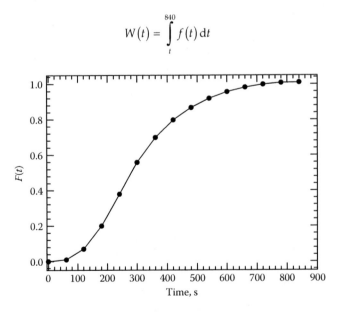

FIGURE 6.7
The cumulative distribution function for the data of Example 6.1. The curve progresses from a value of zero to
one as all of the tracer molecules leave the system.

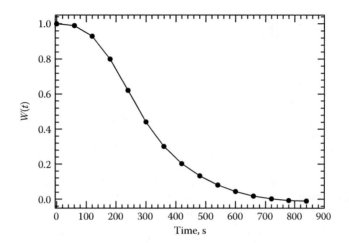

FIGURE 6.8
Washout function for the data of Example 6.1. This line is the mirror image of the cumulative distribution function shown in Figure 6.7.

The washout graph is shown as Figure 6.8. Observe that it is essentially the mirror image of Figure 6.7, the cumulative distribution function.

The data used for the plots generated in this example are shown in the following table:

Time	$C(t)$	$f(t)$	$F(t)$	$W(t)$
0	0	0.00E+00	0.000	1.000
60	1	3.32E−04	0.010	0.990
120	5	1.66E−03	0.070	0.930
180	8	2.66E−03	0.199	0.801
240	10	3.32E−03	0.379	0.621
300	8	2.66E−03	0.559	0.441
360	6	1.99E−03	0.698	0.302
420	4	1.33E−03	0.798	0.202
480	3	9.97E−04	0.868	0.132
540	2.2	7.31E−04	0.920	0.080
600	1.5	4.99E−04	0.956	0.044
660	1.05	3.49E−04	0.982	0.018
720	0.6	1.99E−04	0.998	0.002
780	0.3	9.97E−05	1.007	−0.007
840	0	0.00E+00	1.010	−0.010

6.4 RTD in Ideal Reactors

In Section 6.3, the experimental determination of the RTD for a given system was discussed. If it is possible to write down the exact mole balance equation for the flow vessel in question, it is possible to calculate the RTD from first principles. The RTD in the PFR and the CSTR can be computed easily because the mole balances are known. These mole balance equations are used to calculate the RTD in the following sections.

6.4.1 RTD in Perfectly Mixed CSTR

We consider first a perfectly mixed stirred tank with a constant-density fluid flowing in and out at equal mass flow rates. The RTD can be analyzed using either a step-down or a pulse injection. Assume, in the first instance, that the inlet stream to the vessel contains a tracer at a concentration of C_0, and that the system is at steady state. The tank and effluent concentration of the tracer are thus equal to C_0. At time zero, the tracer concentration in the feed is suddenly reduced to zero. Because there is no more tracer injected, the concentration in the tank begins to decline. The transient mole balance for the tracer is written (using the principles given in Chapter 3) as

$$-QC = V\frac{dC}{dt} \tag{6.33}$$

For a constant-density fluid, the space time is equal to the mean residence time, and Equation 6.33 can be rearranged to give

$$-\frac{Q}{V}dt = -\frac{dt}{t_m} = \frac{dC}{C} \tag{6.34}$$

Equation 6.34 is integrated to give the outlet concentration as a function of time

$$C(t) = C_0 \exp\left(-\frac{t}{t_m}\right) \tag{6.35}$$

Substitute the concentration from Equation 6.35 into Equation 6.12 to obtain the washout function directly:

$$W(t) = \frac{C(t) - C(\infty)}{C(0) - C(\infty)} = \frac{C_0 \exp(-t/t_m)}{C_0} = \exp\left(-\frac{t}{t_m}\right) \tag{6.36}$$

The normalized washout function is then simply given as

$$W(\theta) = \exp(-\theta) \tag{6.37}$$

Furthermore, the RTD function (density function) is obtained by differentiation:

$$f(t) = -\frac{dW(t)}{dt} = \frac{1}{t_m}\exp\left(-\frac{t}{t_m}\right) \text{ and } f(\theta) = \exp(-\theta) \tag{6.38}$$

The normalized density function is equal to the washout function for the perfectly mixed CSTR. The RTD function can also be obtained from a pulse injection. When a pulse of tracer is injected into the tank, it is instantly mixed with the vessel contents to give an initial concentration of C_0. As the material leaves the CSTR, the concentration of the tracer in the effluent begins to decline. It can be seen that this mole balance is identical to that for the step-down case, and the solution is given by Equation 6.35. When the concentration function is substituted into Equation 6.10, we obtain the RTD function:

$$f(t) = \frac{C(t)}{\int_0^\infty C(t)\,dt} = \frac{C_0 \exp\left(-(t/t_m)\right)}{\int_0^\infty C_0 \exp\left(-(t/t_m)\right)dt} = \frac{\exp\left(-(t/t_m)\right)}{t_m} \tag{6.39}$$

The washout function is obtained by the application of Equation 6.4:

$$W(t) = \int_t^\infty \frac{1}{t_m} \exp\left(-\frac{t}{t_m}\right)dt = \exp\left(-\frac{t}{t_m}\right) \tag{6.40}$$

The normalized RTD function is written as

$$f(\theta) = \exp(-\theta) \tag{6.41}$$

Equation 6.41 is the normalized RTD function for all perfectly mixed continuous stirred tanks with a constant-density fluid. The normalized washout function for all perfectly mixed tanks is

$$W(\theta) = \exp(-\theta) \tag{6.42}$$

6.4.2 RTD in Plug Flow System

The RTD function in a tube with plug flow is envisaged most easily in terms of the cumulative distribution or the washout function. Suppose that a steady-state plug flow system is subjected to a step change in the inlet concentration of a tracer (step up or step down). There is no axial mixing in plug flow, so the outlet concentration profile will be exactly the same as the inlet profile, and a sudden step change of the outlet concentration would be observed after a time delay equal to the mean residence time. This pattern is shown in Figure 6.9.

This behavior means that all molecules in a plug flow system have the same residence time. The cumulative distribution function can be expressed as

$$F(t) = 0 \quad t < t_m$$
$$F(t) = 1 \quad t > t_m \tag{6.43}$$

The distribution function is defined in terms of the Dirac delta function, $\delta(x)$. The Dirac function equals zero at all values of x except $x = 0$, and has the further property:

$$\int_0^\infty \delta(x)\,dx = 1 \tag{6.44}$$

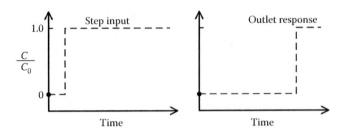

FIGURE 6.9
Response of a plug flow system to a step increase in inlet tracer concentration. The lack of axial mixing implies that the outlet response curve has the same shape as the inlet change.

The RTD function is

$$f(t) = \delta(t - t_m) \tag{6.45}$$

The moments of the normalized distribution all equal one, while the central moments equal zero.

6.5 Modeling RTD

From a reactor analysis perspective, it is usually desirable to use the RTD data to predict reactor performance. In other words, we want to use the RTD to model a reactor. The usual approach is to develop a model flow system that has the same RTD as the existing system. Provided that the conservation equations can be written for this model system, the performance of the real vessel can be predicted. Empirical models with one or more adjustable parameters are often used to model the RTD in vessels. Generally speaking, the more adjustable parameters that are used in a model the better the correlation between a predicted and observed RTD will be. The adjustable parameters in RTD models are often related to the moments of the RTD and are usually expressed in terms of the dimensionless moments about the mean. The first two dimensionless moments about the mean are always given and are therefore not variables, that is, $\mu_0' = 1$ and $\mu_1' = 0$. The moments about the mean are selected so that the mean residence time is not a variable. Quite often the mean residence time is known or may be calculated from other data. For example, for a constant-density fluid, the mean residence time is given by the ratio of the reactor volume to the volumetric flow rate:

$$t_m = \frac{V}{Q} \tag{6.46}$$

The following sections describe some common RTD models.

6.5.1 Multiple Vessel Models

It was seen in Chapter 3 that the mole balance equations for the ideal reactors, PFR and CSTR, were relatively straightforward to write. Many industrial reactors are either tubular reactors or tank reactors, but may not necessarily be operated under conditions of either plug flow or perfect mixing. Many such large-scale reactors are modeled as combinations of ideal reactors, usually in series, sometimes with additional complexities. The following sections describe some of the more common combinations of tanks and tubular reactors that have been used to model industrial reactors. These types of models are sometimes referred to as compartment models (Levenspiel, 1999) or tank and column models (Nauman and Buffham, 1983).

6.5.1.1 Parallel Plug Flow Reactors

It is always possible to devise a combination of ideal reactors to fit an experimental RTD curve. Indeed, any RTD can be closely approximated using a sufficient quantity of plug flow reactors in parallel. For example, consider the washout function shown in Figure 6.10. A set of parallel PFR can be made to fit the curve by adjusting the length of each reactor to fit the curve, as shown in the figure. The more reactors that are used, the better will be the approximation, although usually it is desirable to have as few elements as possible and for the model to have some physical meaning.

6.5.1.2 Fractional Tubularity Model

A simple multiple reactor model is the fractional tubularity model, which consists of a PFR in series with a CSTR, as shown in Figure 6.11. The order of the reactors is not important in regard to the RTD, that is, the same RTD is obtained regardless of which reactor is placed first. The single adjustable parameter in this model is the fractional tubularity, which is the fraction of the total reactor volume that is assigned to the plug flow element. If the PFR volume is V_P and the CSTR volume is V_S, the fractional tubularity is defined as

$$\theta_P = \frac{V_P}{V_P + V_S} \tag{6.47}$$

It was seen in Section 6.2 that the PFR is characterized by a sharp breakthrough of the washout function, while the CSTR exhibits an exponential decay. The cumulative

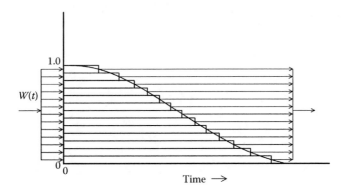

FIGURE 6.10
Any RTD can be approximated using a system of parallel PFR. The length of the reactors is adjusted to match the washout function.

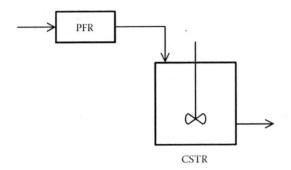

FIGURE 6.11
The fractional tubularity model is one of the simplest multiple vessel models. It models the reactor as a PFR and a CSTR in series. The RTD is the same regardless of the order of the reactors.

distribution function for the two reactors in series thus has a sharp first appearance followed by an exponential tail, as shown in Figure 6.12. The time at which the washout function shows a sharp breakthrough of the tracer compound is equal to the mean residence time of the PFR. The position of the sharp breakthrough is adjusted by changing the fractional tubularity. The dimensional form of the cumulative distribution function is

$$F(t) = 1 - \exp\left(-\frac{t - t_{mP}}{t_{mS}}\right) \quad \text{for} \quad t > t_{mP} \tag{6.48}$$

The mean residence time of the PFR is

$$t_{mP} = \frac{V_P}{Q} \tag{6.49}$$

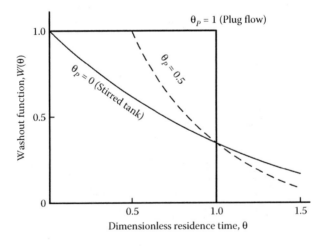

FIGURE 6.12
The washout function for the fractional tubularity model is characterized by a sharp first appearance followed by an exponential tail. The point of breakthrough depends on the relative volume of the PFR.

The mean residence time of the CSTR is

$$t_{mS} = \frac{V_S}{Q} \tag{6.50}$$

The dimensionless form of the cumulative distribution function is

$$F(\theta) = 1 - \exp\left(-\frac{\theta - \theta_P}{\theta_S}\right) \quad \text{for} \quad \theta > \theta_P \tag{6.51}$$

The dimensionless variance is

$$\sigma_\theta^2 = (1 - \theta_P)^2 = \theta_S^2 \tag{6.52}$$

The fractional tubularity can be determined directly from the observed RTD curve or calculated from the variance of the RTD.

6.5.1.3 Tanks-in-Series Model

The tanks-in-series model uses N equally sized perfectly mixed stirred tank reactors in series. The value of N is selected so as to match as closely as possible an observed or expected RTD. Consider, as an illustration, three equally sized tanks in series through which a constant-density fluid flows giving an equal space time (and mean residence time) for each reactor (see Figure 6.13). At time zero, a pulse of the tracer is injected into the first tank. This tracer instantaneously becomes uniformly distributed in the first tank, giving an initial concentration of C_0. The outlet concentration from the first vessel is given by Equation 6.35:

$$C_1(t) = C_0 \exp\left(-\frac{t}{t_{mi}}\right) \tag{6.53}$$

where t_{mi} is the mean residence time in a single tank. The mole balance equation for the second tank is obtained from the perfectly mixed mole balance, without reaction:

$$V\frac{dC_2}{dt} = Q(C_1 - C_2) \tag{6.54}$$

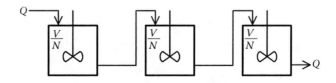

FIGURE 6.13
The tanks-in-series model uses a number of equally sized CSTR placed in series. The volume of each reactor is the total volume divided by the number of reactors.

Substitute for $C_1(t)$ from Equation 6.53 and solve

$$C_2(t) = C_0 \frac{t}{t_{mi}} \exp\left(-\frac{t}{t_{mi}}\right) \tag{6.55}$$

Performing the mole balance on vessel 3 and solving in a similar manner give the outlet concentration with time from the third reactor:

$$C_3(t) = \frac{C_0}{2} \left(\frac{t}{t_{mi}}\right)^2 \exp\left(-\frac{t}{t_{mi}}\right) \tag{6.56}$$

The RTD function based on the outlet from reactor 3 is then obtained using our definition of the RTD function:

$$f(t) = \frac{C_3(t)}{\displaystyle\int_0^\infty C_3(t)\,dt} = \frac{\dfrac{C_0}{2}(t/t_{mi})^2 \exp\left(-(t/t_{mi})\right)}{\displaystyle\int_0^\infty \dfrac{C_0}{2}(t/t_{mi})^2 \exp\left(-(t/t_{mi})\right)dt} = \frac{t^2 \exp\left(-(t/t_{mi})\right)}{2t_{mi}^3} \tag{6.57}$$

In Equation 6.57, the mean residence time (or space time), t_{mi}, is the value for a single vessel, and not for the total system volume. Equation 6.57 can be generalized to a system of N tanks:

$$f(t) = \frac{t^{N-1}}{(N-1)!\,t_{mi}^N} \exp\left(-\frac{t}{t_{mi}}\right) \tag{6.58}$$

Equation 6.58 can be written in terms of the mean residence time for the entire system, t_m, as

$$f(t) = \frac{t^{N-1}N^N}{(N-1)!\,t_m^N} \exp\left(-N\frac{t}{t_m}\right) \tag{6.59}$$

The normalized distribution function is

$$f(\theta) = \frac{\theta^{N-1}N^N}{(N-1)!} \exp\left(-N\theta\right) \tag{6.60}$$

The normalized washout function is

$$W(\theta) = \exp\left(-N\theta\right) \sum_{i=0}^{N-1} \frac{\theta^i N^i}{i!} \tag{6.61}$$

To utilize the tanks-in-series model, we need to determine the number of tanks that will give the same RTD as for the real system. This number may be obtained from the variance of the RTD, which is given by the second normalized moment about the mean (see Equation 6.32):

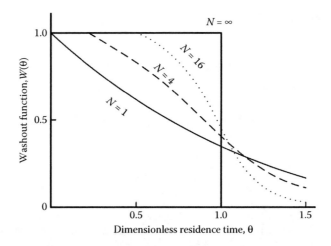

FIGURE 6.14
The shape of the washout function from the tanks-in-series model depends on the number of tanks. For one tank, the washout function is exponential, while as the number of tanks increases, the performance approaches that of a plug flow system.

$$v'_2 = \sigma_\theta^2 = \int_0^\infty (\theta - 1)^2 \frac{\theta^{N-1} N^N}{(N-1)!} \exp(-N\theta)\, d\theta \tag{6.62}$$

Solution of the integral equation gives the result

$$\sigma_\theta^2 = \frac{(N+2-1)!}{N \times N!} - 1 = \frac{(N+1)!}{N \times N!} - 1 = \frac{(N+1)}{N} - 1 = \frac{1}{N} \tag{6.63}$$

The variance of the RTD is seen to be equal to the inverse of the number of tanks used. Therefore, if the RTD curve is known, the appropriate number of tanks in series to use may be readily calculated. The tanks-in-series model is widely used as a reactor model. For example, a tubular reactor with a slight deviation from plug flow may be modeled as a large number of tanks in series. The typical shape of the washout function from a series of CSTR is shown in Figure 6.14 for various numbers of tanks.

Example 6.2

Consider the vessel with the RTD shown in Example 6.1. Calculate the number of tanks in series that should be used to model this vessel.

SOLUTION

To determine the number of tanks in series, we need the variance. The variance is computed as the second normalized moment about the mean residence time. Therefore, we first compute the mean residence time from the formula

$$t_m = \int_0^\infty t f(t)\, dt$$

By numerical integration, $t_m = 310.1$ s. The variance is given by

$$\sigma_\theta^2 = \int_0^\infty (\theta - 1)^2 f(\theta) d\theta = \int_0^\infty \left(\frac{t}{t_m} - 1 \right)^2 f(t) dt$$

The equation is integrated numerically to give $\sigma_\theta^2 = 0.232$. For the tanks-in-series model

$$\sigma_\theta^2 = \frac{1}{N}$$

Solving this equation gives $N = 4.31$. Because only integers are acceptable, the number of tanks is taken as four. The model system would consist of four tanks in series, each with a volume 25% of the volume of the real system. Because the tanks-in-series model is restricted to an integer number of tanks, the match between the model and the observed RTD will not be perfect.

It is worthwhile to consider the output from a series of tanks as the number of tanks increases. The normalized RTD function for a series of perfectly mixed tanks is shown in Figure 6.15 for an increasing number of tanks. This figure represents a response to a pulse input to the system. It can be seen as the number of tanks increases, the shape of the response curve becomes more symmetrical, and the maximum moves close to a dimensionless time of 1. For an infinite number of tanks, the curve would become Gaussian in

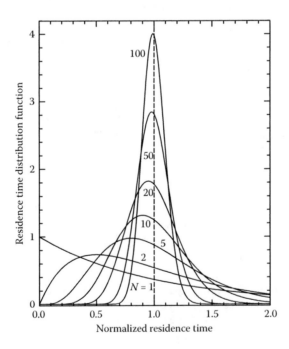

FIGURE 6.15
The shape of the residence time distribution function with an increasing number of tanks. The curve tends to become more symmetrical and, in the limit, becomes a Gaussian distribution.

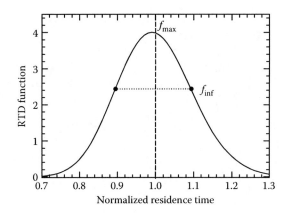

FIGURE 6.16
A portion of the residence time distribution curve for the tanks-in-series model with 100 tanks.

shape, although it approaches this shape for a large but finite number of tanks. A portion of the curve for 100 tanks is given in Figure 6.16. The key features of this curve are summarized by the following equations. The maximum value of the RTD function occurs a dimensionless time of

$$\theta_{max} = \frac{N-1}{N} \tag{6.64}$$

At this value of dimensionless time, the maximum value of the RTD function is

$$f(\theta)_{max} = \frac{N(N-1)^{N-1}}{(N-1)!} \exp(1-N) \tag{6.65}$$

The inflection points on the curve are the positions where the rate of change of the slope changes from positive to negative (or vice versa). The width of the response curve at across these points (marked by the dotted line in the figure) is given by

$$\frac{\Delta\theta}{\theta_{max}} = \frac{2}{\sqrt{N-1}} \tag{6.66}$$

In the limit (or for a very large number of tanks), the width across these points equals twice the standard deviation (square root of the dimensionless variance).

6.5.1.4 Extensions of Tanks in Series: Gamma Function and Fractional Tanks

The standard tanks-in-series model described in the previous section is constrained to have an integer number of equally sized tanks. This constraint is not a serious handicap when large numbers of tanks are used as, for example, when modeling tubular reactors with small deviations from plug flow. However, if N is small, such as the case that may arise if using the model to describe small deviations from ideal mixing in a stirred tank, problems may arise. The constraint can be especially restrictive if the value of N is between 1 and 2. Both the gamma function and the fractional tank extensions have been proposed to increase the flexibility of the tanks-in-series model.

In the gamma function model, the value of N is allowed to assume any value and thus has no physical meaning. The normalized RTD curve is given by

$$f(\theta) = \frac{\theta^{N-1}N^{N}}{\Gamma(\theta)}\exp(-N\theta) \tag{6.67}$$

where the gamma function is defined by

$$\Gamma(\theta) = \int_{0}^{\infty}\exp(-x)x^{N-1}dx \tag{6.68}$$

If N is an integer, Equation 6.67 becomes the same as Equation 6.60 because for integer values the following relation is true:

$$\Gamma(\theta) = (N-1)! \tag{6.69}$$

The cumulative distribution function is

$$F(\theta) = \frac{1}{\Gamma(N)}\int_{0}^{N\theta}\exp(-x)x^{N-1}dx \tag{6.70}$$

In Equation 6.70, the integral is an incomplete gamma function because the upper limit is less than infinity. The variance of the gamma function model is the same as for the standard tanks-in-series model; thus, the only difference is that a noninteger value of N is allowed. The most common use of the gamma function model is to simulate single stirred tanks with less than perfect mixing. The model can account for stagnant regions or bypassing (also called short-circuiting) in stirred tanks. Roughly speaking, stagnant regions tend to increase the residence times of some molecules, while bypassing will tend to reduce it. For a stirred vessel with significant stagnant regions, the variance of the RTD will be less than one, and the value of N in the gamma function model will be greater than 1, while a stirred tank with bypassing will give a variance greater than 1, and hence a value of N that is <1. Typical RTD curves for the gamma function model are shown in Figure 6.17.

Although the gamma function extension offers the advantage of mathematical simplicity, it suffers from a lack of physical meaning, and, for this reason, the fraction tank extension was proposed by Stokes and Nauman (1970). In this extension to the tanks-in-series model, the system is comprised of $I+1$ tanks in series, with I tanks (I must be an integer) the same size and one tank of a smaller size. If V is the total reactor volume, the volumes of the identical tanks, V_{I}, and the volume of the fractional tank, V_{β}, are given by

$$V_{I} = \frac{V}{I+\beta} \quad V_{\beta} = \beta\left(\frac{V}{I+\beta}\right) \tag{6.71}$$

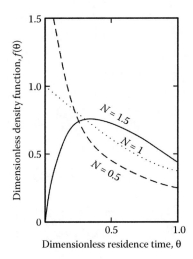

FIGURE 6.17
RTD curves for the gamma function model. The gamma function model can account for stagnant regions and bypassing in CSTR.

Note well that the value of β lies between 0 and 1. If the value of N has a noninteger value, the value of β can be calculated from the system constraint:

$$I + \beta = N \tag{6.72}$$

The normalized washout function of the fractional tank extension model is

$$W(\theta) = \left(\frac{-\beta}{1-\beta}\right)^I \exp\left(-\frac{N\theta}{\beta}\right) + \exp(-N\theta)\sum_{i=0}^{I-1}\left\{\frac{\theta^i N^i}{i!}\left[1 - \left(\frac{-\beta}{1-\beta}\right)^{I-i}\right]\right\} \tag{6.73}$$

The variance of the RTD is different from that for the standard tanks-in-series model, and is

$$\sigma_\theta^2 = \frac{N - \beta + \beta^2}{N^2} = \frac{I + \beta^2}{(I+\beta)^2} \tag{6.74}$$

Figure 6.18 shows a comparison of the dimensionless variance for the gamma function model and the fractional tanks model.

Example 6.3

Consider the vessel with the RTD shown in Example 6.1. Calculate the number and size of the tanks in series that should be used to model this vessel using the fractional tanks model.

SOLUTION

To determine the number of tanks in series, we need the variance. The variance was computed in Example 6.2 to be $\sigma_\theta^2 = 0.232$; therefore

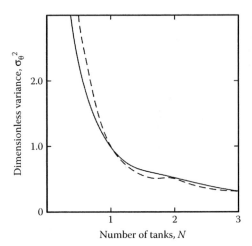

FIGURE 6.18
Comparison of the dimensionless variance for the gamma function model (solid line) and the fractional tanks model (dashed line).

$$\sigma_\theta^2 = \frac{N - \beta + \beta^2}{N^2} = \frac{I + \beta^2}{(I + \beta)^2} = 0.232$$

Now, it was seen in Example 6.2 that the standard tanks-in-series model gave four tanks. The fractional tank size in the fractional tanks model must lie between 0 and 1, and this will occur when $I = 4$. We can then solve for the fractional tank size in the preceding equation, which gives $\beta = 0.167$. Therefore, there will be a total of five tanks (but note that $N = 4.167$), four of which will be equal in volume, and the fifth will be 16.7% of that volume. The total system volume equals the volume of the real system.

6.5.1.5 Backflow, Crossflow, and Side Capacity Models

There are other extensions that can be made to the tanks-in-series model. One of these is the backflow cell model. In this model, the system is once again treated as a series of equally sized tanks; however, a backflow stream is introduced between adjacent vessels, as shown in Figure 6.19. This backflow stream allows material to flow upstream from reactor i to reactor $i - 1$. The backflow model allows for additional axial dispersion in the system. This model is a two-parameter model because both the number of tanks and the backflow ratio must be specified. Note that as the volumetric flow rate of the backflow stream tends

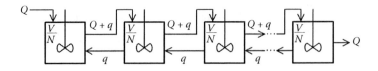

FIGURE 6.19
The backflow cell model is essentially a tanks-in-series, model which allows fluid to flow backwards to upstream tanks.

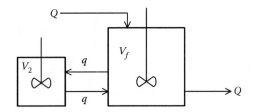

FIGURE 6.20
The crossflow model has a separate stirred tank attached to the main tank. Fluid is transferred between the two at a specified flow rate. This model can be used for a CSTR with stagnant regions.

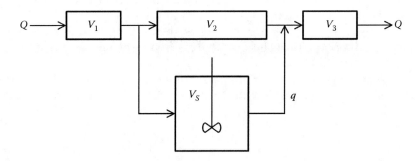

FIGURE 6.21
Side capacity models come in various forms and are another model that can be used to account for stagnant regions.

to zero, the model tends to the standard tanks-in-series model. As the backflow stream q becomes very large compared to Q, the system tends to act like a single stirred tank.

The backflow model allows for backmixing between tanks. A crossflow model is a modification to a stirred tank model that allows for the presence of stagnant regions. In this model, the main tank exchanges mass with a secondary tank at a flow rate q. Mass cannot leave the secondary tank directly, only by exchange with the main reactor. This arrangement is shown in Figure 6.20.

Side capacity models are another method of accounting for stagnancy. A typical example is shown in Figure 6.21. In this figure, the side stream through the CSTR is taken from the inlet of PFR V_2 and passed around to the exit. This pattern corresponds to feed forward. If the same tank arrangement was used but the stream to the CSTR was taken from the exit of PFR V_2 and returned to the entrance of PFR V_2, the situation would be called feedback.

Numerous other possibilities can be envisaged. For example, one could take the backflow cell model shown in Figure 6.19 and add crossflow cells on each of the tanks. Obviously such a model would have a larger number of adjustable parameters. In general, the higher the number of parameters, the better will be the match between the observed and the predicted RTD. However, the predictive ability of the reactor model will not necessarily be better. One should try as much as possible to minimize the number of adjustable parameters in the model.

6.5.2 Theoretical Model for Laminar Flow Tubular Reactor

If the hydrodynamics (velocity pattern) of a vessel are known, the RTD can often be computed analytically. Consider, for example, the fully developed laminar flow of a Newtonian fluid in a circular tube of radius R. The velocity profile is parabolic. In terms of the mean fluid velocity, u_m, and the radial position, r, the axial velocity u is

$$u(r) = 2u_m\left(1 - \frac{r^2}{R^2}\right) \tag{6.75}$$

In laminar flow, the fluid flows along well-defined streamlines. Because of the radial velocity distribution, the fluid velocity increases as the distance from the wall increases, and the residence time of a particle at the center is thus less than that of particles closer to the wall. If a change in the inlet concentration is made, a radial concentration profile will develop as the fluid flows down the tube. Molecular diffusion will act to reduce the radial profile; however, in many systems, especially liquid systems, the rate of diffusion is slow and a radial concentration profile is retained. In the limiting case of zero diffusion, the RTD can be readily derived. The solution of the case where molecular diffusion is included is discussed in Section 6.5.4.

For a particle moving along a streamline in a tube with no diffusion, the residence time of that particle is simply the length of the tube divided by the velocity of the streamline.

$$t(r) = \frac{L}{u(r)} \tag{6.76}$$

Substitute the radial velocity profile and note that the ratio of the mean velocity and the tube length is equal to the mean residence time:

$$t(r) = \frac{L}{2u_m\left(1 - \left(r^2/R^2\right)\right)} = \frac{t_m}{2\left(1 - \left(r^2/R^2\right)\right)} \tag{6.77}$$

From Equation 6.77, it is seen that the residence time of a particle traveling along the axis is one-half the mean residence time, which corresponds to the minimum particle residence time in the tube. The fraction of the fluid that flows in a differential radial increment, dr, is given by

$$\text{Fraction of fluid flowing in } dr = \frac{u(r)\,2\pi r\,dr}{Q} \tag{6.78}$$

The total fraction of the fluid in the tube within the range of $r = 0$ to an arbitrary value of r is found by integration:

$$F(r) = \frac{1}{Q}\int_0^r u(r)\,2\pi r\,dr = \frac{1}{Q}\int_0^r 2u_m\left(1 - \frac{r^2}{R^2}\right)2\pi r\,dr \tag{6.79}$$

Solution of the integral gives the fraction of the fluid in the tube with a radial position less than or equal to r:

$$F(r) = \frac{2r^2R^2 - r^4}{R^4} \tag{6.80}$$

Equation 6.77 relates the residence time to the radial position, and this equation can be substituted into Equation 6.80 to obtain an expression for the fraction of the fluid in the reactor with a residence time less than or equal to t:

$$F(t) = 1 - \frac{t_m^2}{4t^2} \tag{6.81}$$

Equation 6.81 is the cumulative distribution function. The minimum residence time is one-half the mean residence time; thus, the complete cumulative RTD function is

$$F(t) = 0 \quad t < \frac{t_m}{2}$$
$$F(t) = 1 - \frac{t_m^2}{4t^2} \quad t > \frac{t_m}{2} \tag{6.82}$$

The washout function is then defined as

$$W(t) = 1 \quad t < \frac{t_m}{2}$$
$$W(t) = \frac{t_m^2}{4t^2} \quad t > \frac{t_m}{2} \tag{6.83}$$

The RTD function is the derivative of the cumulative distribution function

$$f(t) = 0 \quad t < \frac{t_m}{2}$$
$$f(t) = \frac{t_m^2}{2t^3} \quad t > \frac{t_m}{2} \tag{6.84}$$

The normalized functions are

$$F(\theta) = 0, \quad W(\theta) = 0, \quad f(\theta) = 0 \quad \text{for} \quad \theta < 0.5$$
$$F(\theta) = 1 - \frac{1}{4\theta^2}, \quad W(\theta) = \frac{1}{4\theta^2}, \quad f(\theta) = \frac{1}{2\theta^3} \quad \text{for} \quad \theta > 0.5 \tag{6.85}$$

This result is valid for Newtonian fluids with no diffusion present. Other results apply if the fluid is non-Newtonian. For example, with a power law fluid, the velocity profile is given by

$$u(r) = 2u_m \left[1 - \left(\frac{r}{R} \right)^{(1+(1/n))} \right] n \geq 1 \tag{6.86}$$

It can be shown that the cumulative distribution function is equal to

$$F(t) = \left[1 + \frac{2nt_m}{(3n+1)t} \right] \left[1 - \frac{(n+1)t_m}{(3n+1)t} \right]^{(2n/n+1)} \quad t > \frac{(n+1)}{(3n+1)}t_m \tag{6.87}$$

6.5.3 Dispersion Model for Tubular Reactors

If a step change is made in the inlet concentration of a tracer in a plug flow system, the shape of the axial concentration profile does not change as the fluid flows down the tube. In a system that does not have plug flow, axial mixing tends to smooth out or blur the shape of the concentration profile. Rather than observing a sharp step in the outlet concentration for a step input, the concentration increases more slowly over a measurable period of time, as shown in Figure 6.22.

In turbulent flow, the axial mixing, or backmixing, results primarily from localized velocity fluctuations. In laminar flow, backmixing is caused by molecular diffusion forces. Turbulent axial mixing is referred to as dispersion, and the flow in a tube is modeled in the same manner as molecular diffusion, that is, it is assumed to follow a Fick's law type of relationship. The dispersion model for flow in a tube can be viewed as a plug flow model with a superimposed dispersion. For the transient flow of a nonreacting tracer, the mole balance equation is

$$D_L \frac{\partial^2 C}{\partial z^2} - u_m \frac{\partial C}{\partial z} = \frac{\partial C}{\partial t} \tag{6.88}$$

In Equation 6.88, u_m is the time average mean axial velocity and D_L is the axial (or longitudinal) dispersion coefficient. Equation 6.88 can be made dimensionless by defining the following scaling factors:

$$z^* = \frac{z}{L} \quad \theta = \frac{t}{t_m} \quad C^* = \frac{C}{C_0} \tag{6.89}$$

Here, C_0 is usually taken as the maximum concentration in the system. This concentration would be the initial concentration in the system before a step down or the final concentration

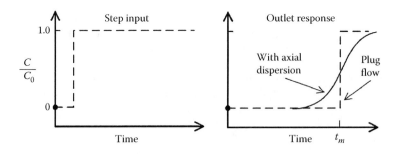

FIGURE 6.22
Comparison of the outlet response to a step input for plug flow and flow with axial dispersion.

after a step up. Note that the mean residence time is given by $t_m = L/u_m$ and therefore the dimensionless equation is, after substitution

$$\left(\frac{D_L}{u_m L}\right)\frac{\partial^2 C^*}{\partial z^{*2}} - \frac{\partial C^*}{\partial z^*} = \frac{\partial C^*}{\partial \theta} \tag{6.90}$$

Equation 6.90 is rewritten using a dimensionless group called the Peclet number, denoted Pe:

$$\left(\frac{1}{\text{Pe}}\right)\frac{\partial^2 C^*}{\partial z^{*2}} - \frac{\partial C^*}{\partial z^*} = \frac{\partial C^*}{\partial \theta} \tag{6.91}$$

Equation 6.91 is a second-order boundary value problem and requires two boundary conditions. If the system is taken as a closed system (which implies that there is no dispersion in the feed line), the dimensionless boundary conditions are

$$
\begin{aligned}
C_{\text{in}}^* &= C^* - \frac{1}{\text{Pe}}\frac{\partial C^*}{\partial z^*} \quad \text{at } z^* = 0 \\
\frac{\partial C^*}{\partial z^*} &= 0 \qquad\qquad\qquad \text{at } z^* = 1
\end{aligned}
\tag{6.92}
$$

These boundary conditions are called the Danckwerts boundary conditions. Note that as the Peclet number tends to infinity (e.g., with a large u_m or small D_L), the flow tends to plug flow, and as the Peclet number becomes very small, the flow tends to be perfectly mixed. The dispersed plug flow reactor performance therefore falls between the limits of a PFR and a CSTR.

The Peclet number can be found experimentally using a pulse injection into the tube inlet and finding the RTD function as described in Section 6.3. The Peclet number depends on the variance of the normalized RTD function (Levenspiel and Smith, 1957):

$$\sigma_\theta^2 = \frac{2}{\text{Pe}} - \frac{2}{\text{Pe}^2}\left[1 - \exp(-\text{Pe})\right] \tag{6.93}$$

The behavior of the washout function with increasing Peclet number is shown in Figure 6.23. Note that as the Peclet number increases the flow approaches plug flow. The behavior of the washout function with increasing Peclet number is similar to that observed with the tanks-in-series model with increasing number of tanks. Either model can be used to model a tubular reactor with dispersion, and can give similar results. An advantage of the tanks-in-series model over the dispersion model is that the tanks-in-series models require the solution of a system of algebraic equations while the dispersion model requires the solution of a differential equation.

Once the Peclet number for the system is calculated from the RTD information, the mole balance for a steady-state system with chemical reaction of a species A is written (in dimensional form) as

$$D_L\frac{\partial^2 C_A}{\partial z^2} - u_m\frac{\partial C_A}{\partial z} - (-r_A) = 0 \tag{6.94}$$

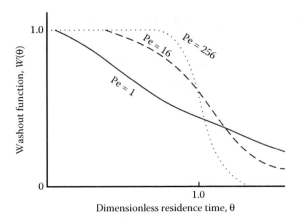

FIGURE 6.23
The washout function depends on the value of the Peclet number. At small Peclet number, the performance is similar to a CSTR, and as the Peclet number increases the behavior approaches plug flow.

When the extent of axial dispersion is relatively small, that is, when the Peclet number is greater than about 100, Equation 6.93 can be approximated by the form

$$\sigma_\theta^2 = \frac{2}{Pe} \tag{6.95}$$

As the Peclet number becomes increasingly large, the shape of the output curve (normalized RTD function) tends toward a symmetrical curve (Gaussian distribution) in the same manner that the TIS model output approaches symmetry as the number of tanks increases. Typically, when Pe > 100, an injection of an idealized pulse of tracer into the feed of the vessel gives a nearly symmetrical output curve. The equation for the Gaussian distribution is

$$f(\theta) = \frac{1}{2\sqrt{\pi/Pe}} \exp\left(-\frac{Pe(1-\theta)^2}{4}\right) \tag{6.96}$$

This equation is plotted in Figure 6.24 for a Pe of 100. The maximum value of $f(\theta)$ occurs at $\theta = 1$, and therefore the maximum value is given by

$$f(\theta)_{max} = \frac{1}{2\sqrt{\pi/Pe}} = \sqrt{\frac{Pe}{4\pi}} \tag{6.97}$$

The inflection points on the plot occur at $0.61 f(\theta)_{max}$ and the width of the curve across the inflection points is equal to $2\sigma_\theta$. The dimensional quantities can give useful information about the system. Recall that

$$f(\theta) = t_m f(t) \quad t_m = \frac{L}{u_m} \quad Pe = \frac{u_m L}{D_L} \tag{6.98}$$

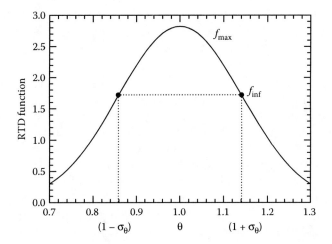

FIGURE 6.24
A near Gaussian-shaped distribution function for the dispersion model with high Peclet number. In this case, the Peclet number is 100.

The dimensional form of the Gaussian curve is thus

$$f(t) = \sqrt{\frac{u_m^3}{4\pi DL}} \exp\left(-\frac{u_m}{4LD_L}(L - tu_m)^2\right) \tag{6.99}$$

The variance of this curve is related to the dimensionless variance, and thence to the physical parameters:

$$\sigma^2 = t_m^2\sigma_\theta^2 = t_m^2 \frac{2}{\text{Pe}} = 2\left(\frac{D_L L}{u_m^3}\right) \tag{6.100}$$

The variance is thus proportional to the distance, which means that the width of the RTD curve is proportional to the square root of the distance.

Another interesting artifact of systems with small amounts of dispersion is the additive property of the nonnormalized variance. Consider, for illustration purposes, a series of tubular vessels in each of which a small amount of dispersion occurs. The overall mean residence time is the sum of the residence times of each vessel, that is

$$(t_m)_{\text{Total}} = (t_m)_1 + (t_m)_2 + (t_m)_3 + \cdots \tag{6.101}$$

The nonnormalized variance in each vessel is given by

$$\sigma^2 = t_m^2\sigma_\theta^2 = t_m^2 \frac{2}{\text{Pe}} = \left(\frac{L}{u_m}\right)^2\left(\frac{2D_L}{u_m L}\right) = 2\left(\frac{D_L L}{u_m^3}\right) \tag{6.102}$$

The variance of the system as a whole is the sum of the variances of the components, or

$$\left(\sigma^2\right)_{\text{Total}} = \left(\sigma^2\right)_1 + \left(\sigma^2\right)_2 + \left(\sigma^2\right)_3 + \cdots \tag{6.103}$$

This surprising result has the consequence that for a system in which there is a small amount of dispersion, the Peclet number for the vessel can be determined from any shape of input pulse, provided that the concentration profiles of both inlet and outlet pulses are known. Equation 6.103 implies that the variance of a system does not depend on the shape of the incoming pulse: in other words, the change in variance across a system is constant. Consider a "sloppy" pulse injection into the vessel, which has a variance σ_{in}^2 and a "mean residence time" $t_{m,\,in}$. This latter value would be computed from the first moment of the injected curve. The outlet variance and "mean residence time" σ_{out}^2 and $t_{m,\,out}$, respectively, are also computed. The Peclet number can then be computed from the difference between the two variances, as follows:

$$\frac{\sigma_{\text{out}}^2 - \sigma_{\text{in}}^2}{\left(t_{m,\text{out}} - t_{m,\text{in}}\right)^2} = \frac{\Delta\sigma^2}{\Delta t_m^2} = \Delta\sigma_\theta^2 = \frac{2}{\text{Pe}} \tag{6.104}$$

For larger axial dispersion (Pe < 100), the preceding simplifications are not valid.

Example 6.4

Consider the vessel with the RTD as shown in Example 6.1. Calculate the Peclet number that would be used with the axial dispersion model.

SOLUTION

The Peclet number is computed from the variance. From Example 6.2 $\sigma_\theta^2 = 0.232$; therefore

$$\sigma_\theta^2 = \frac{2}{\text{Pe}} - \frac{2}{\text{Pe}^2}\left[1 - \exp(-\text{Pe})\right] = 0.232$$

This equation can be solved by iteration to give a value of Pe = 7.47. Note that because the Peclet number is <100 (large amount of dispersion present) the approximate solution of Equation 6.95 would not be valid.

When chemical reaction is included in the dispersion model, a numerical solution is usually required for the resulting differential equation, even at steady state. Consider a reaction with a single reactant A and a simple power law rate expression. Equation 6.94 becomes

$$D_L \frac{\partial^2 C_A}{\partial z^2} - u_m \frac{\partial C_A}{\partial z} - kC_A^n = 0 \tag{6.105}$$

Introducing the dimensionless length coordinate, the Peclet number and the mean residence time (see Equation 6.98), we can write

$$\frac{1}{\text{Pe}} \frac{\partial^2 C_A}{\partial z^{*2}} - \frac{\partial C_A}{\partial z^*} - kt_m C_A^n = 0 \tag{6.106}$$

In terms of the fractional conversion

$$\frac{1}{\text{Pe}}\frac{\partial^2 X_A}{\partial z^{*2}} - \frac{\partial X_A}{\partial z^*} - kt_m C_{A0}^{n-1}(1 - X_A)^n = 0 \qquad (6.107)$$

The fractional conversion is therefore governed by the magnitude of three dimensionless groups, PE, n, and $kt_m C_{A0}^{n-1}$. There is an analytical solution to this equation only for a first-order reaction:

$$\frac{C_A}{C_{A0}} = (1 - X_A) = \frac{4a\exp(\text{Pe}/2)}{(1 + a)^2 \exp(a\text{Pe}/2) - (1 - a)^2 \exp(-a\text{Pe}/2)} \qquad (6.108)$$

where

$$a = \left(1 + \frac{4kt_m}{\text{Pe}}\right)^{0.5} \qquad (6.109)$$

When the Peclet number is greater than 100, the approximate solution is

$$\frac{C_A}{C_{A0}} = (1 - X_A) = \exp\left(-kt_m + \frac{(kt_m)^2}{\text{Pe}}\right) \qquad (6.110)$$

Substituting the variance from Equation 6.100 gives

$$\frac{C_A}{C_{A0}} = (1 - X_A) = \exp\left(-kt_m + \frac{k^2\sigma^2}{\text{Pe}}\right) \qquad (6.111)$$

Recall that for a PFR, the mole balance equation is

$$-\frac{dF_A}{dV} = -Q\frac{dC_A}{dV} = kC_A \qquad (6.112)$$

The solution expressed in terms of the mean residence time is

$$\frac{C_A}{C_{A0}} = (1 - X_A) = \exp(-kt_m) \qquad (6.113)$$

The effect of dispersion is clear on the solution.

6.5.4 Convection–Diffusion Equation for Tubular Reactors

The laminar flow tubular reactor with no diffusion was discussed in Section 6.5.2. In that section, it was seen that the RTD function could be determined analytically. The one-dimensional dispersion model discussed in Section 6.5.3 essentially imposes some axial dispersion on the plug flow model, which preserves the one-dimensional flow profile. Radial concentration gradients are ignored in the axial dispersion model. In fact, the dispersion model can be considered to be a special case of the general convection diffusion equation. For flow in a tube of circular cross section with a one-dimensional velocity profile (the velocity has only an axial component), the convection diffusion equation can be written as a partial differential equation in cylindrical coordinates as follows:

$$\frac{\partial}{\partial z}\left(D_L \frac{\partial C_A}{\partial z}\right) + \frac{1}{r}\frac{\partial}{\partial z}\left(rD_r \frac{\partial C_A}{\partial z}\right) - u_z(r)\frac{\partial C_A}{\partial z} - (-r_A) = \frac{\partial C_A}{\partial t} \tag{6.114}$$

Note that the assumption of zero radial velocity is not strictly valid, especially if there is an entry length in which the flow develops. Equation 6.114 can be applied to either laminar or turbulent flow. In turbulent flow, D_L and D_r would be axial and radial dispersion coefficients, and would be a combination of molecular and eddy diffusivities. The axial velocity in turbulent flow would be a time-averaged value. In laminar flow, D_L and D_r are molecular diffusion coefficients (which, except in special cases, have the same value, here denoted D_A) and Equation 6.114 is simply the fundamental mole balance equation.

For the laminar flow of a Newtonian fluid, the fully developed velocity profile is parabolic, and the convection diffusion equation becomes

$$\frac{\partial}{\partial z}\left(D_A \frac{\partial C_A}{\partial z}\right) + \frac{1}{r}\frac{\partial}{\partial r}\left(rD_A \frac{\partial C_A}{\partial r}\right) - 2u_m\left(1 - \frac{r^2}{R^2}\right)\frac{\partial C_A}{\partial z} - (-r_A) = \frac{\partial C_A}{\partial t} \tag{6.115}$$

Introduce the dimensionless quantities:

$$z^* = \frac{z}{L}, \quad r^* = \frac{r}{R}, \quad \theta = \frac{t}{t_m} = \frac{tu_m}{L} \tag{6.116}$$

If the molecular diffusion coefficient is taken as constant, substitute the dimensionless quantities into Equation 6.115 and rearrange:

$$\frac{D_A}{u_m L}\left[\frac{\partial^2 C_A}{\partial z^{*2}} + \left(\frac{L}{R}\right)^2\left(\frac{\partial^2 C_A}{\partial r^{*2}} + \frac{1}{r^*}\frac{\partial C_A}{\partial r^*}\right)\right] - 2\left(1 - r^{*2}\right)\frac{\partial C_A}{\partial z^*} - t_m(-r_A) = \frac{\partial C_A}{\partial \theta} \tag{6.117}$$

For many tubular reactors, the ratio L/R is very large and thus the axial diffusion term is small compared to the scaled radial diffusion term. Thus, for laminar flow, axial diffusion can often be ignored where the convection diffusion equation is used. Equation 6.117 reduces to

$$\frac{D_A}{u_m R}\left[\left(\frac{L}{R}\right)\left(\frac{\partial^2 C_A}{\partial r^{*2}} + \frac{1}{r^*}\frac{\partial C_A}{\partial r^*}\right)\right] - 2\left(1 - r^{*2}\right)\frac{\partial C_A}{\partial z^*} - t_m(-r_A) = \frac{\partial C_A}{\partial \theta} \tag{6.118}$$

Equation 6.118 must be solved numerically.

6.5.4.1 Taylor–Aris Dispersion

An interesting result for laminar flow was developed by Aris (1956) and Taylor (1953), who showed that the two-dimensional convection diffusion equation for laminar flow could be reduced to a one-dimensional form similar to the 1D dispersion model described in Section 5.5.3. Thus, for laminar flow, the dispersion model is

$$\left(\frac{D_L}{u_m L}\right)\frac{\partial^2 C_A}{\partial z^{*2}} - \frac{\partial C_A}{\partial z^*} - t_m(-r_A) = \frac{\partial C_A}{\partial \theta} \tag{6.119}$$

The axial dispersion coefficient accounts for the radial diffusion, and depends on the velocity. It is given by the formula

$$D_L = D_A + \frac{u_m^2 R^2}{48 D_A} \tag{6.120}$$

This result is valid provided that the following condition is true:

$$\frac{L}{R} > 0.16 \frac{u_m R}{D_A} \tag{6.121}$$

The boundary conditions for this model are the same as for the dispersion model discussed in Section 6.5.3.

6.5.5 Summary of RTD

The previous sections have introduced the concept of the RTD and shows a variety of models for it. Certainly, the wide variety of choices available can lead to confusion when it is time to make a decision as to how to model a deviation from nonideality. It must be emphasized that there are no hard-and-fast rules that govern such a choice, and much is left to the discretion of the designer. For example, the tanks-in-series model and the dispersion model can give essentially identical results, provided that the number of tanks or the Peclet number is selected appropriately, and the choice between the two is a matter of personal preference and computational convenience. In general, when selecting a model, it is preferable to select one that approximates as closely as possible the physical reality. Thus, for a tubular reactor where there is some axial dispersion, the dispersion model might be chosen. For a CSTR with a stagnant region, a crossflow model might be appropriate.

It is important to realize that the RTD function is not necessarily sufficient to describe reactor performance. Many vessel models will give the same RTD, but, when used to calculate conversion in reactors, will lead to different results. This idea is pursed in detail in the next section: suffice to say in summary that the use of the RTD to model reactor behavior is an area that is subject to potential error, and requires a certain amount of experience to use correctly.

6.6 Mixing in Chemical Reactors

The RTD is a key governing factor in the analysis of chemical reactors. However, it is relatively easy to demonstrate that the RTD is, in itself, not sufficient to define uniquely a flow

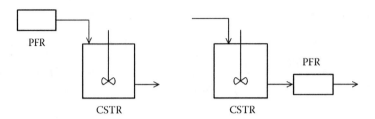

FIGURE 6.25
The fractional tubularity model gives the same RTD function regardless of the order of the reactions. For any reaction other than first order, the conversion from the two reactor systems will be different.

system in which a chemical reaction occurs, except in limited circumstances. Consider, for example, the fractional tubularity model as shown in Figure 6.25. The RTD function is the same regardless of the order in which the PFR and the CSTR are placed. It was seen in Chapter 3 that, for a first-order isothermal reaction occurring in a reactor system comprised of a PFR and a CSTR in series, with a constant-density fluid, the conversion was independent of the reactor order. However, if the reaction is not first order, the conversion will not be the same for the different reactor combinations. The fractional tubularity model is not the only example of a system in which the building blocks of the model can be arranged in different fashions without changing the RTD. Indeed, for any given RTD function it is possible to propose more than one set of vessels that will reproduce exactly the RTD function. For nonlinear reaction kinetics, the RTD is not sufficient to predict the yield and conversion of a chemical reaction. In such cases, it is necessary to consider the degree and intensity of segregation experienced by fluid elements.

Mixing in vessels is a complex subject, especially for multiphase systems, fluids of high viscosity, or reactors in which the reactants enter in separate feed streams. Mixing phenomena in chemical reactors may be considered from the point of view of fluid mechanics using the theory of turbulence, and from a systems approach of macro- and micromixing theory. Both approaches are influenced by turbulence because turbulent mixing affects chemical reactions depending on the kinetics involved, as well as the way in which reactants are transported to the reaction zone.

Mixing is a loose term that encompasses many definitions, depending mainly on the definition of the term *mixture.* A mixture is a combination of two or more ingredients that retain their separate identities however thoroughly comingled. In essence, the extent of mixing refers to the distribution of each species in the mixture with respect to the others. The extent of mixing depends on the scale of view. If the scale of view is very large, then even a very coarse mixture with large clusters of each species may appear to be homogeneous, and only the mixing of large fluid elements must be considered. On the other hand, if the extent of mixing is to be judged on a very small scale, it is necessary to consider mixing by the smallest fluid elements in conjunction with molecular processes. Typically, in chemical reactors, the mixing is important at the molecular level because, except for unimolecular reactions, different reactant molecules must be in physical contact before reaction can occur.

The term *diffusion* refers to the act of spreading out. When this spreading out is caused by relative molecular motion, it is called *molecular diffusion.* In turbulent flow, there is bulk motion of large groups of molecules or eddies. This gives rises to material transport called *eddy diffusion.* When the material transport does not involve molecular or eddy diffusion, it is called *bulk diffusion* or *dispersion.* Bulk diffusion is usually considered to be a result of specific convection mechanisms or large-scale motions that cause dispersion. It is the combination of molecular, eddy, and bulk or convection effects that are best described by the term mixing.

The rate of at which molecular diffusion occurs depends on the relative molecular motion of the diffusing species. In any fluid where there are two or more kinds of molecules, these molecules will intermingle and form a uniform mixture on a submicroscopic level if sufficient time is allowed to elapse. This view is consistent with the definition of a mixture because, at a molecular scale, we can still observe individual molecules of the two kinds; these molecules would always retain their identities. The ultimate in any mixing process would be this submicroscopic homogeneity where the molecules are distributed over the field; however, the molecular diffusion process alone is generally not fast enough for most reaction engineering applications. In some systems, molecular diffusion is so slow as to be completely negligible in any reasonably finite time—for example, in high-molecular-weight polymer processing. Other means must be found to ensure good mixing.

6.6.1 Laminar and Turbulent Mixing

If turbulence can be generated, the eddy-diffusion effects can be used to aid the mixing process. For some cases, however, the generation of high degrees of turbulence might not be practical, for example, solutions of very high viscosity would require very large power inputs to the vessel. Furthermore, some systems may undergo product deterioration under high-energy inputs. In such instances, mechanical means of stretching and folding of intermaterial surfaces are usually employed to promote mixing.

In turbulent flow, the ultimate mixing at the molecular level is achieved by small-scale turbulent eddies, whose size and intensity depend on the local energy dissipation rate. This phenomenon becomes important particularly in the case of multiple simultaneous reactions where it can affect the selectivity of a desired product.

Turbulence is important in fluid mixing operations because it determines to a large extent the microscale or molecular-scale mixing that takes place within the fluid. Turbulence accelerates reaction and promotes uniformity by ensuring that small-scale homogeneity is achieved. It also influences the rate at which chemical reactions occur, and can contribute to the distribution of reaction products formed.

6.6.2 Mixing Process: Micromixing and Macromixing

The process of mixing of two streams of miscible liquid can be described from either the Eulerian or Lagrangian points of view. In the classical Eulerian frame, the observer remains fixed in space, and the contents of the vessel at a single point are described. In the Lagrangian frame perspective, mixing is considered in time, whereby the history of a fluid element is followed as it moves through the vessel, thus allowing the identification and description of elementary processes that constitute mixing. The main concept in this approach is that of the division of mixing into two processes namely, macromixing and micromixing.

6.6.2.1 Macromixing and Macrofluids

Macromixing refers to the large-scale flow characteristics, for example, convection and turbulent dispersion that are responsible for large-scale distributions in the system. Large-scale distributions are characterized by such features as the RTD or age distribution of particles in the vessel. Macromixing depends on the flow pattern in the vessel, which, for a stirred tank, would depend on such factors as the power input. Because the eddies involved in turbulent motion are relative large on the molecular scale, a macrofluid can be viewed as consisting of clusters of molecules circulating in the vessel which tend to retain their identities. An analysis of the concentration distribution of different molecules in the vessel could indicate a perfectly mixed system when the sample size selected is larger than the eddy size.

6.6.2.2 Micromixing and Microfluids

Micromixing is concerned with all features of mixing that help achieve homogeneity at the molecular level. Mixing occurs in three successive or simultaneous stages: (1) inertial–convective disintegration of large eddies that lead to reduction of segregation scale, (2) a viscous–convective process of formation of laminated structures within energy-dissipating vortices caused by fluid engulfment, and (3) molecular diffusion within the deforming-laminated structures. In a microfluid, the individual molecules are free to move and mix in an unrestrained fashion.

6.6.2.3 Mixing and RTD

The RTD is a measure of the level of macromixing; therefore, systems with the same RTD (same level of macromixing) can show different conversions if the level of micromixing is different. The modeling and prediction of the extent of micromixing that can be achieved in a given vessel remains a complex topic in which much research continues to be done. The relationship between RTD, macromixing, micromixing, and conversion in a chemical reaction is illustrated in the following sections for some limiting cases of micromixing. As a first step in constructing a methodology, we introduce a new reactor, the segregated tank reactor.

6.6.3 Segregated Tank Reactor

In Chapter 3, the mole balance equation for the perfectly mixed CSTR was developed. In that analysis, it is implicitly assumed that the perfect mixing extended to the molecular level, that is, perfect micromixing was assumed. We now consider a stirred tank reactor with a different set of assumptions. Following on from the concept of a macrofluid, take the feed to this reactor as being comprised of packages of fluid, packages that are small in comparison to the size of the reactor but, at the same time contain a significantly large number of reactant molecules. These packages could be envisaged as, for example, small hollow spheres containing fluid. When the packages are admitted to a stirred tank, we can envisage two scenarios that involve perfect mixing: one with perfect mixing at the macroscale, and the other with perfect mixing at the microscale. These two scenarios are shown in Figure 6.26.

If the packages are admitted to a stirred tank in which there is a sufficient level of mixing, then the packages of fluid can be perfectly mixed at the macroscale. This process can be viewed as a perfect mixing of the turbulent eddies. However, if we consider the fluid particles as "capsules" of fluid, mass transfer among the different capsules is not allowed. In other words, there is no mixing between fluid elements that have different residence times. The capsules or eddies of fluid will follow the normal RTD for a perfectly mixed

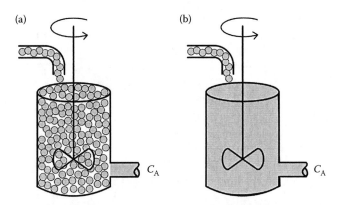

FIGURE 6.26
Different levels of mixing in a stirred tank. (a) The "packages" of fluid are perfectly mixed in the vessel, but each package maintains its integrity as it flows through the vessel. (b) Each package is broken up and mixed with the vessel contents. The RTD for each vessel is the same.

stirred tank, and thus a probe placed on the effluent from such a system would record the exponential function that was developed in Section 6.2. Indeed, provided that the fluid in the tank is perfectly mixed as a macrofluid, the RTD is the same for any level of micromixing. In other words, the RTD can tell us nothing about the level of micromixing that occurs in a system.

A stirred tank reactor that has perfect macromixing but no micromixing is known as a segregated tank reactor. Except for a unimolecular reaction, the conversion in a segregated tank reactor will be different from that obtained in a stirred tank reactor with perfect micromixing. Because the mole balance equations developed in Chapter 3 for the CSTR only apply to a vessel with perfect micromixing, another method must be found for calculating the conversion in the segregated tank reactor.

6.6.3.1 Conversion in Segregated Tank Reactor

The conversion in a segregated tank reactor can be calculated using the batch reactor mole balance and the RTD. Because each package moves through the reactor without intermingling with the fluid in other packages, each package can be seen as a small batch reactor. The time that the "batch reactor" spends in the stirred tank determines its outlet conversion, and this time is given by the RTD. The initial concentration of reactants in the package is the same as the inlet concentration to the stirred tank. The outlet concentration from the tank is the average of the concentration of reactants in all of the exiting packages, which is calculated using the batch reactor mole balance. Consider, for example, a simple second-order reaction occurring in a batch reactor. The batch reactor mole balance is

$$(-r_A) = kC_A^2 = -\frac{dC_A}{dt} \tag{6.122}$$

Integration of Equation 6.122 gives the concentration of A as a function of time:

$$C_A(t) = \left[kt + \frac{1}{C_{A0}} \right]^{-1} \tag{6.123}$$

Therefore, the concentration in any packet is known provided that the residence time of the packet is known. The distribution of concentrations among all of the exiting packages is determined by the RTD. The average outlet concentration of A in the vessel effluent is then computed by adding together the concentrations of all of the packages leaving the reactor at different times, which is obtained by integrating over the entire RTD curve

$$C_{AE} = \int_0^\infty C_A(t) f(t)\, dt \tag{6.124}$$

Substitute Equation 6.123 and the RTD curve for a CSTR into Equation 6.124:

$$C_{AE} = \int_0^\infty \left[kt + \frac{1}{C_{A0}} \right]^{-1} \frac{\exp\left(-(t/t_m)\right)}{t_m}\, dt \tag{6.125}$$

Integration of Equation 6.125 gives the outlet concentration of A. Note that although the upper limit on the integral is infinite, in practice the integration is continued as long as the RTD function has a nonzero value.

Example 6.5

Compare the conversion obtained in a CSTR when it is operated with segregated flow to that obtained with perfect micromixing. The reaction is second order:

$$(-r_A) = kC_A^2 \tag{6.126}$$

Compare the performance for the cases where $t_m k\, C_{A0}$ equals 5, 10, and 20.

SOLUTION

We first calculate the conversion using the segregated flow model. For a stirred tank with perfect macromixing, the outlet concentration for a second-order reaction is given by Equation 6.125. We can introduce the normalized residence time and rearrange the equation to the form

$$C_{AE} = \int_0^\infty \left[kt_m\theta + \frac{1}{C_{A0}} \right]^{-1} \exp(-\theta)\, d\theta \tag{6.127}$$

Substitute for the first case where $t_m k\, C_{A0} = 5$ and rearrange

$$\frac{C_{AE}}{C_{A0}} = \int_0^\infty \left(\frac{\exp(-\theta)}{5\theta + 1} \right) d\theta = 1 - X_{AE} \tag{6.128}$$

The integral must be evaluated numerically. The integral can be computed with increasingly large upper limits until the integral remains essentially unchanged by further increases in the upper limit. For the value $t_m k\, C_{A0} = 5$, we find the fraction conversion of A to be 0.701.

The outlet conversion from the perfectly micromixed CSTR is found using the mole balance equation developed in Chapter 3. For a constant-density system (where the space time equals the mean residence time), the mole balance is

$$C_{A0} - C_{AE} - t_m k C_{AE}^2 = 0 \tag{6.129}$$

Rearranging and substitution of $t_m k C_{A0} = 5$ gives

$$1 - \frac{C_{AE}}{C_{A0}} - 5\left(\frac{C_{AE}}{C_{A0}}\right)^2 = 0 \tag{6.130}$$

The quadratic equation can easily be solved, and thus the outlet conversion computed to be equal to 0.642, which is lower than the value for the segregated tank reactor. Repeating for calculations for the different operating conditions yields the following results:

$t_m k C_{A0}$	5	10	20
X_{AE}, segregated tank (perfect macromixing)	0.701	0.798	0.870
X_{AE}, CSTR (perfect micromixing)	0.642	0.730	0.800

Comparison of the concentrations at the outlet of the two systems shows that there is indeed a difference in conversion. For "normal" kinetics, the conversion from the segregated flow reactor will be higher than that from the CSTR with perfect micromixing. Recall from the comparison of the PFR and the CSTR in Chapter 3 that back mixing reduces the conversion for normal kinetics.

6.6.4 General Segregated Flow Model

The segregated flow model used for the development of the segregated stirred tank can be generalized for any given RTD. The segregation model represents one extreme of micromixing, that is, the extreme of zero micromixing. In this model, there is no mixing between fluid elements of different ages. A PFR is an example of a perfectly segregated reactor because every particle in it has exactly the same residence time. Although there is perfect radial mixing in the PFR, all radial elements have the same residence time. There is no mixing in the axial direction in this system. The RTD for a PFR was developed in Section 6.4 and the desire here is to extend the segregated flow pattern of a PFR to a vessel with an arbitrary RTD, in an analogous manner to that used for the stirred tank. To illustrate how this can be done, we can envisage the vessel as a PFR in which all of the feed enters the reactor at the inlet, but various side streams are taken from the reactor. The positions of the side streams, and the flow rate of fluid at each one, are adjusted as required to give the RTD curve determined for the vessel in question. This arrangement is shown in Figure 6.27. Note that this arrangement is equivalent to a system of parallel PFR of differing lengths. The fluid taken out of the side streams is transported to the reactor exit, where it is mixed with all of the other fluid from the side streams and the reactor effluent. This behavior is also known as late mixing because fluid elements of different residence times are mixed as late as possible in the vessel, in this case at the reactor outlet. The outlet concentration can be determined from the batch reactor kinetics and the RTD function, as done for the segregated stirred tank. That is, the outlet concentration is given by

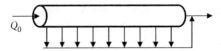

FIGURE 6.27
The segregated flow model for mixing can be envisaged as a plug flow reactor with multiple outlet streams. All the fluid enters at the reactor inlet, and then outlet side streams are added to match the observed residence time distribution.

$$C_{AE} = \int_0^\infty C_A(t)f(t)\,dt \qquad (6.131)$$

Equation 6.131 is the same as used to calculate the outlet concentration from the segregated tank reactor, as illustrated in Example 6.5. However, in the general model, the RTD function used in the solution is the one that is observed in the reactor that is to be analyzed. A numerical integration of the resulting equation is required.

The general segregated flow model can be applied to any vessel, provided that the RTD is known. The integration is usually performed numerically, especially if the RTD function is given as a set of discrete data points.

6.6.5 Maximum Mixedness Model

The segregation model is one extreme of zero micromixing, with no interaction between fluid packets of differing ages. The other extreme of maximum micromixing occurs when the fluid packages of differing residence times are mixed as early as possible in the flow system that is consistent with the observed RTD. The model for this type of behavior is called the maximum mixedness model, or early mixing model. This behavior can be modeled using a PFR with inlet side streams, as shown in Figure 6.28. The feed flow rate is distributed among the side streams in such a way so as to match the required RTD function. It can be seen that in this arrangement, fluid entering the reactor in a given side stream, which has a residence time of zero, will mix with fluid already present in the reactor, that is, fluid that has a finite residence time. Mixing of fluid elements with different ages thus occurs as early as possible. The conversion in a maximum mixedness model can be calculated by performing a mole balance that considers all of the side streams. We start by defining an alternative variable, λ, which is defined as the residual life, or the residence time remaining for an entering particle. Particles that enter at the reactor exit have a residual life of zero. The residual life is nothing more than the residence time for the entering side stream particles. All particles at a given λ have the same time *left* in the reactor, but

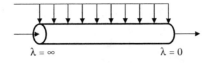

FIGURE 6.28
The maximum mixedness flow model for mixing can be envisaged as a plug flow reactor with multiple inlet streams. The fluid enters at specified inlet side streams, whose positions are added to match the observed residence time distribution.

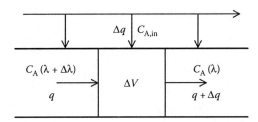

FIGURE 6.29
Control volume used to develop the mole balance for the maximum mixedness model.

they may have different ages. The concentration at the reactor outlet can be derived from a mole balance equation. Take a small control volume of the reactor, as shown in Figure 6.29. The mole balance over the discrete volume element is

$$qC_A(\lambda + \Delta\lambda) + \Delta q C_{A0} - \Delta V(-r_A) = (q + \Delta q)C_A(\lambda) \tag{6.132}$$

Particles with a residual life of λ have a residence time equal to or greater than λ. The washout function gives the fraction of the fluid in the reactor that has a residence time equal to or greater than λ. Therefore, for a given value of λ (at a specified location in the reactor), the flow rate in the reactor at that point can be related to the total feed flow rate using the washout function

$$\frac{q}{Q} = W(\lambda) \tag{6.133}$$

All the fluid that enters a side stream (the incremental flow rate at that point) has the same residence time; therefore, the flow rate of this stream can be expressed in terms of the RTD (density) function

$$\frac{\Delta q}{Q} = f(\lambda)\Delta\lambda \tag{6.134}$$

Dividing Equation 6.134 by Equation 6.133 eliminates the total feed flow rate and gives a relationship between the incremental flow rate and the reactor flow rate at any point in terms of the RTD:

$$\frac{\Delta q}{q} = \frac{f(\lambda)\Delta\lambda}{W(\lambda)} \tag{6.135}$$

The control volume size can be related to the incremental residual life. The incremental residual life is the time it takes for a fluid to pass through the control volume; therefore, we may write

$$\frac{\Delta V}{q} = \Delta\lambda \tag{6.136}$$

Equation 6.132 can be rearranged

$$C_A(\lambda + \Delta\lambda) + \frac{\Delta q}{q}C_{A0} - \frac{\Delta V}{q}(-r_A) = \left(1 + \frac{\Delta q}{q}\right)C_A(\lambda) \qquad (6.137)$$

Substitute the terms from Equations 6.135 and 6.136 to give

$$C_A(\lambda + \Delta\lambda) + \frac{f(\lambda)\Delta\lambda}{W(\lambda)}C_{A0} - \Delta\lambda(-r_A) = \left(1 + \frac{f(\lambda)\Delta\lambda}{W(\lambda)}\right)C_A(\lambda) \qquad (6.138)$$

Rearrange and take the limit as the control volume tends to the differential element

$$\frac{dC_A}{d\lambda} + \frac{f(\lambda)}{W(\lambda)}(C_{A0} - C_A) - (-r_A) = 0 \qquad (6.139)$$

Equation 6.139 is solved to determine the concentration at the reactor outlet. The equation usually requires a numerical solution. An initial condition is also required. In this case, what is known is the inlet concentration at $\lambda = \infty$. From a practical perspective, the solution is started by selecting an arbitrarily large value of λ and integrating numerically until $\lambda = 0$ is reached. The solution of the maximum mixedness model with an exponential RTD curve gives the same solution as the perfectly mixed CSTR. A comparison of the segregated flow and maximum mixedness models is given in the next section.

6.6.6 Bounding Mixing Regions

Two different models for mixing have been considered in the previous sections and the question naturally arises as to the situation prevailing in a given reactor. Although it may not be possible to determine the amount of mixing in a reactor and hence obtain an exact prediction of conversion, it is often possible at least to determine the bounds on the expected conversion. For any given RTD, the completely segregated flow model and the maximum mixedness model give bounds to the level of micromixing in the reactor. This bounding may be represented by the diagram shown in Figure 6.30. The horizontal axis represents the extent of macromixing as defined by the RTD. The vertical axis represents the amount of micromixing, which can vary from zero (segregated flow reactor) to perfect mixing (maximum mixedness model). A plug flow reactor is always a segregated flow reactor while a stirred tank can vary between perfect mixing and segregated flow. The normal region bounds the conversion that can be expected in a real reactor operating without bypassing.

The effect of varying levels of mixing can be illustrated by considering the RTD for the tanks-in-series model. See Zwietering (1959) for more details on this illustration and the maximum mixedness model in general. We shall take the case of second-order kinetics, as used in Example 6.5. The segregated flow model is applied by substituting the tanks-in-series RTD and the batch reactor concentration from Equation 6.123 into Equation 6.131 to give

$$C_{AE} = \int_0^\infty \left[kt + \frac{1}{C_{A0}}\right]^{-1} \frac{t^{N-1}N^N}{(N-1)!t_m^N}\exp\left(-N\frac{t}{t_m}\right)dt \qquad (6.140)$$

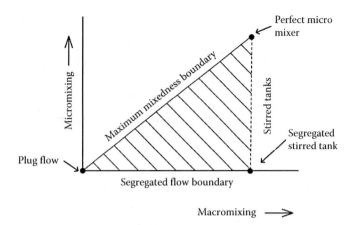

FIGURE 6.30
The bounding of the micro- and macromixing regions are shown. For flow without bypassing, the conversion is bounded by a region delineated by the PFR, the perfectly mixed CSTR, and the segregated tank reactor. (Adapted from Nauman, E.B. and B.A. Buffham, *Mixing in Continuous Flow Systems*, 1983, New York, Copyright Wiley-VCH Verlag GmbH & Co. KGaA.)

Equation 6.140 can be written in dimensionless form using the normalized residence time, which after rearrangement gives

$$\frac{C_{AE}}{C_{A0}} = \int_0^\infty [C_{A0}kt_m\theta + 1]^{-1} \frac{\theta^{N-1}N^N}{(N-1)!} \exp(-N\theta)\,d\theta \qquad (6.141)$$

As in Example 6.5, we characterize the system by the value of $t_m\,k\,C_{A0}$. For example, for a value of 10, Equation 6.141 becomes

$$\frac{C_{AE}}{C_{A0}} = \frac{N^N}{(N-1)!} \int_0^\infty \frac{\theta^{N-1}}{[10\theta + 1]} \exp(-N\theta)\,d\theta \qquad (6.142)$$

Finally, the number of tanks in series is specified and the integral expression can be integrated numerically to determine the outlet concentration and hence the fractional conversion. This operation is repeated for any desired value of $t_m\,k\,C_{A0}$ and number of tanks.

To apply the maximum mixedness model to this system, we first write Equation 6.139 in terms of the normalized residence time. Note that

$$t_m f(\lambda) = f(\theta), \quad W(\lambda) = W(\theta), \quad d\lambda = t_m d\theta \qquad (6.143)$$

Equation 6.139 is then written as

$$\frac{dC_A}{t_m d\theta} + \frac{f(\theta)}{t_m W(\theta)}(C_{A0} - C_A) - (-r_A) = 0 \qquad (6.144)$$

Now substitute the RTD and the washout functions into Equation 6.144 to obtain

$$\frac{dC_A}{d\lambda} + \frac{\left(\theta^{N-1}N^N/(N-1)!\right)\exp(-N\theta)}{t_m\exp(-N\theta)\sum_{i=0}^{N-1}\left(\theta^i N^i/i!\right)}\left(C_{A0}-C_A\right)-(-r_A)=0 \qquad (6.145)$$

The second-order rate expression and the fractional conversion can be introduced directly into Equation 6.145 to give, after simplification

$$\frac{dX_A}{d\theta} - \frac{\left(\theta^{N-1}N^N/(N-1)!\right)}{\sum_{i=0}^{N-1}\left(\theta^i N^i/i!\right)}X_A + k\,C_{A0}\,t_m\left(1-X_A\right)^2=0 \qquad (6.146)$$

Equation 6.146 must be integrated numerically for a specified number of tanks and value of $t_m\,k\,C_{A0}$. Select an arbitrarily large value for θ, for which the conversion is zero. Then integrate the equation numerically until a value of $\theta = 0$, at which point the conversion will correspond to the outlet conversion.

Finally, we can calculate the conversion that would be obtained from a series of equal-sized tanks with perfect micromixing in each tank using the mole balance equations from Chapter 3 for a CSTR. The conversion from the three types of system is shown below for two and three tanks in series (results taken from Zwietering, 1959).

	Two Tanks in Series				Three Tanks in Series			
$t_m\,k\,C_{A0}$	5	10	20	30	5	10	20	30
Complete segregation	0.768	0.860	0.920	0.944	0.791	0.878	0.933	0.954
Maximum mixedness	0.713	0.804	0.868	0.896	0.748	0.834	0.894	0.919
CSTR in series	0.725	0.814	0.878	0.906	0.758	0.845	0.904	0.929

Note that the system of two perfectly micromixed reactors gives a higher conversion than the maximum mixedness model, and less than the segregation model. This result shows that the physical limit on micromixing that can be achieved in CSTR in series is less than the maximum mixedness state. To achieve maximum mixedness with the tanks in series RTD, another reactor configuration must be selected, for example, a plug flow reactor with side entrances.

The effect of micromixing on the conversion is often small. In order for micromixing to have a strong effect, the conversion must be relatively high, the kinetics must be nonlinear and the flow pattern must deviate significantly from plug flow. Most gas-phase systems and liquid systems of low viscosity operate near the limit of maximum mixedness, and in such systems, the RTD may be sufficient to quantify the effects of nonideal behavior.

6.7 Summary

The analysis of nonideal reactors is one of the more challenging aspects of chemical reaction engineering. The material presented in this chapter represents an introduction to this area, and more detail can be found in the references cited. In some cases, the RTD can be used as a tool for the diagnosis of nonideal behavior, and can be combined with a mixing model to estimate reactor conversion. The approaches discussed are most appropriate for flowing systems in turbulent flow where the feed is mixed prior to entering the reactor vessel. In cases such as a fed batch reactor, the RTD approach is not very useful, and in situations where mixing is important, more sophisticated models must be used.

PROBLEMS

6.1 Consider a system of two perfectly mixed tanks connected as shown in Figure 6.P1. Determine the RTD for the system by assuming that a pulse is injected into the first tank.

6.2 Consider a system of three perfectly mixed tanks of volumes V_1, V_2, and V_3 connected as shown in Figure 6.P2. The total volumetric flow rate to the system is denoted Q. The fraction of the total flow rate that flows into vessel 2 is given by α. Calculate the residence time distribution function, $f(t)$, for the system in terms of the space times for the tanks. The three space times are given by

$$\tau_1 = \frac{V_1}{Q}, \quad \tau_2 = \frac{V_2}{\alpha Q}, \quad \tau_3 = \frac{V_3}{(1-\alpha)Q}$$

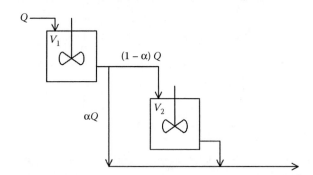

FIGURE 6.P1
Arrangement of tanks and flows for Problem 6.1.

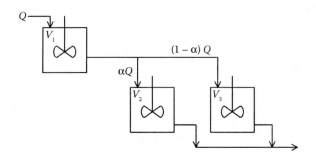

FIGURE 6.P2
Arrangement of tanks and flows for Problem 6.2.

6.3 The data shown below are the concentration response to a pulse input to a vessel. Compute the RTD function and compare its values to a CSTR that has the same mean residence time.

t (s)	0	0.1	0.2	1	2	5	10	30
C (mol/L)	0.50	0.40	0.34	0.30	0.25	0.14	0.04	0.002

6.4 For the system shown in Problem 6.1, calculate the conversion obtained for a second-order reaction with $t_m \, k \, C_{A0} = 10$ when both reactors are the same size and $\alpha = 0.25$. Compare the results obtained using the segregated flow model to that found assuming that each CSTR is perfectly micromixed.

6.5 For the system shown in Problem 6.2, calculate the conversion obtained for a second-order reaction with $t_m \, k \, C_{A0} = 10$ when reactor one is 50% of the total volume and the other two reactors are the same size and $\alpha = 0.25$. Compare the results obtained using the segregated flow model to that found assuming that each CSTR is perfectly micromixed.

6.6 The response to a pulse input to a packed bed reactor is given in the table below. Compute the Peclet number for the system and the number of tanks in series that would be used to model this reactor.

t	C	t	C
7.5	0.0	30.0	11.8
10.0	1.2	32.5	8.5
12.5	4.5	35.0	6.0
15.0	14.2	37.5	4.1
17.5	40.7	40.0	2.6
20.0	46.4	42.5	1.5
22.5	32.9	45.0	0.8
25.0	23.5	47.5	0.4
27.5	16.6	50.0	0.0

6.7 Consider the reactor with the residence time distribution given in Example 6.1. Assume that a second-order reaction in a constant-density system takes place in such a reactor. Calculate the outlet fractional conversion obtained and compare the results for

    a. The standard tanks-in-series model

    b. The fractional tanks model

    c. The segregated flow model

    d. The maximum mixedness model

Use a value of $kt_m \, C_{A0} = 10$ in your calculations.

6.8 Consider a column with a cross-sectional area of 0.5 m² filled with spherical particles to form a packed bed. The porosity (void fraction) of the bed is 0.4 (i.e., 40% of the column volume is available for fluid flow). Liquid of density 1000 kg/m³ flows through the bed with a mass flow rate of 6 kg/s. A tracer is used to measure the extent of dispersion in the system. Two probes are placed 180 cm apart in the bed to record the response of the system to a step change. The first probe records that the variance of the RTD curve at that point is 39 s² and the second probe records that the variance of the RTD curve at that point is 64 s². Compute the Peclet number for the system based on these numbers. State and discuss any assumptions that you make.

6.9 A chemical spill of benzene was recently reported in the North Saskatchewan river. The progress of the pulse of chemical was recorded at two Environment Canada monitoring stations located along the river. At the first station, the pulse was observed to take about 10 h to pass. At the second station, located 170 km downstream from the first one, the pulse arrived 24 h later and took about 15 h to pass by. Estimate how far upstream of the first monitoring station the discharge of benzene occurred. State clearly all of your assumptions.

6.10 Two different vessels of 1 m³ volume have a fluid flowing through it at a volumetric flow rate of 1 m³/min. An RTD for each of the vessels is obtained by allowing a pulse of tracer to flow into the vessel and measuring the outlet response. The outlet concentrations plotted as a function of time for the two different cases are shown in Figure 6.P10a,b, respectively. For each case, propose a flow system model (that is, specify a combination of tanks and columns) that would generate the same RTD as the one shown. Calculate the size of each tank and/or column and specify the flow rates through them.

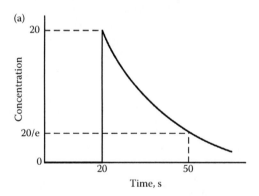

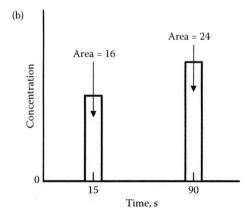

FIGURE 6.P10
Outlet RTD curves for the vessels in Problem 6.10.

References

Aris, R., 1956, On the dispersion of a solute in a fluid flowing through a tube, *Proceedings of the Royal Society*, **A235**, 67–77.

Baldyga, J. and J.R. Bourne, 1999, *Turbulent Mixing and Chemical Reactions*, John Wiley & Sons, New York.

Froment, G.B., K.B. Bischoff, and J. De Wilde, 2011, *Chemical Reactor Analysis and Design*, 3rd Ed., Wiley, New York.

Levenspiel, O., 1999, *Chemical Reaction Engineering*, 3rd Ed., John Wiley & Sons, New York.

Levenspiel, O. and Smith, W.K., 1957, Notes on the diffusion-type model for the longitudinal mixing of fluids in flow, *Chemical Engineering Science*, **6**, 227–233.

Nauman, E.B. and B.A. Buffham, 1983, *Mixing in Continuous Flow Systems*, Wiley, New York.

Ottino, J.M., 1989, *The Kinematics of Mixing: Stretching, Chaos and Transport*, Cambridge University Press, Cambridge.

Stokes, R.L. and Nauman, E.B., 1970, Residence time distribution functions for stirred tanks in series, *Canadian Journal of Chemical Engineering*, **48**, 723–725.

Taylor, G.I., 1953, Dispersion of soluble matter in solvent flowing slowly through a tube, *Proceedings of the Royal Society*, **A219**, 186–203.

Zwietering, T.N., 1959, The degree of mixing in continuous flow systems, *Chemical Engineering Science*, **11**, 1–15.

7

Introduction to Catalysis

In Chapters 3 and 4, we briefly considered some catalytic systems, especially in the example problems, without going into very much, if any, detail. The remainder of this book is devoted to catalytic systems, with the goal of providing the reader with a solid foundation for further study.

In Chapter 1, the plethora of possible products that could be made from a stream of natural gas was discussed (see Figure 1.1). Although not emphasized in that chapter, it is important to state that most of the reactors shown in Figure 1.1 require the presence of a catalyst for optimal efficiency. Indeed, if one were to undertake a survey of all industrial reactors currently used to manufacture useful products, it would be apparent that the vast majority of reactions are conducted in the presence of a catalyst. For most reactions of industrial significance, the noncatalyzed reaction either is not sufficiently speedy or produces an abundance of unwanted side reactions. Indeed, many reactions are simply not feasible without the presence of a catalyst.

To give the simplest definition, a catalyst is a substance that increases the rate of reaction by providing reaction pathways that have an activation energy that is lower than the one for the noncatalytic reaction. Note here that we have specifically stated that a catalyst increases the reaction rate. Although it is actually possible for a catalyst to decrease the reaction rate (the so-called negative catalyst), such catalysts are of little practical interest, and for all intents and purposes, we are interested in increasing reaction rates.

The purpose of this chapter is to provide a general introduction to the field of catalysis, provide some historical perspective, and to show the versatility of catalytic systems. It is also intended to give a somewhat brief overview of some of the complexities of catalytic systems. For more detailed discussion of many of these topics, refer to the books given in the list of suggested reading at the beginning of this book. Subsequent chapters discuss the heat and mass transfer effects that are significant in catalysis, the development of rate expressions, as well as some reactor analysis and modeling methodology.

7.1 Origins of Catalysis: Historical Perspectives

Although, strictly speaking, it may not be considered necessary to know the history of catalysis, it adds somewhat to the appreciation of this field to have some understanding of its origins and the impact of catalytic processes on the modern lifestyle. Indeed, very few of the products that we take for granted today would be possible without catalysis, including the level of food production, most transportation fuels, plastics, and so on. The list is almost endless.

The term "catalysis" was coined in 1835 by Berzelius in a review of some of the pioneers of the area, namely Edmund Davy, J. W. Dobereiner, and several others. Although this is the first use of the actual term, the use of catalysis (usually in an unknowing way) has a history that dates back to prehistorical times. Many of the well-known figures in science have played a part in the development of catalysis over the past nearly 200 years, including Faraday, Nernst, Kirchoff, Ostwald, and many others. Some of these works pre-dated the use of the term "catalysis." For example, Kirchoff observed that acids could promote the hydrolysis of starch, which is an early example of homogeneous catalysis. A classic observation was that of Humphrey Davy, who noted that a platinum wire could sustain a combustion reaction between air and coal gas. Platinum also featured in the later work of Faraday on the combustion of hydrogen and oxygen, and some years later platinum was used as a catalyst in the oxidation of sulfur dioxide to sulfur trioxide, as a first step in the manufacture of sulfuric acid. This reaction was one of the first catalytic processes of industrial significance. Today, platinum is the key ingredient in most automotive catalytic converters, discussed shortly.

Another extremely important reaction, hydrogenation, which involves the addition of hydrogen to molecules, was developed in the late 1800s by Sabatier and Normann. The invention of the process to make a butter substitute, margarine, is considered to be a milestone in the development of hydrogenation processes that have contributed to the production of food and other products.

One of the most important developments in catalysis was the synthesis of ammonia from nitrogen and hydrogen, discovered in 1909 by Fritz Haber. The process was commercialized by BASF in the following years, and this process permitted the production of low-cost fertilizer, which ultimately allowed for the plentiful production of inexpensive food. Other important reactions developed around this time included the Fischer–Tropsch process, in which synthesis gas could be converted into hydrocarbons and alcohols. This process is commercially practiced by SASOL in South Africa today, and is the subject of renewed interest today.

Given the importance of the automobile in today's economy and society in general, it is pertinent to observe that catalysis has played a significant role in the development of transportation fuels. Catalytic cracking appeared in 1937. In this process, large hydrocarbon molecules are broken into smaller ones suitable for inclusion in gasoline and diesel fuel. This reaction uses an acid-based catalyst. Reforming, in which straight-chain hydrocarbons are turned into cyclic ones and dehydrogenated to form aromatics, appeared prior to World War II, and was developed extensively in the 1950s and afterwards. This reaction became especially important in the production of high-octane gasoline after the use of lead-containing additives was banned from gasoline in the 1970s. Over the last few decades, the use of higher-sulfur-content crude oils has become commonplace, which has seen the extensive development of hydrodesulfurization processes, in which the sulfur is reacted with hydrogen to produce hydrogen sulfide, which is easily separated from the reactor effluent.

During the 1960s, concern over air pollution became increasingly prevalent, especially emissions from automobiles in large population centers. Incomplete combustion resulted in large amounts of hydrocarbons and carbon monoxide being emitted, as well as oxides of nitrogen formed by the oxidation of nitrogen present in the combustion air. In what must be considered one of the great triumphs of catalytic reaction engineering, catalytic converters were developed that were able to eliminate virtually all of the unwanted emissions from gasoline engines operated in stoichiometric mode. Lean burn engines, such as the diesel engine, presented greater challenges, but today there are commercial converters for these engines as well.

The preceding paragraphs give a very small sample of the significant catalytic systems that are in use today, and further information can be found in some of the books suggested for further reading.

A final observation to make in this section is that the study of catalysis, including the development of new catalysts, draws upon a wide range of fields, and includes organic and surface chemistry, chemical kinetics, thermodynamics, materials science, and the emerging field of nanotechnology. Both scientists and engineers are involved in catalysis at the fundamental and applied levels. Much of the early work in catalytic systems, and the development of many catalytic processes, was done with a large amount of empirical work. Although there is as of yet no single unified theory of catalysis, progress is being made toward the fundamental design of catalysts for desired processes.

7.2 Definitions and Fundamental Concepts

This section introduces some of the main definitions and concepts that build a foundation for the field of catalysis. Several of these concepts are expanded on in much greater detail in subsequent chapters.

7.2.1 Classification by Number of Phases

The first half of this book focuses on homogeneous reactions, where the reactants and products are all in the same phase. Heterogeneous reactions, such as the combustion of coal, were mentioned but not considered in detail. In catalytic systems, the type of catalyst used is classified according to the number of phases present. Thus, the distinction is made between homogeneous, heterogeneous, and heterohomogeneous catalysts. In homogeneous catalysis, the reactants, products, and catalyst all have the same phase. An example is a liquid-phase reaction that is catalyzed by an acid, where the acid is added in liquid form (e.g., sulfuric acid). Today, there are few industrially significant reactions that use a homogeneous catalyst, and such systems are not considered further in this book. The most common form of a catalyst is a solid, with the reactants and products being liquids or gases (or, infrequently, solids). This system is heterogeneous, and the reaction occurs at the fluid–solid interface. The reaction proceeds through a series of mass transfer and adsorption steps, which are discussed in the following sections and subsequent chapters. A hybrid catalytic system is the heterohomogeneous system. In this case, a solid catalyst can play the role of generating species that promote the reaction in the fluid phase. An example is catalytically stabilized combustion. Standard homogeneous combustion occurs in the gas phase, and involves free radicals that may be formed when larger molecules are broken through collisions. If a solid catalyst is present, it may act to generate free radicals at a lower temperature and pressure than the homogeneous state. These free radicals may migrate from the solid surface to the gas phase and initiate reactions. The remainder of this book is devoted to heterogeneous catalytic systems.

7.2.2 Catalyst Activity and Active Sites

The catalyst activity can be represented in a number of different ways, but it essentially refers to the rate at which the reactants are turned into products. As such, it is related to the reaction rate, which is used to analyze all reactive processes.

A heterogeneous catalytic reaction occurs on the solid surface and, in very simple terms, involves a reaction between the surface and the reactants. This fact has the result that catalytic reactions require specific locations on the surface, called *active sites*, at which the chemical reaction proceeds. Surfaces at the molecular level are not uniform, and their properties at this scale depend on things such as local structure, density of atoms, orientation of crystal faces, among others. Many catalysts (as discussed shortly) consist of a metal spread out on a support, and the dispersion of these metal atoms has an effect on the number and type of active site. The type of surface site and its electronic properties at the atomic level determine its activity for various chemical reactions.

Because the number of active sites on the surface affects the activity, it is common in the literature to quantify catalyst activity in terms of the active sites. Hence, the *turnover number* (or turnover frequency) is defined as the number of molecules that react per active site per unit of time. The drawback of this measure is that it can be difficult to determine accurately the number of active sites present on the surface.

7.2.3 Adsorption

A necessary first step in a heterogeneous catalytic reaction is the *adsorption* of at least one of the reactants on the catalyst surface. Adsorption can be defined as a concentration of species at the fluid–solid interface as a result of unbalanced molecular forces. Adsorption is divided into *physical adsorption*, which results primarily from low-strength forces, and *chemical adsorption* (*chemisorption*), which is akin to a chemical reaction. Chemisorption occurs only at the active sites, while physical adsorption is fairly indiscriminate. Adsorption is discussed in detail in Chapter 8 because it represents such a fundamental role in the catalytic process.

7.2.4 Selectivity and Functionality

If we return to our example of the reactions involving synthesis gas (a mixture of CO and H_2), we will recall that many different products are possible. The actual product distribution will depend on the catalyst used, as well as the operating temperature and pressure. The *selectivity* of the catalyst is a measure of its ability to promote one reaction pathway over another, and thus determine the product distribution. A major benefit of using a catalyst is its ability to change the selectivity of a reaction. When selecting a catalyst for a process, the selectivity and activity will both be important, and in some cases activity will be sacrificed to achieve better selectivity. It is also important to note that selectivity may change with temperature and pressure.

A related concept to the selectivity is the *functionality* of the catalyst. The functionality, in a general sense, describes the role that the active site plays in the reaction. For example, some reaction mechanisms proceed through steps that involve the formation of intermediate products. These intermediates may be formed on one type of site, and then require another type to advance to the final product. One example is the production of dimethyl ether (DME) from synthesis gas. This reaction may be carried out on a single bifunctional catalyst. The synthesis gas is first reacted to form methanol using a metal active site, and then the methanol undergoes further reaction to form DME and water on an acidic site.

7.3 Thermodynamics

A catalyst provides a lower-energy pathway for the reaction than the homogeneous case. A typical potential energy plot for a catalytic reaction is shown in Figure 7.1, which is similar in

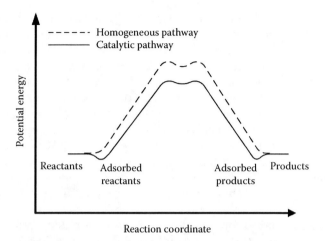

FIGURE 7.1
Energy plot for a heterogeneous catalytic reaction. The catalyst lowers the energy barrier for reaction, compared to the noncatalytic reaction, and thus increases the reaction rate.

concept to the one discussed in Chapter 5. In Figure 7.1, the overall energy barrier for the catalytic reaction is lower than the corresponding barrier for the homogeneous reaction. Thus, for a given energy distribution among reactant molecules, a higher proportion will have enough energy to react, and thus the rate of reaction is increased. The activation energy for both the forward and the reverse reactions is lowered; therefore, the rates of both forward and reverse reactions are affected. The equilibrium position is thus not changed. Another difference for the energy plot associated with the catalytic reaction is the small energy minima that correspond to the adsorbed species. We start with the gas-phase reactants at a given energy level. Then, as mentioned in the previous section, the reactant molecules adsorb on the surface. The adsorbed molecules generally have a lower energy state than the gas-phase molecules. The reactants then must cross the energy barrier to form adsorbed products, which are then desorbed to gas-phase products. As with the adsorbed reactants, the adsorbed products typically have a lower energy state than the gas-phase products.

It should also be emphasized again that the equilibrium for a catalytic reaction should be considered to be a constrained equilibrium, in as much as the product distribution is constrained by the reactions that are promoted by the catalyst. Thus, to compute the equilibrium composition, the list of possible products must be known. Not all catalyst will promote all possible reactions equally; thus, this information is important. Indeed, one of the significant properties of a given catalyst formulation is its ability to promote desired over undesired reactions.

7.4 Catalyst Types and Basic Structure

7.4.1 Basic Catalyst Structure

Heterogeneous catalysts can assume a variety of forms, and the purpose of this section is to give an overview of the more common types. The structure of the catalyst is important because it determines, among other things, the nature of the mass and energy transport that occurs in the reaction. The primary structure types that must be understood include supported and unsupported catalysts, and porous and nonporous catalysts.

Prior to expanding on these terms, we discuss briefly the porous structure of materials. Many natural and man-made materials have a significant fraction of void space contained within the solid. In other words, the material consists of a solid matrix containing pores, which are void space typically filled with fluid. For a given volume of material, a porous one will clearly have a larger fluid/solid interfacial surface area than a nonporous one. For example, a cube of nonporous material measuring 1 cm on a side would have a surface area of 6 cm². If the same cube of material had a porosity of 50% which was composed of pores of 10 nm in diameter, then the surface area can stretch into the hundreds of m². Because heterogeneous catalysis is primarily a surface phenomenon, such a high surface area will tend to make the catalyst more active, although the final activity is determined by the number of active sites on the surface. In catalytic systems, a distinction is often made between the external surface of the catalyst particle, which corresponds to the macroscopic boundary between the particle and the fluid, and the internal surface, which is the fluid–solid interface on the surface of the pores.

Porous microstructures can exist in many forms, and some examples are shown in Figure 7.2. In Figure 7.2a, the structure is in the form of capillaries arranged in a somewhat random way. In Figure 7.2b, the structure that could arise from the use of a selection of spherical microparticles to form structure is shown. This method of formation is quite common in catalysis. In some cases, the spherical microparticles have a porosity of their own. The consequences on the catalytic behavior, specifically the mass transfer within the catalyst, are discussed in Chapter 9. Finally, Figure 7.2c shows a porous structure that contains dead-end pores.

In catalysts that employ relatively high surface area supports, the surface area (measured by physical adsorption; see Chapter 11) is typically of the order of 10–200 m²/g, although areas up to 1000 m²/g are possible. This high surface area is achieved by using small pores in the catalyst support. Typically, catalyst pores are divided into micropores, mesopores, and macropores. Micropores have a diameter <2 nm, mesopores have a diameter between 2 and 50 nm, while macropores have a diameter >50 nm. Other classifications are used, and some care should be exercised.

Many common catalysts have a narrow size distribution centered around 5–10 nm. Some other catalysts show a bimodal type of pore size distribution with mesopores and macropores. Such catalysts are often made by pressing together small particles of catalyst into a larger pellet (see Figure 7.2a). In this case, the mesopores correspond to the porosity of the particles, while the macroporosity results from the voidage left by pressing the particles together. The pore size distribution is often plotted as a function of $\Delta V / \Delta \ln(r)$,

(a)   (b)   (c)

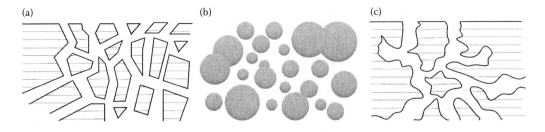

FIGURE 7.2
Examples of typical porous structures. (a) A structure of random capillaries; (b) spheres of different sizes fused together; and (c) a structure of irregular pores with dead-end volume.

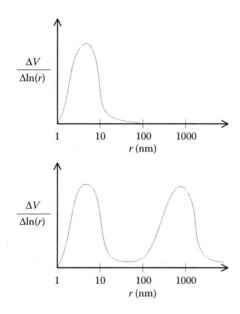

FIGURE 7.3
A typical bimodal pore size distribution.

where ΔV is the fraction of void volume in the incremental radius, $\Delta \ln(r)$. Typical unimodal and bimodal size distributions are shown in Figure 7.3. The porosity is defined as $\varepsilon = \rho_s V_g$, where $\rho_s \equiv$ pellet density and $V_g \equiv$ void volume. The surface area, pore volume, and pore size distribution can be obtained experimentally, as described in Chapter 11.

7.4.2 Supported Catalysts

Many catalytically useful materials do not have a significant porous structure. An example is metals (discussed in the next section), which are very often used as catalysts, as discussed in the following section. When the catalytic material to be used does not have a porous structure, and a high surface area is desired, then the active material may be spread out on the surface of a porous support material. This method is commonly employed for metal catalysts. Such catalysts are called supported catalysts, and the nature and design of the support is often a critical factor in the effectiveness of a catalyst.

7.4.3 Metal Catalysts

Metals comprise some of the most commonly used catalytic materials, and as such they warrant some discussion. They are particularly useful in reactions that involve hydrogen, but they are used in a wide variety of industrial systems. Some of these reactions will be discussed shortly. In some cases, the metal may be used in massive form, for example, as wires or metal gauzes. This type of catalyst would have a relatively low surface area to volume ratio, and thus a low number of active sites. When the catalytic material is used alone to form the catalytic structure, then the catalyst is said to be unsupported. Unsupported catalysts are often used when a very high activity is not desired, or there are serious mass and heat transfer issues involved in the reaction. These issues are further discussed in Chapters 9 and 10.

It is much more common when using metal catalysts to employ a porous support material, as discussed in Section 7.4.2. When metals are supported in this manner, they may be very finely spread out on the support surface. In the most extreme case, all of the metal atoms used in the catalyst will lie on the support surface. The general term for the fraction of the metal that is exposed at the surface is called the dispersion, and the higher the dispersion the more metal will lie on the surface. The degree of dispersion can be one of the significant factors that determine the catalyst activity. Other factors include the crystal structure of the exposed metal atoms, and the interaction between the metal and the support. Some reactions involve single atoms of metal on the surface, while others may require that metal atoms be grouped on the surface with a specific spacing. The methods of catalyst manufacture can be instrumental in determining the ultimate activity. Not surprisingly, the recipes for many industrial catalysts are closely guarded secrets.

7.4.4 Deactivation

It is an unfortunate fact that most catalysts lose their activity with time on stream, which is called *deactivation*. Deactivation processes may occur in a matter of seconds of reaction time (e.g., catalytic cracking) or over many years. The time scale of deactivation is an important consideration for reactor design.

Catalyst deactivation can be caused by a variety of factors. *Poisoning* of the active sites can result when an impurity in the feed adsorbs on the active sites, rendering them inactive. Sulfur compounds, for example, are a common poison for many catalysts, especially platinum and other noble metals. *Fouling* of the catalyst surface occurs when active sites are physically blocked or isolated. An example is when the surface is covered by carbon deposits caused by side reactions. This deposit occurs, for example, in catalytic cracking reactions. Other reasons for deactivation include a physical degradation of the catalyst surface, or a change in the chemical composition, rendering the active sites less active.

Deactivated catalysts may in some cases be reactivated by appropriate treatments. For example, catalytic cracking catalysts are reactivated by burning off the carbon that accumulates on the surface. In other cases, deactivated catalysts must be replaced.

7.5 Classification of Vapor-Phase Reactions

The wide variety of chemical reactions that are used in practice makes classification of them a somewhat arbitrary and over-general process. Nevertheless, there are a number of relatively agreed-upon reaction types. Because of their importance in the chemical process industries, catalytic vapor-phase reactions have perhaps been the most commonly classified, or indeed, studied. Although classification is not an easy task, a systematic arrangement of reaction types makes an orderly precise classification of the multitude of studies on catalytic reactions easier, and thus permits a more reliable generalization where such is possible.

The classification system for catalytic vapor-phase reactions was developed in the 1940s and 1950s, and although it has been changed and perhaps simplified somewhat since then, it remains the basis for reaction classification. Generally speaking, reactions are classified according to the type of bond change, the reaction mechanism, the type of catalyst active for the reaction, and the nature of the product. The primary classes of reaction are given below:

1. Acid-catalyzed reactions
2. Hydrogenation–dehydrogenation
3. Oxidation
4. Hydration–dehydration
5. Halogenation–dehalogenation
6. Combined dehydrogenation–dehydration
7. Isotope exchange
8. Miscellaneous

These groups are discussed briefly in the following sections.

7.5.1 Acid-Catalyzed Reactions

Many reactions that can be promoted by a liquid acid are also catalyzed by some solids, and hence the concept of solids having a surface acidity was developed. Typically, acid-catalyzed reactions proceed via carbonium ions. The acid can be a liquid acid that is absorbed into a solid support, or it can be a solid with Lewis or Bronsted acid sites (i.e., surface that can donate and accept electrons). A Bronsted acid donates a proton to the reactant molecule, while a Lewis acid accepts an electron. Some examples of acid-catalyzed reactions are cracking, alkylation, isomerization, hydrogen transfer, and polymerization.

7.5.1.1 Catalytic Cracking

Catalytic cracking refers to any reaction that involves the rupture of a carbon–carbon bond, which typically results in the splitting of large hydrocarbon molecules into smaller ones. Commercially, gas oil (consisting mostly of paraffins) is cracked into gasoline and diesel range fractions. The general reaction is a dealkylation reaction:

$$R_1CH_2 - CH_2 - R_2 \rightleftharpoons R_1CH{=}CH_2 + R_2H \tag{7.1}$$

The overall reaction is endothermic, although the net heat effect in the reactor is small because of associated isomerization and hydrogenation reactions. The most common catalyst employed in a modern catalytic cracking unit is a Zeolite-based catalyst. The reaction proceeds rapidly, and under typical reactor operating conditions, significant coke laydown occurs on the catalyst, leading to rapid deactivation. Catalytic cracking is one of the largest applications of catalysis in the world, based on a mass of feed stock processed.

7.5.1.2 Alkylation

This reaction involves the replacement of a hydrogen atom in an organic molecule by an alkyl radical. The classic method is Friedel–Crafts alkylation. Usually, an olefin is used to alkylate paraffins or aromatics. A typical paraffin alkylation reaction is shown below.

$$\underset{\text{paraffin}}{\text{iso-}C_4H_{10}} + \underset{\text{olefin}}{C_4H_8} \rightleftharpoons \underset{\text{paraffin}}{\text{iso-}C_8H_1} \tag{7.2}$$

Alkylation can be used to produce higher-octane fuels from lower-octane ones. Benzene can be alkylated to produce ethyl-benzene, which is subsequently dehydrogenated to produce styrene.

7.5.1.3 Isomerization

This type of reaction involves rearrangement of a molecular structure without any change in the numbers of atoms of the various species. In the fuels industry, it is used to make branched hydrocarbons from straight-chained ones, as they have a higher octane number. Branched hydrocarbons may be alkylated to give higher-molecular-weight molecules.

7.5.1.4 Polymerization

Polymerization reactions are used to make larger molecules by combining a large number of small ones. A few polymerization reactions are carried out in the vapor phase. An important one is the polymerization of low-molecular-weight olefins to produce gasoline components.

7.5.1.5 Hydrogen Transfer

Reactions of this type involve a carbon–hydrogen bond rupture and hydrogen is transferred to another hydrocarbon, for example

$$C_2H_4 + C_4H_{10} \rightleftharpoons C_2H_6 + C_4H_8 \tag{7.3}$$

7.5.2 Hydrogenation–Dehydrogenation

These reactions include those in which hydrogen is a reactant or a product, except those where both hydrogen and water are reactants or products. Typical catalysts include (1) transition group and nearby metallic elements, (2) transition group oxides or sulfides, and (3) other oxides.

7.5.2.1 Carbon Bond Saturation or Reverse

Desaturation of a carbon–carbon bond is exothermic. An example is the dehydrogenation of butene to make butadiene:

$$C_4H_8 \rightleftharpoons H_2 + C_4H_6 \tag{7.4}$$

7.5.2.2 Hydrogenation of Aromatics and Aromatization

Aromatization is used to make aromatics from cyclohydrocarbons, for example, cyclohexane is converted to benzene and hydrogen. A common catalyst is platinum. Aromatics have been widely used to increase the octane number of motor fuels, although that use may be declining.

7.5.2.3 Dehydrogenation of Oxyorganic Compounds without Water Removal

In these reactions, oxygen is removed from the reactant and water is not formed. For example, the dehydrogenation of ethanol is

$$CH_3CH_2OH \rightarrow CH_3CHO + H_2 \tag{7.5}$$

Reactions of this type are commonly used for the synthesis of aldehydes, ketones, and esters.

7.5.2.4 Hydrogenolysis of Carbon–Oxygen or Nitrogen–Oxygen Bonds

In these reactions, oxygen is removed from the reactant and water is formed:

$$CH_3CHO + 2H_2 \rightarrow C_2H_6 + H_2O \tag{7.6}$$

7.5.2.5 Hydrogenolysis of Oxides of Carbon

This classification includes the synthesis of hydrocarbons from CO and hydrogen, as well as the methanol synthesis. It also includes the oxo process (hydroformylation). The latter reaction involves the addition of CO and hydrogen to an olefin to form an aldehyde:

$$CO + H_2 + RCH{=}CH_2 \rightarrow RC_2H_4CHO \tag{7.7}$$

The methanol synthesis reactions are one of the most important examples of this class of reactions.

7.5.2.6 Hydrodesulfurization

This reaction is used to remove sulfur atoms from hydrocarbons. Typical catalysts are cobalt–molybdate, used in the sulfided form:

$$C_2H_5SH + H_2 \rightarrow C_2H_6 + H_2S \tag{7.8}$$

Hydrodesulfurization is widely used to reduce the sulfur content in petroleum fractions.

7.5.2.7 Ammonia Synthesis

This reaction is used to make ammonia from hydrogen and nitrogen:

$$\frac{1}{2}N_2 + \frac{3}{2}H_2 \rightarrow NH_3 \tag{7.9}$$

This reaction forms the basis of the modern fertilizer industry. The successful development of a catalyst for the reaction made a major contribution to the chemical industries.

7.5.3 Oxidation

Oxidation reactions involve a transfer of oxygen. The oxygen is usually provided as free oxygen as a reactant, but oxygen may be provided by another source, such as water. If free oxygen is involved, then the breaking of an oxygen–oxygen bond must occur. Catalysts used may either form unstable surface oxides or have lattice oxygen that participates in the reaction. Oxidation reactions are divided into a number of different types, as follows.

7.5.3.1 Oxygen Addition Reactions

Oxygen is added to the reactant molecule, for example

$$C_2H_4 + \frac{1}{2}O_2 \rightarrow C_2H_4O \tag{7.10}$$

7.5.3.2 Oxygenolysis

Oxygen is added and water is also formed, for example

$$C_2H_6 + \frac{1}{2}O_2 \rightarrow C_2H_4 + H_2O \tag{7.11}$$

The above reactions are considered partial oxidation reactions. In complete (or deep) oxidation, hydrocarbons would be converted to carbon dioxide and water. For example, the combustion of methane is given by

$$CH_4 + 3O_2 \rightarrow CO_2 + H_2O \tag{7.12}$$

Deep oxidation reactions are often conducted homogeneously (burning of fuel) but catalytic deep oxidation is also very important in some applications.

7.5.4 Hydration and Dehydration

This class of reactions involves those in which water is transferred, added, or abstracted, and gaseous hydrogen is not involved. Examples include alcohol synthesis, for example

$$CH_2{=}CH_2 + H_2O \rightarrow CH_3CH_2OH \tag{7.13}$$

and aldehyde synthesis

$$HC{=}CH + H_2O \rightarrow CH_3CHO \tag{7.14}$$

Most of the reactions in this class are complicated by the presence of competitive, secondary, and side reactions. Most hydration–dehydration catalysts have a strong affinity for water, a prime example of which is alumina. Alumina has a chemisorbed water content in excess of several percent even at elevated temperatures in the region of 900 K.

7.5.5 Dehydration–Dehydrogenation

This class of reactions involves those in which water and hydrogen appear on the same side of the equation. An example is the synthesis of butadiene from alcohol:

$$2C_2H_5OH \rightarrow C_4H_6 + 2H_2O + H_2 \tag{7.15}$$

This reaction has been used extensively to produce synthetic rubber.

7.5.6 Isotope Exchange

This class of reactions involves the exchange of isotopes of atoms within the molecules. These reactions are primarily used in experimental studies as an aid in determining the reaction mechanism. A reaction mixture that contains a number of isotopically labeled compounds is typically injected into the reactor, and by observing where the isotopes appear in the products, insight can be gained into the mechanism.

7.5.7 Miscellaneous

There are many reactions that do not fall easily into one of the preceding categories, and these reactions are simply lumped into a miscellaneous category until such time as they can be better classified.

7.6 Basic Steps in Heterogeneous Catalytic Reactions

We have emphasized that heterogeneous catalytic reactions occur at the fluid–solid interface. Thus, reactant molecules must be transported to the surface. Also, we have observed that it is necessary for these reactants to adsorb onto the surface. We can formalize the steps that occur in solid-catalyzed reactions as follows:

Step 1: Transport of Reactants to the Catalyst Surface

When a fluid flows past a solid object, the drag force exerted by the solid on the fluid causes the formation of a boundary layer (Figure 7.4). The flow pattern in the boundary layer is different from that in the bulk. Often, the flow in the boundary layer is laminar while the bulk flow may be turbulent. The reactants must be transported from the bulk phase across the boundary layer to the external surface of the catalyst. The driving force for the transport of the reactants is a concentration gradient; therefore, it follows that the reactant concentration at the external catalyst surface is lower than the reactant concentration in the bulk fluid (i.e., outside of the boundary layer). The temperature of the catalyst surface may also be different from that of the bulk fluid if heat is released or absorbed on reaction.

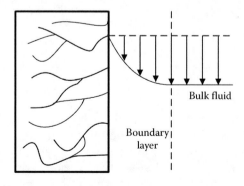

FIGURE 7.4
Simplified diagram of the processes occurring in a solid-catalyzed reaction. Transport effects are often as important as the kinetic effects. The lines in the solid represent pores.

Step 2: Transport within the Porous Catalyst

As noted earlier, many solid catalysts are porous, that is, they contain a high degree of internal void space (much like a sponge). In a porous catalyst, most of the catalytically active surface area is located in the interior of the catalyst; hence, the reactants must diffuse into the pores of the catalyst. Owing to the small size of the pores, this diffusion process may be intrinsically slow. As the reactants diffuse into the pores, some of them react. This combination of diffusion reaction leads to concentration gradients in the catalyst. If the resistance to diffusion is greater than the resistance to reaction, these gradients may be significant. As the reaction occurs at the local concentration and temperature, the rate of reaction varies with position in the catalyst.

Step 3: Adsorption of Reactants

Catalysis is essentially a surface phenomenon, and it is necessary that at least one of the reactants attach itself to the catalyst surface in a process known as adsorption. Reactants adsorb onto active sites of the catalyst. Adsorption is discussed in some detail shortly.

Step 4: Surface Reaction

Once the reactants are adsorbed onto the surface, chemical reaction occurs between adsorbed species. The products of reaction are also usually adsorbed.

Step 5: Desorption of Products

The reaction products desorb from the surface into the fluid phase, the reverse of Step 3.

Step 6: Diffusion of the Products to the External Catalyst Surface

The products diffuse to the external catalyst surface. This step is essentially the reverse of Step 2.

Step 7: Transport of Products to the Bulk Fluid

The products are transported through the boundary layer to the bulk fluid, the reverse of Step 1.

Steps 3 through 5 are considered to be kinetic steps, and are involved in formulating rate expressions and reaction mechanisms. The other steps are transport steps. Steps 3 through 5 are treated in detail in Chapter 8, while the transport steps are discussed at length in Chapter 9.

7.7 Introduction to Catalytic Reactors

Catalysts and catalytic reactors as used in industrial practice have many configurations. Although generalization can lead to oversimplification, there are common features to a number of reactors, and also a number of widely used types. In terms of catalysts used, catalytic reactors in which a solid catalyst is used are one class of heterogeneous reactors that are extremely important in the chemical process industries. Reactors with solid catalysts come in a very wide variety of designs, and a few of the more common ones are introduced in the following paragraphs. The purpose here is to give a brief overview of common reactor configurations, which will set the scene and provide context for Chapters 8 and 9, where the kinetics and transport phenomena of chemical reactors are considered. More detailed discussion of catalytic reactors, including a detailed description of the analysis of fixed beds, is given in Chapter 10.

7.7.1 Fixed Bed Reactor

One common configuration is the *fixed bed reactor*, also known as a packed bed reactor. This type of reactor consists of a bed of stationary solid catalyst particles through which the reacting fluid flows. The fluid may be a gas, a liquid, or a mixture of the two. Probably one of the most common types of industrial reactor is the catalytic fixed bed in which the reactants enter the reactor in gaseous form. The reactor is usually mounted vertically, with gas flow downwards through the bed. The catalyst bed is held in place by some means, so that the bed is stationary in the reactor. Although the flow in such a reactor is fairly complex, the macroscopic flow pattern is often close to plug flow. This configuration is illustrated in Figure 7.5. Some examples of industrial reactions that are carried out in packed bed reactors are the synthesis of methanol, steam reforming of natural gas, and the synthesis of ammonia from hydrogen and nitrogen.

The fixed bed catalytic reactor is possibly the most prevalent reactor type in industry, in large measure because of its apparent simplicity combined with the ease of construction. Fixed bed reactors may be operated with or without external heat transfer. For reactors with heat transfer, single- and multitube designs are used.

7.7.2 Fluidized Bed Reactor

Another common catalytic reactor design contains a bed of small catalyst particles in a vessel. A high-velocity gas passing upwards through the bed causes the bed to "fluidize." The catalyst particles are thus displaced significantly, which ensures a high degree of mixing and heat transfer in the bed. The catalyst bed in this reactor has many of the properties of a fluid, hence the name. The drawback of this type of operation is that the catalyst particles tend to suffer a high degree of attrition, resulting in catalyst loss. A typical fluidized bed reactor is illustrated in Figure 7.6. One of the biggest applications of fluidized catalytic reactors, on the basis of the total worldwide production, is the fluid catalytic cracker (FCC) used to produce hydrocarbon fractions in the gasoline and diesel boiling range from heavier hydrocarbons (typically the bottoms from the vacuum distillation tower in an oil refinery).

There are a number of different modes of operation of fluidized bed reactors, which tend to be classed on the basis of gas velocity. The hydrodynamics of fluidized beds are very complicated, and in turn depend on parameters such as the reactor design and the fluid velocity. Design and modeling of fluid bed reactors is still a very active area of research and there is no consensus as to the best modeling approaches.

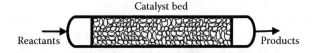

FIGURE 7.5
Illustration of a fixed bed catalytic reactor. A porous bed of catalyst particles is fixed in a tube, and the reactants pass the bed. The fixed bed reactor is simple to build and operate.

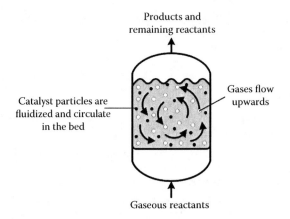

FIGURE 7.6
Illustration of a fluidized bed reactor. The high-velocity gas causes the catalyst bed to behave like a fluid, which gives good heat and mass transfer.

7.7.3 Reactors for Liquids and Gas/Liquid Mixtures

Fixed bed and fluidized bed reactors are typically operated with gas-phase reactants in contact with the solid catalyst. Many industrial reactions are conducted in reactors in which one or more of the reactants are in the liquid phase. Two commonly used reactors for liquid-phase reactions are the *trickle bed reactor* and the *slurry reactor*.

7.7.3.1 Trickle Bed Reactor

The trickle bed reactor is essentially a fixed bed reactor, that is, a fixed bed of solid catalyst is used. At least one of the reactants is in the fluid phase, and most industrial trickle bed reactors contain three phases: solid catalyst plus liquid and gaseous reactants. Trickle bed reactors are often used where it is considered undesirable to heat the feed sufficiently to turn it into a vapor. For example, trickle bed reactors are extensively used in the hydrotreating of petroleum fractions (hydrotreating examples are used later in the text). When higher boiling fractions such as diesel or heavy fuel oil are treated, it is often not possible to boil them with inducing a variety of unwanted chemical reactions, so the reactions are conducted at lower temperatures than the boiling point of the feed. The flow patterns in a trickle bed reactor are complex and only a limited understanding of this reactor type currently exists. A trickle bed reactor is illustrated in Figure 7.7.

7.7.3.2 Slurry Reactor

The slurry reactor is another type of reactor used with liquid reactants and solid catalysts. The slurry reactor usually consists of a stirred tank of liquid, in which a very finely divided solid catalyst is suspended by the action of the impeller. In some operations, a gas-phase reactant is bubbled through the liquid using a sparger. These three-phase reactors are very difficult to model and design. The interactions among the mass transfer and chemical kinetics pose a formidable design problem. A three-phase slurry reactor is illustrated in Figure 7.8. A typical industrial application of a slurry reactor is the hydrogenation of vegetable oil. In this reaction, the saturated fats in liquid vegetable oil are hydrogenated by bubbling hydrogen through liquid oil in the presence of a catalyst. The resulting saturated

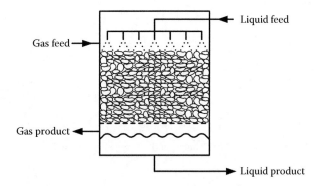

FIGURE 7.7

Illustration of a trickle bed reactor. Typically, the bed is mounted vertically, and both liquid and gaseous reactants flow cocurrently through the bed. This reactor type exhibits complex behavior.

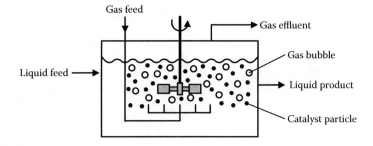

FIGURE 7.8

Illustration of a slurry reactor. The fine catalyst particles are suspended in the liquid by the action of the impeller. In this example, a gas is bubbled through the liquid using a sparger.

fat has a higher melting point than unsaturated fat, and thus, suitably colored and marketed as margarine, can be used as a solid fat replacement for butter.

7.8 Summary

This chapter has introduced, in a very general way, some of the key concepts that are required for an understanding of the area of catalytic reaction engineering. One of the important ideas that must be appreciated is that the role of mass and heat transport at the microscopic level is very important, and can have a significant influence on the reactor performance. The concepts introduced in this chapter are considered in detail in the next three chapters, while Chapter 11 discusses some of the experimental concepts in catalysis.

8

Kinetics of Catalytic Reactions

In Chapter 7, we introduced the field of heterogeneous catalysis, and discussed the steps involved in a catalytic reaction. It was pointed out that when a solid catalyst is used to promote the rate of a chemical reaction, the reaction proceeds at the fluid/solid interface, and involves at least one adsorbed species. The adsorbed species then react on the surface. The dynamics of the adsorption process, in addition to the surface reaction, must therefore have a major impact on the kinetics of the overall process. It is useful, therefore, to consider the theories of adsorption and surface reaction together, and to develop mathematical models for these processes. The combination of the models for these two processes is then used for the development of rate expressions for the reaction. The following sections therefore present first a description of the adsorption process, followed by a development of some popular methods used to derive rate expressions.

8.1 Adsorption

Adsorption results from an imbalance of molecular forces at the fluid–solid interface. This interface is a discontinuity, and as a result the molecular forces at the surface are unsaturated, regardless of whether or not the surface exhibits any catalytic activity. Hence, when a "clean" surface is exposed to a gas, there is a tendency for molecules to accumulate preferentially at the surface, resulting in a higher concentration there. The strength with which the *adsorbate* molecules are bound to the *adsorbent* varies from system to system, and depends on the nature of both. Nonetheless, adsorption may be broadly classified into *physical adsorption* and *chemical adsorption* (the latter is usually referred to as *chemisorption*).

8.1.1 Physical Adsorption

Physical adsorption is caused primarily by van der Waals forces. These forces are relatively low-strength forces, and physical adsorption is similar in principle to condensation. Adsorption is accompanied by a release of heat (the *heat of adsorption*), and in the case of physical adsorption, the amount of heat released is usually of the same order of magnitude as the heat of condensation. Physical adsorption is nonspecific, that is, adsorbing molecules are not particular as to the surface location at which they adsorb. Furthermore, physical adsorption generally requires no activation energy. Physical adsorption plays an important role in many engineering processes, including separations.

8.1.2 Chemisorption

Chemisorption involves a rearrangement of the electronic structure of both adsorbate and adsorbent, and chemical bonds are formed between them. Chemisorption is in effect a chemical reaction between the two and, as a result, the heats of adsorption for

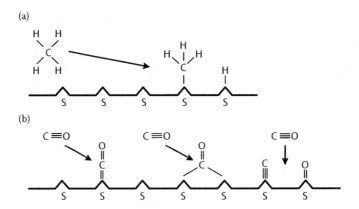

FIGURE 8.1
Chemical adsorption (chemisorption) of species on a surface. In (a), the methane molecule dissociates as it adsorbs. In (b), three possible mechanisms are shown for the adsorption of CO. It can adsorb on a single active site, two sites or it can dissociate as it adsorbs.

chemisorption tend to be much higher than those for physical adsorption: they are of the order of magnitude of chemical reactions. Furthermore, because chemical bonds are formed, the adsorbate molecules will only adsorb at very specific locations on the surface, called *active sites*. Chemisorption requires bonding between the adsorbate and adsorbent; therefore, the maximum amount that can be adsorbed corresponds to a surface layer one molecule in thickness, known as a *monolayer*. Also, because a rearrangement of chemical bonds is required, chemisorption usually requires an activation energy.

Figure 8.1 shows an example of chemisorption, where the points on the surface represent active sites that can form bonds with gas-phase species. In part (a), the methane molecule ruptures before it forms stable adsorbed species; this process is called dissociative chemisorption. In part (b), three possible adsorption mechanisms for carbon monoxide are shown: linear bonding to a single site, bridged bonding to two sites, and dissociative adsorption. This selectivity of the adsorption process governs the nature of the reactions occurring, and hence the product distribution.

8.1.3 Differentiating Chemical and Physical Adsorption

It is not always easy to differentiate between the two types of adsorption. Some of the governing characteristics of the two types of adsorption were discussed in the preceding section, and are elaborated on here. The heat of adsorption is a common criterion, and possibly the best one, for distinguishing between physical and chemical adsorption. The heat liberated during physical adsorption is typically of the same order of magnitude as the heat of condensation, as the two processes are similar. Thus the heat of adsorption is typically of the order of 8–25 kJ/mol, although values as high as 80 kJ/mol have been reported. For example, the heat of adsorption of argon on carbon black is roughly 11.3 kJ/mol, while the heat of liquefaction is 6.5 kJ/mol at the same temperature. The heat of physical adsorption is usually less than two times the heat of liquefaction, although there are exceptions to this rule. In chemisorption, much larger heats of adsorption can be encountered, up to several hundreds of kJ/mol. It is rare for heats of chemisorption to be <80 kJ/mol, but values expected for physical adsorption have been observed, as has endothermic chemisorption.

The second criterion used to distinguish the two adsorption types is the rate at which they occur. As physical adsorption is similar to liquefaction, it should occur rapidly and require no activation energy. Because chemisorption is akin to a chemical reaction that requires the breaking and forming of chemical bonds, it should require an activation energy, and should be a slower process. Care should, however, be used with this criterion. Some chemisorption processes are effectively nonactivated and do occur very rapidly. Physical adsorption in a highly porous solid may also appear to proceed slowly if the diffusion into the solid is very slow: such a process could be mistaken for an activated chemisorption.

The rate of desorption is also used as a criterion to differentiate physical and chemical adsorption. As physical adsorption is a nonactivated process, the activation energy for desorption from the adsorbed state should be equal to the heat of adsorption. This value is typically only a few kJ/mol. For chemisorbed species, the activation energy for desorption should be large, and, in the most common case of activated adsorption, the desorption activation energy (from the adsorbed state) should be higher than the heat of chemisorption.

The temperature range over which adsorption occurs is also a useful guide to which type of adsorption is present. Physical adsorption should only occur at temperatures close to the boiling point of the adsorbate. Chemisorption, involving reactive forces, can occur at much higher temperatures. This criterion can be misleading on highly porous solids which have a very fine porous microstructure. In such a case, kelvin condensation can occur in the pores at much higher temperatures than the boiling point of the adsorbate at the pressure in question.

The specificity of the adsorption can also be used to classify it. Because chemisorption is a chemical reaction, and chemical reactions are specific, then so too is chemisorption. That is, an adsorbate that adsorbs on a given surface of specified cleanliness may not adsorb on a similarly surface of different composition at the same conditions of temperature and pressure. Physical adsorption, on the other hand, being in nature similar to condensation, should not be specific as to the type of surface (i.e., the chemical composition). Physical adsorption, furthermore, can exceed a depth of one molecule on the surface, whereas chemisorption is restricted to a monolayer coverage.

Finally, a physically adsorbed species should not alter the electrical conductivity of the adsorbate. Chemisorption, which alters the electronic structure of the adsorbate, may change the electrical conductivity of the adsorbate.

8.1.4 Energetics of Adsorption

Because adsorption is so critical to catalysis, a brief overview of the energetics of adsorption is discussed in this section. It should be noted that the adsorption process is in reality very complex, and what follows is a somewhat simplistic approximation. The basic ideas described in the following are valid, however.

The energetics of adsorption can be elucidated with the aid of potential energy diagrams, which represent the energy states of molecules as they approach and adsorb on surfaces. A simplified potential energy diagram for physical adsorption is shown in Figure 8.2. The zero energy line represents an infinite separation. The potential energy curve results from a combination of attractive and repulsive forces. The attractive forces cause the adsorbate to approach the surface, resulting in a lowering of the energy (and a concomitant release of heat). As the adsorbate molecules approach more closely, the repulsive forces start to dominate and the energy of the system rises. The combination of attractive and repulsive forces results in a minimum energy at some separation distance r_0.

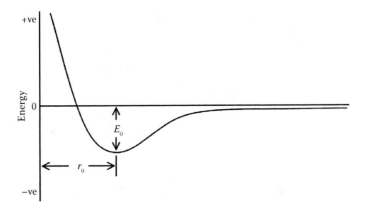

FIGURE 8.2
Simplified potential energy diagram for physical adsorption. The zero potential energy line represents an infinite separation distance. As the adsorbate approaches the surface, a balance is established between the forces of attraction and repulsion.

A simplified analysis of the energetics of physical adsorption is based on the Lennard–Jones relationship. The interaction between gas molecules and the solid surface is expressed in terms of the energy of a molecule and the distance from the surface

$$E = -ar^{-m} + br^{-n} \tag{8.1}$$

where a, b, m, and n are constants. The first term on the right-hand side accounts for attraction and the second term for repulsion. The most commonly used values for m and n are 6 and 12, respectively: this then gives what is known as the Lennard–Jones 6–12 potential. The constant a can be computed from fundamental properties (e.g., atomic susceptibility). The value of b can be computed by setting the experimentally measured heat of adsorption equal to the value of E obtained at the minimum value. At the minimum, the derivative is zero, that is

$$\frac{dE}{dr} = 0 = mar^{-m-1} - nbr^{-n-1} \quad \text{or} \quad b = \frac{m}{n}ar_0^{(n-m)} \tag{8.2}$$

At the minimum distance, it then follows that

$$E_0 = -ar_0^{-m}\left(\frac{m}{n} - 1\right) \tag{8.3}$$

The experimental heat of adsorption is then used to compute r_0 which is then used to compute b.

8.1.4.1 Activated Chemisorption

If only one type of chemisorbed species exists and the chemisorption is nondissociative and nonactivated, then the potential energy diagram would be similar to that shown in Figure 8.2; however, chemisorption processes can be varied. For example, we consider here

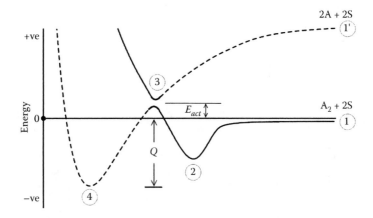

FIGURE 8.3

Simplified potential energy diagram for dissociative activated chemisorption. As molecules approach the surface, they first become physically adsorbed. If the physically adsorbed molecule can attain sufficient activation energy, it can become chemisorbed in the dissociated state.

for illustration purposes potential energy diagrams for dissociative chemisorption. Dissociative chemisorption implies that the adsorbing molecule splits into two or more species as it adsorbs. Consider molecules of A_2. The reference state of zero energy is the undissociated molecules at infinite separation. For reference purposes, the energy line corresponding to dissociated molecules at infinite separation is also shown—the upper line on Figure 8.3.

A theoretically possible route for the dissociative chemisorption of A_2 could follow this upper line, that is, by dissociating the molecules in the gas phase, and then allowing them to approach the surface. In such a case, they would follow the potential energy line commencing at 1′, pass through the point denoted 3 before finally coming to rest in the adsorbed state at point 4. A more likely scenario would be for the dissociative chemisorption to occur without prior dissociation in the gas phase. In this case, the undissociated molecules first become physically adsorbed, as they proceed from position 1 to position 2. Then, if the physically adsorbed molecules of A_2 can acquire sufficient energy to rise to position 3, they can achieve the dissociated state, and fall to the lowest energy level of chemisorbed dissociated A atoms at position 4. The activation energy for adsorption in the dissociated state is given by the height of point 3 above the reference line, while the heat of chemisorption is the distance from point 4 to the reference line. The activation energy for desorption is the height of point 3 above point 4.

8.1.4.2 Nonactivated Chemisorption

Activated chemisorption is common, but nonactivated chemisorption is also possible. For example, if the energy curve corresponding to 1′ in Figure 8.3 is shifted downwards, it is seen how this route is possible. In this case, the dissociation requires a slight activation from the physically adsorbed state, but because point 3 is below the zero energy reference line, the overall process of dissociative chemisorption is nonactivated. This type of nonactivated dissociative chemisorption is not uncommon, and the potential energy diagram is shown in Figure 8.4.

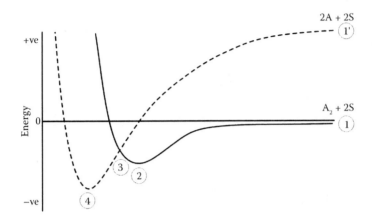

FIGURE 8.4
Simplified potential energy diagram for dissociative nonactivated chemisorption. As molecules approach the surface, they first become physically adsorbed, but no activation energy is required for the physically adsorbed molecules to move to the chemisorbed state.

8.1.4.3 Endothermic Adsorption

For the case of nonactivated chemisorption (Figure 8.4), it was seen that energy curve corresponding to the adsorption of dissociated A atoms was shifted down so that the cross-over at point 3 lay below the zero energy line. For the case of endothermic adsorption, we consider the situation where the energy curve for the adsorption of dissociated A is shifted upwards so that the minimum energy point, point 4, lies above the zero energy line. The resulting potential energy diagram is shown in Figure 8.5. The energetics proceeds in a similar manner as for exothermic adsorption; however, because point 4 now lies above the zero energy reference line, the chemisorption process is endothermic. This type of chemisorption is rare, but has been observed in some systems.

8.1.4.4 Dual Site Adsorption

Often more than one type of adsorbed species can be formed upon adsorption. For example, it was shown in Figure 8.1 that CO can adsorb in more than one way on a surface.

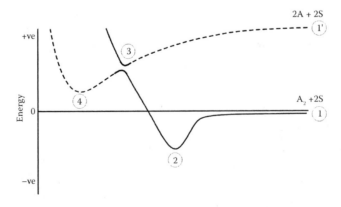

FIGURE 8.5
Potential energy diagram for endothermic dissociative chemisorption. This type of adsorption is rare but has been shown to occur.

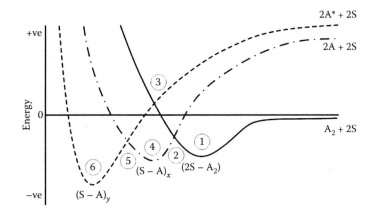

FIGURE 8.6

Potential energy diagram for dual site adsorption. The chemisorption on type x sites is nonactivated. Chemisorption on type y sites may or may not be activated, depending on the path followed.

A Lennard–Jones-type diagram can be used to illustrate qualitatively such a phenomena. Figure 8.6 shows a potential energy diagram for one possible system, where diatomic molecules A_2 can chemisorb on either type x or type y sites. From the shape of the illustrated curve, the dissociative adsorption of A_2 would exhibit the following characteristics:

1. The heat of adsorption for y-type sites is greater than that for x-type sites.
2. The physical adsorption at point 1 is nonactivated. The chemisorption onto type x sites is also nonactivated because position 2 falls below the zero energy reference line. The chemisorption onto type y sites may or may not be activated, depending on the route that is followed. If the chemisorption proceeds directly from the physically adsorbed state, then the adsorption is activated because point 3 lies above the zero energy reference line. However, if the chemisorption onto type y sites proceeds via a route that has as its first step chemisorption onto type x sites, then it is a nonactivated process because both points 2 and 5 lie below the zero energy reference line.

A dual site surface could also exhibit the behavior shown in Figure 8.7. In this case, the chemisorption onto both of site types x and y is activated if chemisorption occurs directly from the physically adsorbed state. Furthermore, the transition from type x to type y requires an activation energy.

8.1.5 Equilibrium Adsorption

The adsorption phenomenon is often quantified for a given adsorbate–adsorbent system by measuring the amount of adsorbed species at equilibrium for a given temperature and pressure. The resulting data can be displayed in a number of ways, both as mathematical models and graphically. For a given adsorbate–adsorbent combination, the amount adsorbed at equilibrium depends on the temperature, the pressure of the adsorbate, and the surface area of the adsorbent. For convenience in data representation, the variation in surface coverage is usually shown when one of these parameters is held constant.

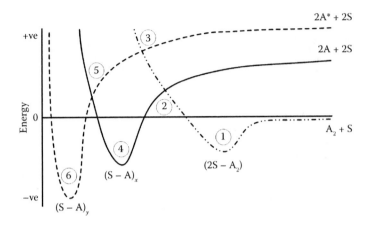

FIGURE 8.7
Potential energy diagram for dual site adsorption. The chemisorption on both type x and y sites is activated.

8.1.5.1 Adsorption Isotherms

The most common method for displaying equilibrium adsorption data is to use isotherms, or lines of constant temperature. An adsorption isotherm shows the relationship between the amount adsorbed at equilibrium and the pressure of the adsorbate in the gas phase at a constant temperature. Typical adsorption isotherms are shown in Figure 8.8. Note that as the pressure of the adsorbate increases, so does the amount adsorbed. As the pressure increases to a large value, the amount adsorbed becomes constant, as all of the adsorption sites become occupied. Usually, the amount adsorbed decreases as the temperature increases, although this is not always true, especially if the adsorption is a mixture of physical and chemical adsorption.

8.1.5.2 Models for Adsorption Isotherms

The main equilibrium adsorption models that have been used in catalysis are briefly discussed in the following paragraphs. These isotherms can be derived either theoretically or empirically, and as shown, the results of such derivations are often shown in terms of an

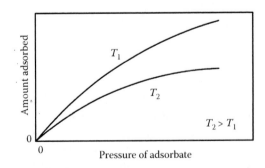

FIGURE 8.8
Typical adsorption isotherms. The amount adsorbed increases with pressure, and typically falls as the temperature increases.

amount adsorbed (uptake) as a function of the equilibrium gas-phase concentration (pressure). The origin and detailed derivation of these equations is beyond the scope of these notes; however, particular attention will be paid to the Langmuir model because of its importance in the development of kinetic rate models. The Langmuir model is widely used for both chemisorption and physical adsorption. The model gives the fraction surface coverage by an adsorbed species A as

$$\frac{v}{v_m} = \theta_A = \frac{KC_A}{1 + KC_A} \tag{8.4}$$

where

v: amount (volume) of gas adsorbed

v_m: amount (volume) of gas adsorbed in one monolayer

θ: fractional coverage, equals one for full monolayer coverage

C_A: equilibrium gas-phase concentration of species A

K: adsorption equilibrium constant

The Freundlich isotherm has been used for physical and chemical adsorption. This model states that the surface coverage varies with the concentration of the adsorbing gas raised to a power

$$v = kC_A^{1/n} \quad (n > 1) \tag{8.5}$$

The Freundlich model predicts that the surface coverage increases in an unbounded manner, which is physically not meaningful. In practice, this isotherm is valid only at relatively small values of surface coverage.

Henry's law for adsorption is the Freundlich model with the power n equal to one.

$$v = kC_A \tag{8.6}$$

Henry's law can be viewed as a limiting case of the Langmuir adsorption isotherm, that is, it is applicable when the concentration of A is very small. Henry's law is also only valid at low values of surface coverage. The final isotherm is the Temkin isotherm, which has been used for chemisorption, which is written as

$$\frac{v}{v_m} = \theta_A = \frac{1}{a}\ln(k_0 C_A) \tag{8.7}$$

The derivation and use of the Langmuir, Freundlich, and Temkin isotherms are shown in more detail shortly.

8.1.5.3 Adsorption Isobars

The adsorption isobar is less commonly used to represent equilibrium adsorption data. In this case, the variation of the amount adsorbed with temperature is shown, while the

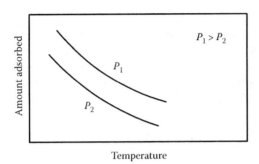

FIGURE 8.9
Typical adsorption isobars. Higher pressures result in an increase in the amount adsorbed at equilibrium. The amount adsorbed often decreases as the temperature increases.

pressure of the adsorbate is held constant. A higher adsorbate pressure usually leads to an increased adsorption. Typical adsorption isobars are shown in Figure 8.9.

8.1.5.4 Adsorption Isoteres

Adsorption isoteres display the temperature and pressure required to produce a surface with a constant amount of adsorbed material. Typical adsorption isoteres are shown in Figure 8.10.

8.1.6 Heat of Adsorption

The heat of adsorption is a key parameter in determining the difference between physical and chemical adsorption. The magnitude of the heat of adsorption determines the strength of the bond formed between the adsorbate and the adsorbent. This factor is obviously important in chemisorption. The heat of adsorption must not be so large so as to result in an irreversible chemisorption. Such a phenomenon would eliminate any catalytic activity.

It has been observed that the heat of adsorption may be a function of the amount of the adsorbed species. Different behavior has been observed for various systems. Some systems have a constant heat of adsorption with coverage, although it is more common for the surface coverage to alter the heat of adsorption. Observed experimental behavior includes a linear decrease in the heat of adsorption with coverage, as well as an exponential decrease. The most likely explanation for this behavior is an energetically heterogeneous

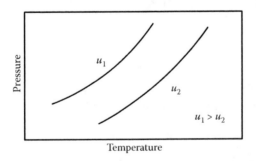

FIGURE 8.10
Typical adsorption isoteres. Each curve represents a line of constant coverage, which changes with temperature and pressure.

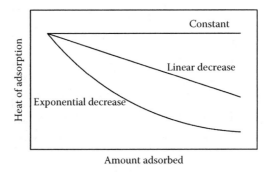

FIGURE 8.11
Behavior of the heat of adsorption as a function of surface coverage. The value may remain constant or decline as coverage increases.

system. The tendency is for molecules to adsorb first on sites with the maximum free energy change (maximum heat of adsorption), so a gradual decrease in heat of adsorption with coverage would result as the higher energy sites become occupied. The dependence of the heat of adsorption on surface coverage has important implications for kinetic modeling. These types of behavior are shown in Figure 8.11.

Heats of adsorption can be estimated from adsorption isotherms using the Clausius–Clapeyron equation. The Clausius–Clapeyron equation is used extensively in vapor–liquid equilibrium, and rests on the assumption of an ideal gas and that the specific volume of the liquid is very small compared to the volume of the gas. The equation can be written as

$$\frac{d(\ln P)}{dT} = \frac{\Delta H}{R_g T^2} \tag{8.8}$$

In vapor–liquid equilibrium, the vapor pressure is not a function of the volume of the system. However, in adsorption equilibrium, the equilibrium pressure is a function of the amount adsorbed as the heat of adsorption is a function of the coverage. The Clausius–Clapeyron analog for an adsorption equilibrium is therefore written in terms of a heat of adsorption at a specified level of coverage (adsorption). The equation is thus written as

$$\frac{d(\ln P)}{dT} = \frac{Q_{iso}}{R_g T^2} \tag{8.9}$$

Q_{iso} is called the isoteric heat of adsorption. Integrating the equation at a constant Q value gives

$$\ln\left(\frac{P_1}{P_2}\right)_{\theta 1} = \frac{Q_{iso}}{R_g}\left(\frac{1}{T_2} - \frac{1}{T_1}\right) \tag{8.10}$$

The heat of adsorption thus calculated would be the value at surface coverage fraction $\theta 1$. The change in heat of adsorption with coverage could be determined by doing the calculation at various levels of surface coverage. The data for calculating isoteric heats of adsorption can be obtained from adsorption isotherms as illustrated in Figure 8.12.

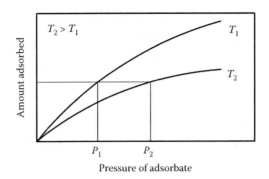

FIGURE 8.12
The heat of adsorption at any given surface coverage can be calculated from two adsorption isotherms.

8.1.7 Langmuir Adsorption Model: Detailed Consideration

Although there are many models for adsorption, the one most commonly used in catalysis is the *Langmuir* adsorption model. This model assumes that

1. All of the active sites are the same.
2. The heat of adsorption is not a function of coverage.
3. The rate of adsorption is proportional to the fraction of empty adsorption sites.
4. The rate of desorption is proportional to the number of molecules adsorbed.
5. At adsorption equilibrium, the rate of adsorption equals the rate of desorption.

8.1.7.1 The Langmuir Model: Single-Component Systems

We begin the analysis of Langmuir adsorption by considering a system in which molecules of A adsorb on active sites on the surface, denoted by an S, as illustrated in Figure 8.13. The adsorption of A is represented by the reversible reaction

$$A + S \rightleftharpoons A \cdot S$$

The rate of adsorption is proportional to the number of vacant sites on the catalyst surface and the gas-phase concentration of the adsorbing species. The rate of desorption is proportional to the concentration of adsorbed species. The net rate of adsorption is written as an elementary reaction

$$r_{ads} = k_a C_A \theta_V - k_d \theta_A \tag{8.11}$$

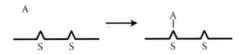

FIGURE 8.13
Pathway for the molecular adsorption of molecule A on a surface site.

In Equation 8.11, the symbols have the following meanings:

$C_A \equiv$ concentration of A in the gas phase
$\theta_V \equiv$ fraction of vacant sites
$\theta_A \equiv$ fraction of sites occupied by A
$k_a \equiv$ rate constant for adsorption
$k_d \equiv$ rate constant for desorption

The active sites on the catalyst surface are either vacant or occupied by an adsorbed molecule. The sum of the fraction of vacant sites and those that are occupied is equal to one; therefore, if A is the only adsorbing species, it follows that

$$\theta_V + \theta_A = 1 \tag{8.12}$$

Substitution of Equation 8.12 into Equation 8.11 eliminates the term for the fraction of vacant sites, and the rate of adsorption can be written as

$$r_{ads} = k_a C_A (1 - \theta_A) - k_d \theta_A \tag{8.13}$$

At equilibrium, the net rate of adsorption is zero. Setting Equation 8.13 equal to zero and rearranging give

$$C_A = \frac{1}{K_A} \left(\frac{\theta_A}{1 - \theta_A} \right) \tag{8.14}$$

The adsorption equilibrium constant, K_A, is the ratio of the adsorption and desorption rate constants:

$$K_A = \frac{k_a}{k_d} \tag{8.15}$$

Equation 8.14 can be rearranged to give an explicit expression for the fractional surface coverage of molecules of A:

$$\theta_A = \frac{K_A C_A}{1 + K_A C_A} \tag{8.16}$$

Equation 8.16 gives the fractional surface coverage of species A for a given gas-phase concentration at a specified temperature. The adsorption equilibrium constant is an exponential function of temperature. Equilibrium adsorption data can be obtained using various experimental techniques. Such results are available in the literature for many gas/solid combinations. The data are usually reported as equilibrium fractional surface coverage plotted as a function of either temperature or pressure.

Some texts use pressure rather than concentration, that is, substitute P_A for C_A in the adsorption rate expressions. This will result in a difference in both the numerical value of

k_a as well as the units. Furthermore, for an ideal gas, the pressure and concentration are related by

$$C_A = \frac{P_A}{R_g T}$$

The temperature dependence of k_a would thus be slightly different if partial pressures are used to define the rate of adsorption.

8.1.7.2 Langmuir Model: Dissociative Chemisorption

Langmuir adsorption can describe the adsorption of a diatomic molecule A_2 that dissociates on adsorption, as shown in Figure 8.14. The adsorption is represented by a reaction between a molecule of A_2 and two surface sites

$$A_2 + 2S \rightleftharpoons 2A \cdot S$$

The adsorption rate would be written as

$$r_{ads} = k_a C_{A_2} \theta_V^2 - k_d \theta_A^2 \tag{8.17}$$

The fraction of vacant sites depends on the fraction of sites occupied by adsorbed A atoms

$$\theta_V + \theta_A = 1 \tag{8.18}$$

Combining Equations 8.17 and 8.18 gives

$$r_{ads} = k_a C_{A_2} \left(1 - \theta_A\right)^2 - k_d \theta_A^2 \tag{8.19}$$

Note that the desorption step requires the recombination of two adsorbed A atoms, so the desorption is a second-order process. At equilibrium, the net rate of adsorption is equal to zero. Setting Equation 8.19 equal to zero and rearranging give

$$C_{A_2} = \frac{1}{K_A} \left(\frac{\theta_A}{1 - \theta_A} \right)^2 \tag{8.20}$$

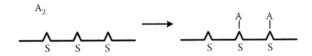

FIGURE 8.14
Pathway for the dissociative adsorption of diatomic molecule A_2 on two surface sites.

We can rearrange Equation 8.20 to give an expression for the adsorbed concentration of A:

$$\theta_A = \frac{\left(K_A C_{A_2}\right)^{0.5}}{1 + \left(K_A C_{A_2}\right)^{0.5}} \tag{8.21}$$

The adsorbed amount is now proportional to the square root of the concentration.

8.1.7.3 Langmuir Model: Multicomponent Systems, Single Site Type

One advantage of the Langmuir adsorption theory is the ease with which it may be extended to multicomponent systems. For example, consider a gas composed of two components, A and B, which compete for adsorption on the same adsorption sites on the adsorbate, as shown in Figure 8.15. The adsorption of A and B are each represented by reversible reactions

$$A + S \rightleftarrows A \cdot S$$

$$B + S \rightleftarrows B \cdot S$$

The rates of adsorption of A and B are given by

$$\left(r_{ads}\right)_A = k_{a(A)} C_A \theta_V - k_{d(A)} \theta_A \tag{8.22}$$

$$\left(r_{ads}\right)_B = k_{a(B)} C_B \theta_V - k_{d(B)} \theta_B \tag{8.23}$$

The active sites may be vacant or occupied by molecules of A or B. Thus, the site balance is

$$\theta_A + \theta_B + \theta_V = 1 \tag{8.24}$$

At equilibrium, the net rates of adsorption of A and B are both equal to zero; therefore

$$K_A C_A = \frac{\theta_A}{1 - \theta_A - \theta_B} \tag{8.25}$$

$$K_B C_B = \frac{\theta_B}{1 - \theta_A - \theta_B} \tag{8.26}$$

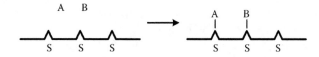

FIGURE 8.15
Pathway for the competitive molecular adsorption of molecules A and B on the surface sites.

Equations 8.25 and 8.26 can be rearranged to give explicit expressions for the surface coverage of molecules of A and B:

$$\theta_A = \frac{K_A C_A}{1 + K_A C_A + K_B C_B} \tag{8.27}$$

$$\theta_B = \frac{K_B C_B}{1 + K_B C_B + K_A C_A} \tag{8.28}$$

In general, for the *i*th component in an *n* component system at equilibrium, the fractional coverage of component *i* is given by

$$\theta_i = \frac{K_i C_i}{1 + \sum_{j=1}^{n} K_j C_j} \tag{8.29}$$

The surface coverages depend on the relative magnitudes of the adsorption equilibrium constants.

Although the Langmuir adsorption isotherm is widely used, especially in the formulation of catalytic rate equations, it has serious limitations. Many experimental data do not follow the Langmuir model, and even for those systems that do, the underlying assumptions are probably wrong. The assumptions that all surface sites are equal, that no interaction occurs between adsorbed molecules, and that adsorption is limited to a monolayer are seldom met exactly. However, these serious limitations notwithstanding, the Langmuir model often fits experimental data reasonable well, especially over a narrow range of surface coverage.

For real surfaces, the inhomogeneity of the surface and the interaction among adsorbed species leads to change (usually a decrease) in the heat of adsorption with coverage. This change in the heat of adsorption with coverage is taken into account in the development of the Freundlich and the Temkin isotherms, which are described in the following section. For multilayer physical adsorption, the BET isotherm is often used. The BET isotherm, and its use in the determination of surface area, is discussed in Chapter 11.

At the limit of low coverages, the Langmuir adsorption isotherm collapses to give Henry's law. The limit on the validity of this expression depends on the value of the gas pressure and the equilibrium constant.

8.1.8 Models Accounting for Variable Heat of Adsorption

There is abundant evidence to support the hypothesis of a fall in the heat of adsorption with surface coverage. As seen earlier, such a decrease may have an exponential or a linear form. In the following paragraphs, the classical derivation for the adsorption on a surface with variable heat of adsorption is briefly elucidated.

Divide the surface into sites of different heat of adsorption. For each type of site, Langmuir adsorption prevails. It is therefore possible to write for each site type, denoted with the subscript *i*, the equilibrium adsorption coverage.

$$\theta_i = \frac{K_i C_A}{1 + K_i C_A} \tag{8.30}$$

Assume that the equilibrium constant is an exponential function of temperature and the heat of adsorption, that is

$$K_i = K_0 \exp\left(\frac{Q_i}{R_g T}\right) \tag{8.31}$$

The equilibrium coverage can be represented by the following temperature-dependent function:

$$\theta_i = \frac{K_0 \exp(Q_i/R_g T) C_A}{1 + K_0 \exp(Q_i/R_g T) C_A} \tag{8.32}$$

The total amount (of component A) adsorbed depends on the numbers of each type of site and the coverage on that type. The relationship can be written in discrete form for a finite number of site types or in integral form for a continuous variation in heat of adsorption, as written below:

$$\theta_A = \frac{\sum_i \theta_i N_i(Q)}{\sum_i N_i(Q)} = \frac{\int_i \theta_i N_i(Q) dQ}{\int_i N_i(Q) dQ} \tag{8.33}$$

Combination of the equations gives an integral expression for the total amount adsorbed. This equation can be used to develop equations for the isotherms provided that a relationship for the variation of the heat of adsorption with coverage is proposed.

$$\theta_A = \frac{\int_i \left[\dfrac{K_0 \exp(Q_i/R_g T) C_A}{1 + K_0 \exp(Q_i/R_g T) C_A}\right] N_i(Q) dQ}{\int_i N_i(Q) dQ} \tag{8.34}$$

The Freundlich isotherm assumes that the change in the heat of adsorption with coverage is exponential, in other words

$$N(Q) = a \exp\left(\frac{-Q}{Q_0}\right) \tag{8.35}$$

Substitution of this exponential expression for the site distribution into the integral equation gives a complex integral that cannot be solved analytically. For small values of coverage, the solution simplifies and it is the Freundlich equation shown previously. Originally, the Freundlich isotherm was considered to be purely empirical, but as seen here, it has a solid theoretical basis. Furthermore, the theoretical derivation shows that the Freundlich isotherm is only valid at low coverages, which explains the apparent inconsistency of the isotherm predicting the unbounded increase in the surface coverage. In several systems in which gases are chemisorbed onto metals, the Freundlich isotherm has been found to be applicable.

It has also been shown that, for some systems, the heat of adsorption can fall in a linear fashion as the surface coverage increases. This type of linear behavior can arise from surface heterogeneities as well as repulsive forces between adsorbed species, even on energetically uniform surfaces. The distribution of surface sites with heat of adsorption is then given by

$$N(Q) = a + b\left(\frac{Q}{Q_0}\right)$$

(8.36)

The derivation of the isotherm proceeds in the same manner as for the Freundlich isotherm, with the substitution of the equation for the coverage dependence of the heat of adsorption into the general integral expression. The solution is

$$\theta_A = \frac{1}{a}\ln(kC_A)$$

(8.37)

One system for which the Temkin isotherm has been shown to be valid is for the adsorption of nitrogen on some iron catalysts. This system is important in the synthesis of ammonia.

8.2 Rate Expressions for Catalytic Reactions

The development of rate equations for solid-catalyzed gas-phase reactions is, generally speaking, a complex process. Catalytic mechanisms can involve many steps, which may be difficult to determine. However, it is not uncommon to make a number of simplifying assumptions, and then derive a rate equation using, in effect, a simplified mechanism. One of the most common techniques gives the LHHW models (Langmuir–Hinshelwood–Hougan–Watson), the development of which are explained in the following section.

8.2.1 Langmuir–Hinshelwood–Hougen–Watson Rate Equations

Langmuir adsorption has been incorporated into a systematic method for generating rate equations for catalytic reactions. The Langmuir–Hinshelwood–Hougen–Watson (LHHW) method of generating rate models based on Langmuir adsorption was systematically developed by Hougen and Watson. It has had some success in the transient case but it works best at steady state. The following steps summarize the LHHW methodology:

Step 1: Propose a mechanism based on: (a) Langmuir adsorption, (b) surface reaction between adsorbed intermediates, and (c) desorption. Proposing a mechanism for the reaction can be difficult and is based on assumptions, intuition, and experimental data. Any information that is available about the nature of the surface intermediates should be used at this stage.

Step 2: Assume that one of the steps in the mechanism is intrinsically slow. This step is said to be the *rate-determining step* (RDS). It must be emphasized that at steady state, the actual rate at which each step in the mechanism proceeds is the same. "Intrinsically slow" implies that the values of the rate constants for this step are small. The other steps are assumed to be in equilibrium. Note that this is not a true equilibrium because actually all steps have the same finite rate, however, if the "equilibrium" steps are intrinsically very fast they will not be far from equilibrium and the assumption of a single rate-determining step will be valid.

Step 3: Write the overall rate of reaction in terms of the rate-determining step. This rate expression will contain surface concentration terms of adsorbed intermediates.

Step 4: Eliminate the surface (adsorbed) concentration terms from the expression using the steps in the mechanism that are assumed to be in equilibrium. The final rate expression is expressed in terms of gas-phase concentrations.

The LHHW method is best illustrated by means of some examples. We show six cases in the following to demonstrate the main features of the method.

LHHW Case 1. Molecular Adsorption with Surface Reaction as RDS

Consider a reaction between reactants A and B, to form products C and D, that is

$$A + B \rightleftarrows C + D$$

A simple reaction mechanism involves the chemisorption of A and B as molecules, reaction between chemisorbed A and B molecules to form C and D molecules, and the subsequent desorption of C and D. This scheme is summarized by the following mechanism (where S represents a surface site). The mechanism is shown also in Figure 8.16.

$A + S \rightleftarrows A \cdot S$	(Step 1, chemisorption of A)
$B + S \rightleftarrows B \cdot S$	(Step 2, chemisorption of B)
$A \cdot S + B \cdot S \rightleftarrows C \cdot S + D \cdot S$	(Step 3, surface reaction)
$C \cdot S \rightleftarrows C + S$	(Step 4, desorption of C)
$D \cdot S \rightleftarrows D + S$	(Step 5, desorption of D)

Let Step 3, the surface reaction, be the rate-determining step. The overall rate of reaction is expressed as an elementary reaction between adsorbed A and adsorbed B, represented by

$$(-r_A) = r_s = k_s \theta_A \theta_B - k_{-s} \theta_C \theta_D \tag{8.38}$$

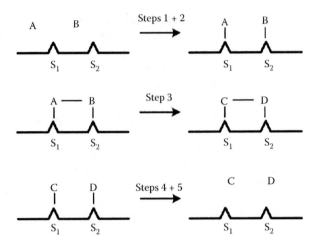

FIGURE 8.16
Sequence of elementary steps for LHHW model, Case 1. A and B adsorb as molecules, react to form adsorbed molecules of C and D, which subsequently desorb.

In Equation 8.38, r_s is the rate of surface reaction, k_s is the rate constant for the forward reaction, and k_{-s} is the rate constant for the reverse reaction. The terms for the adsorbed species are now eliminated by assuming that Steps 1, 2, 4, and 5 are in equilibrium. There is competition for the adsorption sites among molecules of A, B, C, and D. The rate of adsorption of A is given by the Langmuir model as

$$r_{ads(A)} = k_{a(A)}C_A\theta_V - k_{d(A)}\theta_A \tag{8.39}$$

In Equation 8.39, θ_V is the fraction of vacant sites, given by

$$\theta_V = 1 - \theta_A - \theta_B - \theta_C - \theta_D \tag{8.40}$$

At equilibrium, the net rate of adsorption is zero, and from Equation 8.39 the equilibrium surface coverage of molecules of A is thus given by

$$\theta_A = K_A C_A \theta_V \tag{8.41}$$

By applying the same procedure to molecules of B, C, and D, we obtain the equilibrium adsorbed fractions for those species as

$$\theta_B = K_B C_B \theta_V, \quad \theta_C = K_C C_C \theta_V, \quad \theta_D = K_D C_D \theta_V \tag{8.42}$$

Substitution of the adsorbed surface coverage terms into Equation 8.38 gives the rate expression in terms of gas-phase concentrations and θ_V.

$$(-r_A) = k_s K_A C_A K_B C_B \theta_V^2 - k_{-s} K_C C_C K_D C_D \theta_V^2 \tag{8.43}$$

Substitution of Equations 8.4 and 8.42 into Equation 8.40 gives

$$\theta_V = 1 - K_A C_A \theta_V - K_B C_B \theta_V - K_C C_C \theta_V - K_D C_D \theta_V \tag{8.44}$$

Equation 8.44 can be rearranged to the following form:

$$\theta_V = \frac{1}{(1 + K_A C_A + K_B C_B + K_C C_C + K_D C_D)} \tag{8.45}$$

Substituting Equation 8.45 into 8.43 gives

$$(-r_A) = \frac{k_s K_A C_A K_B C_B - k_{-s} K_C C_C K_D C_D}{(1 + K_A C_A + K_B C_B + K_C C_C + K_D C_D)^2} \tag{8.46}$$

Equation 8.46 may be simplified by incorporating the reaction equilibrium constant. If the surface reaction were to proceed to equilibrium, the net reaction rate would be equal to zero and therefore, at equilibrium:

$$k_s \theta_A \theta_B = k_{-s} \theta_C \theta_D \tag{8.47}$$

The equilibrium constant for the surface reaction is defined in terms of the fractional coverage of the reacting species; thus

$$K_s = \left(\frac{\theta_C \theta_D}{\theta_A \theta_B} \right)_{eq} \tag{8.48}$$

It follows from Equations 8.47 and 8.48 that the equilibrium constant for the surface reaction can also be expressed as a ratio of the forward and reverse reaction rate constants, as follows:

$$K_s = \frac{k_s}{k_{-s}} \tag{8.49}$$

Substitution of Equations 8.41 and 8.42 into Equation 8.48 gives

$$K_s = \left(\frac{K_C C_C \theta_V K_D C_D \theta_V}{K_A C_A \theta_V K_B C_B \theta_V} \right) = \left(\frac{K_C C_C K_D C_D}{K_A C_A K_B C_B} \right) = \left(\frac{K_C K_D}{K_A K_B} \right) \left(\frac{C_C C_D}{C_A C_B} \right) \tag{8.50}$$

The equilibrium constant for the overall reaction in terms of concentration (assuming ideal gases) is

$$K = \frac{C_C C_D}{C_A C_B} \tag{8.51}$$

Equation 8.51 can be substituted into Equation 8.50 to give a relationship among the various equilibrium constants

$$K_s = \frac{K_C K_D}{K_A K_B} K \tag{8.52}$$

Note that Equation 8.52 is valid for the sequence of elementary reaction steps proposed, regardless of the rate-determining step. The final rate equation is then obtained by substituting Equation 8.52 into Equation 8.46 and rearranging to give

$$(-r_A) = \frac{k_s K_A K_B \left(C_A C_B - (1/K)C_C C_D\right)}{\left(1 + K_A C_A + K_B C_B + K_C C_C + K_D C_D\right)^2} \tag{8.53}$$

LHHW Case 2. Molecular Adsorption with Adsorption as RDS

Consider the same sequence of elementary steps as proposed in Case 1, but now the adsorption of A (Step 1) is the rate-determining step. Otherwise, the mechanism is the same as Case 1. The surface reaction is at equilibrium, as is the adsorption of B, C, and D. The overall rate of reaction is governed by the rate of adsorption; therefore, the reaction rate is given by

$$(-r_A) = r_{ads(A)} = k_{a(A)}C_A \theta_V - k_{d(A)}\theta_A \tag{8.54}$$

The surface fraction of adsorbed A may not be substituted by the expression used in Case 1, as this step is not in equilibrium. Rather the surface reaction is in equilibrium and its net rate is zero and therefore we can write

$$\theta_A = \frac{\theta_C \theta_D}{K_s \theta_B} \tag{8.55}$$

The adsorption of B, C, and D are in equilibrium, and the fractional coverage of each of these species is

$$\theta_B = K_B C_B \theta_V, \quad \theta_C = K_C C_C \theta_V, \quad \theta_D = K_D C_D \theta_V \tag{8.56}$$

The fractional coverage of A can be expressed in terms of gas-phase concentrations by substituting Equation 8.56 into Equation 8.55 and simplifying

$$\theta_A = \frac{K_C C_C \theta_V K_D C_D \theta_V}{K_s K_B C_B \theta_V} = \frac{K_C K_D}{K_s K_B} \frac{C_C C_D}{C_B} \theta_V \tag{8.57}$$

This result is substituted into Equation 8.54 to give the intermediate result

$$(-r_A) = r_{ads(A)} = k_{a(A)}C_A \theta_V - k_{d(A)} \frac{K_C K_D}{K_s K_B} \frac{C_C C_D}{C_B} \theta_V \tag{8.58}$$

The fraction of vacant sites is given by

$$\theta_V = 1 - \theta_A - \theta_B - \theta_C - \theta_D \tag{8.59}$$

Substituting the fractional coverage terms of Equations 8.56 and 8.57 gives

$$\theta_V = 1 - \frac{K_C K_D}{K_s K_B} \frac{C_C C_D}{C_B} \theta_V - K_B C_B \theta - K_C C_C \theta_V - K_D C_D \theta_V \tag{8.60}$$

Factoring Equation 8.60 to get an explicit expression in θ_V yields

$$\theta_V = \frac{1}{\left(1 + \dfrac{K_C K_D}{K_s K_B} \dfrac{C_C C_D}{C_B} + K_B C_B + K_C C_C + K_D C_D \right)} \tag{8.61}$$

The relationship among equilibrium constants developed in Case 1, Equation 8.52, is still valid. That equation can be rearranged and expressed as

$$\frac{K_A}{K} = \frac{K_C K_D}{K_s K_B} \tag{8.62}$$

Combining Equations 8.58, 8.61, and 8.62 and simplifying give the final result

$$(-r_A) = \frac{k_{a(A)} \left(C_A - \dfrac{1}{K} \dfrac{C_C C_D}{C_B} \right)}{\left(1 + \dfrac{K_A}{K} \dfrac{C_C C_D}{C_B} + K_B C_B + K_C C_C + K_D C_D \right)} \tag{8.63}$$

Note that this rate equation is different from the one obtained in Case 1 when the surface reaction was the rate-determining step.

LHHW Case 3. Dissociative Adsorption with Surface Reaction as RDS

In Case 3, we consider a slightly more complex case, the reaction between two diatomic molecules, A_2 and A_3 to form two molecules of AB:

$$A_2 + B_2 \rightleftarrows 2AB$$

A mechanism where both A_2 and B_2 dissociate on adsorption and the surface reaction proceeds between adsorbed atoms of A and B can be expressed in the following steps. These steps are also illustrated in Figure 8.17.

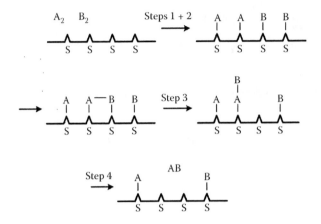

FIGURE 8.17
Steps in the reaction mechanism for LHHW Case 3. The mechanism involves the dissociative chemisorption of two reactants.

$A_2 + 2S \rightleftarrows 2A \cdot S$ \qquad (Step 1, dissociative chemisorption)

$B_2 + 2S \rightleftarrows 2B \cdot S$ \qquad (Step 2, dissociative chemisorption)

$A \cdot S + B \cdot S \rightleftarrows AB \cdot S + S$ \qquad (Step 3, surface reaction)

$AB \cdot S \rightleftarrows AB + S$ \qquad (Step 4, desorption of product)

In this proposed mechanism, Steps 3 and 4 must occur twice for each occurrence of Steps 1 and 2. Let us assume that Step 3, the surface reaction, is the rate-determining step. The overall rate of reaction may be expressed as an elementary reaction between adsorbed atoms of A and B:

$$(-r_{A_2}) = r_s = k_s \theta_A \theta_B - k_{-s} \theta_{AB} \theta_V \tag{8.64}$$

The adsorption of A_2, B_2, and AB are assumed to be at equilibrium. For the dissociative chemisorption of A_2, the rate of adsorption is given by

$$r_{ads(A_2)} = k_{a(A_2)} C_{A_2} \theta_V^2 - k_{d(A_2)} \theta_A^2 \tag{8.65}$$

At equilibrium the net rate of adsorption is equal to zero, and Equation 8.65 can be equated to zero and rearranged to give

$$\theta_A = \sqrt{K_A C_{A_2}}\, \theta_V \tag{8.66}$$

Similarly, we obtain expressions for the adsorbed coverage of B and AB as

$$\theta_B = \sqrt{K_B C_{B_2}}\, \theta_V \quad \text{and} \quad \theta_{AB} = K_{AB} C_{AB} \theta_V \tag{8.67}$$

The fractional coverage of vacant sites is

$$\theta_V = 1 - \theta_A - \theta_B - \theta_{AB} \tag{8.68}$$

Substituting Equations 8.66 and 8.67 into Equation 8.68 and rearranging give

$$\theta_V = \frac{1}{1 + \sqrt{K_A C_{A_2}} + \sqrt{K_B C_{B_2}} + K_{AB} C_{AB}} \tag{8.69}$$

Substituting Equations 8.66, 8.67, and 8.69 into Equation 8.64 gives the rate expression in terms of gas-phase concentrations:

$$(-r_{A_2}) = \frac{k_s \sqrt{K_A C_{A_2}} \sqrt{K_B C_{B_2}} - k_{-s} K_{AB} C_{AB}}{\left(1 + \sqrt{K_A C_{A_2}} + \sqrt{K_B C_{B_2}} + K_{AB} C_{AB}\right)} \tag{8.70}$$

Equation 8.70 may be simplified using the reaction equilibrium constant. The equilibrium constant for the surface reaction expressed in terms of the fractional coverage of the reacting species is

$$K_s = \frac{k_s}{k_{-s}} = \left(\frac{\theta_{AB} \theta_V}{\theta_A \theta_B} \right)_{eq} \tag{8.71}$$

Equations 8.66 and 8.67 may be substituted into Equation 8.71 to give

$$K_s = \frac{K_{AB} C_{AB} \theta_V \theta_V}{\sqrt{K_A C_{A_2}}\, \theta_V \sqrt{K_B C_{B_2}}\, \theta_V} = \frac{K_{AB} \theta_V \theta_V}{\sqrt{K_A K_B}\, \theta_V \theta_V} \frac{C_{AB}}{\sqrt{C_{A_2} C_{B_2}}} \tag{8.72}$$

The overall equilibrium constant for this reaction, assuming ideal gases, is

$$K = \frac{C_{AB}^2}{C_{A_2} C_{B_2}} \tag{8.73}$$

Combination of Equations 8.72 and 8.73 gives

$$K_s = \frac{K_{AB}}{\sqrt{K_A K_B}} \sqrt{K} \tag{8.74}$$

The final rate expression is obtained by substituting Equation 8.74 into Equation 8.70 and rearranging to give

$$(-r_{A_2}) = \frac{k_s \sqrt{K_A K_B} \left(\sqrt{C_{A_2} C_{B_2}} - \frac{1}{\sqrt{K}} C_{AB} \right)}{\left(1 + \sqrt{K_A C_{A_2}} + \sqrt{K_B C_{B_2}} + K_{AB} C_{AB}\right)^2} \tag{8.75}$$

LHHW Case 4. Dissociative Adsorption with Adsorption as RDS

Consider the same reaction and mechanism as in Case 3, except that the adsorption of A is the rate-determining step and the surface reaction is at equilibrium. The adsorption of B_2 and AB are also at equilibrium. The overall rate of reaction is expressed as the rate of dissociative chemisorption of A_2.

$$(-r_{A_2}) = r_{ads(A_2)} = k_{a(A)}C_{A_2}\theta_V^2 - k_{d(A)}\theta_A^2 \tag{8.76}$$

If the rate-determining step is the adsorption of A_2, the surface reaction is in equilibrium. The surface reaction is written as

$$r_s = 0 = k_s\theta_A\theta_B - k_{-s}\theta_{AB}\theta_V \tag{8.77}$$

Rearranging Equation 8.77 leads to the following expression for the adsorbed fraction of A:

$$\theta_A = \frac{k_{-s}}{k_s}\frac{\theta_{AB}\theta_V}{\theta_B} = \frac{1}{K_s}\frac{\theta_{AB}\theta_V}{\theta_B} \tag{8.78}$$

The equations for the adsorbed coverage of B and AB are the same as Case 3

$$\theta_B = \sqrt{K_BC_{B_2}}\,\theta_V \quad \text{and} \quad \theta_{AB} = K_{AB}C_{AB}\theta_V \tag{8.79}$$

The relationship among the equilibrium constants is also the same as in Case 3, which can be expressed in the following form by suitable rearrangement of Equation 8.74:

$$\frac{\sqrt{K_A}}{\sqrt{K}} = \frac{1}{K_s}\frac{K_{AB}}{\sqrt{K_B}} \tag{8.80}$$

Substituting Equations 8.79 and 8.80 into Equation 8.78 gives

$$\theta_A = \frac{1}{K_s}\frac{K_{AB}}{\sqrt{K_B}}\frac{C_{AB}}{C_{B_2}}\theta_V = \frac{\sqrt{K_A}}{\sqrt{K}}\frac{C_{AB}}{\sqrt{C_{B_2}}}\theta_V \tag{8.81}$$

Substituting Equation 8.81 into Equation 8.76 gives

$$(-r_{A_2}) = r_{ads(A_2)} = k_{a(A)}C_{A_2}\theta_V^2 - k_{d(A)}\frac{K_A}{K}\frac{C_{AB}^2}{C_{B_2}}\theta_V^2 \tag{8.82}$$

The fractional coverage of vacant sites is

$$\theta_V = 1 - \theta_A - \theta_B - \theta_{AB} \tag{8.83}$$

Substitution of Equations 8.79 and 8.81 into Equation 8.83 gives

$$\theta_V = \frac{1}{1 + \left(\sqrt{K_A}/K\right)\left(C_{AB}/\sqrt{C_{B_2}}\right) + \sqrt{K_BC_{B_2}} + K_{AB}C_{AB}} \tag{8.84}$$

Finally, substitute Equation 8.84 into Equation 8.82 and simplify to obtain the rate equation

$$(-r_{A_2}) = \frac{k_{a(A)}\left(C_{A_2} - \dfrac{1}{K}\dfrac{C_{AB}^2}{C_{B_2}}\right)}{1 + \left(\sqrt{K_A}/K\right)\left(C_{AB}/\sqrt{C_{B_2}}\right) + \sqrt{K_B C_{B_2}} + K_{AB}C_{AB}} \tag{8.85}$$

LHHW Case 5. Eley–Rideal Mechanism with Surface Reaction as the RDS

A special case of the LHHW model occurs when the surface reaction involves the reaction between an adsorbed species and a component in the gas phase. This is equivalent to a gas-phase molecule adsorbing onto a previously adsorbed species. This type of mechanism is referred to as an Eley–Rideal mechanism. Consider the following reaction:

$$\frac{1}{2}A_2 + B \rightleftarrows AB$$

A possible Eley–Rideal mechanism is given by the reaction steps below, which are also illustrated in Figure 8.18.

$$A_2 + 2S \rightleftarrows 2A \cdot S \qquad \text{(Step 1, dissociative chemisorption)}$$
$$A \cdot S + B \rightleftarrows AB \cdot S \qquad \text{(Step 2, surface reaction with gas-phase B)}$$
$$AB \cdot S \rightleftarrows AB + S \qquad \text{(Step 3, desorption of product)}$$

Assume that Step 2, the surface reaction, is the rate-controlling step. The overall rate of reaction may be expressed as an elementary reaction between adsorbed A and gas-phase B:

$$(-r_{A_2}) = k_s\theta_A C_B - k_{-s}\theta_{AB} \tag{8.86}$$

The adsorption Steps 1 and 3 are assumed to be in equilibrium. The adsorbed surface coverage of A and AB are obtained using the same methods as in Cases 1 through 4 to give

$$\theta_A = \sqrt{K_A C_{A_2}}\,\theta_V \quad \text{and} \quad \theta_{AB} = K_{AB}C_{AB}\theta_V \tag{8.87}$$

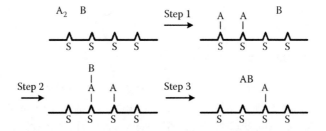

FIGURE 8.18
Eley–Rideal mechanism. Molecules of A adsorb on the catalyst surface, which then react with B molecules in the gas phase. LHHW Case 5.

The surface sites are occupied by either A atoms or AB molecules. Molecules of B do not adsorb on the surface sites. The fraction of vacant sites is therefore given by

$$\theta_V = 1 - \theta_A - \theta_{AB} \tag{8.88}$$

Substituting Equation 8.87 into Equation 8.88 leads to

$$\theta_V = \frac{1}{1 + \sqrt{K_A C_{A_2}} + K_{AB} C_{AB}} \tag{8.89}$$

Substitution of Equations 8.87 and 8.89 into Equation 8.86 gives the rate expression in terms of gas-phase concentrations

$$(-r_{A_2}) = \frac{k_s \left(\sqrt{K_A C_{A_2}} C_B - \left(K_{AB}/K_s \right) C_{AB} \right)}{\left(1 + \sqrt{K_A C_{A_2}} + K_{AB} C_{AB} \right)} \tag{8.90}$$

Equation 8.90 is simplified using the relationship among the equilibrium constants, in a similar fashion as was done in Cases 1 through 4. In this case, at equilibrium, we have

$$K_s = \frac{k_s}{k_{-s}} = \frac{\theta_{AB}}{\theta_A C_B} = \frac{K_{AB}}{\sqrt{K_A}} \frac{C_{AB}}{\sqrt{C_{A_2}} C_B} = \frac{K_{AB} K}{\sqrt{K_A}} \tag{8.91}$$

where K is the equilibrium constant for the overall reaction, as before. Substitution of Equation 8.91 into Equation 8.90 gives the final result

$$(-r_{A_2}) = \frac{k_s \sqrt{K_A} \left(\sqrt{C_{A_2}} C_B - 1/K C_{AB} \right)}{\left(1 + \sqrt{K_A C_{A_2}} + K_{AB} C_{AB} \right)} \tag{8.92}$$

LHHW Case 6. Molecular Adsorption with Surface Reaction as RDS, Dual Site Model
Consider the same overall reaction as in Case 1.

$$A + B \rightleftarrows C + D$$

The mechanism is the same as in Case 1 with one important exception. The catalyst surface has two different types of sites, denoted S_1 and S_2. Molecules of A adsorb on S_1 type of sites and molecules of B adsorb on S_2 type of sites. The steps in the proposed mechanism are shown below and illustrated in Figure 8.19.

$A + S_1 \rightleftarrows A \cdot S_1$	(Step 1, chemisorption of A)
$B + S_2 \rightleftarrows B \cdot S_2$	(Step 2, chemisorption of B)
$A \cdot S_1 + B \cdot S_2 \rightleftarrows C \cdot S_1 + D \cdot S_2$	(Step 3, surface reaction)
$C \cdot S_1 \rightleftarrows C + S_1$	(Step 4, desorption of C)
$D \cdot S_2 \rightleftarrows D + S_2$	(Step 5, desorption of D)

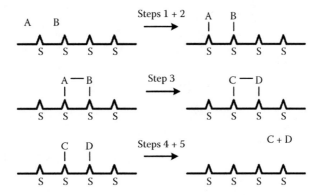

FIGURE 8.19
Steps in a dual site model. The catalyst surface has two different types of sites, which selectively adsorb certain species. LHHW Case 6.

Assume that Step 3, the surface reaction, is the rate-controlling step. The overall rate of reaction is expressed as an elementary reaction between adsorbed A and adsorbed B, which may be represented by

$$(-r_A) = r_s = k_s \theta_A \theta_B - k_{-s} \theta_C \theta_D \tag{8.93}$$

Eliminate the adsorbed species terms by assuming that Steps 1, 2, 4, and 5 are in equilibrium. There will be competition for the adsorption sites between molecules of A and C for type S_1 sites and competition between B and D for type S_2 sites. The adsorption rate of A is

$$(r_{ads})_A = (k_a)_A C_A \theta_{V,1} - (k_d)_A \theta_A \tag{8.94}$$

The fraction of vacant S_1 sites, $\theta_{V,1}$, depends on the extent of adsorption of A and C only, and therefore is given by

$$\theta_{V,1} = 1 - \theta_A - \theta_C \tag{8.95}$$

At equilibrium the net rate of adsorption is equal to zero, and Equation 8.95 can be rewritten in the form

$$\theta_A = K_A C_A \theta_{V,1} \tag{8.96}$$

The adsorption rate of B is

$$(r_{ads})_B = (k_a)_B C_B \theta_{V,2} - (k_d)_B \theta_B \tag{8.97}$$

The fraction of vacant S_2 sites, $\theta_{V,2}$, depends on the coverage of B and D only, and is therefore

$$\theta_{V,2} = 1 - \theta_B - \theta_D \tag{8.98}$$

To summarize, the equilibrium coverage of the four species are

$$\theta_A = K_A C_A \theta_{V,1}, \quad \theta_C = K_C C_C \theta_{V,1}$$
$$\theta_B = K_B C_B \theta_{V,2}, \quad \theta_D = K_D C_D \theta_{V,2}$$

(8.99)

Substitution of Equation 8.99 into Equation 8.93 gives the rate expression in terms of gas-phase concentrations and θ_V.

$$(-r_A) = k_s K_A C_A K_B C_B \theta_{V,1} \theta_{V,2} - k_{-s} K_C C_C K_D C_D \theta_{V,1} \theta_{V,2}$$

(8.100)

The fraction of each type of vacant site is obtained by substituting the terms of Equation 8.99 into Equations 8.95 and 8.98 appropriately to give

$$\theta_{V,1} = 1 - K_A C_A \theta_{V,1} - K_C C_C \theta_{V,1}$$

(8.101)

for type 1 sites and

$$\theta_{V,2} = 1 - K_B C_B \theta_{V,2} - K_D C_D \theta_{V,2}$$

(8.102)

for type 2 sites. Equations 8.101 and 8.102 can be rearranged to the forms

$$\theta_{V,1} = \frac{1}{(1 + K_A C_A + K_C C_C)}$$

(8.103)

$$\theta_{V,2} = \frac{1}{(1 + K_B C_B + K_D C_D)}$$

(8.104)

Equations 8.103 and 8.104 are then substituted into Equation 8.100 to give the rate equation

$$(-r_A) = \frac{k_s K_A C_A K_B C_B - k_{-s} K_C C_C K_D C_D}{(1 + K_A C_A + K_C C_C)(1 + K_B C_B + K_D C_D)}$$

(8.105)

Equation 8.105 may be simplified by incorporating the reaction equilibrium constant in the same manner as in Case 1. This gives the final rate equation

$$(-r_A) = \frac{k_s K_A K_B \left(C_A C_B - \frac{1}{K} C_C C_D \right)}{(1 + K_A C_A + K_C C_C)(1 + K_B C_B + K_D C_D)}$$

(8.106)

8.2.1.1 General Form of LHHW Reaction Models

In the preceding six cases, several LHHW rate expressions were developed for various reaction scenarios. There are many more possible reaction mechanisms, and the complexity of the rate expression increases as the complexity of the proposed reaction mechanism increases. However, it is possible to make some general observations about the nature of all LHHW models.

The common feature of all LHHW models is that the rate equation is expressed as a ratio of two driving forces. For example, let us return to the solution for Case 1. The numerator in Equation 8.53 is

$$k_s K_A K_B \left(C_A C_B - \frac{1}{K} C_C C_D \right) \qquad (8.107)$$

Equation 8.107 is referred to as the kinetic-driving force. The dominator in Equation 8.53 is

$$(1 + K_A C_A + K_B C_B + K_C C_C + K_D C_D)^2 \qquad (8.108)$$

Equation 8.108 is an inhibition force that arises from the adsorption of species on the surface. If we consider a reaction mixture in the absence of products, it can be seen that, at any given temperature, an increase in reactant concentration will result in an increase in the kinetic-driving force, which has the tendency to increase the rate of reaction. However, an increase in C_A and C_B will also increase the value of the adsorption inhibition term, which will tend to decrease the rate of reaction. There is in essence a competition between the two driving forces, and it is therefore not obvious for any given set of operating conditions whether an increase in reactant concentration will result in an increase or a decrease in the reaction rate. An increase in product concentration decreases the value of the kinetic-driving force and increases the value of the adsorption inhibition term, so the overall effect is to decrease the reaction rate.

The presence of the inhibition term can give rise to interesting phenomena. Consider the magnitude of the rate of reaction as the reactant concentration is increased. Initially, the numerator term will increase faster than the denominator, and the rate of reaction will increase. At some point, however, the denominator will increase faster than the numerator and, after this point, the rate will decrease as the pressure of reactants increases. A plot of reaction rate versus reactant concentration, with everything else the same, will result in a graph of the type illustrated in Figure 8.20. This type of curve is commonly known as a "volcano" graph. Note that in regions 1 or 2 (marked on the figure), the rate behavior could be modeled by a power law-type expression

$$(-r_A) = k' C_A^\alpha$$

where in region 1, α would be positive, and in region 2, it would be negative. This behavior is typical of reactions that follow LHHW-type mechanisms.

For some reactions, under certain operating conditions, some of the terms in the LHHW rate expression may be negligibly small. Furthermore, some species may be so weakly adsorbed to the surface that their corresponding term in the denominator effectively vanishes. This is especially true as temperature increases, as the adsorption equilibrium constants tend to decrease in value as the temperature increases. Naturally, at sufficiently small concentrations of reactant and product, the denominator term is essentially equal to one. In this case, the rate equation resembles the power law type of model often encountered in catalytic reactions.

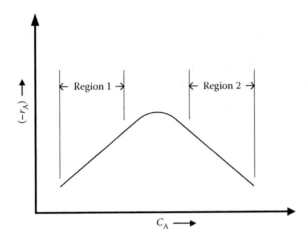

FIGURE 8.20
Reaction rate as a function of concentration of A for a typical LHHW-type rate expression. The volcano shape results from a competition between the kinetic-driving force and the adsorption inhibition term.

There is some controversy in the literature regarding the validity and relevance of LHHW models. Critics point out that some of the assumptions made in the LHHW approach are questionable, or simply incorrect. Some authors will go so far as to say that the LHHW method should simply be discarded. Supporters suggest that in many cases the validity of the assumptions involved in the development of the LHHW method is not important. In fact, LHHW models can and do work. There are several commercial processes that use LHHW models for reactor design and simulation purposes, with very good results. In some cases, the models work well in spite of the fact that the mechanism used to generate the model is not correct. The following points should be considered when deriving and using LHHW models:

1. They can provide a rate equation that is capable of predicting accurately the reaction rate, at least over the required range of operating conditions.

2. The mechanism upon which an LHHW model is based is not necessarily correct, or is a simplification of what occurs.

3. In the event that a proposed LHHW model is observed to fit experimental data sufficiently well over the required temperature range, it may be used as a reaction rate expression for design purposes. It must be emphasized, however, that such an equation should not be extrapolated to regions outside those at which it was experimentally verified.

LHHW models should be used with caution and considered more as an empirical model rather than a mechanistic equation.

Example 8.1

The synthesis of methanol from carbon monoxide and hydrogen is an important industrial reaction, as methanol is the feedstock for many other processes. The synthesis reaction may be represented by the following overall reaction:

$$CO + 2H_2 \rightleftharpoons CH_3OH$$

This reaction has been widely studied, and there is still a certain amount of controversy regarding the mechanism. However, it is necessary to have an appropriate rate model to design reactors. One proposed mechanism following LHHW principles is based on the molecular adsorption of carbon monoxide and hydrogen, followed by a surface reaction among two adsorbed hydrogen molecules and an adsorbed carbon monoxide molecule. This mechanism can be summarized in the following steps:

$$CO + S \rightleftarrows CO \cdot S \tag{8.109}$$

$$H_2 + S \rightleftarrows H_2 \cdot S \tag{8.110}$$

$$2H_2 \cdot S + CO \cdot S \underset{k_{-s}}{\overset{k_s}{\rightleftarrows}} CH_3OH \cdot S + 2S \tag{8.111}$$

$$CH_3OH \cdot S \rightleftarrows CH_3OH + S \tag{8.112}$$

Derive the LHHW-type rate equation for this proposed mechanism, assuming that the surface reaction is the rate-determining step and that the adsorption steps are in equilibrium.

SOLUTION

The surface reaction, Equation 8.111, is the rate-determining step, therefore the rate of reaction is governed by this step. The rate of the surface reaction is written as an elementary reaction among two adsorbed hydrogen molecules and one adsorbed carbon monoxide molecule. The reverse reaction occurs between one adsorbed methanol molecule and two vacant sites.

$$r_s = k_s \theta_{CO} \theta_{H2}^2 - k_{-s} \theta_{CH3OH} \theta_V^2 \tag{8.113}$$

The adsorption steps, Equations 8.109, 8.110, and 8.112, are assumed to be in equilibrium. Following the principles described in Case 1 for molecular adsorption, the equilibrium surface coverage of the three adsorbed species are given by

$$\theta_{CO} = K_{CO} C_{CO} \theta_V$$

$$\theta_{H2} = K_{H2} C_{H2} \theta_V \tag{8.114}$$

$$\theta_{CH3OH} = K_{CH3OH} C_{CH3OH} \theta_V$$

The equilibrium adsorbed surface coverage can be eliminated from the surface reaction rate equation, Equation 8.113, using these terms. That is, substitute the terms of Equation 8.114 into Equation 8.113 to give

$$r_s = k_s K_{CO} K_{H2}^2 C_{CO} C_{H2}^2 \theta_V^3 - k_{-s} K_{CH3OH} C_{CH3OH} \theta_V^3 \tag{8.115}$$

The active sites may be either vacant or occupied by an adsorbed molecule. The sum of the fractions of all occupied and vacant sites must equal one, therefore the fraction of vacant sites is

$$\theta_V = 1 - \theta_{H2} - \theta_{CO} - \theta_{CH3OH} \tag{8.116}$$

Substitute the terms of Equation 8.114 into Equation 8.116, and rearrange to give

$$\theta_V = \frac{1}{\left(1 + K_{CO}C_{CO} + K_{H2}C_{H2} + K_{CH3OH}C_{CH3OH}\right)^3} \tag{8.117}$$

The overall reaction equilibrium constant can be expressed in terms of the equilibrium constants for each elementary step following the principles given in Case 1. The resulting expression is

$$K = K_s \frac{K_{CO}K_{H2}^2}{K_{CH3OH}} \tag{8.118}$$

Substitution of Equations 8.117 and 8.118 into Equation 8.115 gives the final result

$$r_S = \frac{k_s K_{CO}K_{H2}^2 \left[C_{CO}C_{H2}^2 - \dfrac{1}{K}C_{CH3OH}\right]}{\left(1 + K_{CO}C_{CO} + K_{H2}C_{H2} + K_{CH3OH}C_{CH3OH}\right)^3} \tag{8.119}$$

Note: This proposed mechanism is unlikely to represent what actually occurs because of the third-order surface reaction. It was observed earlier that third-order elementary reactions tend to be slow. Furthermore, there is some experimental evidence that the reaction proceeds via a different route. However, this rate equation is used in the design of methanol synthesis reactors with a few minor changes. As the pressure in the commercial reactors is of the order of 50 to 100 atm, the gas is very nonideal, and C is replaced by the fugacity. The rate constants for the catalyst to be used in the reactor are determined experimentally.

8.2.2 Models for Oxidation Reactions

The LHHW approach is arguably the most common systematic method used for the development of rate equations. However, there are many cases where the LHHW approach is not used, and to illustrate this idea we briefly discuss some rate equations for oxidation and partial oxidation reactions. Oxidation reactions involve reactions with oxygen, and are divided into complete (or deep) oxidation and partial oxidation. In deep oxidation, the end products are carbon dioxide and water, while in partial oxidation, other products are desired. Industrially, reactions involving oxygen and hydrocarbons are very common.

There are various permutations possible. In some cases, oxygen appears in the final product, such as with the oxidation of propylene to produce acrolein.

$$CH_3CH{=}CH_2 + O_2 \rightarrow CH_2{=}CHCHO + H_2O \tag{8.120}$$

In this reaction, oxygen is added to the hydrocarbon and water is formed. In other cases, oxygen does not appear in the main product, but rather water is produced. An example reaction is the oxidative dehydrogenation of butene to butadiene.

$$CH_2{=}CH{-}CH_2{-}CH_3 + \frac{1}{2}O_2 \rightarrow CH_2{=}CH{-}CH{=}CH_2 + H_2O \tag{8.121}$$

Oxidation reactions tend to be highly exothermic and the reaction mixtures can potentially be explosive. Typical catalysts used are the transition metal oxides (iron, vanadium, bismuth), which contain oxygen in the lattice. Noble metals, such as silver and platinum, are also used. These catalysts can chemisorb oxygen on their surfaces.

The reaction mechanism can be expressed as an oxidation reduction (redox) of the catalyst surface. The classical redox mechanism proposed by Mars and van Krevelen (1954) consists of reduction and oxidation steps. In the first step, the oxidized catalyst reacts with a hydrocarbon

$$S \cdot O + R \rightarrow RO + S \tag{8.122}$$

where $S \cdot O$ is an oxidized catalyst site and R represents a hydrocarbon molecule. In the second step, the catalyst is reoxidized by reaction with oxygen.

$$2S + O_2 \rightarrow 2S \cdot O \tag{8.123}$$

Define the variables

$\theta \equiv$ fraction of sites in the reduced state

$\beta \equiv$ moles O_2 consumed per mole of R reacted

$P_R \equiv$ partial pressure of hydrocarbon

$P_{O_2} \equiv$ partial pressure of oxygen

$k \equiv$ rate constant for oxidation of R

$k^* \equiv$ rate constant for oxidation of catalyst surface

Mars and van Krevelen proposed that at steady state the rates of reduction and oxidation of the catalyst were equal. The rate of reduction of the catalyst (hydrocarbon oxidation) was given by the simple expression

$$(-r) = kP_R(1 - \theta) \tag{8.124}$$

The oxidation rate of the catalyst depends on the number of reduced sites and the oxygen partial pressure

$$(-r) = \frac{k^*}{\beta} P_{O_2}^n \theta \tag{8.125}$$

Equating the rates of oxidation and reduction gives

$$kP_R (1 - \theta) = \frac{k^*}{\beta} P_{O_2}^n \theta \tag{8.126}$$

Rearranging to obtain an expression for θ

$$\theta = \frac{kP_R}{kP_R + \left(\dfrac{k^*}{\beta} \right) P_{O_2}^n} \tag{8.127}$$

Substitution of θ into the rate equation gives

$$(-r) = \cfrac{1}{\left(\cfrac{\beta}{k^*} P_{O_2}^n\right) + \cfrac{1}{kP_R}} \qquad (8.128)$$

If the oxidation of the surface were the rate-limiting step, then

$$kP_R \gg \frac{k^*}{\beta} P_{O_2}^n \qquad (8.129)$$

The rate simplifies to

$$(-r) = \frac{k^*}{\beta} P_{O_2}^n \qquad (8.130)$$

This reaction is zero order in hydrocarbon partial pressure. This behavior has been observed for many catalytic oxidation reactions. If, on the other hand, the oxidation of the hydrocarbon is the rate-limiting step, then

$$\frac{k^*}{\beta} P_{O_2}^n \gg kP_R \qquad (8.131)$$

and

$$(-r) = kP_R \qquad (8.132)$$

In this case, only a small oxygen partial pressure is required to reoxidize the catalyst surface.

8.3 Mechanisms and Models

In the previous sections of this chapter, we explained the development of rate functions from proposed mechanisms. It is worthwhile to summarize the key points of these sections, prior to a description of methods used to estimate reaction rate parameters. In point form, these are:

1. All reactions proceed via a sequence of steps at the molecular level which, taken together, comprise the reaction mechanism.
2. The mechanism may comprise a single step (elementary reaction) or a sequence of steps.
3. The reaction mechanism, if known, can be used to develop a rate function, with the reaction rate expressed as a function of reactant and product concentration and temperature. Usually the development of the rate equation from the mechanism involves making some assumptions.

4. The rate function, whether for an elementary reaction or one based on a complex reaction mechanism, contains one or more temperature-dependent parameters. The numerical value of these parameters cannot be calculated from first principles and must be determined by experimental measurement.

The significance of these points should not be underestimated, especially point 4. The fact that rate functions contain experimentally determined parameters, obtained by statistical analysis of rate data, means that a certain degree of caution should be exercised when using them. This is especially true when it comes to making statements about the validity of the assumptions made in deriving the rate function. We have seen earlier that it is possible for more than one mechanism to give the same rate function. It is also possible for two different rate functions, derived from different mechanisms, to give an acceptable correlation of the experimental data. It is often difficult to determine exactly the reaction mechanism. Furthermore, even if the reaction mechanism is reasonably well established, it may be so complex that its use in a practical reactor model is too unwieldy. In such a case, simplifying assumptions or "lumped" kinetics are often used. Although this latter point is of less concern as the power of computers continues to grow, it is still valid. The net result is that the rate function used to design or analyze a reactor should be regarded more as a model than a mechanistic description of the reaction process.

A model is developed using experimental data collected over a range of operating conditions. The model should not be used outside of this range of conditions. Even a model that is an exact description of the mechanism may not be correct if extrapolated to new operating conditions, as the reaction mechanism may change. For example, a rate model that assumes that a certain step in the reaction mechanism is the rate-determining step would not be appropriate under a new set of operating conditions where the rate-determining step changed.

8.4 Summary

In this chapter, we have discussed the processes involved in gas-phase reactions catalyzed by solids. Appropriate rate expressions are needed for the analysis and design of catalytic reactors, and special attention always needs to be paid to this aspect of modeling. It is important to remember at all times that rate equations always contain some element of empiricism, which must be accounted for in design. The complexity of most catalytic reactions means that it is rare to analyze a reactor using a detailed mechanism, and most rate equations used in industrial practice have a large dose of the empirical. Notwithstanding this limitation, these rate equations are often sufficient for design and scale-up purposes.

PROBLEMS

8.1 The dissociative chemisorption of hydrogen on glass at 300 K is endothermic. The heat of adsorption is approximately 60 kJ/mol, and the activation energy is approximately 100 kJ/mol. The equilibrium distance between the chemisorbed hydrogen and the glass surface is 10 nm. Sketch a potential energy diagram for this surface.

8.2 Oxygen chemisorbs on platinum in two ways (Type 1 and Type 2 adsorption). The following observations have been made:

    a. Both types are exothermic.

    b. Type 1 is nonactivated.

    c. Type 2 is activated.

    d. Type 1 does not change to Type 2 on the surface.

    e. The energy for desorption for Type 1 is greater than for Type 2.

    Sketch a potential energy diagram that is consistent with the above observations.

8.3 The following adsorption data were obtained for the adsorption of ethyl chloride on charcoal:

Pressure (Torr)	Uptake, U (g C_2H_5Cl/g of charcoal)	Pressure (Torr)	Uptake, U (g C_2H_5Cl/g of charcoal)
10	0.133	250	0.446
30	0.240	300	0.462
50	0.278	350	0.473
70	0.318	400	0.485
100	0.357	450	0.494
150	0.394	500	0.500
200	0.426		

    Test the data for adherence to the Langmuir isotherm. Evaluate b and V_m using linear regression (three forms) and nonlinear regression.

8.4 The following data were recorded for the adsorption of a gas on a solid:

Volume Adsorbed (cm^3/g)	Equilibrium Pressure		
	$T = 0°C$	$T = 25°C$	$T = 100°C$
2	0.12	0.31	5.8
3.5	0.21	0.54	10.3
5	0.31	0.8	14
7	0.48	1.2	21
10	0.8	2.0	32
15	1.7	4.3	56
20	3.1	7.8	91
30	7.7	19	210
40	15	38	380

    a. Plot the three isotherms on a log–log scale.

    b. Plot the heat of adsorption as a function of coverage.

    c. Is the Langmuir adsorption isotherm a good assumption?

8.5 Consider the methanation of synthesis gas, represented by the overall reaction:

$$3H_2 + CO \rightleftharpoons CH_4 + H_2O$$

A proposed LHHW-type mechanism is shown below.

$$CO + 2S \rightleftharpoons C \cdot S + O \cdot S \tag{8.133}$$

$$H_2 + 2S \rightleftharpoons 2H \cdot S \tag{8.134}$$

$$\text{RDS} \qquad C \cdot S + 3H \cdot S \rightleftharpoons CH_3 \cdot S + 3S \tag{8.135}$$

$$CH_3 \cdot S + H \cdot S \rightleftharpoons CH_4 + 2S \tag{8.136}$$

$$2H \cdot S + O \cdot S \rightleftharpoons H_2O + 3S \tag{8.137}$$

Derive a rate expression for the disappearance of CO if step (3) is the rate-determining step.

8.6 Consider the following overall reaction:

$$\frac{1}{2}A_2 + B \rightarrow C + D$$

The mechanism has the following features:
a. Molecules of A_2 dissociatively adsorb on two type 1 sites (one A atom on each site).
b. Molecules of B adsorb on type 2 sites.
c. The rate-determining step is the surface reaction between adsorbed A and adsorbed B.
d. Molecules of C adsorb on type 1 sites and molecules of D adsorb on type 2 sites.

Derive an LHHW-type rate expression that is consistent with this mechanism.

8.7 Consider the essentially irreversible oxidation of carbon monoxide:

$$CO + \frac{1}{2}O_2 \rightarrow CO_2$$

The reaction proceeds at 1 atmosphere pressure and 246°C on a supported platinum catalyst.
a. Derive a rate expression using LHHW principles assuming that (a) CO_2 is adsorbed very weakly so that the coverage is close to zero, (b) O_2 adsorbs in the dissociated state, and (c) the rate-determining step is the surface reaction between adsorbed CO and adsorbed O.
b. Derive the rate expression for an Eley–Rideal mechanism assuming that the rate-limiting step is the reaction between adsorbed O and gas-phase CO.
c. The following set of experimental data were taken under conditions of a large excess of oxygen, and 1 atm and 246°C. Determine which type of model

gives the best correlation for the data from the following three choices. A power law model of the form

$$(-r_{CO}) = kC_{CO}^n$$

The rate model developed in part (a) and the rate model developed in part (b).

Reaction Rate (mol/h · gcat)	% CO
2.5	0.1
4.5	0.2
5.65	0.3
7.5	0.48
6.0	1.0
5.0	1.25
4.0	1.6

Reference

Mars, P., and D.W. van Krevelen, 1954, Oxidations carried out by means of vanadium oxide catalysts, *Chemical Engineering Science*, **3** (Special supplement), 41–59.

9

Transport Processes in Catalysis

In Chapter 7, we introduced the basic steps that occur in a solid-catalyzed reaction. The three kinetic steps are detailed in Chapter 8. The remaining steps are those that involve the transport of mass and energy. As seen in Chapter 7, these steps can be divided into external transfer, which involves the transport of mass and energy through the external boundary layer that separates the bulk fluid from the solid, and internal transport, which occurs inside the catalyst volume if it is porous. The external transport is normally referred to as interphase transport, whereas that occurring within the catalyst is called intraphase transport. Both transport processes can lead to significant changes in the observed rate, and, in extreme cases, may be the dominant rate-determining factors in the reactor. Reactors that are analyzed or designed without consideration of transport effects will in most cases fail to meet expectations.

In the following sections, an overview of diffusion is first presented in the context of catalytic reactions. This section is then followed by a detailed discussion of both external and internal diffusion effects.

9.1 Diffusion in Bulk Phase

9.1.1 Basics of Bulk Diffusion: Diffusion Coefficient

In catalysis, diffusion plays a role in the mass transfer process, both in the transfer across the external boundary layer between the bulk fluid and the catalyst surface, and within the porous catalyst. The mechanism of diffusion is related to molecular motion and essentially involves collisions, either between molecules in the fluid phase or between fluid phase molecules and the walls of a container (or other solid object). The governing factor that determines the type of diffusion is the mean free path of a molecule, which is simply the distance that a molecule travels before it hits something. The mean free path depends on the concentration of the fluid and the size of the confining vessel.

In a bulk gas phase in a relatively large container, the diffusion process is governed by bulk diffusion or molecular diffusion (two common terms for the same thing). Diffusion, especially in multicomponent mixtures, is a far from simple phenomenon. Readers are referred to, for example, the book by Taylor and Krishna (1993) for a complete treatment. For multicomponent diffusion in gas-phase systems, a rigorous treatment of diffusion requires the solution of the Maxwell–Stefan equations. Typically, diffusion is analyzed using Fick's law, in one of its various permutations. We now consider the basics of molecular diffusion.

For a fluid moving with a mole average velocity $v*$, the diffusion velocity is defined as the difference between the velocity of the species and the average fluid velocity. The molar diffusion flux of species is given by

$$J_1 = C_1 \left(v_1 - v* \right) \tag{9.1}$$

where C_1 is the molar concentration of species 1. The mole average velocity for a set of n species, each of which has a molecular velocity of v_i, is defined as

$$v* = \sum_{i=1}^{n} Y_i v_i \qquad (9.2)$$

For a binary system, that is, one containing two species only, Fick's first law of diffusion states that the diffusion flux is proportional to the mole fraction gradient:

$$J_1 = C_1 \left(v_1 - v* \right) = -D_{12} C \nabla Y_1 \qquad (9.3)$$

where C is total molar concentration, Y_1 is the gas-phase mole fraction, and D_{12} is the molecular diffusion coefficient, which has units of area per time, for example, m^2/s. A similar expression for component 2 in the binary mixture can be written as

$$J_2 = C_2 \left(v_2 - v* \right) = -D_{21} C \nabla Y_2 \qquad (9.4)$$

It follows that $J_1 + J_2 = 0$ and $Y_1 + Y_2 = 1$; thus, $D_{12} = D_{21}$. For a binary mixture, therefore, there is a single value for the diffusion coefficient that describes the diffusion fluxes. Methods for estimating values for binary diffusion coefficients are given shortly.

The diffusion flux of species 1 can be also defined using a mass flux, which incorporates the mass average velocity and the mass fractions, w_i. The mass average velocity is

$$v = \sum_{i=1}^{n} w_i v_i \qquad (9.5)$$

In general, mole average and mass average velocities are not the same. Fick's law written in terms of the mass fraction gradients is

$$j_1 = \rho_1 \left(v_1 - v \right) = -D_{12} \rho \nabla w_1 \qquad (9.6)$$

Equation 9.6 is a convenient form of Fick's law to use when a mass balance equation is solved simultaneously with the equation of motion. Note that any reference velocity may be used to define the diffusion flux. For example, the volume average reference velocity defined in terms of the volume fraction, ϕ_i, is

$$v^V = \sum_{i=1}^{n} \phi_i v_i \qquad (9.7)$$

The diffusion flux is

$$J_1^V = C_1 \left(v_1 - v^V \right) = -D_{12} \nabla C_1 \qquad (9.8)$$

When the volume average velocity is used, the diffusion-driving force is the concentration fraction. This form is actually not very convenient, especially in nonisothermal system, because the concentration changes with temperature. Thus, for an ideal gas

$$\nabla C_1 = \nabla(CY_1) = C\left(\nabla Y_1 - \frac{Y_1}{T}\nabla T\right) \tag{9.9}$$

Note that the diffusion coefficients in Equations 9.3, 9.6, and 9.8 are the same.

9.1.2 Calculating Molecular Diffusion Coefficients in Binary Systems

The calculation of binary molecular diffusion coefficients in gases has been the subject of much research. Two relationships that are widely used are the Chapman–Enskog formula and the Fuller formula. The Chapman–Enskog formulation is classical, and details can be found in, for example, Bird et al. (2002). The formula is

$$D_{12} = \frac{1.858 \times 10^{-7} T^{1.5}[(1/M_1) + (1/M_2)]^{0.5}}{P\sigma_{12}^2 \Omega_D} \tag{9.10}$$

where T is the temperature in kelvin, M_1 and M_2 are the molar mass of 1 and 2, respectively, P is the total pressure in atm, Ω_D is the collision integral, and σ_{12} is a force constant. The values of Ω_D and σ_{12} are tabulated in many references for a number of species. In the above equation, D_{12} is expressed in m^2/s.

The Fuller correlation can also be used for calculating bulk diffusion coefficients (Reid et al., 1987). With pressure P in Pascal and temperature T in kelvin, the diffusion coefficient in m^2/s is

$$D_{12} = \frac{1.013 \times 10^{-2} T^{1.75}[(1/M_1) + (1/M_2)]^{0.5}}{P\left[\left(\Sigma v_i\right)_1^{1/3} + \left(\Sigma v_i\right)_2^{1/3}\right]^2} \tag{9.11}$$

The variables v_i are diffusion volumes. Values of v_i for simple molecules are

H_2	7.07	CO	18.9	NH_3	14.9
N_2	17.9	CO_2	26.9	CH_4	24.42
O_2	16.6	N_2O	35.9	Ar	16.1
Air	20.1	H_2O	12.7	Kr	22.8
				He	2.88

For complex molecules, the diffusion volume can be estimated by adding incremental volumes based on the species present. The values are

C	16.5	Cl	19.5
H	1.98	S	17.0
O	5.48	N	5.69

For example, the diffusion coefficient of methane in air at 120 kPa and 500 K is

$$D_{12} = \frac{1.013 \times 10^{-2}(500)^{1.75}\left((1/16) + (1/28.96)\right)^{0.5}}{120,000\left[(24.42)^{\frac{1}{3}} + (20.1)^{\frac{1}{3}}\right]^2}$$

$$D_{12} = 4.4 \times 10^{-5}\ \text{m}^2/\text{s}$$

This magnitude is typical of diffusion coefficients in gases at low pressure.

9.1.3 Total Molar Diffusive Flux in Binary System

Equation 9.3 defines the diffusion flux of component 1. The total flux, N_1, of component 1 is defined by rearrangement:

$$N_1 = C_1 v_1 = C_1 v^* - D_{12}C\nabla Y_1 = CY_1 v^* - D_{12}C\nabla Y_1 \tag{9.12}$$

The overall total flux is

$$N_t = N_1 + N_2 = Cv^* \tag{9.13}$$

Combining these equations gives the result

$$N_1 = Y_1(N_1 + N_2) - D_{12}C\nabla Y_1 \tag{9.14}$$

Equation 9.12 can be used with a basic shell balance to define the mole balance of species 1 in a system, thus

$$\nabla \cdot \left(CY_1 v^* - D_{12}C\nabla Y\right) = \nabla \cdot \left(Cv^* Y_1\right) - \nabla \cdot \left(D_{12}C\nabla Y\right) = 0 \tag{9.15}$$

For a constant-density nonreacting system, Equation 9.15 reduces to

$$D_{12}\nabla^2 Y_1 - v^*\nabla Y_1 = 0 \tag{9.16}$$

9.1.4 Diffusion in Multicomponent Mixtures: Generalized Fick's Law

For a multicomponent mixture comprised of n species, there are $n-1$ independent fluxes and concentration gradients. The diffusion flux of any species is equal to

$$J_i = -C\sum_{k=1}^{n-1} D_{ik}\nabla Y_k \tag{9.17}$$

For example, for a ternary system (three components), the fluxes for 1 and 2 are

$$J_1 = -C\left(D_{11}\nabla Y_1 + D_{12}\nabla Y_2\right) \tag{9.18}$$

$$J_2 = -C\left(D_{21}\nabla Y_1 + D_{22}\nabla Y_2\right) \tag{9.19}$$

There are some significant differences between the D_{ik} in the above equation and the binary diffusion coefficients discussed earlier. The values for D_{ik} are the multicomponent diffusion coefficients. They can be either positive or negative, and are not generally symmetric, that is $D_{12} \neq D_{21}$. Furthermore, it was noted earlier that in a binary system the diffusion of both species was characterized by a single diffusion coefficient. The value of this binary diffusion coefficient does not depend on the reference velocity, and hence is independent of the gradient. That is, the mole fraction, concentration, or mass fraction gradients use the same value of diffusion coefficient. It should be noted that this uniformity is no longer true for multicomponent systems, and the matrix of diffusion coefficients will assume different numerical values depending on the choice of reference velocity and gradient type.

In general, the diffusion in a multicomponent system is governed by an $(n-1) \times (n-1)$ matrix of coefficients, $[D]$. The challenge is to obtain data for the matrix coefficients for the system of interest. These values could be measured, but this step would be impractical in many cases. In practice, multicomponent diffusion coefficients are computed from the Stefan–Maxwell binary diffusion pairs, as described in the following. It can be shown that the matrix of Fickian diffusion coefficients, $[D]$, can be related to the Stefan–Maxwell diffusion coefficients

$$[D] = [B]^{-1}[\Gamma] \tag{9.20}$$

$[B]$ is a matrix built using Maxwell–Stefan binary diffusion pairs and $[\Gamma]$ is a thermodynamic matrix that contains the activity coefficients. For an ideal solution, which includes most gas mixtures, $[\Gamma] = 1$ and we can write $[D] = [B]^{-1}$. The terms in the matrix $[B]$ are

$$B_{ii} = \frac{Y_i}{D_{in}} + \sum_{\substack{k=1 \\ i \neq k}}^{n} \frac{Y_k}{D_{ik}} \quad (\text{main diagonal terms}) \tag{9.21}$$

$$B_{ij} = -Y_i \left(\frac{1}{D_{ij}} - \frac{1}{D_{in}} \right) \quad (\text{off diagonal terms}) \tag{9.22}$$

Note that n represents the last component. In this case, D_{ij} represents a binary Maxwell–Stefan diffusion coefficient. For ideal solutions, the Maxwell–Stefan diffusion coefficients are equal to the Fick's diffusion coefficients.

9.1.5 Multicomponent Diffusion in Dilute Mixtures

In many cases, approximate values of the diffusion coefficient are used for simplicity. For example, for multicomponent diffusion in which all of the binary diffusivity pairs have similar values, the off-diagonal terms in the Fick's matrix are very small compared with the main diagonal terms. Furthermore, the main diagonal terms are nearly equal, and it is safe to use an average diffusion coefficient for the entire mixture. On the other hand, if one component is present in large excess, then this component is assigned as the nth component, and the diffusion of the other components are

$$D_{in} = \left[\left(\frac{1}{1 - Y_i} \right) \sum_{\substack{j=1 \\ j \neq i}}^{n} \frac{Y_j}{D_{ij}} \right]^{-1} \tag{9.23}$$

where D_{ij} is the binary diffusion coefficient for the pair ij.

9.2 External Mass and Heat Transfer Effects

Whenever a fluid flows over a surface, a boundary layer forms near the interface. To illustrate the concept of boundary layer development for fluid flow over a surface, we examine the case of flow over a flat plate. Consider the plate shown in Figure 9.1, where a fluid with an approach velocity u_∞, a temperature of T_∞, and a concentration of C_∞ flows over the plate. A velocity boundary layer starts to develop at the leading edge, and gradually grows in thickness as the distance over the plate increases. We suppose further that an exothermic chemical reaction occurs at the surface, which results in the generation of thermal energy and a diminution of the concentration, giving rise to a surface temperature T_S and a surface concentration C_S. The result will be temperature and concentration profiles of some magnitude over the boundary layer, as shown in Figure 9.1.

9.2.1 Mass and Heat Transfer Coefficients

Although the catalyst geometry is rarely as simple as a flat plate, there will always be present a boundary layer of some type, with gradients of velocity, temperature, and concentration.

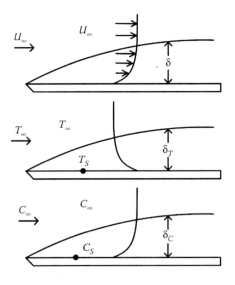

FIGURE 9.1
The development of velocity, temperature, and concentration boundary layers for flow across a flat plate. Regardless of the shape, a boundary layer will be present at the fluid–solid boundary.

The flow patterns and temperature and concentration profiles in the boundary layer are determined by a mixture of advection (bulk flow) and diffusion. The magnitude of the gradients is determined by the relative size of these two forces. For simple geometries with laminar flow, the gradients may be calculated either analytically or numerically. In practice, because of the presence of turbulent flow, complex geometry, or simply a desire for computational simplicity, use is usually made of the heat and mass transfer coefficients to describe the rates of heat and mass transfer. These coefficients are in most cases determined from an empirical correlation that includes the local conditions of flow, temperature, and concentration.

We recall now from the basics of momentum, mass and energy transfer in fluids that these transport phenomena are governed by three dimensionless groups, the Reynolds, Prandtl, and Schmidt numbers. These three numbers are defined as

$$\text{Re} = \frac{u_\infty L_C \rho}{\mu}, \quad \text{Pr} = \frac{C_P \mu}{k}, \quad \text{and} \quad \text{Sc} = \frac{\mu}{\rho D_{AB}}$$

The transfer rates of both energy and mass are equal to the rates of diffusion at the solid surface. For a constant-density fluid, the transfer of mass equals the rate of diffusion at the surface

$$N_A = -D_A \left. \frac{\partial C_A}{\partial y} \right|_{y=0} \tag{9.24}$$

where N_A is the molar flux of species A and D_A is the bulk diffusion coefficient of species A. The rate of mass transfer across the boundary layer is then typically defined in terms of a mass transfer coefficient and (for a constant-density fluid) an overall driving force equal to the concentration difference, which gives the following rate of mass transfer for species A:

$$N_A = k_m \left(C_{A,S} - C_{A,f} \right) \tag{9.25}$$

The subscripts f and S denote the bulk fluid- and solid-phase mole fractions, respectively. For gas-phase systems, the driving force may be written in terms of the gas-phase mole fraction (or sometimes the partial pressure) and the bulk-phase molar density, thus

$$N_A = k_m C_f \left(Y_{A,S} - Y_{A,f} \right) \tag{9.26}$$

By definition, the two rates must be equal; therefore, we write

$$N_A = -D_A \left. \frac{\partial C_A}{\partial y} \right|_{y=0} = k_m \left(C_{A,S} - C_{A,f} \right) \tag{9.27}$$

or, in terms of mole fraction

$$N_A = -D_A C_S \left. \frac{\partial Y_A}{\partial y} \right|_{y=0} = k_m C_f \left(Y_{A,S} - Y_{A,f} \right) \tag{9.28}$$

We can define a dimensionless coordinate and a dimensionless mole fraction

$$y^* = \frac{y}{L_C}, C^* = \left(\frac{C_A - C_{A,f}}{C_{A,S} - C_{A,f}} \right) \quad \text{and} \quad \frac{\partial C^*}{\partial y} = \left(\frac{1}{C_{A,S} - C_{A,f}} \right) \frac{\partial C_A}{\partial y} \tag{9.29}$$

Substituting these two terms into Equation 9.27 and rearrangement give

$$-\frac{1}{\left(C_{A,S} - C_{A,f}\right)} \frac{\partial C_A}{\partial y^*}\bigg|_{y=0} = \left(\frac{k_m L_C}{D_A} \right) = -\frac{\partial C^*}{\partial y^*} \tag{9.30}$$

It is seen that the dimensionless concentration gradient at the surface depends on a dimensionless group, which is known as the *Sherwood* number, denoted Sh:

$$\frac{k_m L_C}{D_A} = -\frac{\partial C^*}{\partial y^*}\bigg|_{y=0} = \text{Sh} \tag{9.31}$$

The mass transfer through the boundary layer depends on the Reynolds and Schmidt numbers, as noted earlier, and thus it follows that

$$\text{Sh} = f(\text{Re}, \text{Sc}) \tag{9.32}$$

This functional dependence is often an empirical relationship.

By analogy, a similar analysis can be made for the rate of heat transfer across the boundary layer. The rate of heat transfer (heat flux) into the boundary layer is given by the rate of conduction at the surface:

$$q'' = -k \frac{\partial T}{\partial y}\bigg|_{y=0} \tag{9.33}$$

The rate of heat transfer is also defined in terms of an overall driving force and a heat transfer coefficient; thus

$$q'' = h\left(T_S - T_f\right) \tag{9.34}$$

A dimensionless temperature and distance are defined

$$\theta = \left(\frac{T - T_f}{T_S - T_f} \right), y^* = \frac{y}{L_C} \tag{9.35}$$

Combination of Equations 9.33, 9.34, and 9.35 gives

$$\frac{h L_C}{k} = -\frac{\partial \theta}{\partial y}\bigg|_{y=0} = \text{Nu} \tag{9.36}$$

The dimensionless surface temperature gradient is the *Nusselt number*, whose magnitude governs the heat transfer rate. The Nusselt number depends on the Reynolds and Prandtl numbers; thus

$$\mathrm{Nu} = f(\mathrm{Re}, \mathrm{Pr}) \tag{9.37}$$

The equations for both the Nusselt and Sherwood numbers depend on local conditions, including catalyst shape and flow regime. For a few laminar flow situations, the values can be calculated from the solution of the governing conservation equations. In most situations, an empirical correlation is used. The form of these correlations will be discussed in more detail with the appropriate reactor type when they are introduced in Chapter 10.

The relative thickness of each boundary layer depends on the Reynolds, Schmidt, and Prandtl numbers. Let δ, δ_C, and δ_T be the thickness of the hydrodynamic, concentration, and temperature boundary layers, respectively. Then, the following ratios have been observed to hold:

$$\frac{\delta}{\delta_C} = \mathrm{Sc}^n \quad \frac{\delta}{\delta_T} = \mathrm{Pr}^n \quad \frac{\delta_T}{\delta_C} = \frac{\mathrm{Sc}^n}{\mathrm{Pr}^n} = \mathrm{Le}^n = \left(\frac{\alpha}{D_A}\right)^n \tag{9.38}$$

For laminar flow, the value of n is about 0.33. The ratio of the Schmidt number to the Prandtl number is the Lewis number, and is the ratio of the thermal- and molecular-diffusion coefficients (although other definitions of the Lewis number have been reported in the literature).

9.2.2 Effect of External Mass Transfer on Observed Reaction Rate

The effect of mass transfer through the boundary layer is to give a lower reactant concentration at the external catalyst surface than would otherwise be present. The result, for reactions with normal kinetics, is to lower the observed rate of reaction. Consider, for the moment, an isothermal system (i.e., with no temperature gradients) in which a first-order reaction occurs at the surface of the catalyst. The reaction rate per volume of catalyst is given by

$$(-R_A) = k_V C_{A,S} \tag{9.39}$$

The rate of mass transfer to the catalyst surface depends on the concentration gradient, the mass transfer coefficient, and the area available for mass transfer, as noted earlier. Thus

$$(-R_A) = k_m a_v \left(C_{A,f} - C_{A,S}\right) \tag{9.40}$$

where a_v is the surface area per unit volume of catalyst. At steady state, the rate of mass transfer is equal to the rate of reaction, and thus

$$k_m a_v \left(C_{A,f} - C_{A,S}\right) = k_V C_{A,S} \tag{9.41}$$

Equation 9.41 can be rearranged to give an explicit relationship between the bulk and the surface concentration:

$$C_{A,S} = \frac{k_m a_v}{\left(k_V + k_m a_v\right)} C_{A,f} \tag{9.42}$$

The surface concentration given by Equation 9.42 can then be substituted into Equation 9.39 to give an expression for the reaction rate in terms of the bulk concentration:

$$\left(-R_A\right) = k_V \frac{k_m a_v}{\left(k_V + k_m a_v\right)} C_{A,f} \tag{9.43}$$

The reaction rate can therefore be written in terms of an apparent rate constant, which contains terms for both kinetics and mass transfer:

$$\left(-R_A\right) = k_{app} C_{A,f} \tag{9.44}$$

where

$$k_{app} = k_V \frac{k_m a_v}{\left(k_V + k_m a_v\right)} = \left(\frac{1}{k_V} + \frac{1}{k_m a_v}\right)^{-1} \tag{9.45}$$

Examining the form of Equation 9.45, we can consider two limiting cases. In the first, the mass transfer coefficient is very large compared with the rate constant, or $k_m a_v \gg k_V$. The surface and bulk concentrations are nearly equal and the rate is controlled only by the kinetics, or

$$\left(-R_A\right) \approx k_V C_{A,f} \tag{9.46}$$

Alternatively, in the limiting case where the reaction rate constant is very large, or $k_m a_v \ll k_V$, the rate is controlled by the rate of mass transfer, or

$$\left(-R_A\right) \approx k_m a_v C_{A,f} \tag{9.47}$$

In the latter case, the surface concentration is essentially equal to zero and the kinetics, including the form of the rate expression, become unimportant.

The introduction of mass transfer across the boundary layer also introduces changes in the behavior of the apparent reaction rate constant with temperature. The intrinsic rate constant typically varies exponentially with temperature, according to the Arrhenius relationship, whereas the mass transfer coefficient has a much smaller variation in temperature, which is typically linear or less. The temperature dependence of the apparent rate constant can therefore be written as

$$k_{app} = \frac{A \exp\left(-E/R_g T\right) k_m a_v}{\left(A \exp\left(-E/R_g T\right) + k_m a_v\right)} \tag{9.48}$$

Then, when the mass transfer coefficient is large, and Equation 9.46 prevails, the apparent rate constant will exhibit an apparent exponential temperature dependence, whereas when the mass transfer coefficient is relatively small such that Equation 9.47 dominates, then the apparent rate constant will show little temperature dependence. This type of behavior is shown on the Arrhenius-type plot illustrated in Figure 9.2.

For the first-order reaction in concentration, we have just shown that the apparent rate constant shows a significant deviation from the Arrhenius behavior as mass transfer limitation of the reaction rate begins. However, for the first-order reaction, the apparent reaction order remains unchanged by the change in controlling factor. Let us now repeat the calculation for a second-order reaction. We shall again assume an isothermal system for convenience. The reaction rate per volume of catalyst is

$$(-R_A) = k_V C_{A,S}^2 \quad \text{mol/m}^3\text{s} \tag{9.49}$$

The rate of mass transfer to the catalyst surface is, as before

$$(-R_A) = k_m a_v \left(C_{A,f} - C_{A,S} \right) \tag{9.50}$$

At steady state, the rate of mass transfer is equal to the rate of reaction, and thus

$$k_m a_v \left(C_{A,f} - C_{A,S} \right) = k_V C_{A,S}^2 \tag{9.51}$$

Equation 9.41 can be rearranged to give a quadratic equation

$$k_V C_{A,S}^2 + k_m a_v C_{A,S} - k_m a_v C_{A,f} = 0 \tag{9.52}$$

The solution is

$$C_{A,S} = \frac{-k_m a_v + \sqrt{\left(k_m a_v \right)^2 + 4 k_V k_m a_v C_{A,f}}}{2 k_V} \tag{9.53}$$

For more complex kinetics, it is not possible to obtain explicit relationships between the surface and bulk concentrations. Consider, for example, the reaction

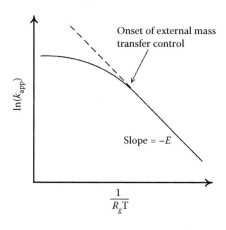

FIGURE 9.2
The apparent activation for a reaction changes as the external mass transfer effects become significant.

$$A + B \rightleftharpoons C + D \tag{9.54}$$

Assume, for illustration purposes, that we have LHHW kinetics, an irreversible reaction, the surface reaction, the rate controlling step, and insignificant adsorption inhibition by the products, C and D. The reaction rate for such a case is

$$(-R_A) = \frac{k_V K_A K_B C_{A,S} C_{B,S}}{\left(1 + K_A C_{A,S} + K_B C_{B,S}\right)^2} \tag{9.55}$$

For an isothermal system, the rate of mass transfer of A and B depends on their respective mass transfer coefficients; thus

$$(-R_A) = (k_m)_A \, a_v \left(C_{A,f} - C_{A,S}\right) \tag{9.56}$$

and

$$(-R_B) = (k_m)_B \, a_v \left(C_{B,f} - C_{B,S}\right) \tag{9.57}$$

The surface concentrations must then be determined by a numerical solution of two nonlinear equations, obtained by equating the rates of mass transfer and reaction. Noting that A and B react in a one-to-one ratio, we have

$$(k_m)_A \, a_v \left(C_{A,f} - C_{A,S}\right) = \frac{k_V K_A K_B C_{A,S} C_{B,S}}{\left(1 + K_A C_{A,S} + K_B C_{B,S}\right)^2} \tag{9.58}$$

and

$$(k_m)_B \, a_v \left(C_{B,f} - C_{B,S}\right) = \frac{k_V K_A K_B C_{A,S} C_{B,S}}{\left(1 + K_A C_{A,S} + K_B C_{B,S}\right)^2} \tag{9.59}$$

Assuming that $C_{A,f}$ and $C_{B,f}$ are known, Equations 9.58 and 9.59 can be solved to determine the values of $C_{A,S}$ and $C_{B,S}$, which in turn would be substituted into Equation 9.55 to evaluate the rate.

9.2.3 Combined Heat and Mass Transfer

As noted earlier, the release or absorption of thermal energy at the catalyst surface will result in the development of a thermal boundary layer through which the energy is transferred by convection. The rate of transfer at steady state is equal to the rate of generation, and therefore the following energy balance is appropriate:

$$h a_v \left(T_f - T_S\right) = (-R_A)(\Delta H_R) \tag{9.60}$$

In the general case, the energy balance and the mole balance must be solved simultaneously to obtain the surface temperature and concentration and hence the reaction rate. Typically, temperature effects are strongest for gas-phase reactions, where

$$(-R_A) = k_m a_v C_f \left(Y_{A,f} - Y_{A,S}\right) \tag{9.61}$$

Regardless of the kinetic expression, a relationship between the temperature and mole fractions can be generated by equating Equations 9.60 and 9.61 to give

$$\frac{h a_v \left(T_f - T_S\right)}{(\Delta H_R)} = k_m a_v C_f \left(Y_{A,f} - Y_{A,S}\right) \tag{9.62}$$

Equation 9.62 can be rearranged:

$$T_S = T_f + (-\Delta H_R)\frac{k_m}{h} C_f \left(Y_{A,f} - Y_{A,S}\right) \tag{9.63}$$

The maximum temperature difference across the boundary layer occurs when the surface mole fraction goes to zero; thus

$$\left(T_S - T_f\right)_{max} = +(-\Delta H_R)\frac{k_m}{h} C_f Y_{A,f} \tag{9.64}$$

We substitute from Equations 9.31 and 9.36:

$$Sh = \frac{k_m L_C}{D_A} \quad \text{and} \quad Nu = \frac{h L_C}{k} \tag{9.65}$$

If the Nusselt and Sherwood numbers are equal, then substitution into Equation 9.63 and rearrangement gives

$$\left(T_S - T_f\right)_{max} = (-\Delta H_R)\frac{D_A}{k} C_f Y_{A,f} \tag{9.66}$$

9.2.4 Effect of External Transport on Selectivity

It was seen in Section 9.2.2 that mass transfer through the external boundary layer affects the reaction rate. For normal kinetics, in an isothermal system, the surface concentration of reactants is reduced, compared with the bulk, which causes a reduction in the observed reaction rate. With systems of parallel and/or series reactions, the situation is more complicated. Mass transfer rates across the boundary layer are not the same for all species; thus, they may not all experience the same concentration difference. Furthermore, in series reactions, the product of one reaction is the reactant for another. Because the concentration of this species will also be different from the bulk, the relative rates of the reactions will change. Clearly, the potential exists for mass transfer to affect the product distribution. To quantify the effects of mass transfer, we will define point selectivity as the ratio of reaction rates at a particular place in the reactor, and show how this selectivity can be affected by mass and heat transfer for a few simple cases.

9.2.4.1 Isothermal Series Reaction

Consider a system of simple consecutive reactions in an isothermal system:

$$A \xrightarrow{k_1} B \xrightarrow{k_2} C \tag{9.67}$$

Both reactions are assumed to be first order and irreversible. The ratio of the two reaction rates (here we use the rates of formation), which defines the selectivity, can be expressed in terms of the surface concentrations:

$$S_P = \frac{\text{rate of formation of B}}{\text{rate of formation of C}} = \frac{r_B}{r_C} = \frac{k_1 C_{A,S} - k_2 C_{B,S}}{k_2 C_{B,S}} \tag{9.68}$$

As discussed in Section 9.2.2, the surface concentration for a first-order irreversible reaction is related to the bulk by Equation 9.42:

$$C_{A,S} = \frac{(k_m a_v)_A}{(k_1 + (k_m a_v)_A)} C_{A,f} \tag{9.69}$$

To derive the relationship for species B, we write the mass balance across the boundary layer:

$$(k_m a_v)_B (C_{B,S} - C_{B,f}) = k_1 C_{A,S} - k_2 C_{B,S} \tag{9.70}$$

Equation 9.70 can be rearranged to give

$$C_{B,S} = \frac{k_1 C_{A,S} + (k_m a_v)_B C_{B,f}}{(k_m a_v)_B + k_2} \tag{9.71}$$

If we now substitute the surface concentration from Equation 9.69, we obtain

$$C_{B,S} = \frac{(k_1 (k_m a_v)/(k_1 + (k_m a_v)_A)) C_{A,f} + (k_m a_v)_B C_{B,f}}{(k_m a_v)_B + k_2} \tag{9.72}$$

The selectivity in terms of the surface concentrations is thus given by

$$S_P = \frac{k_1 C_{A,f}}{k_2 C_{B,f}} \left[\frac{1 + (k_2/(k_m a_v)_B)}{(k_1/(k_m a_v)_B)(C_{A,f}/C_{B,f}) + (k_1/(k_m a_v)_A) + 1} \right] - 1 \tag{9.73}$$

As the mass transfer coefficients increase to infinity, the quantity in square brackets approaches zero, and the selectivity in the absence of mass transfer limitations is

$$S_P = \frac{k_1 C_{A,f}}{k_2 C_{B,f}} - 1 \tag{9.74}$$

This result is the same as obtained by substituting the bulk concentrations into the definition of selectivity, Equation 9.68. The quantity in square brackets, which depends on both the kinetic constants and the mass transfer coefficients, determines the overall selectivity of the system.

9.2.4.2 Parallel Isothermal Reactions

The selectivity effects for parallel reactions depend on the reaction network. For a simple system of parallel first-order reactions with a common reactant

$$A \xrightarrow{\ k_1\ } B$$
$$A \xrightarrow{\ k_2\ } C \tag{9.75}$$

The selectivity is not affected because both reactions are diminished by the same amount. For parallel independent reactions

$$A \xrightarrow{\ k_1\ } B$$
$$C \xrightarrow{\ k_2\ } D \tag{9.76}$$

The change in selectivity depends on the relative magnitudes of the mass transfer coefficients. If both these reactions are first order, then the selectivity defined as the rate of disappearance of A divided by the rate of disappearance of C is

$$S_P = \frac{\left((1/k_1) + \left(1/(k_m a_v)_A \right) \right)^{-1} C_{A,f}}{\left((1/k_2) + \left(1/(k_m a_v)_C \right) \right)^{-1} C_{C,f}} \tag{9.77}$$

Again, the local selectivity is determined by a combination of the mass transfer coefficients and the kinetic constants. In one extreme, where the mass transfer coefficients of all reactants are extremely large, the selectivity is governed only by the kinetics:

$$S_P = \frac{k_1 C_{A,f}}{k_2 C_{C,f}} \tag{9.78}$$

At the other extreme of extremely fast reactions, the selectivity depends only on the mass transfer coefficients:

$$S_P = \frac{(k_m a_v)_A \, C_{A,f}}{(k_m a_v)_C \, C_{C,f}} \tag{9.79}$$

9.2.4.3 Nonisothermal Reactions

When temperature effects are included, the change in selectivity will occur as a result of both temperature and concentration changes. For exothermic reactions, the surface temperature is higher than the bulk, which increases the reaction rate constants. For an endothermic reaction, the surface temperature is lower than the bulk and the rate constants decrease. Consider again the simple system of first-order parallel reactions with a common reactant. The selectivity is

$$S_P = \frac{k_1}{k_2} \frac{C_{A,S}}{C_{A,S}} = \frac{A_1 \exp\left(-E_1/R_g T_S \right)}{A_2 \exp\left(-E_2/R_g T_S \right)} \tag{9.80}$$

Although in this case the mass transfer coefficients do not affect the selectivity, the temperature differences may. For example, for an exothermic reaction, we have $T_S > T_f$, and the difference in activation energies will determine the change in selectivity. For the case where $E_1 > E_2$, the rate of reaction 1 increases more than that of reaction 2, resulting in an increase in selectivity. For the case where $E_1 < E_2$, reaction 2 increases by a larger margin than reaction 1, and the selectivity goes down.

The previous cases are relatively simple; however, they serve to illustrate the concept. For complex reaction networks with non-first-order kinetics, it is not possible to derive such simple relationships, and numerical solutions are required.

9.3 Diffusion in Porous Catalysts

In Chapter 7, the porous nature of many catalysts was discussed. As noted there, in a porous catalyst, much of the active site catalytic surface area is located within the interior of the catalyst particle, and most of the reaction occurs inside the pores on their surface. The reactants must therefore diffuse into the catalyst pores before reaction can proceed. As the reactants diffuse into the catalyst, they encounter active sites and some of the reactants react. Therefore, the reactant concentration declines as the distance into the catalyst increases, giving rise to a concentration gradient inside the catalyst. If the diffusion velocity is much lower than the intrinsic reaction rate, the concentration gradients can be quite large. Therefore, it is necessary to have a means of calculating the effective rate of reaction in a catalyst with a variable concentration profile.

Most chemical reactions also have associated with them heat effects, which involve the release or absorption of thermal energy. These heat effects cause local changes in temperature, forming temperature gradients inside the catalyst. The reaction rate occurring at a local position within the catalyst will be different (either higher or lower) than the rate at the external surface of the catalyst particle. These differences must be considered when analyzing catalytic systems.

In the following sections, we consider first the case of concentration gradients in an isothermal system, and then expand the analysis to nonisothermal systems.

9.3.1 Diffusion in Catalyst Pores

Diffusion in catalyst pores is more complicated than simple diffusion in bulk systems. First, the complex tortuous nature of the catalyst pores means that the exact length of the diffusion path is unknown. Second, because the pores are usually quite small, there is the potential of a significant number of collisions between gas molecules and the walls of the pores. The typical pore size in active catalysts is of the same order of magnitude as the mean free path for gas-phase molecules, especially at low-to-moderate pressure, and therefore in gas-phase reactions, collisions between molecules and the walls predominate. For liquid-phase reactions, the mean free path remains small, so molecular collisions are the dominating diffusion-driving force.

We begin the discussion by considering the case of diffusion in a single ideal cylindrical pore. For a binary system, the diffusion flux can be expressed using Fick's law, as described earlier:

$$J_1 = D_{P1} C \nabla Y_1 \qquad (9.81)$$

D_{P1} is used to denote the pore scale diffusion coefficient of component 1. The diffusion coefficient for large pores can be computed using the methods discussed earlier for molecular diffusion. However, as the pores become smaller, collisions between the molecules and the pore walls become significant for gas-phase systems. The collisions between the diffusing molecules and the wall gives rise to *Knudsen diffusion*. The Knudsen diffusion coefficient, in m²/s, in a straight round pore can be calculated from the formula:

$$D_K = 48.5 \, d_p \left(\frac{T}{M} \right)^{\frac{1}{2}} \text{m}^2/\text{s} \tag{9.82}$$

where d_p is the pore diameter in m, T is the temperature in kelvin, and the molar mass of the diffusion species, M, is in kg/kmol or g/mol. The Knudsen diffusion coefficient varies with the pore diameter, which implies that a range of diffusion coefficients applies to a catalyst with a significant pore size distribution. Frequently, the equivalent pore diameter is taken as the average diameter of the pores in the pellet, which gives a reasonable approximation provided that there is a relatively narrow pore size distribution.

Depending on the size of the catalyst pore, the diffusion phenomenon depends primarily on bulk diffusion, Knudsen diffusion, or a combination of the two. For a binary system, it has been shown that the pore diffusion coefficient can be expressed as a function of the two quantities using the following relationship:

$$\frac{1}{D_{\text{pore}}} = \frac{1 - \alpha Y_1}{D_{12}} + \frac{1}{D_K} \tag{9.83}$$

where α depends on the ratio of the molar fluxes of the two species:

$$\alpha = 1 + \frac{N_2}{N_1} \tag{9.84}$$

For equimolar counterdiffusion, $N_1 = -N_2$, $\alpha = 0$ and then

$$\frac{1}{D_{\text{pore}}} = \frac{1}{D_{12}} + \frac{1}{D_K} \tag{9.85}$$

Equation 9.85 is sometimes referred to as the Bosanquet formula. The concentration dependence of the pore diffusion coefficient contained in Equation 9.83 leads to a complex problem. It has been shown that, generally speaking, the concentration dependence of the pore diffusion coefficient is not large, and in many cases, Equation 9.85 is used to compute the pore diffusion coefficient, even for cases when the diffusion is not equimolar.

9.3.2 Effective Diffusivity

The diffusion of the reactants occurs only in the pores of the catalyst; however, a mathematical description of the diffusion process based on a pore by pore analysis would be very complex, and furthermore, would require a detailed knowledge of the porous microstructure. It is therefore convenient to develop an analysis of diffusion based on the total cross-sectional area of the catalyst, effectively treating the catalyst as a continuum. This requires that we modify the diffusion coefficient of the gas in the pores to account for the following factors:

1. The void fraction of the catalyst. This is done using the catalyst porosity, which is defined as the ratio of catalyst void volume to total volume.
2. The tortuosity of the pores. The pores do not run straight, but rather follow a tortuous path as they pass into the catalyst. The tortuosity can be defined as the ratio of the distance that a molecule travels to get between two points to the shortest distance between those points.
3. The pores vary in cross-sectional area.

An *effective diffusivity*, D_{eff}, is defined based on the entire cross-sectional area of the catalyst (i.e., in the direction orthogonal to diffusion), both solid and pores. It includes the catalyst porosity and tortuosity, either explicitly or implicitly. The effective diffusivity of a catalyst can be measured experimentally (see Chapter 11). In the absence of experimental data, its value may be estimated from pore size distribution and catalyst porosity data.

9.3.3 Parallel Pore Model

An early pore model is the parallel pore model of Wheeler (1955). In this model, the porous material is considered to be a bundle of capillaries through which species diffuse. The volume available for diffusion is the pore volume. An additional factor is introduced to account for the fact that the pores are not straight, but rather follow a tortuous path. The effective diffusivity is given by

$$D_{eff,A} = \frac{\varepsilon D_{pore,A}}{\tau} \tag{9.86}$$

where ε is the porosity (fractional pore volume), τ is the tortuosity factor, and $D_{pore, A}$ is the diffusion coefficient of A in the pores. Estimating the tortuosity can present problems. Satterfield (1970) has suggested a value of about 4 for typical catalyst pellets, with a range of 2–12 being reported by various researchers. Satterfield (1970) reports that an equivalent pore diameter can be computed from the porosity, ε, pellet density, ρ_p, and the surface area (normally the BET surface area determined from measurements of physical adsorption; see Chapter 11), S_g, by using the parallel pore model as follows:

$$d_p = \frac{4\varepsilon}{S_g \rho_p} \tag{9.87}$$

The parallel pore model works best for catalysts with a unimodal pore size distribution that has a narrow range of pore diameters.

Example 9.1

Estimate the effective diffusivity for methane in air at 500 K and 120 kPa in a catalyst with a porosity of 0.4, a pellet density of 1400 kg/m³, and a BET surface area of 180,000 m²/kg.

SOLUTION

The first step is to calculate the bulk and Knudsen diffusion coefficients. The bulk diffusion coefficient for methane in air at this temperature and pressure was calculated in Section 9.1.2 to be equal to 4.40×10^{-5} m²/s. The Knudsen diffusion coefficient requires the equivalent pore radius, which may be calculated from Equation 9.87:

$$d_p = \frac{4\varepsilon}{S_g \rho_p} = \frac{(4)(0.4)}{(180{,}000\,\text{m}^2/\text{kg})(1400\,\text{kg/m}^3)} = 6.35 \times 10^{-9}\ \text{m} \tag{9.88}$$

The Knudsen diffusion coefficient is calculated from Equation 9.82:

$$D_K = 48.5\, d_p \left(\frac{T}{M}\right)^{\frac{1}{2}} = (48.5)(6.35 \times 10^{-9}\,\text{m})\left(\frac{500\,\text{K}}{16\,\text{kg/kmol}}\right)^{0.5} = 1.72 \times 10^{-6}\ \text{m}^2/\text{s} \tag{9.89}$$

Combining the Knudsen and bulk diffusion coefficients using Equation 9.85 gives

$$D_V = \left(\frac{1}{D_{12}} + \frac{1}{D_K}\right)^{-1} = \left(\frac{1}{4.4 \times 10^{-5}\,\text{m}^2/\text{s}} + \frac{1}{1.72 \times 10^{-6}\,\text{m}^2/\text{s}}\right)^{-1} = 1.66 \times 10^{-6}\ \text{m}^2/\text{s} \tag{9.90}$$

With a tortuosity factor of 4, the effective diffusivity calculated from Equation 9.86 is

$$D_\text{eff} = \frac{\varepsilon D_V}{\tau} = \frac{0.4(1.66 \times 10^{-6}\,\text{m}^2/\text{s})}{4} = 1.66 \times 10^{-7}\ \text{m}^2/\text{s} \tag{9.91}$$

9.3.4 Random Pore Model

Another method for estimating effective diffusion coefficients is the *random pore model*, initially proposed by Wakao and Smith (1962) for bidisperse systems. A bidisperse catalyst has two distinct regions of pore sizes (see Chapter 7), small pores referred to as micropores or mesopores, and larger pores referred to as macropores. A bidisperse system may result when a catalyst is produced by compressing an assembly of small porous particles. A microporous or mesoporous structure is present in the small particles, while the space between the particles gives macroporosity. To use the random pore model, it is necessary to have data on the average pore size and the porosity of each region. In each region, a diffusion coefficient is calculated from the bulk and Knudsen diffusion coefficients, and the two coefficients are then combined to give the final result. For the macroregion, we define a macrodiffusion coefficient according to

$$\frac{1}{D_M} = \frac{1}{D_{12}} + \frac{1}{(D_K)_M} \tag{9.92}$$

For the microregion, the equation is

$$\frac{1}{D_\mu} = \frac{1}{D_{12}} + \frac{1}{(D_K)_\mu} \tag{9.93}$$

Let ε_M and ε_μ represent the porosity of the macro- and microregions, respectively, then the effective diffusivity is

$$D_{\mathrm{eff}} = D_M \varepsilon_M^2 + \left[\frac{\varepsilon_\mu^2 (1 + 3\varepsilon_M)}{1 - \varepsilon_M} \right] D_\mu \qquad (9.94)$$

For monodisperse systems, the model may also be used; simply set $\varepsilon_M = 0$ if there are no macropores and $\varepsilon_\mu = 0$ if there are no micropores.

Example 9.2

For an alumina pellet, the following data are available:

Macroregion: $\varepsilon_M = 0.20$	Average $r_1 = 500$ nm
Mesoregion: $\varepsilon_\mu = 0.35$	Average $r_2 = 10$ nm

Calculate the effective diffusion coefficient using the random pore model for methane diffusing in air at a temperature of 500 K and a pressure of 120 kPa.

SOLUTION

The bulk diffusion coefficient, D_{AB}, for methane in air was computed in Section 9.1.2 to be 4.4×10^{-5} m²/s. The Knudsen diffusion coefficient for each pore size region can be calculated from Equation 9.82, using the average pore radius for each region. The molar mass of methane is 16 kg/kmol, which gives

$$\left(D_K \right)_M = 2.711 \times 10^{-4}\,\mathrm{m^2/s} \quad \text{and} \quad \left(D_K \right)_\mu = 5.422 \times 10^{-6}\,\mathrm{m^2/s}$$

at the temperature given. Using Equations 9.92 and 9.93, we then obtain

$$D_M = 3.786 \times 10^{-5}\,\mathrm{m^2/s} \quad \text{and} \quad D_\mu = 4.827 \times 10^{-6}\,\mathrm{m^2/s}$$

These values and the porosities are then substituted into Equation 9.94 to give

$$D_{\mathrm{eff}} = 2.70 \times 10^{-6}\,\mathrm{m^2/s}$$

Finally, we note the pressure effect on the effective diffusion coefficient. Pressure has no effect on the Knudsen diffusion, but an increase in pressure will decrease the bulk diffusion coefficient, D_{12}.

Example 9.3

Consider the diffusion of methane in air occurring inside a porous catalyst with a porosity of 0.4, tortuosity factor of 4.0, and mean pore radius of 6.3 nm. Plot the molecular diffusion coefficient, D_{12}, the Knudsen diffusion coefficient, D_K, the pore diffusion coefficient, D_P, and the effective diffusion coefficient, D_{eff} over the temperature range 300–1300 K at pressures of 1 and 20 bar, respectively.

SOLUTION

The bulk diffusion coefficient, D_{12}, for methane in air can be computed using Equation 9.11, with the result left in terms of temperature and pressure as variables; thus

$$D_{12} = 9.991 \times 10^{-5} \, \frac{T^{1.75}}{P} \tag{9.95}$$

The Knudsen diffusion coefficient at any temperature can be computed using Equation 9.82 and the average pore radius. The molar mass of methane is 16 kg/kmol.

$$D_K = 48.5 \times 12.6 \times 10^{-9} \left(\frac{T}{16} \right)^{\frac{1}{2}} = 1.53 \times 10^{-7} \sqrt{T} \tag{9.96}$$

The diffusion coefficient in the pores is then computed using Equation 9.85:

$$D_{\text{eff,A}} = \left(\frac{1}{9.991 \times 10^{-5} \left(T^{1.75}/P \right)} + \frac{1}{1.53 \times 10^{-7} \sqrt{T}} \right)^{-1} \tag{9.97}$$

Then, D_{eff} is calculated from Equation 9.86:

$$D_{\text{eff,A}} = 0.1 D_P \tag{9.98}$$

The resulting diffusion coefficients are plotted as a function of temperature for 1 and 20 bar pressure in Figure 9.3. It can be seen from Figure 9.3 that at 1 bar D_{12} is much larger than D_K at 1 bar; in fact, the diffusion in the pores is governed almost entirely by the Knudsen diffusion, that is, $D_P \approx D_K$. At 20 bar, however, D_{AB} is of the same order as D_K which leads to a reduction in the value of D_P.

9.4 Diffusion with Reaction in Porous Catalysts

9.4.1 Effectiveness Factor

Having discussed the nature of diffusion in porous catalyst, it is now time to introduce the combined problem of diffusion with reaction. As noted earlier, the combination of diffusion with reaction gives rise to concentration gradients within the particle. The reaction rate thus varies continuously within the catalyst volume. To analyze a catalytic reactor, it is necessary to have an expression for the reaction rate of the entire catalyst pellet, and this quantity is usually expressed in terms of the average reaction rate. In turn, the average reaction rate for the particle is expressed in terms of the rate calculated at the temperature and concentration at the external surface using an *effectiveness factor* to quantify the effect of the diffusion in the catalyst. This factor is defined as

$$\eta = \frac{\text{average rate for a catalyst pellet}}{\text{rate evaluated at surface conditions}} \tag{9.99}$$

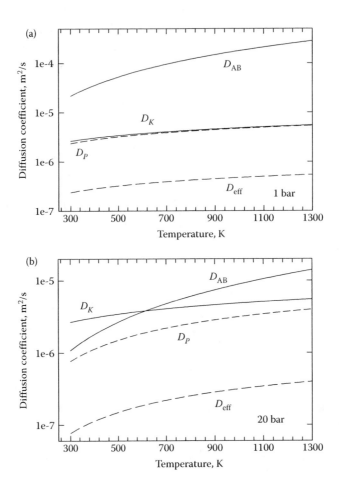

FIGURE 9.3
Effect of temperature and pressure on the diffusion coefficients of methane. Top graph shows the pressure of
1 bar. The short dashed line represents the pore diffusion coefficient. The bottom graph shows the results from
a pressure of 20 bars. Note the difference in scales for the diffusion coefficient axis in the two graphs.

When the potential for reaction is low and the potential for diffusion is high, the concentration gradients are small and the effectiveness factor approaches unity. In other circumstances, the effectiveness factor may be significantly less than or greater than unity.
The determination of the effectiveness factor is therefore of paramount importance in the
design of a catalytic system. Its value depends on the form of the rate expression and the
catalyst geometry. In the following sections, the derivation of the effectiveness factor will
be shown for three common catalyst geometries, flat plate, cylinder, and sphere, first for
the isothermal case and then the nonisothermal one. These three geometries can be considered as having only one space dimension in which the concentration varies, which
makes the analysis much easier.

9.4.2 Effectiveness Factor for Isothermal Flat Plate

Consider a flat plate of catalyst, as illustrated in Figure 9.4. We assume that the plate is
sealed on the edge faces, and that no diffusion occurs at these faces, and thus consider the

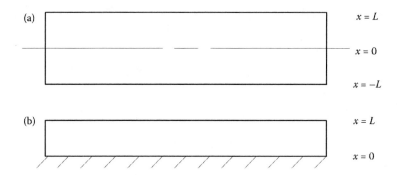

FIGURE 9.4
Representation of a flat slab of catalyst. In (a), the diffusion occurs from each side of the slab, while in (b), the diffusion occurs from one side only with the other side sealed. The situations are mathematically the same.

variation of the concentration in the x direction only. The problem thus only has one space dimension. Two mathematically identical cases are shown in Figure 9.4. In case (a), we see a porous slab of catalyst of thickness $2L$ with diffusion of reactants occurring from both exposed faces. The center of the slab is denoted with $x = 0$. In case (b), the slab is of thickness L, and diffusion into the slab occurs only at the face where $x = L$, with the face at $x = 0$ being impermeable. The catalyst surface at $x = \pm L$ is exposed to reactant at concentration C_S and temperature T_S. Assume that reactant A is diffusing into the catalyst, where it undergoes simultaneous reaction as the reactants encounter active sites. This reaction will cause a diminution of the reactant, and, as a result, a concentration gradient will develop along the catalyst depth. The magnitude of the gradient will depend on the relative speed of the diffusion and reaction processes. At any point in the catalyst, the rate of reaction is given by the *intrinsic* rate expression evaluated at local concentration and temperature, which will clearly vary within the catalyst. It is necessary to compute the average or observed reaction rate in the total catalyst volume, and to this end we now develop the mole balance equations for this situation. To this end we use the effective diffusivity, and base the analysis on the entire cross-sectional area.

The diffusive flux based on the effective diffusivity is assumed to follow Fick's law. For a reactant denoted by subscript A, the shell balance on an elementary element, Δx, of the flat plate gives the following mole balance equation:

$$[\text{Flow in}] = [\text{Flow out}] + [\text{Disappearance by reaction}] + [\text{Rate of accumulation}] \qquad (9.100)$$

For a steady-state process the last term is equal to zero and

$$(N_A)_x = (N_A)_{x+\Delta x} + (-R_A)\Delta x \qquad (9.101)$$

Dividing by Δx and rearranging gives

$$\frac{(N_A)_{x+\Delta x} - (N_A)_x}{\Delta x} = -(-R_A) \qquad (9.102)$$

We take the limit as $\Delta x \to 0$ to give

$$\frac{(N_A)_{x+\Delta x} - (N_A)_x}{\Delta x} = \frac{dN_A}{dx} \qquad (9.103)$$

therefore

$$\frac{dN_A}{dx} = -(-R_A) \qquad (9.104)$$

Note the signs on the quantities associated with the variables in the above equations. If A is a reactant, it diffuses in the negative x direction, that is from $x = L$ to $x = 0$ and therefore, N_A is negative. Because A is disappearing by reaction as it moves into the catalyst, it follows that the magnitude (absolute value) of the molar flow at x is less than the value at $x + \Delta x$. Because both quantities are negative, it follows that dN_A/dx is also a negative number, see Equation 9.103. This is consistent with Equation 9.104 because $(-R_A)$ is positive. The concentration of A drops as the x coordinate becomes smaller (as we move deeper into the slab); hence, the gradients, dC_A/dx and dY_A/dx are positive. The molar flux results from diffusion only and is given by

$$N_A = D_{\text{eff}} C \frac{dY_A}{dx} + Y_A N_T \qquad (9.105)$$

For equimolal counterdiffusion, the net molar flux rate is zero, that is, $N_T = 0$. For example, for a combustion reaction such as

$$CH_4 + O_2 \to CO_2 + 2H_2O \qquad (9.106)$$

the relationship between the diffusive fluxes would be

$$-(N_{CH_4} + N_{O_2}) = N_{CO_2} + N_{H_2O} \qquad (9.107)$$

and hence

$$N_A = -D_{\text{eff}} C \frac{dY_A}{dx} \qquad (9.108)$$

Substituting Equation 9.108 into Equation 9.104 gives

$$\frac{d}{dx}\left(D_{\text{eff}} C \frac{dY_A}{dx}\right) - (-R_A) = 0 \qquad (9.109)$$

If D_{eff} is constant and $C_A = C Y_A$ (valid for an isothermal system), then

$$D_{\text{eff}} \frac{d^2 C_A}{dx^2} - (-R_A) = 0 \qquad (9.110)$$

Equation 9.110 is the mole balance in a flat plate for simultaneous diffusion and reaction. The equation requires two boundary conditions. At the surface, the concentration is given by the surface concentration, while at $x = 0$ the diffusive flux is zero, because it is either a line of symmetry or an impermeable boundary. Thus, we write

$$C_A = C_{A,S} \text{ at } x = L, \quad \text{and} \quad \frac{dC_A}{dx} = 0 \quad \text{at } x = 0 \tag{9.111}$$

Note that the rate, $(-R_A)$, has units of mol/m$^3 \cdot$ s, that is, the rate is measured in terms of volume of catalyst. The rate per mass of catalyst can be converted to the rate per catalyst volume by multiplying by the density of the catalyst slab. It is convenient to put the equation into dimensionless form as an aid in identifying the important dimensionless groups which govern the diffusion/reaction process. Consider a first-order reaction

$$\left(-R_A\right) = k_V C_A \tag{9.112}$$

In common with much engineering analysis, Equation 9.110 can be written in dimensionless form by defining the dimensionless concentration as the ratio of the concentration to the surface concentration, and the dimensionless coordinate as the ratio of the position to the length:

$$C^* = \frac{C_A}{C_{A,S}} \quad \text{and} \quad x^* = \frac{x}{L} \tag{9.113}$$

The dimensionless form of the differential equation is

$$\frac{d^2 C^*}{dx^{*2}} - \left(L^2 \frac{k_V}{D_{\text{eff}}}\right) C^* = 0 \tag{9.114}$$

The dimensionless boundary equations are

$$C^* = 1 \text{ at } x^* = 1 \quad \text{and} \quad \frac{dC^*}{dx^*} = 0 \quad \text{at } x^* = 0 \tag{9.115}$$

The solution of Equations 9.114 and 9.115 can be used to generate the concentration profile in the catalyst, and hence to compute the average reaction rate. First, we introduce a new dimensionless parameter, the Thiele modulus.

9.4.2.1 Thiele Modulus for First-Order Reaction

Equation 9.114 can be simplified by introducing a dimensionless group, called the Thiele modulus, ϕ, which is defined (for a first-order reaction in a flat slab) as

$$\phi = L\sqrt{\frac{k_V}{D_{\text{eff}}}} \tag{9.116}$$

Substitute into Equation 9.114:

$$\frac{d^2C^*}{dx^{*2}} - \phi^2 C^* = 0 \qquad (9.117)$$

The Thiele modulus is defined differently for different rate equations, and is sometimes defined differently by various authors even for the same rate equation. Regardless of the definition, it is a standard dimensionless parameter in diffusion-reaction analysis in porous catalysts. Equation 9.114 with the boundary conditions of Equation 9.115 has an analytical solution (Thomas and Thomas, 1967; Satterfield, 1970; Fogler, 2006; Froment et al., 2011):

$$C^* = \frac{\cosh\left(\phi x^*\right)}{\cosh\left(\phi\right)} \qquad (9.118)$$

Equation 9.118 gives the concentration profile in the catalyst and hence the reaction rate at any point can be calculated.

9.4.2.2 Effectiveness Factor for First-Order Reaction in Flat Plate

We shall now derive the effectiveness factor. The effectiveness factor was defined earlier as the ratio of the average intrinsic rate in the catalyst to the rate evaluated at surface conditions. The average intrinsic rate could be found by calculating the rate of reaction at every point in the pellet using the concentration profile and averaging the results. This method is known as the integral approach and, although not widely reported in textbooks, is a useful technique. We first note that we can substitute the local concentration given by Equation 9.118 to compute the local rate in terms of the surface concentration, viz.

$$\left(-R_A\right) = k_V C_{A,S} \frac{\cosh\left(\phi x^*\right)}{\cosh\left(\phi\right)} \qquad (9.119)$$

Integration over the thickness L (1 in the dimensionless domain) gives the average rate in the slab

$$\left(-\bar{R}_A\right) = \int_0^1 k_V C_{A,S} \frac{\cosh\left(\phi x^*\right)}{\cosh\left(\phi\right)} dx^* \qquad (9.120)$$

The integrated result is

$$\left(-\bar{R}_A\right) = \frac{k_V C_{A,S}}{\cosh\left(\phi\right)} \left[\frac{1}{\phi} \sinh\left(\phi x^*\right)\right]_0^1 = k_V C_{A,S} \frac{\tanh\left(\phi\right)}{\phi} \qquad (9.121)$$

The reaction rate evaluated at the surface conditions is simply

$$\left(-R_A\right)_S = k_V C_{A,S} \qquad (9.122)$$

The ratio of the average rate to the rate calculated at the surface conditions is the effectiveness factor. Dividing Equation 9.121 by Equation 9.122 gives the result

$$\eta = \frac{\left(-\bar{R}_A\right)}{\left(-R_A\right)_S} = \frac{\tanh\left(\phi\right)}{\phi} \tag{9.123}$$

This result is valid for a first-order reaction occurring in an isothermal flat plate of catalyst with a uniform catalyst activity. The same result may be derived using a material balance approach. At steady state, the rate at which reactants diffuse into the plate across the entire surface area is equal to the total rate of reaction in the catalyst. The rate of reaction in terms of the rate of diffusion across the pellet external surface area, A_S, gives

$$V\left(-\bar{R}_A\right) = A_S D_{\text{eff}} \left(\frac{dC_A}{dx}\right)_{x=L} \tag{9.124}$$

where V is the total volume of the slab, the product of the surface area and the thickness. Rearranging

$$\left(-\bar{R}_A\right) = \frac{A_S}{V} D_{\text{eff}} \left(\frac{dC_A}{dx}\right)_{x=L} = \frac{1}{L} D_{\text{eff}} \left(\frac{dC_A}{dx}\right)_{x=L} \tag{9.125}$$

The concentration gradient at the surface can be determined by differentiating Equation 9.118:

$$\frac{dC^*}{dx^*} = \frac{\phi \sinh\left(\phi x^*\right)}{\cosh\left(\phi\right)} \tag{9.126}$$

At the surface, $x^* = 1$. Using the trigonometric relationships, Equation 9.126 becomes

$$\left(\frac{dC^*}{dx^*}\right)_{x^*=1} = \phi \tanh\left(\phi\right) \tag{9.127}$$

or, rewriting in dimensional quantities

$$\left(\frac{dC_A}{dx}\right)_{x=L} = \frac{C_{A,S}\phi}{L} \tanh\left(\phi\right) \tag{9.128}$$

The average rate in the catalyst is therefore given by

$$\left(-\bar{R}_A\right) = \frac{1}{L} D_{\text{eff}} \frac{C_{A,S}\phi}{L} \tanh\left(\phi\right) \tag{9.129}$$

The rate of reaction for the plate at surface conditions is given in Equation 9.122. The effectiveness factor, η, is calculated by dividing Equation 9.129 by Equation 9.122 to give

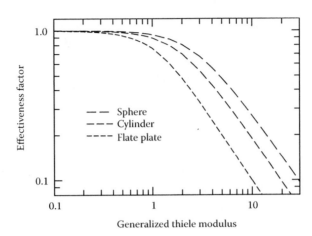

FIGURE 9.5
Effectiveness factor plots for the three one-dimensional geometries.

$$\eta = \frac{\left(-\bar{R}_A\right)}{\left(-R_A\right)_S} = \frac{(1/L)D_{eff}\left(C_{A,S}\phi/L\right)\tanh(\phi)}{k_V C_{A,S}} = \frac{D_{eff}}{k_V}\frac{\phi}{L^2}\tanh(\phi) \qquad (9.130)$$

Substituting the definition of the Thiele modulus gives the final result

$$\eta = \frac{\tanh(\phi)}{\phi} \qquad (9.131)$$

Note that the effectiveness factor is independent of the value of the concentration at the surface. This result is true only for first-order isothermal reactions. A plot of Equation 9.131 is shown in Figure 9.5. Once the Thiele modulus has been determined for the catalyst slab, then the effectiveness factor is computed and hence the actual reaction rate for the slab is

$$\left(-R_A\right)_{observed} = \eta\left(-R_A\right)_{intrinsic} \qquad (9.132)$$

9.4.2.3 Asymptotic Solutions for Effectiveness Factor

An interesting feature of this solution for the effectiveness factor is the limiting or *asymptotic* solutions, obtained when the Thiele modulus becomes very large or very small. As the Thiele modulus becomes large (greater than about 3), Equation 9.131 gives the approximate solution $\eta \approx 1/\phi$. At small values of Thiele modules (less than about 0.2), the asymptotic solution is $\eta \approx 1$. The importance of having a small Thiele modulus is therefore evident. Because the effective diffusivity and the reaction rate constant are fixed for a given catalyst formulation, operating temperature, and rate expression, the only parameter within the control of the reactor designer is the catalyst size. Thus, a thin catalyst slab has a higher effectiveness than a thick one, for an isothermal first-order reaction, everything else being equal.

Example 9.4

Consider the oxidation of propane in excess air, which has the rate expression

$$(-R_{C3H8}) = 5.0 \times 10^9 \exp\left(\frac{-89{,}791}{R_g T}\right) C_{C3H8} \; \text{mol/m}^3 \cdot \text{s} \qquad (9.133)$$

Calculate the effectiveness factor for a flat catalyst plate 30 μm thick with diffusion from one side only. The effective diffusivity of propane is $D_{\text{eff}} = 2.0 \times 10^{-7}$ m²/s. The surface concentration of propane is 0.1 mol% and the total pressure is 1 bar. Perform the calculation over the temperature range 500–900 K assuming that D_{eff} is not a function of temperature.

SOLUTION

The reaction is first order and, as seen in the previous derivation, the effectiveness factor does not depend on the surface concentration. The rate constant is given by

$$k_V = 5.0 \times 10^9 \exp\left(\frac{-89{,}791}{R_g T}\right) \text{s}^{-1} = 5.0 \times 10^9 \exp\left(\frac{-10{,}800}{T}\right) \text{s}^{-1} \qquad (9.134)$$

The Thiele modulus is computed from Equation 9.114:

$$\phi = L\sqrt{\frac{k_V}{D_{\text{eff}}}} = 30 \times 10^{-6} \text{ m} \sqrt{\frac{5.0 \times 10^9 \exp(-10{,}800/T) \; \text{s}^{-1}}{2.0 \times 10^{-7} \; (\text{m}^2/\text{s})}} = 4743 \exp\left(\frac{-5400}{T}\right) \qquad (9.135)$$

The effectiveness factor is computed from Equation 9.131. Some results are given below.

T (K)	Φ	η
500	0.0968	0.997
600	0.5853	0.900
700	2.117	0.459
800	5.554	0.180
900	11.76	0.085

Note how the effectiveness factor falls rapidly with temperature due to the exponential rise in the value of the reaction rate constant with temperature.

9.4.3 Effectiveness Factor for Isothermal Spherical Catalyst Pellet

A similar analysis can be applied to a spherical catalyst particle of radius R. For this geometry, we assume that there is symmetry about the center point, so that the only dimension for which the concentration varies is the radial one. The mole balance equation for species A in spherical coordinates in an isothermal system is

$$\frac{D_{\text{eff}}}{r^2} \frac{d}{dr}\left(r^2 \frac{dC_A}{dr}\right) - (-R_A) = 0 \qquad (9.136)$$

The boundary conditions are

$$C_A = C_{A,S} \quad \text{at } r = R \quad \text{and} \quad \frac{dC_A}{dr} = 0 \quad \text{at } r = 0 \qquad (9.137)$$

Equation 9.136 can also be written in dimensionless form. Define a dimensionless distance as

$$r^* = \frac{r}{R} \qquad (9.138)$$

We examine the case of a first-order reaction, and write the mole balance equation as

$$\frac{1}{r^{*2}} \frac{d}{dr^*}\left(r^{*2} \frac{dC^*}{dr^*}\right) - \frac{R^2 k_V}{D_{eff}} C^* = \frac{1}{r^{*2}} \frac{d}{dr^*}\left(r^{*2} \frac{dC^*}{dr^*}\right) - \phi^2 C^* = 0 \qquad (9.139)$$

The dimensionless boundary conditions are

$$C^* = 1 \quad \text{at } r^* = 1 \quad \text{and} \quad \frac{dC^*}{dr^*} = 0 \quad \text{at } r^* = 0 \qquad (9.140)$$

The analytical solution to Equation 9.139 gives the dimensionless concentration profile

$$C^* = \frac{\sinh\left(\phi r^*\right)}{r^* \sinh\left(\phi\right)} \qquad (9.141)$$

The solution of this equation can be used to develop the equation for an effectiveness factor for a first-order isothermal reaction in the same manner as was done for the flat plate. The resulting expression for the effectiveness factor is

$$\eta = \frac{3}{\phi}\left[\frac{1}{\tanh\left(\phi\right)} - \frac{1}{\phi}\right] \qquad (9.142)$$

As noted in Equation 9.139, the Thiele modulus is defined as

$$\phi = R\sqrt{\frac{k_V}{D_{eff}}} \qquad (9.143)$$

A plot of the effectiveness factor as a function of the Thiele modulus is also given in Figure 9.5. For values of Thiele modulus greater than about 7, Equation 9.142 gives the approximate (asymptotic) solution $\eta \approx 3/\phi$, while at small values of Thiele modules, less than about 0.2, the asymptotic solution is $\eta \approx 1$. All else being equal, decreasing the pellet size decreases the Thiele modulus, thus increasing the catalyst effectiveness, just as decreasing the thickness for the flat slab of catalyst does. Smaller pellet sizes, however, may lead to other problems, for example, they give a higher pressure drop in fixed bed reactors, which may not be desirable, and compromises must be made.

9.4.4 Effectiveness Factor for Isothermal Cylindrical Catalyst Pellet

The third geometry to consider is the cylindrical catalyst pellet. The 1D analysis for a cylindrical pellet requires the assumption that there is no diffusion through the ends of the cylinder, and thus there is symmetry about the axis. The mole balance equation is

$$\frac{D_{eff}}{r}\frac{d}{dr}\left(r\frac{dC_A}{dr}\right) - (-R_A) = 0 \tag{9.144}$$

The boundary conditions are

$$C_A = C_{A,S} \quad \text{at} \quad r = R \quad \text{and} \quad \frac{dC_A}{dr} = 0 \quad \text{at} \quad r = 0 \tag{9.145}$$

Using the same definition for dimensionless radius, the dimensionless form of the equation for the cylinder with a first-order reaction is

$$\frac{1}{r^*}\frac{d}{dr^*}\left(r^*\frac{dC^*}{dr^*}\right) - \frac{R^2 k_V}{D_{eff}}C^* = \frac{1}{r^*}\frac{d}{dr^*}\left(r^*\frac{dC^*}{dr^*}\right) - \phi^2 C^* = 0 \tag{9.146}$$

The boundary conditions are

$$C^* = 1 \quad \text{at} \quad r^* = 1 \quad \text{and} \quad \frac{dC^*}{dr^*} = 0 \quad \text{at} \quad r^* = 0 \tag{9.147}$$

The effectiveness factor is derived in the same manner as before for a first-order isothermal reaction:

$$\eta = \frac{2}{\phi}\frac{I_1(\phi)}{I_0(\phi)} \tag{9.148}$$

I_0 and I_1 are zero- and first-order Bessel functions of the first kind. The Thiele modulus is

$$\phi = R\sqrt{\frac{k_V}{D_{eff}}} \tag{9.149}$$

The plot of effectiveness factor is also shown in Figure 9.5. For values of Thiele modulus greater than about 5, Equation 9.148 gives the approximate (asymptotic) solution $\eta \approx 2/\phi$, while at small values of Thiele modules, less than about 0.2, the asymptotic solution is $\eta \approx 1$.

9.4.5 Generalized Thiele Modulus and Effectiveness Factors

The three geometrical shapes presented in the previous sections each give rise to a different definition of the Thiele modulus. To give a consistency to the definition, and also to aid in the analysis of other shapes using an approximate one-dimensional analysis, a generalized Thiele modulus has been proposed. The characteristic length for the catalyst particle is defined as the total particle volume divided by the external surface area through which diffusion occurs. Thus, we can define a generalized Thiele modulus as

$$\phi = L_C \sqrt{\frac{k_V}{D_{eff}}} \quad \text{where } L_C = \frac{V}{A} \tag{9.150}$$

For each of the three geometries (plane wall, infinite cylinder, and sphere) considered thus far, the generalized Thiele modulus becomes, respectively

$$\phi = L \sqrt{\frac{k_V}{D_{eff}}}, \quad \phi = \frac{R}{2} \sqrt{\frac{k_V}{D_{eff}}}, \quad \phi = \frac{R}{3} \sqrt{\frac{k_V}{D_{eff}}} \tag{9.151}$$

With this new definition of the Thiele modulus, the solution must also change, and the three new equations for the effectiveness factor are

$$\eta = \frac{\tanh(\phi)}{\phi} \quad \text{(plane wall)} \tag{9.152}$$

$$\eta = \frac{1}{\phi} \frac{I_1(2\phi)}{I_0(2\phi)} \quad \text{(cylinder)} \tag{9.153}$$

$$\eta = \frac{1}{\phi} \left[\frac{1}{\tanh(3\phi)} - \frac{1}{3\phi} \right] \quad \text{(sphere)} \tag{9.154}$$

A plot of effectiveness factor as a function of the generalized Thiele modulus is given Figure 9.6. It can be seen that the solutions for the three equations are nearly coincident.

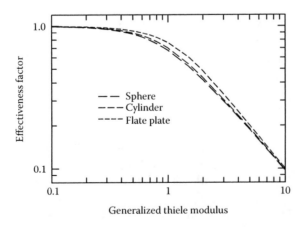

FIGURE 9.6
Effectiveness factor plots for the three one-dimensional geometries using the generalized Thiele modulus.

Thus, for a catalyst particle that does not conform to one of the three idealized shapes discussed, one can simply define a characteristic length as the ratio of volume to surface area and then any of the three analytical solutions. Furthermore, the asymptotic solutions for each of Equations 9.152, 9.153, and 9.154 at high Thiele modulus are the same, being equal to $\eta \approx 1/\phi$.

9.4.6 Effectiveness Factors for Non-First-Order Reactions

The results presented thus far pertain to first-order reactions. As noted earlier, an interesting result for the first-order case is that the effectiveness factor does not depend on the value of the surface concentration, but only on the temperature. Effectiveness factors can also be derived for other types of kinetics, and we start by considering a reaction of order n; thus, let us consider a rate equation of the form

$$(-R_A) = k_V C_A^n \tag{9.155}$$

The mole balance equation for the flat plate is

$$D_{eff} \frac{d^2 C_A}{dx^2} - k_V C_A^n = 0 \tag{9.156}$$

Using the same dimensionless quantities as previously

$$C^* = \frac{C_A}{C_{A,S}} \quad \text{and} \quad x^* = \frac{x}{L} \tag{9.157}$$

we can write the mole balance in dimensionless form

$$\frac{d^2 C^*}{dx^{*2}} - \left(L^2 \frac{k_V C_{A,S}^{n-1}}{D_{eff}} \right)(C^*)^n = \frac{d^2 C^*}{dx^{*2}} - \phi^2 (C^*)^n = 0 \tag{9.158}$$

It can be seen that the Thiele modulus becomes

$$\phi = L \sqrt{\frac{k_V C_{A,S}^{n-1}}{D_{eff}}} \tag{9.159}$$

The value of the Thiele modulus includes the surface concentration, which in turn means that the effectiveness factor depends on the surface concentration of reactant. In this case, the mole balance equation becomes a nonlinear differential equation and requires a numerical solution. A unique solution for Equation 9.158 can be obtained; however, the computation of the effectiveness factor now requires the value of the surface concentration.

For simple power law kinetics, the solution for the mole balance equation depends on a unique parameter, the Thiele modulus. As kinetic expressions become more complicated, the determination of the effectiveness factor becomes more involved. Consider, for example, the oxidation of carbon monoxide in an excess of oxygen:

$$CO + \frac{1}{2}O_2 \rightarrow CO_2 \tag{9.160}$$

A reported rate equation for this reaction over a platinum catalyst in an excess of oxygen has the LHHW model form:

$$(-R_{CO}) = \frac{k_V C_{CO}}{(1 + K_{CO}C_{CO})^2} \tag{9.161}$$

The mole balance equation in an isothermal system could be written as

$$D_{eff}\frac{d^2C_{CO}}{dx^2} - \frac{k_V C_{CO}}{(1 + K_{CO}C_{CO})^2} = 0 \tag{9.162}$$

We again write the equation in dimensionless form using the scaling factors

$$C^* = \frac{C_{CO}}{C_{CO,S}} \quad \text{and} \quad x^* = \frac{x}{L} \tag{9.163}$$

which gives the result

$$\frac{d^2C^*}{dx^{*2}} - L^2\frac{k_V}{D_{eff}}\frac{C^*}{(1 + K_{CO}C_{CO,S}C^*)^2} = 0 \tag{9.164}$$

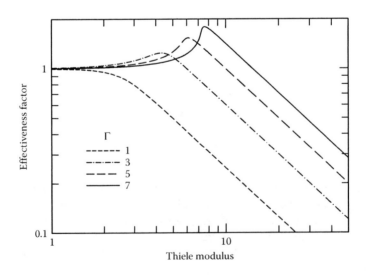

FIGURE 9.7
Effectiveness factor plots for a reaction with LHHW-type kinetics. The value of Γ varies (see Equation 9.151).

Now, it is not possible to obtain a result in terms of only two dimensionless groups. Rather, the equation has the form

$$\frac{d^2C^*}{dx^{*2}} - \phi^2 \frac{C^*}{\left(1 + \Gamma C^*\right)^2} = 0 \tag{9.165}$$

The effectiveness factor will be a function of the two groups, ϕ and Γ. Equation 9.165 is highly nonlinear and may be difficult to solve. Further, we recall from Chapter 8 that LHHW kinetics do not exhibit "normal" behavior, which implies that a reduction in concentration can lead to an increase in reaction rate. Therefore, for LHHW kinetics, effectiveness factors greater than one are possible, even for the isothermal case. A plot of the effectiveness factors generated for Equation 9.165 is shown in Figure 9.7. Note the region where the effectiveness factor exceeds unity, thus enhancing the reaction rate.

9.4.7 Nonisothermal Effectiveness Factors

We have seen that the bulk of the chemical reaction occurs inside the catalyst pellet. Unless the enthalpy change of reaction is zero, the progress of the reaction will result in the absorption or release of thermal energy. This thermal energy change requires the transport of thermal energy either into or out of the catalyst pellet by conduction. The conduction occurs through both the void space and the solid matrix. Thermal conductivities of catalyst pellets can be low, of the order of 0.5–1 W/m · K, and therefore, depending on the relative potential rates of heat release and heat conduction, significant temperature gradients may arise inside the catalyst. As the effect of a temperature increase is generally to increase the rate of reaction, this may counterbalance the decrease in rate due to a decrease in concentration. The increase in rate due to a temperature rise may result in effectiveness factors greater than one.

To determine the effectiveness factor in a nonisothermal case requires the simultaneous solution of the energy balance and mole balance equations for the catalyst layer. The resulting differential equations for the three 1D catalyst geometries in which a single reaction occurs are (assuming that the effective thermal conductivity is not a function of temperature)

Flat plate:

$$k_{\text{eff}} \left(\frac{d^2T}{dx^2} \right) - \Delta H_R \left(-R_A \right) = 0 \tag{9.166}$$

Cylindrical pellet:

$$\frac{k_{\text{eff}}}{r} \frac{d}{dr} \left(r \frac{dT}{dr} \right) - \Delta H_R \left(-R_A \right) = 0 \tag{9.167}$$

Spherical pellet:

$$\frac{k_{\text{eff}}}{r^2} \frac{d}{dr} \left(r^2 \frac{dT}{dr} \right) - \Delta H_R \left(-R_A \right) = 0 \tag{9.168}$$

In each case, k_{eff} is the *effective* thermal conductivity of the porous matrix, which is assumed to be independent of temperature. Equations 9.166, 9.167, and 9.168 can be derived following the usual shell balance methods. The boundary conditions frequently used are

$$T = T_S \quad \text{at} \quad x = L \quad \text{or} \quad r = R$$

$$\frac{dT}{dx} = 0 \quad \text{at} \quad x = 0 \quad \text{or} \quad \frac{dT}{dr} = 0 \quad \text{at} \quad r = 0 \tag{9.169}$$

The mole and energy balance equations form a system of coupled ordinary differential equations which must be solved numerically.

Regardless of which 1D geometry is taken, a simple relationship between temperature and conversion can be derived, provided that some simplifying assumptions are made. Consider the flat plate. The mole balance, Equation 9.110, and the energy balance, Equation 9.166, both contain the rate $(-R_A)$. Writing each equation explicitly in terms of the rate gives

$$D_{eff} \frac{d}{dx}\left(C\frac{d^2 Y_A}{dx^2}\right) = (-R_A) \tag{9.170}$$

$$\frac{k_{eff}}{\Delta H_R}\left(\frac{d^2 T}{dx^2}\right) = (-R_A) \tag{9.171}$$

Combining these two equations gives the following relationship:

$$D_{eff} \frac{d}{dx}\left(C\frac{dY_A}{dx}\right) = \frac{k_{eff}}{\Delta H_R}\left(\frac{d^2 T}{dx^2}\right) \tag{9.172}$$

Note that Equation 9.172 is independent of the kinetic rate expression and assumes that both D_{eff} and k_{eff} are not functions of temperature. We shall now assume that ΔH_R is also constant. We can integrate both sides of Equation 9.172 to give

$$D_{eff}C\frac{dY_A}{dx} + \Gamma_1 = \frac{k_{eff}}{\Delta H_R}\frac{dT}{dx} + \Gamma_2 \tag{9.173}$$

where Γ_1 and Γ_2 are constants. Equation 9.173 must be satisfied at all points in the solution domain. At $x = 0$, we have the boundary conditions

$$\frac{dY_A}{dx} = \frac{dT}{dx} = 0 \tag{9.174}$$

It follows that $\Gamma_1 = \Gamma_2$; thus

$$D_{eff}C\frac{dY_A}{dx} = \frac{k_{eff}}{\Delta H_R}\frac{dT}{dx} \tag{9.175}$$

For small temperature changes, the values of C and ΔH_R can be assumed to be essentially constant, and then integrating again subject to boundary conditions of specified temperature and concentration at $x = L$ gives

$$T - T_S = (-\Delta H_R)\frac{D_{\text{eff}}}{k_{\text{eff}}}C_S(Y_{A,S} - Y_A) \tag{9.176}$$

It follows that the largest temperature difference will occur when the concentration goes to zero. This can be restated in the form

$$\frac{(\Delta T)_{\max}}{T_S} = \frac{(-\Delta H_R)D_{\text{eff}}C_{A,S}}{k_{\text{eff}}T_S} = \beta \tag{9.177}$$

The parameter β is called the *Prater number*. Its value determines the importance of temperature gradients in the catalyst pellet.

9.4.7.1 Solving Coupled Mole and Energy Balance Equations

The mole and energy balance equations given by, for example, Equations 9.170 and 9.171 are a coupled system of nonlinear equations. Furthermore, these equations are boundary value problems (BVP). The solution of the coupled diffusion reaction system is discussed by a number of authors, and Finlayson (1980) gives a good discussion of the appropriate numerical methods for this problem. One of the original methods was presented by Weisz and Hicks (1962), and is widely cited. The relationship between temperature and concentration may be expressed as (ignoring the temperature dependence of the concentration):

$$T = T_S + (-\Delta H_R)\frac{D_{\text{eff}}}{k_{\text{eff}}}(C_{A,S} - C_A) \tag{9.178}$$

Using the definition of the Prater number from Equation 9.177 gives

$$T = T_S + \beta\frac{T_S}{C_{A,S}}(C_{A,S} - C_A) = T_S\left[1 + \beta(1 - C^*)\right] \tag{9.179}$$

For a first-order reaction in a flat plate, the dimensionless mole balance is

$$\frac{d^2C^*}{dx^{*2}} - \left(L^2\frac{k_V}{D_{\text{eff}}}\right)C^* = 0 \tag{9.180}$$

Define a reference value for the rate constant at the surface temperature; thus

$$k_V = k_0 \quad \text{at } T = T_S \tag{9.181}$$

Thus

$$k_V = k_0\ \exp\left\{\frac{-E}{R_g}\left(\frac{1}{T} - \frac{1}{T_S}\right)\right\} \tag{9.182}$$

Equation 9.180 can thus be written to include the temperature effects:

$$\frac{d^2C^*}{dx^{*2}} - \left(L^2 \frac{k_0}{D_{eff}}\right)C^* \exp\left\{\frac{-E}{R_g}\left(\frac{1}{T} - \frac{1}{T_S}\right)\right\} = 0 \tag{9.183}$$

If we introduce the Thiele modulus evaluated at the surface temperature, ϕ_S

$$\frac{d^2C^*}{dx^{*2}} - \phi_S^2 C^* \exp\left\{\frac{-E}{R_g}\left(\frac{1}{T} - \frac{1}{T_S}\right)\right\} = 0 \tag{9.184}$$

Substitute the temperature from Equation 9.179 into Equation 9.184:

$$\frac{d^2C^*}{dx^{*2}} - \phi_S^2 C^* \exp\left\{\frac{-E}{R_g T_S}\left(\frac{1}{\left[1 + \beta\left(1 - C^*\right)\right]} - 1\right)\right\} = 0 \tag{9.185}$$

Define a new variable

$$\gamma = \frac{-E}{R_g T_S} \tag{9.186}$$

Substitute into Equation 9.185 and rearrange

$$\frac{d^2C^*}{dx^{*2}} - \phi_S^2 C^* \exp\left(\frac{\gamma\beta\left(1 - C^*\right)}{\left[1 + \beta\left(1 - C^*\right)\right]}\right) = 0 \tag{9.187}$$

The boundary conditions are the same as before:

$$C^* = 1 \quad \text{at } x^* = 1 \quad \text{and} \quad \frac{dC^*}{dx^*} = 0 \quad \text{at } x^* = 0 \tag{9.188}$$

Equation 9.187 must be solved numerically using an appropriate method. Once the concentration profile is computed, the average rate in the catalyst is then calculated either by computing the rate of diffusion into the catalyst or by integrating the rate over the thickness.

Note that there are three governing parameters, ϕ_S, γ, and β. Typically, results are presented as effectiveness factor plotted as a function of Thiele modulus at various values of γ and β (see, e.g., Weisz and Hicks, 1962). A value of β less than zero implies an endothermic reaction, a value of zero is the isothermal case, and a value greater than zero is the exothermic case. A typical result is shown in Figure 9.8 for a value of $\gamma = 20$. Several observations can be made. The first is that for $\beta = 0$ the solution is the same as before (i.e., Equation 9.123). An endothermic reaction results in an internal temperature lower than the surface, which reduces the reaction rate. Thus, the effectiveness factor is lower than the

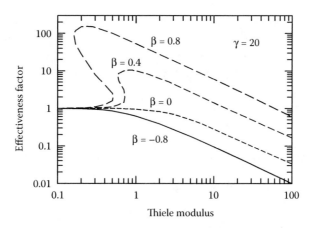

FIGURE 9.8
Effectiveness factor plots for a nonisothermal catalyst pellet with different values of β and γ = 20.

one for the isothermal case. For an exothermic reaction, the temperature inside the catalyst is higher than the surface, and as a result the effectiveness factor is higher than that for the isothermal case. Indeed, for significant temperature gradients, the effectiveness factor may exceed unity, which implies an enhancement of the conversion. This artifact is not necessarily beneficial because in some cases the elevated temperatures may cause catalyst damage. Another artifact of the exothermic case is the existence of multiple steady states over a small range of Thiele modulus. Theoretically, these multiple steady states can result in unstable reactor performance. Note that when there are three possible steady-state solutions, the middle solution is unstable and cannot be realized in practice.

9.4.7.2 Solution Algorithm Using Only Initial Value Problems

Equation 9.187 can be solved as a boundary value problem or, alternatively, a coordinate transformation can be used to recast the problem as an initial value problem that can be solved using a standard IVP solver, such as the Runga–Kutta method. First, define a new variable

$$x^* = a\xi \tag{9.189}$$

Equation 9.187 becomes

$$\frac{d^2C^*}{d\xi^2} - a^2\phi_S^2 C^* \exp\left(\frac{\gamma\beta(1 - C^*)}{\left[1 + \beta(1 - C^*)\right]}\right) = 0 \tag{9.190}$$

The solution algorithm is then as follows:

0. Select a value for β and γ. These values remain fixed.
1. Select a value for $a^2\phi_S^2$.

2. To start the solution at $\xi = 0$, assume a value of C^*. Then, integral the equation along ξ until a value of $C^* = 1$ is reached. Whatever the value of ξ at which this occurs corresponds to the surface, or $x^* = 1$.

3. Because $x^* = 1$, it follows that $a = 1/\xi$ and hence calculate

$$\phi_S = \left[a^2 \phi_S^2 \right]^{0.5} \xi$$

where the $\left[a^2 \phi_S^2 \right]$ corresponds to the value assumed in step (1).

4. Calculate the value of the gradient at the surface; thus

$$\left. \frac{dC^*}{dx^*} \right|_{x^* = 1} = \frac{1}{a} \left. \frac{dC^*}{d\xi} \right|_{\xi = \frac{1}{a}} \tag{9.191}$$

5. The effectiveness factor for the corresponding value of ϕ_S is then computed:

$$\eta = \frac{1}{\phi_S^2} \left. \frac{dC^*}{dx^*} \right|_{x^* = 1} = \frac{1}{\phi_S^2} \frac{1}{a} \left. \frac{dC^*}{d\xi} \right|_{\xi = \frac{1}{a}} \tag{9.192}$$

6. Repeat steps (1) through (5) until the entire range of ϕ_S is covered.

9.4.8 Effect of Internal Diffusion Limitation on Observed Kinetics

Strong internal diffusion resistance affects the observed, or apparent, kinetics. The observed kinetics will be the average rate in the catalyst; thus, for an isothermal first-order reaction

$$\left(-\bar{R}_A \right) = \eta k_V C_{A,S} \tag{9.193}$$

In the asymptotic region, we can write

$$\left(-\bar{R}_A \right) = \frac{k_V C_{A,S}}{\phi} = \frac{1}{L} \sqrt{\frac{D_{eff}}{k_V}} k_V C_{A,S} = \frac{1}{L} \sqrt{D_{eff} k_V} \, C_{A,S} \tag{9.194}$$

If the reaction rate constant has an Arrhenius-type temperature dependence

$$\left(-\bar{R}_A \right) = \frac{1}{L} \sqrt{D_{eff}} \, A_0 \, \exp\left(\frac{-E}{2R_g T} \right) C_{A,S} \tag{9.195}$$

In the diffusion-free (kinetically controlled) region

$$\left(-\bar{R}_A \right) = A_0 \, \exp\left(\frac{-E}{R_g T} \right) C_{A,S} \tag{9.196}$$

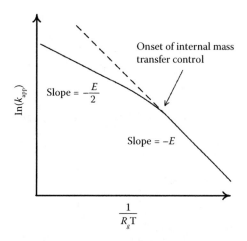

FIGURE 9.9

Effect of internal mass transfer limitation on the apparent activation energy of reaction. For a first-order reaction, the apparent activation energy falls by one-half as the reaction becomes completely mass transfer controlled.

A plot of the apparent rate constant versus temperature on an Arrhenius plot will thus be as shown in Figure 9.9. It can be seen that the slope of the Arrhenius line decreases by a factor of two, which means that the apparent activation energy in the asymptotic region will be one half of the true value. For a first-order reaction, however, the reaction will still appear to be first order. For a power law expression, the apparent reaction order will also change.

9.4.9 Combined Internal and External Mass Transfer Resistance

Having reviewed the resistances due to both external mass transfer, which uses the concept of a mass transfer coefficient, and internal mass transfer, which uses the concept of an effectiveness factor, it is worthwhile to discuss briefly the combination of the two resistances. For a first-order isothermal reaction, the addition of the resistances is straightforward. The solution obtained previously for the effectiveness factor is still valid, except that the surface concentration is unknown, which implies that the observed rate of reaction cannot be found from this value alone. An overall effectiveness factor which incorporates both internal and external mass transfer resistance can be obtained by changing the boundary conditions in the mole balance equation. Replace the boundary conditions for Equation 9.110 (i.e., Equation 9.111) by

$$k_m\left(C_{A,f} - C_{A,S}\right) = D_{\text{eff}}\left(\frac{dC_A}{dx}\right)_{x=L} \quad \text{at } x = L$$

$$\frac{dC_A}{dx} = 0 \quad \text{at } x = 0$$

(9.197)

The solution of the mole balance equation gives the following equation for the concentration inside the catalyst in terms of the bulk concentration:

$$C_A = C_{A,f} \left[\frac{\cosh\left(\dfrac{\phi x}{L}\right)}{\cosh\phi + \dfrac{D_{eff}\phi}{Lk_m}\sinh\phi} \right] \tag{9.198}$$

The local reaction rate at any point inside the catalyst is therefore

$$(-R_A) = k_V C_{A,f} \left[\frac{\cosh\left(\dfrac{\phi x}{L}\right)}{\cosh\phi + \dfrac{D_{eff}\phi}{Lk_m}\sinh\phi} \right] \tag{9.199}$$

The average rate in the catalyst can be obtained by integration over the thickness, as before

$$(-\bar{R}_A) = \frac{1}{L}\int_0^L k_V C_{A,f} \left[\frac{\cosh\left(\dfrac{\phi x}{L}\right)}{\cosh\phi + \dfrac{D_{eff}\phi}{Lk_m}\sinh\phi} \right] dx \tag{9.200}$$

The result is

$$(-\bar{R}_A) = k_V C_{A,f} \left[\frac{\dfrac{\sinh\phi}{\phi}}{\cosh\phi + (D_{eff}\phi/Lk_m)\sinh\phi} \right] = k_V C_{A,f} \left[\frac{\dfrac{\tanh\phi}{\phi}}{1 + (D_{eff}\phi/Lk_m)\tanh\phi} \right] \tag{9.201}$$

If we now define a global effectiveness factor as the ratio of the average rate in the catalyst divided by the rate evaluated at bulk conditions, it is evident that the global effectiveness factor is simply

$$\eta_G = \frac{\dfrac{\tanh\phi}{\phi}}{1 + (D_{eff}\phi/Lk_m)\tanh\phi} = \frac{\eta}{1 + (D_{eff}\phi^2/Lk_m)\eta} \tag{9.202}$$

We can define the mass Biot number, Bi_m, for the catalyst as

$$Bi_m = \frac{k_m L}{D_{eff}} \tag{9.203}$$

The Biot number is a ratio of convective to diffusive mass transfer. Substitute Equation 9.203 into Equation 9.202 and take the reciprocal

$$\frac{1}{\eta_G} = \frac{1}{\eta} + \frac{\phi^2}{Bi_m} \tag{9.204}$$

For reactions that are not isothermal and first order, a numerical solution to the mole and energy balance equations is required, and the results cannot be presented as neatly as in the preceding equations. In the asymptotic region, Equation 9.204 becomes

$$\frac{1}{\eta_G} \approx \phi + \frac{\phi^2}{Bi_m} \tag{9.205}$$

Example 9.5

The combustion of methane in a washcoated monolith reactor was studied by Kolaczkowski and Worth (1995), who reported the rate expression

$$\left(-r_{CH_4,S} \right) = 2.84 \times 10^8 \eta \exp\left(\frac{-15,757}{T_S} \right) Y_{CH_4,S}^{0.72} \tag{9.206}$$

Leung et al. (1996) studied the diffusion in the washcoat for this reactor and derived an expression for the effectiveness factor (for this rate expression) as

$$\eta = \left[1 + \left(\frac{1.122}{\phi^{0.991}} \right)^{-1.3406} \right]^{\frac{-1}{1.3406}} \tag{9.207}$$

The Thiele modulus was defined as

$$\phi = \left(\frac{L_c k_s R_g T_S}{D_{eff} P Y_{CH_4}^{0.28}} \right)^{\frac{1}{2}} \tag{9.208}$$

The characteristic length, L_c, for the washcoat was 45.8 μm. Note that the effectiveness factor is a function of the surface mole fraction of methane because this is not a first-order reaction.

Consider a point in the reactor downstream from the entrance where the entrance effect is negligible. The mean bulk fluid mole fraction of methane is 0.01 with the remainder air, the mean bulk temperature is 750 K, and the total pressure is 100 kPa. Compute the value of the surface temperature, methane mole fraction, washcoat effectiveness factor, and global effectiveness factor if the Reynolds number at this point in the reactor is

 a. 200 (laminar flow case)
 b. 10,000 (turbulent flow case)

Assume that the reactor tubes are circular in cross section with a diameter of 1 mm. Take the effective diffusivity of CH_4 in the washcoat to be 3.0×10^{-6} m^2/s.

SOLUTION

Part (a)
The key to this problem is to recognize that the rate of mass transfer to the surface equals the rate of reaction, and that the rate of heat transfer equals the rate of

heat generation. Using the principles outlined earlier, we can write the following equations:

$$k_m \frac{P}{R_g T_b}\left(Y_{CH4,b} - Y_{CH4,S}\right) = \eta k_s Y_{CH4,S}^{0.72} \tag{9.209}$$

$$h\left(T_S - T_b\right) = -\Delta H_R \eta k_s Y_{CH4,S}^{0.72} \tag{9.210}$$

The effectiveness factor is given by Equation 9.207 and the Thiele modulus by Equation 9.208. Note that T and Y_{CH4} in k_s and ϕ are the unknown values at the catalyst surface.

To solve this problem, we require heat and mass transfer coefficients. For illustration purposes, we shall use the results for nonreacting flows. For fully developed laminar flow (Re < 2100), take the values of Sh and Nu for constant wall temperature:

$$Nu = \frac{hD_T}{k_f} = 3.67 \quad \text{and} \quad Sh = \frac{k_m D_T}{D_{AB}} = 3.67 \tag{9.211}$$

The thermal conductivity of air is 0.0549 W/m · K at 750 K which gives

$$h = \frac{Nu k_f}{D_T} = \frac{3.67\left(0.0549\,\text{W/m} \cdot \text{K}\right)}{1 \times 10^{-3}\,\text{m}} = 201\,\text{W/m}^2\text{K} \tag{9.212}$$

The diffusion coefficient of CH_4 in air at 750 K is 1.07×10^{-4} m²/s, giving

$$k_m = \frac{Sh D_{AB}}{D_T} = \frac{3.67\left(1.07 \times 10^{-4}\,\text{m}^2/\text{s}\right)}{1 \times 10^{-3}\,\text{m}} = 0.394 \ \text{m/s} \tag{9.213}$$

ΔH_R at this temperature is approximately −802 kJ/mol. Substitution of the numerical values into Equations 9.208, 9.209, and 9.210 gives the following equations for the mole balance:

$$\left(0.01 - Y_{CH4,S}\right) = 4.50 \times 10^7 \eta \, \exp\left(\frac{-15{,}757}{T_S}\right) Y_{CH4,S}^{0.72} \tag{9.214}$$

for the energy balance

$$\left(T_S - 750\right) = 1.13 \times 10^2 \, \eta \, \exp\left(\frac{-15{,}757}{T_S}\right) Y_{CH4,S}^{0.72} \tag{9.215}$$

for the Thiele modulus

$$\phi = 600\left[\frac{T_S \exp\left(-15{,}757/T_S\right)}{Y_{CH4,S}^{0.28}}\right]^{0.5} \tag{9.216}$$

These three equations plus Equation 9.207 are a system of four nonlinear equations which must be solved simultaneously using an appropriate numerical method to give values of surface temperature and concentration. The solution is

$$Y_{CH4,S} = 0.008 \quad \text{and} \quad T_S = 789 \text{ K}$$

Then, we calculate $\phi = 1.518$ and $\eta = 0.506$, respectively. We can now calculate a global effectiveness factor. This is defined as the observed rate divided by the rate evaluated at the mean bulk conditions. The first step is to calculate the reaction rate at the surface conditions, that is, at $Y_{CH4,S} = 0.008$ and $T_S = 789$ K.

$$\begin{aligned}
(-R_{CH4})_S &= 2.84 \times 10^8 \exp\left(\frac{-131,004}{R_g T_S}\right) Y_{CH4,S}^{0.72} \\
&= 2.84 \times 10^8 \exp\left(\frac{-131,004}{(8.314)(789)}\right) 0.008^{0.72} = 0.0186 \text{ mol/m}^2\text{s}
\end{aligned} \tag{9.217}$$

The actual rate in the catalyst washcoat is this surface rate multiplied by the effectiveness factor:

$$(-R_{CH4})_{actual} = \eta(-R_{CH4})_S = (0.506)(0.0186) = 9.43 \times 10^{-3} \text{ mol/m}^2 \cdot \text{s} \tag{9.218}$$

The rate evaluated at the bulk conditions is:

$$(-R_{CH4})_b = 2.84 \times 10^8 \exp\left(\frac{-131,004}{(8.314)(750)}\right) 0.01^{0.72} = 0.00775 \text{ mol/m}^2 \cdot \text{s} \tag{9.219}$$

The global effectiveness factor is then

$$\eta_G = \frac{\eta(-R_{CH4})_S}{(-R_{CH4})_b} = \frac{(0.506)(0.0186)}{7.75 \times 10^{-3}} = \frac{9.43 \times 10^{-3}}{7.75 \times 10^{-3}} = 1.22 \tag{9.220}$$

Note that the global effectiveness factor is greater than one, which implies that the rate at bulk conditions is lower than the observed rate. This occurs because the increase in reaction rate due to the increase in catalyst temperature more than offsets the decrease in rate resulting from the decrease in concentration.

Part (b)
In the turbulent region, the Dittus–Boelter correlation for Sh and Nu might be used:

$$\text{Sh} = 0.023 \text{Re}^{0.8} \text{Sc}^{0.33} \tag{9.221}$$

$$\text{Nu} = 0.023 \text{Re}^{0.8} \text{Pr}^{0.33} \tag{9.222}$$

For methane in air at 750 K, the Schmidt and Prandtl numbers are $\text{Sc} = 0.449$ and $\text{Pr} = 0.702$, respectively. These values in turn give $\text{Sh} = 28$ and $\text{Nu} = 32.4$ at $\text{Re} = 10,000$. From these values, we calculate the mass and heat transfer coefficients:

$$k_m = 4.76 \quad \text{and} \quad h = 1781$$

Note that these values are about an order of magnitude higher than the laminar flow case. Substitution into Equations 9.209 and 9.210 gives

$$\left(0.01 - Y_{CH4,S}\right) = 3.72 \times 10^6 \eta \exp\left(\frac{-15,757}{T_S}\right) Y_{CH4,S}^{0.72} \tag{9.223}$$

$$\left(T_S - 750\right) = 1.28 \times 10^{11} \eta \exp\left(\frac{-15,757}{T_S}\right) Y_{CH4,S}^{0.72} \tag{9.224}$$

These two equations, plus Equations 9.207 and 9.208 are a system of four nonlinear equations which can be solved to give

$$Y_{CH4,S} = 0.0099, \quad T_S = 752 \text{ K}, \quad \phi = 0.8904, \quad \text{and} \quad \eta = 0.6631$$

The global effectiveness factor is $\eta_G = 0.696$. The results for the two cases are given below.

Re	200	10,000
Y_b	0.01	0.01
Y_S	0.008	0.0099
T_b	750 K	750 K
T_S	789 K	752 K
η	0.506	0.663
η_G	1.22	0.696

In summary, increasing the flow rate reduces the external mass and heat transfer resistances, with the result that the surface conditions are closer to the bulk conditions. However, in both cases, the effectiveness factor, η, is low and significant errors would occur in a modeling study if this was not considered.

9.5 Summary

This chapter has considered the internal and external heat and mass transfer effects related to solid-catalyzed fluid reactions. There is usually some degree of transfer resistance associated with these systems, which results in a change in the reaction rates from those predicted by the intrinsic kinetic model. It is important to quantify the magnitude of these effects, either to know if they are small enough to be safely ignored or must be included in the analysis. Failure to account properly for these resistances can lead to a large discrepancy in predicted and observed conversions.

PROBLEMS

9.1 A series reaction $A \xrightarrow{k_1} B \xrightarrow{k_2} C$ occurs over a solid catalyst. The temperature of the bulk gas is 500 K and that of the catalyst surface is 525 K. The mass transfer coefficients for all of the species are equal to 1.25 m/s and the kinetic rate constants are

$$k_1 = 7.75 \times 10^6 \, \exp\left(\frac{-10{,}000}{T}\right) \text{m/s} \quad k_2 = 2.375 \times 10^4 \, \exp\left(\frac{-7541}{T}\right) \text{m/s}$$

At a certain point in the reactor, the concentration of A in the bulk divided by the concentration of B in the bulk is equal to 2. The selectivity for this reaction is defined as the net rate of formation of B divided by the net rate of formation of C. Calculate the selectivity at the bulk conditions and the selectivity at the surface conditions.

9.2 Consider the fully developed laminar flow of a gas mixture down a cylindrical duct which has a catalyst coated on the inside wall, as shown in Figure 9.P2. The Nusselt and Sherwood numbers have a value of 4.0. The gas is a mixture of propane in air, which is oxidized by the catalyst at the surface, according to

$$C_3H_8 + 5O_2 \rightarrow 3CO_2 + 4H_2O$$

The reaction rate expressed in terms of the catalyst external surface area is

$$\left(-r_{C_3H_8}\right) = 2.40 \times 10^5 \, \exp\left(\frac{-89{,}791}{R_g T}\right) C_{C_3H_8} \text{ mol/m}^2\text{s}$$

For a gas at 1 bar pressure and 700 K that is 0.1 mol% propane (average bulk value):

a. Assume isothermal operation and calculate the observed reaction rate for these conditions.

b. Assume nonisothermal operation and calculate the temperature and concentration at the surface.

Data:

Bulk diffusion coefficient of propane in air is $5 \times 10^{-5} \text{ m}^2/\text{s}$.
Enthalpy of reaction is $(-\Delta H_R) = 2044$ kJ/mol.
Thermal conductivity of air is 0.055 W/m · K.

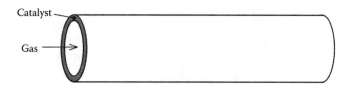

FIGURE 9.P2
Tube wall reactor used in Problem 9.2.

The characteristic length that is used in the Sherwood and Nusselt numbers is the duct diameter, which is 1×10^{-3} m.

Ignore internal diffusion resistance and assume that the properties are not functions of temperature

9.3 Consider a porous catalyst pellet with a diameter of 6 mm. The porosity of the pellet is 0.55, the tortuosity factor is 4.0, the internal surface area is 200 m²/g, and the pellet density is 2.0 g/cm³. The following first-order reaction occurs in the pellet at a temperature of 475 K:

$$A \rightarrow B \quad \text{where } (-r_A) = kC_A$$

The rate constant at the reaction temperature has a value of 0.1 s⁻¹. The molecular weight of A and B is 16 g/mol. The bulk diffusion coefficient of A in the reacting mixture is 5×10^{-5} m²/s. Determine the value of the effectiveness factor in the pellet.

9.4 Consider the platinum-catalyzed oxidation of carbon monoxide:

$$CO + \frac{1}{2}O_2 \rightleftharpoons CO_2$$

Take a catalyst in the form of a flat plate with a thickness of 40 μm with diffusion from one face only, that is, the back face is impermeable to mass transfer. The catalyst is uniformly distributed throughout the catalyst. The effective diffusivity of CO in the catalyst is 5×10^{-6} m²s. The thermal conductivity of catalyst is 0.5 W/m · K. The reaction rate in terms of catalyst volume is given by an LHHW-type expression (Y is the mole fraction)

$$(-r_{CO}) = \frac{k_s Y_{CO} Y_{O_2}}{(1 + K_A Y_{CO})^2} \text{ mol/m}^3\text{s}$$

where

$$k_s = 9 \times 10^{15} \exp\left(-\frac{12,600}{T}\right) \quad \text{and} \quad K_A = 65.5 \exp\left(+\frac{961}{T}\right)$$

a. Prepare a plot of effectiveness factor versus external surface temperature, for a temperature range of 500–900 K. The surface boundary composition is 2% CO in air at 1 atm total pressure.

b. Determine the error involved in the plot if the assumption of isothermal catalyst is made. Compute the maximum temperature rise in the catalyst.

9.5 Consider a first-order irreversible reaction that occurs in a packed bed reactor

$$A \rightarrow B (-r_A) = kC_A$$

where C_A has units of mol/m^3. At a point in the reactor, the following conditions exist:

Superficial mass flow rate	$0.30 \ kg/m^2s$
Diameter of spherical catalyst pellets	$0.01 \ m$
Intrinsic rate constant k (based on catalyst solid volume)	$20 \ s^{-1}$
Catalyst pellet density	$2000 \ kg/m^3$
Molecular mass of A and B	50
Molecular mass of inert	32
Mole percent of inert	80%
Mole percent of A	12%
Total pressure	100 kPa
Temperature	400 K
Schmidt number of the gas	1.0
Viscosity of the gas	$3.0 \times 10^{-5} \ Pa \cdot s$
Effective diffusion coefficient in the catalyst	$1.0 \times 10^{-6} \ m^2/s$
Porosity of the bed	0.36

Compute the observed reaction rate at these conditions and express the result in terms of moles reacted per second per cubic meter of reactor volume.

9.6 Consider the reaction

$$A + B \rightarrow C + D$$

where B is present in excess and thus the rate is represented by

$$(-r_A) = kY_A^{0.72}$$

where Y_A is the mole fraction of A. Assume this rate is valid at atmospheric pressure. Consider that the reaction is occurring in a long rod of catalyst with a square cross section, with diffusion from all four faces (symmetry about the axis).

a. Plot the effectiveness factor as a function of Thiele modulus for the isothermal case.

b. Determine a single expression for η versus ϕ using the asymptote matching technique.

9.7 Heat transfer correlations for flow around spherical catalyst pellets sometimes have the form

$$Nu = 2.0 + a \ Re^n$$

At a Reynolds number of zero, the Nusselt number has a value of 2.0. Prove that the value 2.0 is the appropriate constant.

References

Bird, R.B., W.E. Stewart, and E.N. Lightfoot, 2002, *Transport Phenomena*, 2nd Ed., John Wiley & Sons, New York.

Finlayson, B.A., 1980, *Non-Linear Analysis in Chemical Engineering*, McGraw-Hill, New York.

Fogler, H.S., 2006, *Elements of Chemical Reaction Engineering*, 4th Ed., Prentice-Hall, Englewood Cliffs.

Froment, G.B, K.B. Bischoff, and J. De Wilde, 2011, *Chemical Reactor Analysis and Design*, 3rd Ed., Wiley, New York.

Kolaczkowski, S.T. and D. Worth, 1995, Modelling channel interactions in a non-adiabatic multichannel catalytic combustion reactor, *Catalysis Today*, **26**, 275–282.

Leung, D., R.E. Hayes, and S.T. Kolaczkowski, 1996, Diffusion limitation in the washcoat of a catalytic monolith reactor, *Canadian Journal of Chemical Engineering*, **74**, 94–103.

Reid, R.C., J.M. Prausnitz, and B.E. Poling, 1987, *The Properties of Gases and Liquids*, 4th Ed., McGraw-Hill, New York.

Satterfield, C.N., 1970, *Mass Transfer in Heterogeneous Catalysis*, MIT Press, Cambridge, Massachusetts.

Taylor, R. and R. Krishna, 1993, *Multicomponent Mass Transfer*, John Wiley & Sons, New York.

Thomas, J.M. and W.J. Thomas, 1967, *Introduction to the Principles of Heterogeneous Catalysis*, Academic Press, London.

Wakao, W. and J.M. Smith, 1962, Diffusion in catalyst pellets, *Chemical Engineering Science*, **17**, 825–834.

Weisz, P.B. and J.S. Hicks, 1962, The behaviour of porous catalyst particles in view of internal mass and heat diffusion effects, *Chemical Engineering Science*, **17**, 265–275.

Wheeler, A. in P.H. Emmett (ed.), 1955, *Catalysis*, Vol. II, Chapter 2, 105, Reinhold Publishing Corporation, New York.

10

Analysis of Catalytic Reactors

Chapter 7 gave an introduction to some of the types of catalytic reactor used by industry. There are an extremely wide variety of reactor types used, and it would not be possible even to mention them all, let alone discuss the conservation equations that govern their performance. The choice of reactor type depends on many variables, including the reaction rate, the heat effects, the catalyst selectivity, the phase of the process stream, the operating conditions (especially temperature and pressure), the longevity of the catalyst, and so on. It is not possible to give general rules for choosing an appropriate reactor for any given process. Indeed, even for the same reaction, it will be seen that different reactor types are often employed by different companies. In the following sections, the analysis of catalytic reactors will be discussed. We will present the packed-bed reactors in most detail, both because of their importance in industry and their relative ease of analysis. Following this detailed presentation, other types will be introduced at varying levels of complexity. As usual, if more detail is desired on these reactor types, reference books or the open literature should be consulted as necessary.

10.1 Packed-Bed Reactor: Introduction and Overview

It is probably fair to say that the majority of industrial chemical reactors are based on the fixed-bed concept. As a result, they have been the subject of much investigation, and discussions of packed-bed reactors can be found in most reaction engineering texts and also in review papers. A good example of the latter is Feyo de Azevedo et al. (1990). At its simplest, this reactor consists of a vessel, usually circular in cross section, which is filled with solid catalyst particles, through which the process stream flows. Packed beds are favored as a reactor type because of their relative simplicity, which makes them easier to design, build, and operate. In the sections that follow, the detailed conservation equations for the packed-bed reactor will be developed. For the moment, we focus on some general concepts that must be considered in design.

As with any fluid flow, a pressure drop results from the flow through the packed bed. Generally speaking, the pressure drop increases as the size of the particles decreases, everything else being equal. For operating reasons, the pressure drop consideration results in relatively large pellets being used in the reactor; typically of the order of 3–6 mm in size. As we have seen in Chapter 9, the internal diffusion limitation increases with increasing particle size, and in commercial scale reactors mass and heat transfer limitation are usually a fact of life, and therefore must be considered in design.

Another significant design decision is whether the reactor is to be operated adiabatically or with heat transfer through the reactor walls. Adiabatic operation is preferred from the standpoint of simplicity, both in construction and operation. This choice is determined by the magnitude of the enthalpy changes that occur with the reaction. If the enthalpy change of reaction is large, then there is a correspondingly large amount of thermal energy that is either released or absorbed. Therefore, there will be a significant change in the temperature of the process stream. If the reactor is operated adiabatically, there is a danger that the temperature may fall sufficiently far that the reaction rate becomes negligible (endothermic reaction), or that it rises so high that the catalyst is damaged, or equilibrium constraints become significant (exothermic reaction). In such a case, there are two options. The first is to use multiple adiabatic beds of catalyst, with heating or cooling of the process stream between beds. This concept was discussed in Chapter 4. The alternative is to have continuous addition or removal of heat along the length of the reactor. This concept was discussed in Chapter 4 for homogeneous reactors.

When a reactor is operated adiabatically, the constraints on the reactor diameter are fairly minimal. However, when it is desired to transfer heat with the process stream, the reactor diameter should be kept fairly small, typically of the order of 10 particle diameters or so. This constraint arises from the relatively small thermal conductivity of packed beds. Because the desired reactor volume is usually quite large, the solution is to use a large number of parallel reactors of small diameter. This concept was also mentioned in Chapter 4, Figure 4.8. It is not unusual to have 20,000 tubes in a multitubular reactor.

When it comes to the analysis of packed-bed reactors, there are a number of model choices to be made. In the following sections we start by presenting the momentum balance equations used for calculating the velocity and pressure drop in a packed bed. This presentation is followed by a discussion of the different forms of the mole and energy balance that can be employed. Both steady-state and transient models are presented.

10.2 Conservation Equations for Packed Beds

10.2.1 Momentum Balance in Packed Beds

In this section, the pressure/velocity relations for forced flow in a porous medium consisting of a packed bed are developed. In these porous media, there is bulk flow due to an imposed pressure gradient. A packed bed is a container filled with particles of catalyst. The particles are not bonded together, and are held in place by the walls of the vessel. A packed bed consists of two parts, (i) the solid matter and (ii) the void space, through which the gas flows. The *porosity*, ϕ, is defined as the ratio of void volume to the total volume.

$$\text{Bed porosity}, \phi = \frac{\text{Void volume in bed}}{\text{Total reactor volume}} \tag{10.1}$$

10.2.1.1 Velocities in Packed Beds

The velocity pattern in a packed bed is complex at the pellet or pore scale, because the fluid must wend its way through a tortuous pore network formed by the randomly packed particles. The time-averaged axial velocity profile at a macroscopic level is approximately flat in the central portion of the bed, and rises to a higher value in the boundary layer at the reactor wall. This rise near the wall is caused by local ordering of the particles induced by the presence of the wall (Schlünder, 1978). Usually this rise near the wall is ignored, and the profile is considered to be flat along the entire cross-section of the bed. In all of the models described here, an appropriate pressure/velocity equation must be used in conjunction with the mole and energy balances.

Two velocities are used to describe flow in porous media. The *interstitial* velocity, v, is the average velocity of the fluid in the void space. The *superficial* velocity, v_s, is the velocity based on the total cross-sectional area. This velocity is computed by dividing the volumetric flow rate of the fluid by the cross-sectional area of the bed, and is the velocity which would exist in the duct if the packing were absent. The superficial and interstitial velocities are related by

$$v_s = \phi v \tag{10.2}$$

The superficial velocity is commonly used to define the Reynolds number in packed beds.

10.2.1.2 One-Dimensional Flow

The macroscopic flow in a packed bed is one-dimensional if all of the flow is parallel to the axis. The simplest one-dimensional equation for describing the fluid flow in a porous medium was proposed by Darcy (1856), and is called Darcy's law. The linear relationship between velocity and pressure drop for Newtonian fluids is

$$-\frac{dP}{dz} = \frac{\mu}{K} v_s \tag{10.3}$$

K is the permeability of the medium, and is a measure of the ease with which a fluid may pass through it. This linear relationship between pressure drop and velocity is only valid at low flow rate (Reynolds numbers being less than one). At higher flow rate, fluid inertial effects become important. Darcy's law was extended by Forchheimer (1901) to account for these inertial effects. Equations of the Forchheimer type have the following general form:

$$-\frac{dP}{dz} = \left(\frac{\mu}{K} + \frac{C\rho}{\sqrt{K}} v_s \right) v_s \tag{10.4}$$

The constant, C, may assume various values depending on the type of medium.

The permeability of porous media is usually measured experimentally, although some empirical correlations exist for packed beds. A commonly used permeability for packed beds of uniformly sized smooth spherical particles comes from the Blake–Kozeny friction factor analysis, supplemented with experimental results (Bird et al., 2002), which gives

$$K = \frac{D_p^2 \phi^3}{150(1 - \phi)^2} \tag{10.5}$$

The equivalent particle diameter, D_p, is defined as

$$D_p = \frac{6}{a_v} = 6\frac{V}{A} \tag{10.6}$$

a_v is the ratio of particle surface area, A, to volume, V. For a sphere, D_p is the particle diameter.

Example 10.1

Calculate the equivalent diameter for a cylindrical pellet, 4 mm long and 3 mm in diameter.

SOLUTION

The surface area of a cylinder consists of the area of the curved face plus the area of the two end surfaces. The area of the curved surface is

$$2\pi RL = (2)(3.1416)(1.5 \text{ mm})(4 \text{ mm}) = 37.7 \text{ mm}^2$$

The area of each end is given by

$$\pi R^2 = (3.1416)(1.5 \text{ mm})^2 = 7.07 \text{ mm}^2$$

The total surface area is

$$A = 37.7 + 2 \times 7.07 = 51.84 \text{ mm}^2$$

The volume of the cylinder is

$$V = \pi LR^2 = (3.1416)(4 \text{ mm})(1.5 \text{ mm})^2 = 28.27 \text{ mm}^3$$

Therefore,

$$a_v = 51.84 \text{ mm}^2 / 28.27 \text{ mm}^3 = 1.833 \text{ mm}^{-1}$$

and

$$D_p = 6/a_v = 6/1.833 = 3.27 \text{ mm}$$

The equivalent diameter lies between the values of the particle diameter and length.

10.2.1.3 Ergun Equation

A classical value for the constant C in Equation 10.4 is

$$C = \frac{0.143}{\phi^{1.5}} \tag{10.7}$$

Substitution of this value and the Blake–Kozeny permeability (from Equation 10.5) into Equation 10.4 gives an equation for the differential pressure drop:

$$-\frac{dP}{dz} = \frac{v_s}{D_p} \frac{(1-\phi)}{\phi^3} \left[\frac{150\mu(1-\phi)}{D_p} + 1.75\rho v_s \right] \tag{10.8}$$

Equation 10.8 is the Ergun equation (Ergun, 1952), and is widely used for pressure drop prediction in packed beds. The constants 150 and 1.75 were determined by experiment, and other values have been proposed. MacDonald et al. (1979) suggested values of 180 and 1.8 for the two constants. Note that these constants only apply for randomly packed beds of uniformly sized smooth spherical particles. The values depend on surface roughness, size distribution, shape, and packing homogeneity. In critical applications, it is best to determine the permeability of the bed experimentally. This measurement is performed by measuring the pressure drop across the packed bed as a function of the flow rate over a wide range of Reynolds number.

The Reynolds number in a packed bed is usually defined in terms of the equivalent particle diameter and the superficial velocity, namely,

$$\text{Re}_b = \frac{D_p \rho v_s}{\mu} \tag{10.9}$$

The reader should note that other definitions of the Reynolds number are also used in packed-bed correlations.

The Ergun and MacDonald equations are only valid for Reynolds numbers <300. Above this value, they will predict higher pressure drops than are observed experimentally. Other pressure drop equations have been proposed for packed beds (e.g., Hayes et al., 1995).

10.2.1.4 Deriving the Ergun Equation

Permeability relationships can be derived using a model microstructure for the porous medium. In general, flow in porous media is modeled using one of two types of structure models:

1. *Capillary models.* The bed is viewed as a network of interconnected capillaries, of varying cross-section.
2. *Swarm of particles.* The bed is viewed as a swarm of submerged particles around which the fluid moves.

The Ergun equation was developed using a capillary model. To derive the equation, we start from the definition of the friction factor for flow in a tube, which is

$$f = \frac{1}{4} \frac{D}{L} \left(\frac{P_0 - P_L}{\frac{1}{2} \rho v_m^2} \right) \tag{10.10}$$

where v_m is the average velocity. By analogy, we can define a friction factor for a packed bed:

$$f = \frac{1}{4} \frac{D_P}{L} \frac{P_0 - P_L}{\left(\frac{1}{2} \rho v_s^2 \right)} \tag{10.11}$$

where v_s is the superficial velocity and D_P is the equivalent particle diameter. For laminar flow in a circular tube of diameter D, the mean velocity is given by the Hagan–Pouiseile equation:

$$v_m = \frac{(P_0 - P_L)D^2}{32\mu L} \tag{10.12}$$

The packed bed can be treated like a tube of complex cross-section, with a hydraulic diameter, D_H, defined as

$$D_H = 4\frac{\text{Cross-section for flow}}{\text{Wetted perimeter}} = 4\frac{\text{Volume available for flow}}{\text{Total wetted surface}} \tag{10.13}$$

Divide the top and bottom by the total volume of the bed:

$$D_H = 4\frac{(\text{Volume of voids})/(\text{Volume of bed})}{(\text{Wetted surface})/(\text{Volume of bed})} = 4\frac{\phi}{a_m} \tag{10.14}$$

For a single particle, we have defined the ratio of the particle area to the volume as

$$a_v = \frac{\text{Particle surface area}}{\text{Volume of particle}} \tag{10.15}$$

To relate the wetted surface to the volume of the bed, we can use the porosity, thus

$$a_m = a_v(1 - \phi) \tag{10.16}$$

Recalling from Equation 10.6, the equivalent particle diameter is

$$D_P = \frac{6}{a_v} \tag{10.17}$$

The Hagan–Poiseuille flow in the packed bed can be described in terms of D_H by substitution into Equation 10.12:

$$v_m = \frac{(P_0 - P_L)D_H^2}{32\mu L} \tag{10.18}$$

In terms of the superficial velocity:

$$v_s = \frac{(P_0 - P_L)D_H^2\phi}{32\mu L} \tag{10.19}$$

Substitute for D_H from Equation 10.14:

$$v_s = \frac{(P_0 - P_L)}{2\mu L a_m^2}\phi^3 \tag{10.20}$$

Substitute for a_m and a_v in succession:

$$v_s = \frac{(P_0 - P_L)\phi^3}{2\mu L a_v^2 (1 - \phi)^2} = \frac{(P_0 - P_L)}{2\mu L}\left(\frac{D_P}{6}\right)^2 \frac{\phi^3}{(1 - \phi)^2} \tag{10.21}$$

Simplification gives an expression for the superficial velocity:

$$v_s = \left(\frac{P_0 - P_L}{L}\right)\frac{D_P^2}{72\mu}\frac{\phi^3}{(1 - \phi)^2} \tag{10.22}$$

This equation is valid for straight capillaries. In a packed bed, the flow pathways are not straight, but rather have a tortuous nature. This tortuous flow has a longer fluid path, and hence a higher pressure drop. These effects are compensated for by the addition of a tortuosity factor. After introducing an experimentally determined tortuosity factor, the equation for pressure drop in a packed bed with laminar flow is

$$v_s = \left(\frac{P_0 - P_L}{L}\right)\frac{D_P^2}{150\mu}\frac{\phi^3}{(1 - \phi)^2} \tag{10.23}$$

This equation is known as the Blake–Kozeny equation and is valid for

$$\frac{D_P \rho v_s}{\mu(1 - \phi)} < 10 \tag{10.24}$$

The packed-bed friction factor can be computed by substituting v_s into the definition of friction factor for a packed bed, Equation 10.11.

$$f = \frac{(1 - \phi)^2}{\phi^3} \cdot \frac{75\mu}{D_P \rho v_s} \tag{10.25}$$

Rearrange and substitute for the definition of the Reynolds number:

$$f\,\mathrm{Re}_b = \frac{(1 - \phi)^2}{\phi^3} \cdot 75 \tag{10.26}$$

For high Reynolds turbulent flow, the friction factor becomes constant. To develop a relation for high Reynolds flow, we start from the friction factor for an empty tube:

$$\frac{P_0 - P_L}{L} = \frac{1}{D}\frac{1}{2}\rho v_m^2 4 f_0 \tag{10.27}$$

Substitute the packed parameters as before:

$$D_P = \frac{6}{a_v}; \quad v_s = \phi v_m; \quad D_H = 4\frac{\phi}{a}; \quad a_m = a_v(1 - \phi) \tag{10.28}$$

The following result is obtained for the pressure drop:

$$\frac{P_0 - P_L}{L} = 6 f_0 \frac{1}{D_P} \left(\frac{1}{2} \rho v_s^2 \right) \frac{1 - \phi}{\phi^3} \tag{10.29}$$

The term $6 f_0$ was determined experimentally to be 3.5, giving the Burke–Plummer equation:

$$\frac{P_0 - P_L}{L} = 3.5 \frac{1}{D_P} \left(\frac{1}{2} \rho v_s^2 \right) \frac{1 - \phi}{\phi^3} \tag{10.30}$$

Substitution into the definition of the packed-bed friction factor gives

$$f = 0.875 \frac{1 - \phi}{\phi^3} \tag{10.31}$$

The Blake–Kozeny and the Burke–Plummer equations give the pressure drop for the low and high Reynolds number regions. It has been shown that simple addition of these two equations gives a relationship that is applicable over the entire range of Reynolds number.

$$\frac{P_0 - P_L}{L} = \frac{150 \mu v_s}{D_P^2} \frac{(1 - \phi)^2}{\phi^3} + \frac{1.75 \rho v_s^2}{D_P} \frac{1 - \phi}{\phi^3} \tag{10.32}$$

Equation 10.32 is the Ergun equation, which can be written in differential form:

$$-\frac{dP}{dz} = \frac{v_s}{D_P} \left(\frac{1 - \phi}{\phi^3} \right) \left[150 \frac{\mu (1 - \phi)}{D_P} + 1.75 \rho v_s \right] \tag{10.33}$$

10.2.1.5 Continuity Equation

Darcy's law and the Forchheimer-type equations relate pressure drop to superficial velocity. For incompressible fluid flow (most liquids can be assumed incompressible at low-to-moderate pressure), the density of the fluid is constant. However, when the fluid is a compressible gas, the density depends on the temperature and the pressure. The equation of continuity can be applied to one-dimensional flow in a packed tube of constant cross-sectional area:

$$\frac{d(\rho v_s)}{dz} = 0 \tag{10.34}$$

It is seen that the velocity must change if density changes, that is, an increase in density must have a concomitant decrease in velocity to preserve the equality. There are various ways to rewrite the pressure drop/velocity equations to reflect the change in density with pressure or temperature. One approach is to use the *superficial mass velocity*, denoted G, which is defined as the mass of fluid flowing through a unit cross-section of bed per unit time (i.e., the product of density and superficial velocity). In SI units, this quantity would be measured in $kg/(m^2 \, s)$. Because of the conservation of mass, in a one-dimensional flow the superficial mass velocity is constant at every location in the bed. That is,

$$\frac{d(\rho v_s)}{dz} = \frac{dG}{dz} = 0 \tag{10.35}$$

Consider the flow of an ideal gas in a packed bed. The mass density of the gas can be expressed in terms of the ideal gas law, as follows:

$$\rho = \frac{PM_m}{R_g T} \tag{10.36}$$

Substitution of this definition of density into the Ergun equation gives

$$-\frac{dP}{dz} = \frac{GR_g T}{PM_m D_p} \frac{(1-\phi)}{\phi^3} \left[\frac{150\mu(1-\phi)}{D_p} + 1.75G \right] \tag{10.37}$$

If we consider a packed bed through which a nonreacting gas is flowing at constant temperature, the only variables in the Ergun equation are the pressure and axial distance. We can combine all of the parameter values into a constant, E, defined as

$$E = \frac{GR_g T}{M_m D_p} \frac{(1-\phi)}{\phi^3} \left[\frac{150\mu(1-\phi)}{D_p} + 1.75G \right] \tag{10.38}$$

The final equation that defines the pressure profile is then:

$$-\frac{dP}{dz} = \frac{E}{P} \tag{10.39}$$

The solution to this simple differential equation is

$$P^2 = P_0^2 - 2Ez \tag{10.40}$$

where P_0 is the inlet pressure (i.e., the pressure at $z = 0$). This equation defines a quadratic pressure distribution. Sometimes the pressure profile in a packed bed is (incorrectly) computed using the gas density at the bed inlet. This assumption is equivalent to substituting the inlet pressure into Equation 10.37 to give a linear pressure profile:

$$P = P \frac{E}{P_0} z \tag{10.41}$$

Generally, the temperature and viscosity are also a function of z, so the quadratic pressure profile will not apply exactly. Moreover, if there is a change in the number of moles on reaction, the average molar mass of the mixture will also change as the fluid moves through the reactor.

Example 10.2

Consider a packed bed with a 10 cm inside diameter, length of 2 m, packed with 3 mm diameter smooth spheres. An air stream at an approach velocity of 1 m/s enters the bed at a temperature of 500 K and 300 kPa pressure. For isothermal operation, calculate the pressure at the bed outlet and the superficial velocity. The bed porosity is 0.36. The molar mass of air is 28.96.

SOLUTION

The first step is to calculate the bed Reynolds number to determine which bed pressure drop equation should be used. This calculation is based on the superficial velocity or superficial mass flow rate in the bed. The approach velocity is the velocity of the nitrogen before it enters the bed, which is therefore the superficial velocity at the bed inlet. The superficial velocity will change as the pressure drops through the bed, so we shall work in terms of superficial mass velocity. The gas density is computed from the ideal gas law. Therefore, the density is given by

$$\rho = \frac{PM_m}{R_g T} = \frac{(300{,}000 \text{ Pa})(28.96 \text{ g/mol})(0.001 \text{ kg/g})}{(8.314 \text{ J/mol} \cdot \text{K})(500 \text{ K})} = 2.09 \text{ kg/m}^3 \qquad (10.42)$$

The superficial mass velocity is the product of the density and the approach velocity. Thus, $G = 2.09P^2 = P_0^2 - 2Ez \text{ kg/m}^2 \cdot \text{s}$. The viscosity of air at 500 K and atmospheric pressure is obtained from Incropera et al. (2006). The viscosity is not a strong function of pressure, so this value was used throughout the calculation. The Reynolds number is

$$\text{Re}_b = \frac{\rho v_s D_p}{\mu} = \frac{G D_p}{\mu} = \frac{(2.09(\text{kg/m}^2\text{s}))(0.003 \text{ m})}{2.577 \times 10^{-5} \text{ Pa} \cdot \text{s}} = 243 \qquad (10.43)$$

which is <300, so the Ergun equation is valid. Substituting into Equation 10.37 give

$$-\frac{dP}{dz} = \frac{(2.09 \text{ kg/m}^2\text{s})(8.314 \text{ J/mol} \cdot \text{K})(500 \text{ K})}{P(0.003 \text{ m})(0.02896 \text{ kg/mol})} \frac{(1 - 0.36)}{0.36^3}$$

$$\times \frac{150(2.577 \times 10^{-5} \text{ Pa} \cdot \text{s})(1 - 0.36)}{0.003 \text{ m}} + 1.75(2.09 \text{ kg/m}^2 s) \qquad (10.44)$$

Performing the arithmetic and integrating give the result:

$$P^2 = P_0^2 - \left(1.2 \times 10^{10} \frac{\text{Pa}^2}{\text{m}}\right) z \qquad (10.45)$$

The inlet pressure is 300,000 Pa. Therefore, at $z = 2$, the pressure is

$$P = \left[\left(300,000\right)^2 - \left(1.2 \times 10^{10} \frac{\text{Pa}^2}{\text{m}}\right) 2\,\text{m} \right]^{\frac{1}{2}} = 256,905\,\text{Pa} \tag{10.46}$$

The outlet density can be calculated from the ideal gas law:

$$\rho = \frac{PM_m}{R_g T} = \frac{\left(256,095\ \text{Pa}\right)\left(28.96 \times 10^{-3}\ \text{kg/mol}\right)}{\left(8.314\ \text{mol J/mol} \cdot \text{K}\right)\left(500\ \text{K}\right)} = 1.78\ \text{kg/m}^3 \tag{10.47}$$

The total molar flow rate and the temperature are constant. Therefore,

$$Q_v P = v_s A_C P = \text{Constant} \tag{10.48}$$

The outlet velocity is therefore given by

$$v_s = \frac{\left(P v_s\right)_0}{P} = \frac{(300)(1)}{256.9} = 1.17\ \text{m/s} \tag{10.49}$$

For comparison purposes, we can calculate the pressure drop in the bed using the inlet density. Substitution into Equation 10.41 gives

$$P = P - \frac{E}{P_0} z = 300,000\ \text{Pa} - \frac{1.2 \times 10^{10}(\text{Pa}^2/\text{m})}{300,000\ \text{Pa}} 2\,\text{m} = 260,000\ \text{Pa} \tag{10.50}$$

The outlet pressure of 260,000 Pa is about 1% higher than the value computed previously. This is insignificant in the case, although in other instances the difference may be more significant.

The Reynolds numbers at which most reactors operate are within the range of the Ergun equation. It should be noted that although the catalyst particles in packed beds are often of uniform size, they are not always spherical in shape.

10.2.2 Mole and Energy Balances in Packed Beds

Because packed-bed reactors are so widely used, it is not surprising to find a large number of models for them in the literature. In the following sections, models based on the Fickian approach are described. Such models are also called dispersion models or dispersed plug flow models. The cell model approach, in which the reactor is approximated as a series of mixed reactors, is described by Feyo et al. (1990), among others.

Both one- and two-dimensional models have been used to model packed-bed reactors. The 1D model assumes that there are no gradients of temperature or concentration in the radial direction, and considers only the axial variations. In adiabatic reactors, this approximation is usually valid. The lack of heat transfer through the wall combined with the essentially flat velocity profile implies that there is no mechanism for radial gradients of temperature and concentration to be initiated so, at the macroscopic level, the radial gradients will not exist. The only exception will be the thin boundary layer at the wall but, in most adiabatic reactors, this layer will contribute only a minor error unless the reactor diameter is less than about 5 particle diameters. For nonadiabatic reactors, the situation is different. The thermal conductivity of a typical packed bed is low and, therefore, the removal of heat through the wall causes radial temperature gradients to form. Because the reaction rate is an exponential function of temperature, the reaction rates can vary significantly along a radius, causing concentration gradients to develop in the radial direction as well. In this case, a 2D model is more appropriate than a 1D model. Because 2D models are much more complex than 1D models, sometimes the nonadiabatic reactor is also modeled using a 1D model. This is an approximation because it ignores the very real gradients. It may not lead to significant error, however, depending on the relative resistances to heat transfer in the reactor. If the thermal conductivity of the bed is high compared to the overall heat transfer coefficient at the reactor wall, most of the resistance to heat transfer in the reactor will be located at the wall, and the resulting temperature profile in the reactor may not differ significantly from a flat profile.

Packed-bed reactor models may be classified as pseudo-homogeneous or heterogeneous. The pseudo-homogeneous model treats the reactor bed as a continuous phase (i.e., it assumes that the solid and fluid temperatures are the same at any given point in the reactor). The heterogeneous model accounts separately for the solid and fluid phases, with heat and mole balance equations being written explicitly for each phase. The fluid and solid equations are coupled through mass and heat transfer coefficients. This model allows for temperature and concentration differences between the fluid and solid phases. In many industrial applications, it has been shown that the pseudo-homogeneous model gives a reasonable prediction of reactor performance. However, reactions with high heats of reaction may give significant differences between fluid and solid conditions, which makes the use of the heterogeneous model preferable. In transient modeling of packed-bed reactors, greater differences between fluid and solid conditions can exist, even under mild operating conditions and, in this instance, the heterogeneous model is also more appropriate.

10.3 One-Dimensional Steady-State Plug Flow Models

The simplest one-dimensional model considers only transport by bulk flow in the axial direction in plug flow. Axial diffusion/dispersion of both mass and heat are thus ignored. Both pseudo-homogeneous and heterogeneous models have been proposed, with the former being the simpler. Both types are developed in the following sections. We start with the adiabatic case and then move to the nonadiabatic reactor. Steady-state models will be developed for all of these cases.

10.3.1 Pseudo-Homogeneous Model for an Adiabatic-Packed Bed

The mole and energy balance equations are developed by performing shell balances on an elemental reactor volume element, ΔV, as shown in Figure 10.1. The molar flow rate of reactant A is denoted F_A and the enthalpy of the process stream is denoted as H. In a catalytic reactor, both homogeneous and catalytic reactions may occur. In the following analysis, the homogeneous reaction is neglected.

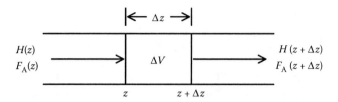

FIGURE 10.1
Elemental reactor volume element used for developing the mole and energy balance equations for an adiabatic-packed bed.

10.3.1.1 Steady-State Mole Balance Equation

The mole balance in words is

$$\begin{bmatrix} \text{Moles} \\ \text{in} \end{bmatrix} - \begin{bmatrix} \text{Moles} \\ \text{out} \end{bmatrix} - \begin{bmatrix} \text{Moles reacted} \\ \text{in catalytic} \\ \text{reaction} \end{bmatrix} = 0 \tag{10.51}$$

here moles reacted include both homogeneous and catalytic reaction. Consider a reactant denoted by A. The change in the molar flow rate across the volume element is

$$\begin{bmatrix} \text{Moles} \\ \text{in} \end{bmatrix} - \begin{bmatrix} \text{Moles} \\ \text{out} \end{bmatrix} = (F_A)_z - (F_A)_{z+\Delta z} = -\Delta F_A \tag{10.52}$$

The mole balance on the control volume requires a value for the rate of reaction in terms of reactor volume, $\text{mol/m}^3\text{s}$. In a catalytic, it is common to quote the reaction rate in terms of the catalyst mass. Typical units are mol/(kgcat·s). The reaction rate is then related to the catalyst volume through the density. Two densities are important, the density of the solid catalyst and the bulk density of the bed:

$$\text{Bulk density, } \rho_b = \frac{\text{Catalyst mass}}{\text{Reactor volume}} \tag{10.53}$$

$$\text{Catalyst density, } \rho_C = \frac{\text{Mass of a catalyst particle}}{\text{Volume of a catalyst particle}} \tag{10.54}$$

The relationships are

$$\rho_b \left(-r_A\right)_{\text{Mass basis}} = \rho_C(1 - \phi)\left(-r_A\right)_{\text{Mass basis}} = \left(-r_A\right)_{\text{Volume basis}} \tag{10.55}$$

As usual, it is necessary to pay close attention to units and the basis for specifying the reaction rate. In the following equations, the catalytic reaction rate has units of mol/(kgcat·s). Using this basis, the loss of reactant in the elementary volume due to catalytic reaction is

$$\begin{bmatrix} \text{Moles reacted in} \\ \text{catalytic reaction} \end{bmatrix} = \eta(-R_A)_S \Delta V \rho_b \qquad (10.56)$$

The catalytic reaction rate is here evaluated at the temperature and concentration at the surface of the catalyst pellets, signified by the subscript S. Therefore, this rate must be multiplied by the effectiveness factor, η. The steady-state mole balance then becomes:

$$-\Delta F_A - \eta(-R_A)_S \Delta V \rho_b = 0 \qquad (10.57)$$

Simplify by writing the elemental volume in terms of incremental reactor length, as follows:

$$\Delta V = A_C \Delta z \qquad (10.58)$$

A is the cross-sectional or flow area. Equation 10.57 can thus be written as

$$-\Delta F_A - \eta(-R_A)_S A_C \Delta z \rho_b = 0 \qquad (10.59)$$

Dividing through by $A_C \Delta z$ gives

$$-\frac{1}{A_C} \frac{\Delta F_A}{\Delta z} - \eta(-R_A)_S \rho_b = 0 \qquad (10.60)$$

When the volume element becomes small, it approaches the differential volume element and the final mole balance equation is

$$-\frac{1}{A_C} \frac{dF_A}{dz} - \eta(-R_A)_S \rho_b = 0 \qquad (10.61)$$

This mole balance can also be written in terms of superficial bed velocity. The molar flow rate can be related to the volumetric flow rate and hence the velocity, as follows:

$$F_A = Q_V C_A = v_s A_C C_A = v_s A_C C Y_A \qquad (10.62)$$

If we then substitute and simplify, and assume that the change in total molar flow rate with reaction is negligible, we obtain the final result:

$$-C v_s \frac{dY_A}{dz} - \eta(-R_A)_S \rho_b = 0 \qquad (10.63)$$

Equation 10.63 is a first-order initial value problem and requires that the values of Y_A be known at $z = 0$ in order for a solution to be obtained. We emphasize that the total molar concentration, C, as well as the component concentrations depend on temperature and pressure.

10.3.1.2 Steady-State Energy Balance Equation

The energy balance for the volume element must account for changes in energy of the flowing mixture. The energy balance is written in terms of enthalpy changes resulting from changes in temperature and composition (owing to reaction). Changes in kinetic and potential energy are ignored. For an adiabatic reactor, there is no heat transfer to the surroundings. The energy balance over ΔV is

$$\begin{bmatrix} \text{Enthalpy} \\ \text{in} \end{bmatrix} - \begin{bmatrix} \text{Enthalpy} \\ \text{out} \end{bmatrix} = 0 \tag{10.64}$$

The enthalpy balance may then be written as

$$-Q_V \rho C_P \Delta T - \Delta V \rho_b \eta \left(-R_A \right)_S \Delta H_R = 0 \tag{10.65}$$

Take the limit as $\Delta V \to dV$ and express the result using the reactor length. The differential volume is

$$dV = A_C \, dz \tag{10.66}$$

where A_C is the cross-sectional area of the reactor. Thus we obtain:

$$-\frac{Q_V}{A_C} \rho C_P \frac{dT}{dz} - \rho_b \eta \left(-R_A \right)_S \Delta H_R = 0 \tag{10.67}$$

We can express the equation in terms of the superficial velocity:

$$-v_s \rho C_P \frac{dT}{dz} - \rho_b \eta \left(-R_A \right)_S \Delta H_R = 0 \tag{10.68}$$

If multiple reactions occur, the energy balance becomes:

$$-v_s \rho C_P \frac{dT}{dz} - \rho_b \sum_{i=1}^{n} \eta_i \left(-R_i \right)_S \left(\Delta H_R \right)_i = 0 \tag{10.69}$$

The superficial mass velocity $(G = \rho v_s)$ is constant. Recall that the heat capacity, C_P, of the reacting stream is a function of both temperature and composition. The temperature at $z = 0$ must be known before Equation 10.69 may be solved.

Solution of the mole and energy balance equations gives the temperature and concentration profiles in the reactor. No allowance is made for a temperature difference between fluid and solid and, if this difference is thought to be important, the heterogeneous model described in the next section is appropriate.

10.3.2 Heterogeneous Model for an Adiabatic-Packed Bed

The pseudo-homogeneous model does not differentiate between the fluid and solid phases, and hence the temperature and concentration in each phase are the same. In some reactors, however, the external transport resistance causes significant differences in temperature and concentration between the two phases, and hence use of the pseudo-homogeneous model can lead to error in reactor simulation. The heterogeneous model treats the fluid and solid phases separately, and thus requires a mole and energy balance for each phase. The mole balance is performed on the small volume element shown in Figure 10.1. In describing the various heterogeneous packed-bed models, we use the subscript f to denote the fluid phase and S the solid (catalyst) phase.

10.3.2.1 Steady-State Mole Balance Equation for the Fluid

We start with the species balance for the gas phase. The catalytic reaction does not occur in the fluid phase, and therefore there is no loss of mass by reaction. Rather, the disappearance of a reactant from the fluid phase is represented as a loss by mass transfer to the solid, that is, the rate of disappearance of A by catalytic reaction is equal to the rate of transfer of species A from the fluid to the solid phase. The mole balance on the fluid is therefore:

$$
\begin{bmatrix} \text{Moles} \\ \text{in} \end{bmatrix} - \begin{bmatrix} \text{Moles} \\ \text{out} \end{bmatrix} - \begin{bmatrix} \text{Moles transported} \\ \text{to catalyst} \\ \text{surface} \end{bmatrix} = 0 \tag{10.70}
$$

The change in molar flow rate of A across the volume element is

$$
\begin{bmatrix} \text{Moles} \\ \text{in} \end{bmatrix} - \begin{bmatrix} \text{Moles} \\ \text{out} \end{bmatrix} = (F_A)_z - (F_A)_{z+\Delta z} = -\Delta F_A \tag{10.71}
$$

The moles transported to the catalyst surface is expressed in terms of a mass transfer coefficient:

$$
\begin{bmatrix} \text{Moles transported to} \\ \text{catalyst surface} \end{bmatrix} = k_m \Delta S C_f \left(Y_{A,f} - Y_{A,S} \right) \tag{10.72}
$$

where ΔS is the external catalyst surface area contained in the volume element. The steady-state mole balance then becomes

$$
-\Delta F_{A,f} - k_m \Delta S C_f \left(Y_{A,f} - Y_{A,S} \right) = 0 \tag{10.73}
$$

Dividing through by $\Delta V = A_C \Delta z$ gives

$$
-\frac{1}{A_C} \frac{\Delta F_{A,f}}{\Delta z} - k_m \frac{\Delta S}{\Delta V} C_f \left(Y_{A,f} - Y_{A,S} \right) = 0 \tag{10.74}
$$

The effective area for mass transfer, a_m, is the particle surface area per unit bed volume:

$$
a_m = \frac{\Delta S}{\Delta V} = \frac{S}{V} \tag{10.75}
$$

As the volume element becomes small, it approaches the differential volume element and the final mole balance equation is

$$-\frac{1}{A_C}\frac{dF_{Af}}{dz} - k_m a_m C_f \left(Y_{Af} - Y_{A,S}\right) = 0 \tag{10.76}$$

In terms of superficial velocity, the fluid phase mole balance equation is

$$-C_f v_s \frac{dY_{Af}}{dz} - k_m a_m C_f \left(Y_{Af} - Y_{A,S}\right) = 0 \tag{10.77}$$

The units of k_m are m/s. The inlet mole fraction of A must be known to solve this differential equation.

10.3.2.2 Steady-State Mole Balance Equation for the Solid Catalyst

The steady-state mole balance equation for species A in the solid catalyst is

$$\begin{bmatrix} \text{Moles transported to} \\ \text{catalyst surface} \end{bmatrix} = \begin{bmatrix} \text{Moles reacted in} \\ \text{catalytic reaction} \end{bmatrix} \tag{10.78}$$

The balance can be expressed in terms of the incremental surface area as

$$k_m \Delta S C_f \left(Y_{Af} - Y_{A,S}\right) = \eta \left(-R_A\right)_S \Delta V \rho_b \tag{10.79}$$

Using Equation 10.75, this can be simplified to

$$C_f k_m a_m \left(Y_{Af} - Y_{A,S}\right) - \left(-R_A\right)_S \rho_C \left(1 - \phi\right) = 0 \tag{10.80}$$

Note that this equation is not a differential equation.

10.3.2.3 Steady-State Energy Balance Equation for the Fluid

The enthalpy balance for the gas phase is a nonadiabatic energy balance because heat is transferred to the solid from the fluid. Thus,

$$\begin{bmatrix} \text{Enthalpy} \\ \text{in} \end{bmatrix} - \begin{bmatrix} \text{Enthalpy} \\ \text{out} \end{bmatrix} + \begin{bmatrix} \text{Heat added from} \\ \text{catalyst surface} \end{bmatrix} = 0 \tag{10.81}$$

The enthalpy of the stream changes as a result of temperature and composition changes. The enthalpy change over the volume element, ignoring viscous work is

$$\begin{bmatrix} \text{Enthalpy} \\ \text{in} \end{bmatrix} - \begin{bmatrix} \text{Enthalpy} \\ \text{out} \end{bmatrix} = Q_V \rho C_P \Delta T_f \tag{10.82}$$

The heat transfer from the solid to the gas depends on the heat transfer coefficient, h_{fs}:

$$\left[\begin{matrix} \text{Heat added from} \\ \text{catalyst surface} \end{matrix}\right] = h_{fs}\Delta S\left(T_S - T_f\right) \tag{10.83}$$

Therefore, the energy balance over ΔV is

$$-Q_V\rho C_P\Delta T_f + \Delta Sh_{fs}\left(T_S - T_f\right) = 0 \tag{10.84}$$

Dividing through by $\Delta V = A_C\,\Delta z$, note again that

$$a_m = \frac{\Delta S}{\Delta V} = \frac{S}{V} \quad \text{and} \quad \frac{Q_V}{A_C} = v_s \tag{10.85}$$

Substitution into Equation 10.84 gives

$$-v_s\rho C_P\frac{\Delta T_f}{\Delta z} + h_{fs}a_m\left(T_S - T_f\right) = 0 \tag{10.86}$$

Taking the limit as $\Delta V \rightarrow dV$, we obtain the ordinary differential equation:

$$-v_s\rho C_P\frac{dT_f}{dz} + h_{fs}a_m\left(T_S - T_f\right) = 0 \tag{10.87}$$

The reactor inlet temperature must be known to solve this initial value problem.

10.3.2.4 Steady-State Energy Balance Equation for the Solid Catalyst

The solid consists of the solid catalyst which is not flowing and the gas which fills the pore volume of the catalyst. We can consider the solid as an open system with diffusion of mass in and out. The gas reacts and therefore undergoes a change in enthalpy due to reaction. The enthalpy change of the gas must be balanced by the heat that is added to the solid, thus:

$$\left[\begin{matrix} \text{Enthalpy} \\ \text{in} \end{matrix}\right] - \left[\begin{matrix} \text{Enthalpy} \\ \text{out} \end{matrix}\right] + \left[\begin{matrix} \text{Heat added from} \\ \text{catalyst surface} \end{matrix}\right] = 0 \tag{10.88}$$

In the PFR model, heat is added to the solid by convective heat transfer from the fluid. The change in enthalpy of the gas is due to reaction. Thus,

$$\left[\begin{matrix} \text{Enthalpy} \\ \text{in} \end{matrix}\right] - \left[\begin{matrix} \text{Enthalpy} \\ \text{out} \end{matrix}\right] = -\rho_b\eta\left(-R_A\right)_S\Delta H_R \tag{10.89}$$

The energy transfer by convection is

$$\begin{bmatrix} \text{Heat added from} \\ \text{catalyst surface} \end{bmatrix} = h_{fs}a_m\left(T_S - T_f\right) = 0 \tag{10.90}$$

The energy balance equation is therefore

$$\rho_b\eta\left(-R_A\right)_S \Delta H_R + h_{fs}a_m\left(T_S - T_f\right) = 0 \tag{10.91}$$

Note that Equation 10.91 is not a differential equation.

Example 10.3

Consider a bed packed with cylindrical pellets 3.2 mm in diameter and 3.2 mm long. The pellet density is 900 kg/m³. The overall bed density is 600 kg/m³. Compute the value for a_m, the external pellet surface area per unit volume of bed.

SOLUTION

The external surface area of a single cylindrical pellet of length, L, and diameter, D, is

$$a_p = 2\frac{\pi D^2}{4} + \pi DL \tag{10.92}$$

Substitution of the numerical values gives an area of 4.83×10^{-5} m². Compute the mass of one pellet by multiplying the volume by the density:

$$m_p = \rho_c V_p = \rho_c\pi\frac{D^2}{4}L \tag{10.93}$$

The pellet density is 900 kg/m³, so the pellet mass is 2.316×10^{-5} kg. The surface area per unit mass of catalyst is $a_p/m_p = 2.08$ m²/kg. The bed density is given as 600 kg/m³. Typically the bed density is measured with the void space filled with air at atmospheric pressure. The mass of gas is negligible compared to the mass of solid, so take the bed density as the mass of solid per volume of bed. Thus multiply the surface area per mass of catalyst by the bed density to give

$$a_m = 2.08 \text{ m}^2/\text{kg} \times 600 \text{ kg/m}^3 = 1250 \text{ m}^{-1}$$

10.3.3 Pseudo-Homogeneous Model for a Nonadiabatic-Packed Bed

In a nonadiabatic reactor, there is heat transfer through the reactor wall. The energy equation for the fluid phase must be modified to account for this external heat transfer. The volume element for the nonadiabatic case is represented in Figure 10.2. Note that the main difference between Figures 10.2 and 10.1 is the addition of the convective energy transfer term at the wall.

For the pseudo-homogeneous model, the energy balance for a volume element, ΔV, given in Equation 10.64 for the adiabatic model is modified to include external energy transfer:

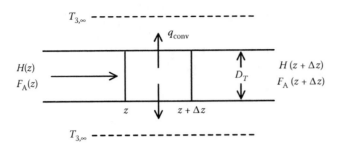

FIGURE 10.2
Elemental reactor volume element used for developing the mole and energy balance equations for a nonadiabatic-packed bed.

$$\begin{bmatrix} \text{Enthalpy} \\ \text{in} \end{bmatrix} - \begin{bmatrix} \text{Enthalpy} \\ \text{out} \end{bmatrix} + \begin{bmatrix} \text{Heat added from} \\ \text{surroundings} \end{bmatrix} = 0 \qquad (10.94)$$

The area for heat transfer in the elemental volume is the circumference of the channel multiplied by the length. That is,

$$\Delta A_T = \pi D_T \Delta z \qquad (10.95)$$

In Equation 10.95, D_T is the reactor diameter. The heat exchange with the surroundings depends on an overall heat transfer coefficient, which includes heat transfer at the inside reactor wall, outside reactor surface and conduction through the wall, and the driving force, which is the temperature difference between the fluid in the reactor and the surroundings. The heat loss from the elemental volume element is

$$\begin{bmatrix} \text{Heat added from} \\ \text{surroundings} \end{bmatrix} = q_{\text{conv}} = \pi D_T \Delta z U \left(T_{3,\infty} - T \right) \qquad (10.96)$$

This value is positive if heat is added to the reactor. U is the overall heat transfer coefficient, and is elaborated on shortly. $T_{3,\infty}$ is the temperature of the heat transfer fluid surrounding the reactor.

Addition of the external heat transfer term to the pseudo-homogeneous model energy balance for the adiabatic reactor, Equation 10.65 gives

$$-Q_V \rho C_P \Delta T - \Delta V \rho_b \eta \left(-R_A \right)_S \Delta H_R + \pi D_T \Delta z U \left(T_{3,\infty} - T \right) = 0 \qquad (10.97)$$

Simplify as for the adiabatic reactor model. The elemental volume is

$$\Delta V = \frac{\pi}{4} D_T^2 \Delta z \qquad (10.98)$$

Dividing Equation 10.97 by ΔV and simplifying give

$$-v_s \rho C_P \frac{dT}{dz} - \Delta H_R \rho_b \eta (-R_A)_S + \frac{4}{D_T} U \left(T_{3,\infty} - T \right) = 0 \qquad (10.99)$$

If multiple reactions occur, the energy balance becomes

$$-v_s \rho C_P \frac{dT}{dz} - \rho_b \sum_{i=1}^{n} \eta_i \left(-R_i \right)_S \left(\Delta H_R \right)_i + \frac{4}{D_T} U \left(T_{3,\infty} - T \right) = 0 \qquad (10.100)$$

10.3.4 Heterogeneous Model for a Nonadiabatic-Packed Bed

A similar modification of the energy balance equation can be applied to the heterogeneous model. In this case, the fluid phase energy balance equation must be adjusted. The energy balance for the fluid is

$$\begin{bmatrix} \text{Enthalpy} \\ \text{in} \end{bmatrix} - \begin{bmatrix} \text{Enthalpy} \\ \text{out} \end{bmatrix} + \begin{bmatrix} \text{Heat added} \\ \text{from catalyst} \\ \text{surface} \end{bmatrix} + \begin{bmatrix} \text{Heat added} \\ \text{from} \\ \text{surroundings} \end{bmatrix} = 0 \qquad (10.101)$$

Then we can proceed as before to obtain the result for a single reaction:

$$-v_s \rho C_P \frac{dT_f}{dz} + h_{fs} a_m \left(T_S - T_f \right) + \frac{4}{D_T} U \left(T_{3,\infty} - T_f \right) = 0 \qquad (10.102)$$

The other model equations are unchanged.

10.3.4.1 Heat Transfer Coefficient for Nonadiabatic Reactors

In a nonadiabatic reactor, the extent of the heat transfer with the surroundings is quantified using the heat transfer coefficient. The heat transfer between the reactor side fluid and the heat transfer fluid surrounding the reactor is comprised of three resistances, which can be combined using the resistance in series concepts. There is convection heat transfer between the reactor fluid and the inside wall, which is governed by the inside convection heat transfer coefficient, $h_{w,\text{eff}}$. There is conduction through the reactor wall that is proportional to the thermal conductivity of the wall, k_w. Finally, there is heat transfer between the outside reactor wall and the heat transfer fluid, governed by the outside heat transfer coefficient, h_o. The overall heat transfer coefficient, U, combines these three parameters. For a reactor tube of outside diameter D_o, the relationship is

$$\frac{1}{U} = \frac{1}{h_{w,\text{eff}}} + \frac{D_T}{2k_w} \ln \left(\frac{D_o}{D_T} \right) + \frac{D_T}{D_o} \frac{1}{h_o} \qquad (10.103)$$

The outside heat transfer coefficient may be estimated from standard heat transfer correlations. The best value for the effective inside heat transfer coefficient is still a matter of some debate, and various correlations have been proposed (e.g., De Wasch and Froment (1972), Froment et al. (2011), Schlünder (1978), Dixon and Cresswell (1979)). The model of Dixon and Cresswell (1979) is given in Section 10.6.

The heat transfer fluid may have a uniform temperature along the entire length of the reactor. This situation could exist if the heat transfer fluid is at its boiling point, such that the addition of energy to it causes a phase change at constant temperature. Alternatively, the temperature of the heat transfer fluid may vary along z. In this case, a separate energy balance must be written for the heat transfer fluid. Consider, for example, a tubular packed-bed reactor with reactants entering at $z = 0$ and leaving at $z = L$. The reactor is surrounded by an annular jacket through which the heat transfer fluid flows, parallel to the reactor. Assume that this fluid enters the annulus also at $z = 0$ and flows at some mass flow rate, m_{HTF}, with heat capacity, $C_{P, HTF}$. The temperature of the heat transfer fluid depends on the rate of heat transfer from the reactor, which in turn depends on the local temperature difference between the reactor and the heat transfer fluid. Thus,

$$\frac{\mathrm{d}T_{\mathrm{HTF}}}{\mathrm{d}z} = \frac{UD_T\pi}{m_{\mathrm{HTF}}C_{P,\mathrm{HTF}}}(T_{\mathrm{HTF}} - T) \tag{10.104}$$

The inlet heat transfer fluid temperature must be known. The temperature, T, in this equation is the reactor temperature, and this equation must be solved simultaneously with the reactor mole and energy balance equations.

10.4 One-Dimensional Steady-State Axial Dispersion Models

We have already observed that plug flow is an approximation to the real flow pattern and that the presence of concentration gradients will induce some axial mixing. In the case of a packed bed, the mixing is caused by molecular diffusion down a concentration gradient as well as larger scale fluid mixing induced by the particles. Both of these mixing effects act to reduce concentration and temperature gradients. Various types of models have been proposed to incorporate mixing effects into packed-bed models, with one of the popular models being the so-called Fickian models. In these types of models, the mixing is assumed to follow a Fickian type of relationship. Thus, the axial dispersive flux would be given by

$$J_A = -D_{ea}C\frac{\mathrm{d}Y_A}{\mathrm{d}z} \tag{10.105}$$

where D_{ea} is a dispersion coefficient for axial flow.

The dispersion of heat is assumed to follow Fourier's law. The dispersive heat flux is

$$q = -k\frac{\mathrm{d}T}{\mathrm{d}z} \tag{10.106}$$

The thermal conductivity, k, is an effective thermal conductivity that depends on conduction contributions, as well as local fluid mixing. There may also be a radiation component, which will be discussed in Section 10.7.3. When dealing with heat transfer, the situation is somewhat more complex than with mass transfer because transfer of heat also occurs through the solid phase. The form of k will therefore depend on whether a heterogeneous or pseudo-homogeneous model is used.

In the following sections, the pseudo-homogeneous and heterogeneous 1D dispersion models are introduced.

10.4.1 Pseudo-Homogeneous Axial Dispersion Model

The one-dimensional dispersion model for packed beds simply adds the dispersion term to the plug flow model to give the dispersed plug flow model.

10.4.1.1 Mole Balance Equation

The mole balance equation is derived using standard shell balance techniques. The result expressed for a single reaction is

$$\frac{d}{dz}\left(D_{ea}C\frac{dY_A}{dz}\right) - Cv_s\frac{dY_A}{dz} - \eta(-R_A)_s\rho_b = 0 \tag{10.107}$$

To use this equation, it is necessary to have values for the axial dispersion coefficient. The value of the effective axial dispersion coefficient does not depend strongly on the type of species. Therefore, all species will be dispersed at approximately the same rate. This means that, for the single reaction case, the mole balance equation can be written for one reacting species with the other mole fractions being evaluated from the stoichiometry. For multiple reactions, at least one mole balance equation must be written for every independent reaction. Techniques for obtaining dispersion coefficients are discussed in Section 10.6.

10.4.1.2 Energy Balance Equation

The energy balance equation is also obtained by adding an axial conduction term to the PFR model. Again, using the same techniques applied elsewhere, the energy balance for a single reaction for an adiabatic reactor is

$$\frac{d}{dz}\left(k_{a,\text{eff}}\frac{dT}{dz}\right) - v_s\rho_f C_p\frac{dT}{dz} - \Delta H_R\eta(-R_A)_S\rho_b = 0 \tag{10.108}$$

The energy balance for a single reaction for a nonadiabatic reactor is

$$\frac{d}{dz}\left(k_{a,\text{eff}}\frac{dT}{dz}\right) - v_s\rho_f C_p\frac{dT}{dz} - \Delta H_R\eta(-R_A)_S\rho_b + \frac{4}{D_T}U(T_{3,\infty} - T) = 0 \tag{10.109}$$

The effective axial thermal conductivity, $k_{a,\text{eff}}$, combines an effective fluid thermal conductivity and a solid thermal conductivity (see Section 10.6 for details).

10.4.1.3 Boundary Conditions for the 1D Axial Dispersion Model

The following boundary conditions may be applied for the mole and energy balance equations:

$$\left(-D_{ea} \frac{dT_A}{dz} \right)_{0^+} = v_s \left[Y_{Ao} - Y_A(0^+) \right] \quad \text{at } z = 0 \tag{10.110}$$

$$\left(-k_{a,\text{eff}} \frac{dT}{dz} \right)_{0^+} = \rho v_s C_P \left[T_o - T(0^+) \right] \quad \text{at } z = 0 \tag{10.111}$$

$$\frac{dY_A}{dz} = \frac{dT}{dz} = 0 \quad \text{at } z = L \tag{10.112}$$

These boundary conditions are based on the Danckwerts boundary conditions.

10.4.2 Heterogeneous Axial Dispersion Model

The heterogeneous model has equations for each phase, which makes the model a little more complex, because we have to consider axial dispersion in each phase. For the mole balance, dispersion only occurs in the fluid phase. However, for the energy balance, both phases are considered because both phases can conduct heat.

10.4.2.1 Mole Balance Equations

The fluid phase mole balance is the PFR equation with a superimposed dispersion term:

$$\frac{d}{dz} \left(D_{ea} C_f \frac{dY_{A,f}}{dz} \right) - C_f v_s \frac{dY_{A,f}}{dz} - k_m a_m C_f \left(Y_{A,f} - Y_{A,S} \right) = 0 \tag{10.113}$$

The effective dispersion coefficient in the heterogeneous model is the same as in the pseudo-homogeneous model, because there is no mass diffusion in the solid phase.

The solid phase mole balance equation for the heterogeneous dispersion model is the same as for the PFR model:

$$k_m a_m C_f \left(Y_{A,f} - Y_{A,S} \right) - \eta (-R_A)_S \rho_b = 0 \tag{10.114}$$

10.4.2.2 Energy Balance Equation in the Fluid Phase

The fluid phase energy balance for the adiabatic reactor is

$$\frac{d}{dz} \left(k_{af} \frac{dT_f}{dz} \right) - v_s \rho C_P \frac{dT_f}{dz} + h_{fs} a_m \left(T_S - T_f \right) = 0 \tag{10.115}$$

The fluid phase energy balance for the adiabatic reactor is

$$\frac{d}{dz}\left(k_{af}\frac{dT_f}{dz}\right) - v_s\rho C_P\frac{dT_f}{dz} + h_{fs}a_m(T_S - T_f) = 0 \tag{10.116}$$

The fluid phase energy balance for the nonadiabatic reactor is

$$\frac{d}{dz}\left(k_{af}\frac{dT_f}{dz}\right) - v_s\rho C_P\frac{dT_f}{dz} + h_{fs}a_m(T_S - T_f) + \frac{4}{D_T}U(T_{3,\infty} - T_f) = 0 \tag{10.117}$$

The axial fluid thermal conductivity, k_{af}, depends on the thermal conductivity of the fluid plus fluid-mixing effects. Its value is a function of the fluid velocity, and is discussed in Section 10.6.

10.4.2.3 Energy Balance Equation for the Solid Phase

The solid phase energy balance equation is

$$\frac{d}{dz}\left(k_{as}\frac{dT_S}{dz}\right) - \Delta H_R\eta(-R_A)_S\rho_b - h_{fa}a_m(T_S - T_f) = 0 \tag{10.118}$$

where k_{as} is the axial solid phase thermal conductivity. Thermal conduction through the solid consists of conduction through the particles, conduction between particles at their points of contact and radiation across the void space.

10.4.2.4 Boundary Conditions

The following boundary conditions may be applied to the conservation equations:

$$\left(-D_{ea}\frac{dY_{A,f}}{dz}\right)_{0^+} = v_s\left[Y_{A,fo} - Y_{A,f}(0^+)\right] \quad \text{at } z = 0 \tag{10.119}$$

$$\left(-k_{af}\frac{dT_f}{dz}\right)_{0^+} = \rho v_s C_P\left[T_{fo} - T_f(0^+)\right] \quad \text{at } z = 0 \tag{10.120}$$

$$\frac{dT_S}{dz} = 0 \quad \text{at } z = 0 \tag{10.121}$$

$$\frac{EY}{dz} = \frac{dT_f}{dz} = \frac{dT_S}{dz} = 0 \quad \text{at } z = 0 \tag{10.122}$$

These boundary conditions are based on the Danckwerts boundary conditions.

10.5 Two-Dimensional Steady-State Models

A more rigorous packed-bed model is the two-dimensional model, which allows for gradients in both radial and axial directions, and includes both axial and radial dispersion of heat and mass. The 2D model should be used where critical results are required, or where a large diameter reactor is used (>10–15 particle diameters). As in the 1D formulation, both pseudo-homogeneous and heterogeneous forms of the model have been used. The conservation equations for the 2D model are based on the elemental volume shown in Figure 10.3. The mole and energy balances are developed using the same techniques used for the 1D models. Note that for the elemental volume element shown in Figure 10.3b, the elemental volume is

$$\Delta V = 2\pi r \Delta r \Delta z \tag{10.123}$$

The flow area in the axial direction is

$$A_z = 2\pi r \Delta r \tag{10.124}$$

The flow area in the radial direction is

$$A_r = 2\pi r \Delta z \tag{10.125}$$

10.5.1 Pseudo-Homogeneous 2D Model

We consider first the pseudo-homogeneous model. In the axial direction, the transport mechanisms in the fluid phase are dispersion and bulk flow. In the radial direction, the fluid transport mechanisms are dispersion only.

10.5.1.1 Fluid Transport Mechanisms Are Dispersion Only Mole Balance Equation

The mole balance equation is written over the elemental volume represented in Figure 10.3. It includes axial bulk flow, homogeneous and catalytic reaction, and radial and axial dispersion.

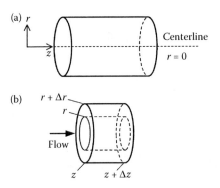

FIGURE 10.3
Elemental reactor volume element used for developing the mole and energy balance equations for the two-dimensional packed-bed reactor model. (a) Shows a typical cylindrical reactor and coordinate system, while (b) shows the elemental volume element.

$$\Delta\left[2\pi r\Delta z CD_{er}\frac{\partial Y_A}{\partial r}\right] + \Delta\left[2\pi r\Delta r CD_{ea}\frac{\partial Y_A}{\partial z}\right] - (2\pi r\Delta r)\Delta\left(v_s^*CY_A\right)$$

$$- \eta(-R_A)_S(1-\phi)\rho_c 2\pi r\Delta r\Delta z = 0 \tag{10.126}$$

D_{ea} is the axial dispersion coefficient (see Equation 10.105) and D_{er} is the radial dispersion coefficient. We divide by $\Delta V = 2\pi\, r\Delta\, r\Delta z$, take the limit as $\Delta V \to dV$ and rearrange to give

$$\frac{1}{r}\frac{\partial}{\partial r}\left(rD_{er}C\frac{\partial Y_A}{\partial r}\right) + \frac{\partial}{\partial z}\left(D_{ea}C\frac{\partial Y_A}{\partial z}\right) - Cv_s^*\frac{\partial Y_A}{\partial z} - \eta(-R_A)_S\rho_b = 0 \tag{10.127}$$

Dispersion coefficients for both radial and axial directions are required (see Section 10.6.4). The value of the dispersion coefficient does not depend on the type of species.

10.5.1.2 Energy Balance Equation

The energy balance equation with a single homogeneous and single heterogeneous reaction is obtained by performing an energy balance over the elemental volume to give

$$\Delta\left(2\pi r\Delta z k_{r,\text{eff}}\frac{\partial T}{\partial r}\right) + \Delta\left(2\pi r\Delta r k_{a,\text{eff}}\frac{\partial T}{dz}\right) - 2\pi r\Delta r v_x\rho C_P\Delta T$$

$$- \Delta H_R\eta(-R_A)_S\rho_b 2\pi r\Delta r\Delta z = 0 \tag{10.128}$$

We divide by $\Delta V = 2\pi r\Delta r\Delta z$, take the limit as $\Delta V \to dV$ and rearrange to give

$$\frac{1}{r}\frac{\partial}{\partial r}\left(rk_{r,\text{eff}}\frac{\partial T}{\partial r}\right) + \frac{\partial}{\partial z}\left(k_{a,\text{eff}}\frac{\partial T}{dz}\right) - v_s\rho C_P\frac{\partial T}{\partial z} - \Delta H_R\eta(-R_A)_S\rho_b = 0 \tag{10.129}$$

This energy balance is valid in both the adiabatic and nonadiabatic case. Unlike the 1D model, in the 2D model, the heat transfer with the surroundings (if present) appears in the boundary conditions, as shown next. The effective thermal conductivities in the axial direction, $k_{a,\text{eff}}$, are discussed in Section 10.6.

10.5.1.3 Boundary Conditions

In the 2D model, boundary conditions are required for both r and z directions. The boundary conditions at the reactor entrance are based on the Danckwerts conditions. Therefore, for the mole balance equation:

$$\left(-D_{ea}\frac{dY_A}{dz}\right)_{0^+} = v_s^*\left[Y_{Ao} - Y_A(0^+)\right] \quad \text{at } z = 0 \tag{10.130}$$

and for the energy balance equation:

$$\left(-k_{a,\text{eff}}\frac{dT}{dz}\right)_{0^*} = \rho v_s C_P\left[T_o - T(0^*)\right] \quad \text{at } z = 0 \tag{10.131}$$

At the reactor outlet, the zero flux condition is applied for mole and energy balance equations:

$$\frac{\partial Y_A}{\partial z} = \frac{\partial T}{\partial z} = 0 \quad \text{at } z = L \tag{10.132}$$

At the reactor centerline, the symmetry condition also gives a zero flux and thus:

$$\frac{\partial Y_A}{\partial r} = \frac{\partial T}{\partial r} = 0 \quad \text{at } r = 0 \tag{10.133}$$

The reactor wall is impermeable to mass. Thus, there is a zero mole flux at the wall and, therefore, for the mole balance equation we have

$$\frac{\partial Y_A}{\partial r} = 0 \quad \text{at } r = R \tag{10.134}$$

The boundary condition for the energy balance equation depends on whether or not there is heat transfer with the surroundings. For an adiabatic reactor, the zero flux condition also applies and:

$$-k_{r,\text{eff}} \frac{\partial T}{\partial r} = 0 \quad r = R \tag{10.135}$$

For a nonadiabatic reactor, the heat flux at the wall balances with the heat transfer to the surroundings, and we have the condition:

$$-k_{r,\text{eff}} \frac{\partial T}{\partial r} = U\left(T - T_{3,\infty}\right) \quad \text{at } r = R \tag{10.136}$$

The overall heat transfer coefficient, U, is given by Equation 10.103.

10.5.2 Heterogeneous 2D Model

The heterogeneous model equations are derived using an analogous procedure to the derivation of the homogeneous model, except that the catalytic reaction is represented by a rate of transfer to the solid phase.

10.5.2.1 Mole Balance Equation for the Fluid

The mole balance equation with both axial and radial dispersion is

$$\frac{1}{r}\frac{\partial}{\partial r}\left(rD_{er}C_f \frac{\partial Y_{Af}}{\partial r}\right) + \frac{\partial}{\partial z}\left(D_{ea}C_f \frac{\partial Y_{Af}}{\partial z}\right) - C_f v_s^* \frac{\partial Y_{Af}}{\partial z} - k_m a_m C\left(Y_{Af} - Y_{A,S}\right) = 0 \tag{10.137}$$

where a_m is the surface area of the catalyst packing per unit volume of reactor.

10.5.2.2 Mole Balance Equation for the Solid

The solid phase mole balance equation is a balance between mass transfer and reaction rate, and is the same as for the 1D model, namely,

$$k_m a_m C_f \left(Y_{A,f} - Y_{A,S} \right) - \eta \left(-R_A \right)_S \rho_b = 0 \tag{10.138}$$

The presence of multiple species and reactions is dealt with in the same manner as described in the 1D heterogeneous model description.

10.5.2.3 Energy Balance Equation for the Fluid

The fluid phase energy balance is also derived in a manner analogous to that used for the homogeneous model to give

$$\frac{1}{r} \frac{\partial}{\partial r} \left(r k_{rf} \frac{\partial T_f}{\partial r} \right) + \frac{\mathrm{d}}{\partial z} \left(k_{af} \frac{\partial T_f}{\partial z} \right) - v_s \rho C_P \frac{\partial T_f}{\partial z} - h_{fs} a_m \left(T_f - T_S \right) = 0 \tag{10.139}$$

The radial and axial thermal conductivity of the fluid is required: see Section 10.6.

10.5.2.4 Energy Balance Equation for the Solid

The solid phase energy balance equation is:

$$\frac{1}{r} \frac{\partial}{\partial r} \left(r k_{rs} \frac{\partial T_S}{\partial r} \right) + \frac{\partial}{\partial z} \left(k_{as} \frac{\partial T_S}{\partial z} \right) - \Delta H_R \eta \left(-R_A \right)_S \rho_b + h_{fs} a_m \left(T_f - T_S \right) = 0 \tag{10.140}$$

Radial and axial thermal conductivities of the solid are required: see Section 10.6.

10.5.2.5 Boundary Conditions

Boundary conditions at the reactor entrance are required for the fluid phase mole and energy balance equations, and the solid phase energy balance. These can be based on the Danckwerts conditions. For the mole balance equation:

$$\left[-D_{ea} \frac{\partial Y_{A,f}}{\partial z} \right]_{0^+} = v_s^* \left[Y_{A,fo} - Y_{A,f} \left(0^+ \right) \right] \quad z = 0 \tag{10.141}$$

For the fluid phase energy balance:

$$\left[-k_{af} \frac{\partial Y_f}{\partial z} \right]_{0^+} = \rho v_s C_P \left[T_{fo} - T_f \left(0^+ \right) \right] \quad z = 0 \tag{10.142}$$

For the solid phase energy balance:

$$\frac{\partial T_S}{\partial z} = 0 \quad \text{at } z = 0 \tag{10.143}$$

At the reactor exit, a condition is needed for the fluid phase mole and energy balances, and for the solid phase energy balance. The zero flux condition can also be applied to each of these equations to give

$$\frac{\partial Y_{Af}}{\partial z} = \frac{\partial T_f}{\partial z} = \frac{\partial T_S}{\partial z} = 0 \quad \text{at } z = L \tag{10.144}$$

At the centerline, the symmetry condition gives the zero flux condition. Therefore,

$$\frac{\partial Y_{Af}}{\partial r} = \frac{\partial T_f}{\partial r} = \frac{\partial T_S}{\partial r} = 0 \quad \text{at } r = 0 \tag{10.145}$$

At the reactor wall, the molar flux is zero because the wall is impermeable, and therefore the concentration gradient is zero, that is,

$$\frac{\partial Y_{Af}}{\partial r} = 0 \quad \text{at } r = R \tag{10.146}$$

For an adiabatic reactor, the heat flux at the wall is also zero and:

$$\frac{\partial T_f}{\partial r} = \frac{\partial T_S}{\partial r} = 0 \quad \text{at } r = R \tag{10.147}$$

If the reactor is nonadiabatic, then the heat flux at the wall equals the heat transfer to the surroundings. The total flux at the wall is made up of transfer from both fluid and solid phases. The following conditions may then be applied for the fluid phase:

$$-k_{rf} \frac{\partial T_f}{\partial r} = U_f \left(T_f - T_{3,\infty} \right) \quad \text{at } r = R \tag{10.148}$$

The condition for the solid phase is:

$$-k_{rs} \frac{\partial T_S}{\partial r} = U_S \left(T_S - T_{3,\infty} \right) \quad \text{at } r = R \tag{10.149}$$

The overall heat transfer coefficients contain inside wall coefficients for fluid and solid phases, respectively. The overall coefficients are given by

$$\frac{1}{U_f} = \frac{1}{h_{wf}} + \left(\frac{D_T}{2k_W} \right) \ln \left(\frac{D_o}{D_T} \right) + \frac{D_T}{D_o} \frac{1}{h_o} \tag{10.150}$$

$$\frac{1}{U_S} = \frac{1}{h_{ws}} + \left(\frac{D_T}{2k_W} \right) \ln \left(\frac{D_o}{D_T} \right) + \frac{D_T}{D_o} \frac{1}{h_o} \tag{10.151}$$

Correlations for the various thermal conductivities and heat transfer coefficients required in the 2D model are given in Section 10.6.

10.6 Transient Packed-Bed Models

Thus far, we have restricted the discussion to steady-state models. Many, if not most, packed-bed reactors are operated at steady state. However, during start-up and shutdown cycles transient operation is inevitable, and some reactors are deliberately operated in a transient fashion. Some examples of these types of reactors will be given later. Transient operation introduces some complex control issues, which will not be discussed here. Rather, we will limit the presentation to a description of the additional complexities introduced by the time-dependent terms.

Transient operation can result from a change in feed concentration or temperature, a change in surrounding temperature for a nonadiabatic system, or a change in flow rate. Transients may be the result of a step change, or a gradual change over time. The new versions of the mole and energy balance equations are given in the following. We will not explore these models very far; the main purpose here is to present the models so that they are available if needed. Note that all of these models represent challenging systems of nonlinear PDE, which require a numerical solution. To achieve these solutions, it is recommended to use commercial software which has been developed for this purpose.

10.6.1 One-Dimensional Reactor: Pseudo-Homogeneous Case

We start by writing the mole and energy balances for the pseudo-homogeneous case, using the most general case with axial dispersion and including terms for the heat transfer in the energy equation. For simplicity in writing the energy balances, we consider the case of a single reaction only. The transient mole balance is

$$\frac{\partial}{\partial z}\left(D_{ea}C\frac{\partial Y_A}{\partial z}\right) - Cv_s\frac{\partial Y_A}{\partial z} - \eta\left(-R_A\right)_s\rho_b = C\frac{\partial Y_A}{\partial t} \tag{10.152}$$

For the plug flow case, the dispersion term is simply dropped. For the energy balance, we have

$$\frac{\partial}{\partial z}\left(k_{a,\text{eff}}\frac{\partial T}{\partial z}\right) - v_s\rho_f C_{p,f}\frac{\partial T}{\partial z} - \Delta H_R\eta\left(-R_A\right)_S\rho_b + \frac{4}{D_T}U\left(T_{3,\infty} - T\right)$$

$$= \left[\phi\rho_f C_{p,f} + (1-\phi)\rho_S C_{p,S}\right]\frac{\partial T}{\partial t} \tag{10.153}$$

Again, in the case of plug flow, the term for axial conduction is eliminated, and for an adiabatic reactor the term for external heat transfer is dropped. In the accumulation term, the thermal is a weighted average (using the porosity) of the thermal mass of the fluid and solid. For gas phase systems, the thermal mass of the fluid is usually much smaller than the solid.

Notice that both of the conservation equations are now partial differential equations (PDE). The boundary conditions for the coupled mole and energy balances are the same as for the steady-state case. The initial condition must also be specified, that is, the temperature and concentration must be known everywhere in the reactor at time equals zero.

If we consider the time scales at which changes occur in the reactor, a few observations can be made. The space time in most packed-bed systems is small, of the order of a few seconds or less. Practically, this means that changes in the fluid phase happen much more rapidly than in the solid, and, as a result, the two coupled PDE represent a stiff system. To help the solution, sometimes the pseudo-steady-state assumption is used for the mole balance. In this case, the transient term is only retained on the energy balance equation, and the steady-state version of the mole balance is used.

10.6.2 One-Dimensional Reactor: Heterogeneous Case

For the heterogeneous model, we nominally have four new equations, two each for the mole and energy balance. However, the accumulation of species in the catalyst is rarely considered, and this balance retains its steady-state form. The transient mole balance for the fluid is

$$\frac{\partial}{\partial z}\left(D_{ea}C_f \frac{\partial Y_{A,f}}{\partial z}\right) - C_f v_s \frac{\partial Y_{A,f}}{\partial z} - k_m a_m C_f \left(Y_{A,f} - Y_{A,S}\right) = C_f \frac{\partial Y_{A,f}}{\partial t} \tag{10.154}$$

As with the pseudo-homogeneous model, in the case of plug flow the term for axial dispersion is dropped. The energy balance for the fluid is

$$\frac{\partial}{\partial z}\left(k_{af}\frac{\partial T_f}{\partial z}\right) - v_s\left(\rho_f C_{p,f}\right)\frac{\partial T_f}{\partial z} + h_{fs}a_m\left(T_S - T_f\right) + \frac{4}{D_T}U\left(T_{3,\infty} - T_f\right)$$

$$= \phi\rho_f C_{p,f}\frac{\partial T_f}{\partial t} \tag{10.155}$$

The energy balance for the solid is

$$\frac{\partial}{\partial z}\left(k_{as}\frac{\partial T_S}{\partial z}\right) - \Delta H_R \eta\left(-R_A\right)_S \rho_b - h_{fa}a_m\left(T_S - T_f\right) = (1-\phi)\rho_S C_{p,S}\frac{\partial T_S}{\partial z} \tag{10.156}$$

The axial dispersion and heat transfer terms can be eliminated if desired to give the PFR and adiabatic reactor equations. As for the pseudo-homogeneous model, the pseudo-steady-state assumption can be made. Both fluid equations are used in their steady-state forms, and only the energy accumulation term for the solid is retained. Boundary and initial conditions are required.

10.6.3 Two-Dimensional Reactor: Pseudo-Homogeneous Case

For completeness, we also give the transient two-dimensional conservation equations. The pseudo-homogeneous mole balance is

$$\frac{1}{r}\frac{\partial}{\partial r}\left(rD_{er}C\frac{\partial Y_A}{\partial r}\right) + \frac{\partial}{\partial z}\left(D_{ea}C\frac{\partial Y_A}{\partial z}\right) - Cv_s^*\frac{\partial Y_A}{\partial z} - \eta\left(-R_A\right)_S \rho_b = C\frac{\partial Y_A}{\partial t} \tag{10.157}$$

The energy balance equation is

$$\frac{1}{r}\frac{\partial}{\partial r}\left(rk_{r,\text{eff}}\frac{\partial T}{\partial r}\right) + \frac{\partial}{\partial z}\left(k_{a,\text{eff}}\frac{\partial T}{\partial z}\right) - v_s\rho_f C_{p,f}\frac{\partial T}{\partial z} - \Delta H_R\eta(-R_A)_S\rho_b$$

$$= \left[\phi\rho_f C_{p,f} + (1-\phi)\rho_S C_{p,S}\right]\frac{\partial T}{\partial t} \tag{10.158}$$

The accumulation is again a weighted average of the solid and fluid terms. Boundary and initial conditions are the same as before. The use of the pseudo-steady-state assumption would also see the steady-state version of the mole balance equation used.

10.6.4 Two-Dimensional Reactor: Heterogeneous Case

Finally, the full 2D heterogeneous model is presented. The mole balance equation for the fluid is

$$\frac{1}{r}\frac{\partial}{\partial r}\left(rD_{er}C_f\frac{\partial Y_{Af}}{\partial r}\right) + \frac{\partial}{\partial z}\left(D_{ea}C_f\frac{\partial Y_{Af}}{\partial z}\right) - C_f v_s^*\frac{\partial Y_{Af}}{\partial z}$$

$$- k_m a_m C\left(Y_{Af} - Y_{AS}\right) = C_f\frac{\partial Y_{Af}}{\partial t} \tag{10.159}$$

The solid is usually assumed to be at steady state, as before. The energy balance equation for the fluid is

$$\frac{1}{r}\frac{\partial}{\partial r}\left(rk_{rf}\frac{\partial T_f}{\partial r}\right) + \frac{\partial}{\partial z}\left(k_{af}\frac{\partial T_f}{\partial z}\right) - v_s\rho_f C_{p,f}\frac{\partial T_f}{\partial z} - h_{fs}a_m\left(T_f - T_S\right)$$

$$= \phi\rho_f C_{p,f}\frac{\partial T_f}{\partial t} \tag{10.160}$$

The solid phase energy balance equation is

$$\frac{1}{r}\frac{\partial}{\partial r}\left(rk_{rs}\frac{\partial T_S}{\partial r}\right) + \frac{\partial}{\partial z}\left(k_{as}\frac{\partial T_S}{\partial z}\right) - \Delta H_R\eta(-R_A)_S\rho_b + h_{fs}a_m\left(T_f - T_S\right)$$

$$= (1-\phi)\rho_S C_{p,S}\frac{\partial T_S}{\partial z} \tag{10.161}$$

As before, it is common practice only to retain the transient terms for the solid energy balance.

10.7 Transport Properties in Packed Beds

The fixed-bed reactor models require values for one or more transport properties. Correlations for these parameters are presented in this section.

10.7.1 Fluid/Solid Mass and Heat Transfer Coefficients

There are many correlations for heat and mass transfer coefficients in packed beds in the literature. Many of these are presented in terms of dimensionless j factors, denoted j_D for mass transfer and j_H for heat transfer. For gases, these quantities are given by

$$j_D = \frac{k_m \rho}{G} Sc^{\frac{2}{3}} \quad \text{and} \quad j_H = \frac{h_{fs}}{C_p G} Pr^{\frac{2}{3}} \tag{10.162}$$

The Schmidt number, Sc, and Prandtl number, Pr, are

$$Sc = \frac{\mu}{\rho D} \quad \text{and} \quad Pr = \frac{C_p \mu}{k_f} \tag{10.163}$$

The mass transfer coefficient in Equation 10.162 has units of m/s. Commonly it is assumed that $j_D = j_H$. Correlations which fit most of the available experimental data are (Bird et al., 2002)

$$j_D = j_H = 0.91 Re_j^{-0.51} S_f \quad \text{at} \, Re_j < 50 \tag{10.164}$$

$$j_D = j_H = 0.61 Re_j^{-0.41} S_f \quad \text{at} \, Re_j > 50 \tag{10.165}$$

S_f is a shape factor, which equals 1.0 for a sphere, 0.91 for a short cylinder, and 0.79 for Rashig rings. The Reynolds number in Equations 10.164 and 10.165 is defined as

$$Re_j = \frac{GD_p}{(1-\phi)\mu S_f} \tag{10.166}$$

Recall that D_p is the equivalent particle diameter, G is the superficial mass velocity, ϕ is the porosity and μ is the viscosity. More recently, a general correlation has been proposed (Bird et al., 2002), applicable over a broad range of Reynolds numbers:

$$j_D = j_H = 2.91 Re_j^{-2/3} + 0.78 Re_j^{-0.381} \tag{10.167}$$

At a Reynolds number of zero, the j factor correlations predict a zero rate of mass transfer which is not realistic. Wakao and Kaguei (1982) analyzed a large amount of packed-bed data using the dispersed plug flow model, and proposed correlations for Sh and Nu:

$$Sh = \frac{k_m D_p}{D} = 2 + 1.1 Sc^{\frac{1}{3}} Re_b^{0.6} \tag{10.168}$$

$$\text{Nu} = \frac{h_{fs}D_P}{k_f} = 2 + 1.1\text{Pr}^{\frac{1}{3}}\text{Re}_b^{0.6} \tag{10.169}$$

The Reynolds number is

$$\text{Re}_b = \frac{D_p G}{\mu} \tag{10.170}$$

Example 10.4 illustrates the calculation of mass and heat transfer coefficients.

Example 10.4

Consider a bed of porosity 0.37, packed with cylinders 3 mm in diameter and 4 mm in length. A gas mixture at a temperature of 500 K and a pressure of 100 kPa consisting of 1 mol% CH_4 in air flows through the bed with a superficial velocity of $v_s = 1$ m/s. Calculate k_m and h_{fs} using:

    a.   Equation 10.162.
    b.   Equations 10.168 and 10.169.

    Note that the equivalent particle diameter of these particles is 3.79 mm. The molecular diffusion coefficient of CH_4 in air is 4.4×10^{-5} m²/s.

SOLUTION

The physical properties will be taken as those of air. At the stated conditions:

    Density = 0.7048 kg/m³
    Viscosity = 2.671×10^{-5} Pa · s
    Heat capacity = 1029.5 J/kg · K
    Thermal conductivity = 0.04038 W/m · K

From these physical property data, we compute:

$$\text{Pr} = 0.680 \quad \text{and} \quad \text{Sc} = 0.862$$

Part (a)

To calculate the j factors, we first compute Re_j using Equation 10.166. For a porosity of 0.37 and $S_f = 0.91$ (short cylinder value), we obtain:

$$\text{Re}_j = 17.7$$

From Equation 10.164, we calculate:

$$j_D = j_H = 0.192$$

From Equation 10.162:

$$k_m = 0.213 \text{ m/s} \quad \text{and} \quad h_{fs} = 180.2 \text{ W/m}^2\text{K}$$

Part (b)

The bed Reynolds number is defined by

$$\text{Re}_b = D_p G / \mu = 60.7$$

Substitution of Re_b, Sc, and Pr into Equations 10.168 and 10.169 gives

$$k_m = 0.166 \text{ m/s} \quad \text{and} \quad h_{fs} = 142.4 \text{ W/m}^2\text{K}$$

These values are about 20% lower than those calculated using the j factors. This degree of uncertainty is typical when calculating heat and mass transfer coefficients in packed beds.

After the mass and heat transfer coefficients have been calculated, the reaction rate at the pellet surface can be computed.

10.7.2 Dispersion Coefficients

Analysis of dispersion in packed beds is not a trivial matter, as the vast literature on the subject indicates. The subject is covered by Wakao and Kaguei (1982), Carberry (1976), among others. The following analysis is based on the work of Wilhelm (1962), as presented in Carberry (1976). Data for radial and axial dispersion coefficients were presented in terms of Peclet numbers for mass transfer. Define a hydraulic diameter as

$$D_B = \frac{D_T}{\frac{3}{2}(D_T/D_P)(1 - \phi) + 1} \tag{10.171}$$

Define a Reynolds number using this hydraulic diameter as

$$\text{Re}_H = \frac{D_B \rho v_s}{\mu} \tag{10.172}$$

The radial and axial Peclet numbers for mass transfer are defined as

$$(\text{Pe}_m)_r = \frac{D_B v_s}{D_{er}} \quad \text{and} \quad (\text{Pe}_m)_z = \frac{D_B v_s}{D_{ea}} \tag{10.173}$$

For Re_H greater than about 40, the value of $(\text{Pe}_m)_r$ varies between about 8 and 12 (usually a value of 10 is used). For beds of particles more than about 50 particle diameters deep and a Reynolds number greater than about 10, a value of $(\text{Pe}_m)_z = 2$ is a good approximation. Most packed-bed reactors fall into this range. At a lower Reynolds number, the Peclet numbers decrease in value. See Carberry (1976) for more details.

Example 10.5

Consider the packed bed discussed in Example 10.4. Calculate the effective radial and axial dispersion coefficients, D_{er} and D_{ea}. The bed diameter is 3 cm.

SOLUTION

First calculate the hydraulic radius given by Equation 10.171. The equivalent particle diameter of the given particles was calculated in Example 10.1 to be 3.79 mm. The value of D_B is

$$D_B = \frac{D_T}{\frac{3}{2}(D_T/D_P)(1-\phi)+1} = \frac{30\ \text{mm}}{\frac{3}{2}(30\ \text{mm}/3.79\ \text{mm})(1-0.37)+1} = 3.54\ \text{mm} \tag{10.174}$$

This value can be used with the viscosity and density given in Example 10.3 to calculate Re_H.

$$\text{Re}_H = \frac{D_B \rho v_s}{\mu} = \frac{(3.54 \times 10^{-3}\ \text{m})(0.7048\ \text{kg/m}^3)(1\ \text{m/s})}{2.671 \times 10^{-5}\ \text{Pa} \cdot \text{s}} = 93.3 \tag{10.175}$$

This value is greater than 40. Therefore, the Peclet numbers are

$$\left(\text{Pe}_m\right)_r = \frac{D_B v_s}{D_{er}} = 10 \quad \text{and} \quad \left(\text{Pe}_m\right)_z = \frac{D_B v_s}{D_{ea}} = 2 \tag{10.176}$$

Substitution of D_B and v_s gives

$$D_{er} = 3.54 \times 10^{-4}\ \text{m}^2/\text{s}$$

$$D_{ea} = 1.77 \times 10^{-3}\ \text{m}^2/\text{s}$$

10.7.3 Thermal Conductivity

Dixon and Cresswell (1979) attempted a theoretical approach to analyzing the values of transport properties in packed beds. We shall not explain the derivation of the model equations, but will simply report them and give one or two numerical examples. The energy balance equations for the 1D and 2D pseudo-homogeneous models require effective thermal conductivities. These depend on the conductivities of the fluid, k_f, and solid, k_p, the fluid/wall heat transfer coefficient, h_{wf}, the fluid/solid heat transfer coefficient, h_{fs}, the particle diameter and the tube diameter/particle diameter ratio. The equations are

$$k_{r,\text{eff}} = k_{rf} + k_{rs}\left[\frac{1 + 8k_{rf}/h_{wf}D_T}{1 + \frac{16}{3}k_{rs}\left((1/h_{fs}D_p)+(0.1/k_p)\right)/(1-\phi)(D_T/D_p)^2}\right] \tag{10.177}$$

$$k_{a,\text{eff}} = k_{af} + \frac{k_{as}}{\left[1 + \frac{16}{3}k_{as}\left((1/h_{fs}D_p)+(0.1/k_p)\right)/(1-\phi)(D_T/D_p)^2\right]^2} \tag{10.178}$$

For $\text{Re}_b > 50$ and $D_T/D_p > 10$, we can approximate Equations 10.177 and 10.178 by

$$k_{r,\text{eff}} \approx k_{rf} + k_{rs} \tag{10.179}$$

$$k_{a,\text{eff}} \approx k_{af} + k_{as} \tag{10.180}$$

The radial and axial conductivity of the solid, k_{rs} and k_{as}, respectively, are given by

$$k_{rs} = k_{as} = \frac{2k_f(1-\phi)^{0.5}}{(1-(k_fB/k_p))}\left[\frac{(1-(k_f/k_p))B}{(1-(k_fB/k_p))^2}\ln\left(\frac{k_p}{Bk_f}\right) - \frac{B+1}{2} - \frac{B-1}{(1-(k_fB/k_p))}\right] \qquad (10.181)$$

where

$$B = C\left(\frac{1-\phi}{\phi}\right)^{10/9} \qquad (10.182)$$

and $C = 1.25$ for spheres and 1.4 for crushed particles.

The conductivities, k_{rf} and k_{af}, are correlated using Peclet numbers for heat transfer. Define the Peclet numbers as

$$(\mathrm{Pe}_H)_{rf} = \frac{GC_PD_P}{k_{rf}} \quad \text{and} \quad (\mathrm{Pe}_H)_{af} = \frac{GC_PD_P}{k_{af}} \qquad (10.183)$$

Then

$$\frac{1}{(\mathrm{Pe}_H)_{rf}} = 0.1 + \frac{0.66\phi}{\mathrm{Re}_b\mathrm{Pr}} \qquad (10.184)$$

$$\frac{1}{(\mathrm{Pe}_H)_{af}} = \frac{0.73\phi}{\mathrm{Re}_b\,\mathrm{Pr}} + \frac{0.5}{(1+(9.7\phi/\mathrm{Re}_b\,\mathrm{Pr}))} \qquad (10.185)$$

Thus, provided that the thermal conductivities of the fluid and the catalyst particles, the bed porosity, equivalent particle diameter and reactor diameter are known, the thermal conductivities required for the reactor models can be computed. An example is given in the next section.

The Dixon and Cresswell models do not account for radiation among particles. Radiation tends to increase the transmission of thermal energy, thus reducing the temperature gradients. One technique used to account for radiation is to add its effect to the thermal conductivity. Feyo de Azevedo et al. (1990) proposed equations for a modified effective solid thermal conductivity:

$$k'_{rs} = k_{rs} + 4\psi_r\sigma D_pT_S^3 \qquad (10.186)$$

$$k'_{as} = k_{as} + 4\psi_r\sigma D_pT_S^3 \qquad (10.187)$$

as modified effective thermal conductivities. These modified conductivities are used in place of k_{rs} and k_{as} in the preceding analysis. Ψ_r is a radiation transfer factor defined as

$$\psi_r = \frac{2}{(1/\varepsilon) - 0.264} \qquad (10.188)$$

where ε is the particle emissivity.

10.7.4 Bed-to-Wall Heat Transfer Coefficients

There are three important wall heat transfer coefficients that are used in the various models described in the previous sections. In the heterogeneous 2D model, the heat transfer coefficient from the solid to the wall, h_{ws}, and from the fluid to the wall, h_{wf}, is used explicitly. In the pseudo-homogeneous model and the 1D heterogeneous model, the combined effective heat transfer coefficient at the wall, $h_{w,eff}$, is used. Defining a Nusselt number at the wall:

$$\text{Nu}_{fw} = \frac{h_{wf}D_p}{k_f} \tag{10.189}$$

Yagi and Wakao (1959) reported:

$$\text{Nu}_{fw} = 0.6\,\text{Pr}^{\frac{1}{3}}\text{Re}_b^{0.5} \quad 1 < \text{Re}_b < 40$$
$$\text{Nu}_{fw} = 0.2\,\text{Pr}^{\frac{1}{3}}\text{Re}_b^{0.8} \quad 40 < \text{Re}_b < 2000 \tag{10.190}$$

Dixon and Cresswell (1979) gave the following relationship:

$$\frac{h_{wf}}{k_{rf}} = \frac{h_{w,\text{eff}}}{k_{r,\text{eff}}} \tag{10.191}$$

Finally, the third significant heat transfer coefficient, the wall to solid coefficient, is given by

$$h_{ws} = 2.12\frac{k_{rs}}{D_p} \tag{10.192}$$

As stated previously, other models have been published for the transport properties in packed beds. There is not general agreement as to the best models to use, and the applicability of the different proposed models may depend on the nature of the packed bed and the operating conditions. It is important to appreciate that all parameter values are subject to a measure of uncertainty, and the area of packed-bed modeling remains one of active research. The calculation of a set of transport parameters for a packed bed is illustrated in Example 10.6.

Example 10.6

Consider the packed bed in Examples 10.4 and 10.5. Calculate the bed thermal conductivities heat transfer coefficients which are required in both the 1D and 2D models. The thermal conductivity of the particles is 0.3 W/m K.

SOLUTION

The fluid properties will be those of air. The fluid thermal conductivity is thus $k_f = 0.04038$ W/m · K. From Equation 10.182, with $C = 1.25$ and $\phi = 0.37$, we obtain:

$$B = C\left(\frac{1-\phi}{\phi}\right)^{10/9} = 1.25\left(\frac{1-0.37}{0.37}\right)^{10/9} = 2.26 \tag{10.193}$$

The value of $k_{rs} = k_E$ is then computed from Equation 10.181 by substituting into that equation the following values:

$$k_f = 0.04038 \text{ W/m} \cdot \text{K} \qquad k_p = 0.3 \text{ W/m} \cdot \text{K}$$

$$B = 2.26 \quad \phi = 0.37$$

The substitution then gives the following values:

$$k_{rs} = k_{as} = 0.126 \text{ W/m} \cdot \text{K}$$

The fluid phase conductivities will be calculated next, using Equations 10.184 and 10.185. The bed Reynolds number, Re_b, was calculated in Example 10.4 to be 60.7 and the Prandtl number was 0.68. The two Peclet numbers can therefore be computed to give

$$\left(\text{Pe}_H\right)_{rf} = \frac{GC_P D_P}{k_{rf}} = 0.106 \quad \text{and} \quad \left(\text{Pe}_H\right)_{af} = \frac{GC_P D_P}{k_{af}} = 0.467 \tag{10.194}$$

Now $G = 0.7048 \text{ kg/m}^2 \text{ s}$ and $C_P = 1029.5 \text{ J/kg} \cdot \text{K}$, therefore:

$$k_{rf} = 0.0252 \text{ W/m} \cdot \text{K} \quad \text{and} \quad k_E = 0.00572 \text{ W/m} \cdot \text{K}$$

These four thermal conductivities are used in the heterogeneous model. The effective values for use in the pseudo-homogeneous model can also be computed from them. However, the effective radial conductivity depends on the wall heat transfer coefficient, which is calculated next.

We can use the high Reynolds version of Equation 10.190 to obtain:

$$\text{Nu}_{fw} = \frac{h_{wf} D_P}{k_f} = 4.696 \tag{10.195}$$

which gives a value of $h_{wf} = 50.0 \text{ W/(m}^2\text{K)}$. The value of h_{ws} is calculated from Equation 10.192 to be 70.5 W/(m²K).

Next, we compute the effective thermal conductivities for the pseudo-homogeneous model. We require the fluid/solid heat transfer coefficient, which was calculated in Example 10.4 to be

$$h_{fs} = 142.4 \text{ W/m}^2 \cdot \text{K}$$

Thus, from Equation 10.177, the effective radial thermal conductivity is

$$k_{r,\text{eff}} = 0.163 \text{ W/m} \cdot \text{K}$$

From Equation 10.178, the effective axial thermal conductivity is

$$k_{a,\text{eff}} = 0.123 \text{ W/m} \cdot \text{K}$$

Finally, the effective wall heat transfer coefficient used in the pseudo-homogeneous model can be calculated from Equation 10.191 to be

$$h_{w,\text{eff}} = 323 \text{ W/m}^2\text{K}$$

10.8 Autothermal Operation of Packed Beds

An exothermic reaction is accompanied by the release of thermal energy, or heat. In practical operation, this thermal energy may be removed continuously, as in nonadiabatic operation, or the reactor may be allowed to run in adiabatic mode with an increase in temperature of the reacting stream. Regardless of the mode of operation, some thermal energy is ultimately removed from the process stream, and, from a purely economic perspective, it is advantageous to recover this energy for use elsewhere. It is often necessary to preheat the feed to a reactor so that the reaction rate will be sufficient to ensure the desired conversion with a practical reactor volume. An obvious choice is to use the thermal energy generated in an exothermic reaction to preheat the reactor feed. If the energy added to the reactor feed stream is all obtained from the thermal energy generated by the reaction, then the reactor is said to operate in an autothermal mode.

There are many possibilities for exchanging energy between the reactor feed and the choice depends on the nature of the reaction and the operating conditions. The simplest choice is to let the reactor operate in the adiabatic mode and then to exchange heat between the reactor effluent and the feed. For this choice to be appropriate, the adiabatic temperature rise expected in the reactor must be small enough so that the catalyst is not damaged and unwanted side reactions are not initiated. This type of arrangement is illustrated in Figure 10.4. Note that this figure illustrates an arrangement in which part of the feed is allowed to bypass the heat exchanger. This arrangement gives more control over the reactor. Coupling the reactor effluent and feed can lead to some complex operational phenomena, including the possibility of multiple steady states. This aspect of autothermal operation is discussed in detail by Froment et al. (2011).

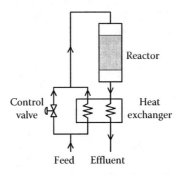

FIGURE 10.4
Adiabatic-packed-bed reactor with feed preheat using the reactor effluent. The amount of preheat can be controlled by bypassing the heat exchanger with a portion of the feed. This design can have multiple steady-state operating points.

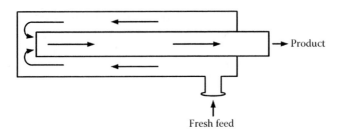

FIGURE 10.5
Reactor with feed preheat in a nonadiabatic arrangement. This reactor can also exhibit multiple steady states for some operating conditions.

Another possible method of exchanging the thermal energy produced by the reaction with the feed is shown in Figure 10.5. This figure shows an arrangement analogous to the double pipe heat exchanger, operated in counter current flow, where the shell side fluid is the fresh feed. This arrangement of internal heat exchange can lead to other interesting phenomena, such as a reactor temperature that exceed the adiabatic temperature rise by a significant amount. There are many types of reactors combined with heat exchangers. See Churchill (2007) for additional examples.

10.9 Fluidized-Bed Reactors

Fluidized-bed reactors were briefly introduced in Chapter 7. Consider a packed bed of catalyst particles through which we pass a stream of gas in an upwards direction. We assume that the top of the bed is constrained by any wire mesh or equivalent obstacle. If we monitor the pressure drop across the bed, we will observe an approximately linear increase in pressure with increasing velocity, in accordance with the Ergun equation. At some velocity, however, called the minimum fluidization velocity, the particles will start to move and the bed is said to become fluidized. This behavior is illustrated in stylized form in Figure 10.6 (adapted from Carberry, 1976). In region (a) the bed is fixed, and we see the continuous pressure increase. At the minimum fluidization velocity, the bed starts to move and there is a slight drop in pressure. At this point there will be a large number of gas bubbles moving upwards through the bed, and this region is generally referred to as a bubbling fluidized bed. As the velocity increases, the bubbles tend to increase in size. If they increase to the reactor diameter, the flow is called slug flow. Finally, as the gas velocity becomes sufficiently large, the particles are carried along with the fluid stream, and are lifted out of the unit. The latter form of operation is called a transport reactor. Mechanical devises must be present to avoid the loss of catalyst to the environment. Fluidized beds offer some advantages compared to fixed beds, as well as suffering from some drawbacks.

Fluidized beds tend to use small particles, because these are easier to be fluidized. As a result, the internal diffusion resistance tends to be quite small. As will be discussed shortly, it is fairly easy to add and remove catalyst from the reactor while it is operating. If a catalyst deactivates quickly, this mode of operation allows for the removal of spent catalyst. Indeed, this advantage was the driving force in the development of fluidized beds for catalytic

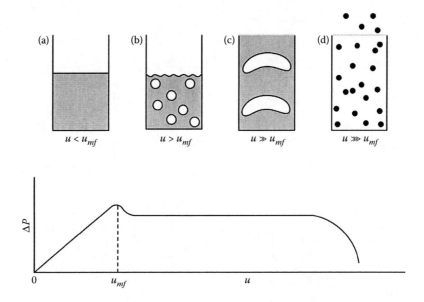

FIGURE 10.6
Pressure drop across the bed as a function of fluid velocity, with the associated flow regimes. (a) Fixed bed, (b) bubbly flow, (c) slug flow, and (d) complete particle entrainment. (Adapted from Carberry, J.J., 1976, *Chemical and Catalytic Reaction Engineering*, McGraw-Hill, New York.)

cracking. The motion induced by the flowing fluid tends to induce a large amount of radial and axial mixing, therefore the reactor bed has a tendency to be isothermal, at least in the bubbling region. Generally there is a good rate of transfer between the fluid and solid phases, and the pressure drop is also independent of the flow rate.

On the negative side, the high level of catalyst motion leads to catalyst breakage and subsequent loss to the environment. Their use is generally not recommended if the catalyst cost is a significant factor, for even with the best systems for preventing catalyst loss, rates of attrition can be high, of the order of 1% of the catalyst mass or more each day of operation.

Fluidized beds were first used in the 1920s for the gasification of lignite, but the real development push came with their use for the catalytic cracking of hydrocarbons (as discussed in Chapter 7). The primary demand for refined petroleum products is in the gasoline and diesel range, and only a small fraction of crude oil naturally occurs in this region. To increase yields of these products, longer chain hydrocarbons molecules must be broken into smaller fragments by catalytic cracking. The typical feed to a catalytic-cracking reactor is gasoil, which is a relatively high-molecular weight petroleum fraction. The products of cracking have a higher hydrogen-to-carbon ratio than the gasoil feed, and as a result the reaction is accompanied by the deposition of carbon on the catalyst surface, which renders it inactive. Therefore, the catalyst must be continuously removed from the reactor for regeneration, which is achieved by burning the carbon from the surface of the catalyst. The development of the fluid-bed reactor was an elegant solution to this problem. There are currently a number of designs extant for the fluid catalytic cracker (FCC), and one design is illustrated in Figure 10.7.

Referring to Figure 10.7, it is seen that this design consists of two large vessels placed side by side. These vessels are called the regenerator and the reactor, and the two are connected using large U-shaped pipes (Ubends). The role of the regenerator is to burn the carbon

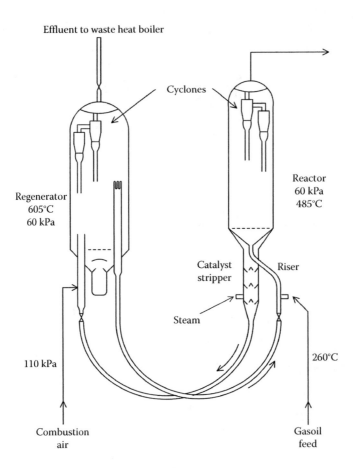

Effluent to waste heat boiler

Cyclones

Reactor
60 kPa
485°C

Regenerator
605°C
60 kPa

Catalyst
stripper

Riser

Steam

110 kPa

260°C

Combustion
air

Gasoil
feed

FIGURE 10.7
Schematic of a fluidized-bed catalytic cracker. This design is one of many that exist.

from the deactivated catalyst, thus rendering it active. The regenerator is also a fluid-bed reactor. Vaporized gasoil is fed into the reactor through the Ubend that brings the reactivated catalyst from the regenerator. The catalyst and feed are mixed as they flow through the riser and into the reactor proper. The catalyst bed sits above a distributor grid. The catalyst in the reactor deactivates quickly with reaction, and falls through a standpipe into the bottom of the reactor. There the catalyst surface is stripped of adsorbed hydrocarbon molecules by injected steam before it moves into the Ubend that will transport it to the regenerator. The catalyst is moved along by the injection of air required for the combustion of the surface carbon. The deactivated catalyst moves into the regenerator, where the carbon burns from the surface. The regenerated catalyst leaves the regenerator through a standpipe, and flows through the Ubend into the reactor, thus repeating the process. To reduce catalyst losses in the effluent streams, two stage cyclones are used in both reactor and regenerator.

The unit shown is an early design that is still in wide use. Prior to the 1960s, the unit was operated with a significant catalyst bed in the reactor. During and after the 1960s, highly active zeolite catalysts were introduced, which reduced the amount of catalyst used. With this catalyst, most of the reaction occurs in the riser, and thus a very small catalyst bed is used. Later designs, developed specifically for the new highly active catalysts, have much smaller reactors. See, for example, Froment et al. (2011). Catalytic crackers come in various sizes. One example, of the type illustrated in Figure 10.7, processes 25,000 bbl a day of gas-oil. The unit contains about 72 tonnes of catalyst, and the catalyst loss rate is of the order of 1 tonne/day.

10.9.1 Minimum Fluidization Velocity

The velocity at minimum fluidization is calculated from a force balance between gravitational and buoyancy forces. At minimum fluidization, this force balance can be represented by a pressure drop. If we consider a bed with cross-sectional area A and length L containing particles of density ρ_S, then the pressure drop is given by

$$\left(\rho_S - \rho_f\right)AL(1 - \phi)g = \Delta PA \tag{10.196}$$

This equation gives the force required to move the bed in the vertical direction. This pressure drop can be obtained from the Ergun equation, developed previously:

$$\frac{\Delta P}{L} = \frac{150(1 - \phi)}{d_p^2\phi^3}\mu u_{mf} + 1.75\frac{\rho_f}{d_p\phi^3}u_{mf}^2 \tag{10.197}$$

Combine Equations 10.196 and 10.197 to give the minimum fluidization velocity.

$$1.75\frac{\rho_f}{d_p\phi^3}u_{mf}^2 + \frac{150(1 - \phi)}{d_p^2\phi^3}\mu u_{mf} - \left(\rho_S - \rho_f\right)(1 - \phi)g = 0 \tag{10.198}$$

10.9.1.1 Entrainment Velocity

As the velocity increases from the minimum fluidization velocity, it reaches a point where the particles become entrained by the flowing gas. Entrainment happens when the terminal velocity of the particle is reached. This value can be computed from Stokes' law, which for a spherical particle gives the entrainment velocity, u_t as

$$u_t = d_p^2\frac{\left(\rho_S - \rho_f\right)g}{18\,\mu} \tag{10.199}$$

A rule of thumb for operation as a fluidized bed (as opposed to a transport reactor) is to operate at about one-half of the entrainment velocity.

10.9.1.2 Fluid-Bed Reactor Modeling

Early models for fluid beds were based on plug flow or complete mixing assumptions. Neither of these approaches works very well, and there is a large literature on the modeling and analysis of fluidized beds. The reader is referred to Froment et al. (2011), Carberry (1976), Levenspiel (1999), or Kunii and Levenspiel (1991) for further reference material.

10.10 Metal Gauze Reactors

Some reactions occur so quickly that the reaction is completely controlled by the rate of external mass transfer. For such a case the use of a porous catalyst is not effective, and their use would result in much wasted catalytic material. One reactor design that can be used in these cases is the wire screen or gauze reactor. The use of gauze as a catalyst is common in the chemical industry, with the oxidation of ammonia being the main application. The gauze reactor consists of a series of wire screens, stacked one on top of the other. The wire, composed of the requisite metal (platinum, silver, etc.), acts as the catalyst. The diameter of the wire may vary from 0.4 to 1.0 mm. For the oxidation of ammonia at reaction temperatures of 750°C to 950°C, reaction rates are so fast that the rate is generally considered to be limited by external mass transfer effects. In this reaction, the catalysts used are platinum or a platinum–rhodium alloy. Other applications include the oxidation of methanol to produce formaldehyde with silver gauze and some catalytic combustion reactions.

The parameter that controls conversion is the number of gauzes used, rather than the actual length of the bed. This number varies from 2 up to 30 or more. Even with a high number, the thickness of the bed is low, and the total contact time is of the order of milliseconds. The reactor diameter can be quite large for industrial scale units, with diameters of 6 m or more.

10.11 Counter Diffusive Reactor: Radiant Heaters

The catalytic reactors considered so far have one feature in common—all reactants enter at the reactor entrance and proceed together as a reacting flow along the reactor, with or without back mixing. Another method of operation is used in the counter diffusive reactor, in which one reactant enters at one end of the reactor, and the other reactant enters at the reactor exit (Radcliffe and Hickman, 1975). The latter reactant must diffuse against the bulk fluid flow to react. This arrangement is not common, and can be considered to be a specialized system that is primarily used in catalytic radiant heaters. A design for a counter flow radiant heater is illustrated in Figure 10.8. The main components are identified in the figure, and comprise the catalyst pad, the insulation blanket, an electric heating element and a fuel distributor. A picture of the pads used in the reactor is given in Figure 10.9. The fuel is a gaseous hydrocarbon, either methane or propane. Methane is usually used where a mains supply is available, and propane is used for portable units. Propane is generally easier to combust catalytically than methane.

The fuel is admitted to the back of the heater unit (the reactor inlet) through a flow distributor, whence it flows by forced convection toward the reactor exit (the front of the heater). It flows first through the insulation blanket, and then into the catalyst pad. The role of the insulation blanket is to reduce energy loss through the back of the unit. The catalyst is usually platinum, and is supported on a fibrous material in the shape of a pad. Oxygen diffuses in a counter-current direction from the front face of the heater (reactor exit). Catalytic combustion occurs in the catalytic pad, causing it heat to a relatively high temperature. Heat is transferred from the front face by radiation to the surroundings. Convection heat losses by the exit of hot exhaust gas are much lower, primarily

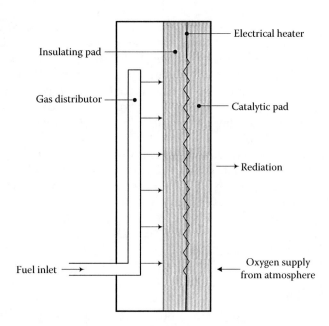

FIGURE 10.8
Schematic of a counter diffusive radiant heater.

because the combustion air is not admitted with the fuel, thus the flow rate of effluent gas is low. Because the fuel is admitted at ambient temperature and the surrounding air is also at ambient, an electrical element is used to start the reactor by heating it sufficiently such that the combustion reaction is initiated. Once this occurs, the heater is switched of and the reactor operates autothermally, provided that the fuel flow rate is adjusted correctly. The advantages of the counter diffusive design are that the reactor operates in an autothermal mode with ambient temperature feed and little heat is lost in the exhaust gas.

FIGURE 10.9
Photograph of the pads used in a counter diffusive radiant heater. On the left is a pad impregnated with catalyst, and on the right is an insulation pad.

With a sufficiently active catalyst, the rate-limiting step is the transport of oxygen across the natural convection boundary layer at the front of the heater. The conversion can be increased by imposing some forced convection at this surface.

There are many instances where heat transfer by radiation is desirable. Examples range from applications in domestic and commercial space heaters to manufacturing processes where paint is dried or materials are cured. There are a number of commercial designs for radiant heaters in operation giving low heat flux (from 10 to 30 kW/m^2), although higher heat flux designs (100–200 kW/m^2) are more difficult to achieve. The units are common in the oil and gas industry. The factor in analyzing this system is that the governing phenomenon is diffusion, so careful attention must be paid to the way in which this term is handled. The conservation equations should be written in terms of mass balances, rather than mole balances for systems where diffusion dominates.

10.12 Monolith Reactors

The packed bed of catalyst pellets is one of the most common types of reactor and for many applications it is an ideal choice. However, there are some applications where it is not suitable to support the catalyst on particles or pellets. For example, packed beds potentially suffer from fluid channeling at the wall, and the possibility of catalyst attrition if the reactor is moved or the fluid velocity is high. There are also potential problems with large pressure drops in some applications. These disadvantages can be overcome by the use of structured catalysts. The development of structured catalyst supports was advanced during the 1970s with the advent of the catalytic converter for the mitigation of harmful automobile exhaust gases. In this application, it is necessary to have a minimal pressure drop, structural rigidity, and an absence of channeling and bypassing of fluid. These requirements were met using monolith supports, also known as honeycomb monoliths.

10.12.1 Structure and Properties of Monoliths

The honeycomb monolith consists of a single block containing a series of straight, uniformly sized and nonconnected channels through which the reacting gas flows. The surface of the monolith support is coated with a layer of high surface area material, called the washcoat, in which the catalyst is dispersed. The washcoat thickness may vary from 10 to 150 μm and usually does not have a uniform thickness around the cell perimeter. The fluid flows down the channels, the reactants are transported to the surface of the washcoat by diffusion, and then diffusion and reaction occur in the porous catalytic washcoat. Monolith channels cross-sectional shapes include circular, hexagonal, square, or sinusoidal. Monolith structures can be manufactured to have a specified size of channel, cell density, wall thickness, and hence a known free cross-sectional flow area. Materials for the support vary, ranging from ceramics to metallic alloys. Typically, monoliths are classified by the number of channels (cells) in a given frontal area, usually in terms of the cells per square inch (CPSI). Commercial units range from 100 to 1200 CPSI.

Figure 10.10 shows the end view of a 400 CPSI ceramic monolith with square channels. A scanning electron micrograph showing the details of the channels is given in Figure 10.11. The catalytic washcoat can clearly be seen. Note that the layer does not have a uniform

FIGURE 10.10
End view of a monolith honeycomb type reactor. The square ceramic substrate has 400 CPSI.

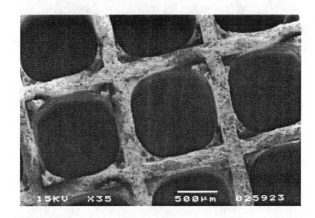

FIGURE 10.11
Close up view of the 400 CPSI ceramic monolith showing the catalytic washcoat. Note that the washcoat is not uniformly distributed.

thickness. This lack of uniformity means that the diffusion path is not the same in all locations, which adds complexity to the computation of effectiveness factors. A monolith formed from corrugated sheets of thin metal is shown in Figure 10.12, with a close up picture of the cells given in Figure 10.13. The monolith core is formed by rolling two sheets of metal, one flat and one corrugated, into the form of a spiral. The resulting channels have a shape resembling a sine wave, and the washcoat also has a nonuniform thickness.

Honeycomb monolith type reactors have been used for both gas phase, liquid phase and multiphase reactions. The analysis that is presented in this book is limited to gas phase systems with laminar flow in the channels, such as used in the automobile catalytic converter. This latter application is the largest application of honeycomb monolith reactors.

FIGURE 10.12
End view of a metal monolith honeycomb type reactor. The spiral wound monolith is made from thin sheets of corrugated metal.

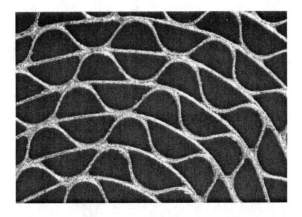

FIGURE 10.13
Close up view of the metal monolith showing the catalytic washcoat. Note that the washcoat is also not uniformly distributed and there is a large accumulation in the corners.

For the automobile catalytic converter application, the inside dimension of the cells is of the order of 0.7–1 mm. Notwithstanding the fact that the gas velocities in the channels is high (of the order of several m/s), the flow in the channels is laminar because of the small size. A typical monolith has an open frontal area of the order of 65% or higher. Furthermore, because the fluid flows down straight channels, the pressure drops is much lower than from a packed bed. Back pressure is particularly critical in an exhaust gas after treatment device.

10.12.2 Analysis of Gas Phase Monolith Reactors

In the discussion of packed-bed reactor modeling, the concepts of discrete and continuous models, as well as pseudo-homogeneous and heterogeneous models, were presented. These same concepts apply to monolith reactor models, with some additional complexities.

There is a fairly large literature on monolith reactor modeling, and a good overview is provided in Hayes and Kolaczkowski (1997). A relatively brief discussion will be presented in the following sections.

Models of monoliths can be divided into two types, each of which has many variants. The first main type is a model of the entire monolith reactor, often taking into account the piping associated with the inlet and outlet streams. In our packed-bed analogy, this model is analogous to modeling a packed bed. The second major type of monolith reactor model is the single channel model, in which one of the channels is modeled for the entire length of the reactor.

10.12.2.1 Complete Converter Models

Models for the complete converter resemble models for packed beds in many respects, in as much as the equations are similar. In both cases, the reactor is effectively treated as a porous medium. For a typical monolith converter, the flow inside the channels is laminar, and therefore Darcy's law is a good approximation of the flow and pressure drop. The flow is constrained by the channels, so there is no transverse flow. Mole and energy balance equations are written for the fluid and solid phases as needed. If the monolith is uniform in temperature, concentration, and velocity in the transverse to flow direction, then a single channel model can be used. However, when modeling the entire monolith, a one-dimensional model is rarely used, and the transverse gradients of temperature and concentration are considered. Detailed description of modeling a monolith using a continuum model can be found in Hayes and Kolaczkowski (1997).

10.12.2.2 Single Channel Models

Prior to discussion on the development of the single channel model (SCM), we illustrate the physical and chemical phenomena occurring inside the channel. Consider the stylized picture of a cross section of a monolith channel, taken along the axial flow direction, as shown in Figure 10.14. The main features are the flow of the gas along the channel in the axial direction. For most applications, especially those conducted at near ambient pressure, the flow in the channels is laminar. The implication of laminar flow is that there

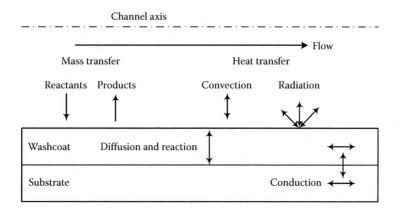

FIGURE 10.14
Cross-sectional representation of a monolith channel showing the heat and mass transfer phenomena that occur. Drawing is not to scale.

will be a well-defined velocity profile in the channel, which will remain near constant after an initial development length has passed. The reaction occurs in the porous wash-coat; therefore, the reactants in the bulk gas phase must first be transferred to the inter-face between the gas and washcoat surface, which occurs by molecular diffusion. Once at the washcoat surface, the reactants diffuse into the porous washcoat, with concomitant reaction. After reaction, the process is reversed, with diffusion back to the washcoat sur-face and then diffusion into the bulk gas. The overall nature of these processes is the same as for any catalytic reactor using a porous catalyst, and the analysis draws on the concepts discussed in Chapters 8 and 9.

There are many levels at which a single channel model can be built. We will start by describing the most realistic (and hence most complex) and then show how the model can be progressively simplified by making more and more approximations. The detailed conservation equations will be developed for one of the simpler, and commonly used, models.

If we consider the geometry of a washcoated channel, as shown in Figures 10.11 and 10.13, it is evident that axial symmetry does not exist, and therefore a model in three spatial dimensions (3D SCM) is required for the most accurate representation. For the square channels of Figure 10.11, the size of the problem can be reduced by assuming symmetry, and thus only one-eighth of the channel must be modeled. The analysis of the channel requires that the conservation equations for momentum, energy, and species must be solved for the fluid, washcoat, and substrate. Although easy to write, the equa-tions are computationally intensive to solve. The solution is done numerically, usually with the help of commercially available software. The availability of such software means that the solution is relatively straight forward, assuming appropriate knowledge of the software, but is still expensive to achieve. The first level of simplification is to use geometric transformation to represent the monolith channels as right cylinders with symmetry about the axis, as represented in Figure 10.15. The washcoat and substrate are defined as annular rings of uniform thickness. If it is desired to model correctly the per-formance of the real channel, it is important to map correctly the relative dimensions of the two systems. This is especially important in transient systems, where slight differ-ences in thermal mass can cause large differences in performance. Ideally one should preserve:

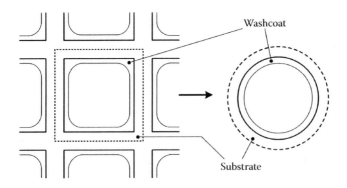

FIGURE 10.15
Geometric transformation used to represent the real geometry as a right circular cylinder with annular layers of washcoat and substrate. The area inside the dashed square is mapped. Drawing not to scale.

The volume (cross-sectional area) of the channel

The volume (cross-sectional area) of the washcoat

The volume (cross-sectional area) of the substrate

The thickness of the washcoat

Hydraulic diameter of the channel

It is not possible to preserve all of these features, and the best compromise is to conserve the three cross-sectional areas. Note that this is the best choice when reaction rates are expressed in terms of the washcoat volume. Preserving the cross-sectional area of the channels also ensures that the mass flow rate per channel is the same in the model system as in the real system. The conservation equations for the 2D model are essentially the same as for the 3D model, except that they can be solved using cylindrical coordinates.

The 2D and 3D SCM are both called distributed parameter models, in that the parameters of interest (velocity, concentration, and temperature) vary spatially in the domain, and such variations are accounted for. The next level of simplification requires the move to a lumped parameter model. The lumped parameter model averages the values of the parameters in one or more of the space dimensions, and thus sacrifices some accuracy for the benefit of model simplicity. We can assume plug flow in the channel, and thus the variations of velocity, temperature, and concentration in the transverse to flow direction are ignored. As will be seen shortly, some dispersion can be added in the axial (flow) direction to correct for the plug flow assumption. By assuming a uniform temperature and concentration in the fluid, equal to the mixing cup values, a discontinuity in temperature and concentration is created at the wall—the wall temperatures and concentrations are different from the average fluid values. This discontinuity of the wall is accounted for using the appropriate heat and mass transfer coefficients that define the rate of heat and mass transfer between the fluid and the solid. The key factor when using a 1D model is the correct evaluation of the heat and mass transfer coefficients. If this is done properly, then close agreement can be obtained between 1D and 2D or 3D models.

Because all radial resolution is lost, it is not possible to include the radial variations of temperature and concentration in the washcoat directly in the model, but rather a submodel has to be used to account for diffusion effects in the washcoat.

10.12.2.3 A 1D Single Channel Monolith Model

The axial dispersion model was discussed in Chapter 6, and its development in the context of an SCM is explained in detail in Hayes and Kolaczkowski (1997). The derivation is not explained in detail; rather the main equations are given. The axial dispersion is superimposed on to a bulk flow term giving the advection dispersion equation. Consider flow down a channel with catalyst located on or in the wall. Species are transported by bulk flow and dispersion, are formed or removed by reaction. The reaction occurs in the solid, thus the reaction is represented by the rate of mass transfer to the surface, using a mass transfer coefficient. Assuming that there is no homogeneous reaction, the steady-state gas phase mole balance equation can be written in terms of the mole fraction as

$$\frac{d}{dz}\left(D_l C_f \frac{dY_{j,f}}{dz}\right) - C_f u_m \frac{dY_{j,f}}{dz} + a_m k_{m,j} C_f \left(Y_{j,f} - Y_{j,S}\right) = 0 \qquad (10.200)$$

The constant a_m is the effective area for mass transfer from the bulk gas to the surface. It is defined as the ratio of the gas volume to the gas/solid interfacial area, which is equivalent to the channel perimeter, P, divided by the cross-sectional area of the channel, A_C. This ratio is related to the hydraulic diameter, D_H, for a duct of arbitrary shape by

$$D_H = 4\frac{A_C}{P} = \frac{4}{a_m} \tag{10.201}$$

The mole balance can thus be expressed in terms of the hydraulic diameter:

$$\frac{d}{dz}\left(D_l C_f \frac{dY_{j,f}}{dz}\right) - C_f u_m \frac{dY_{j,f}}{dz} + \frac{4}{D_H} k_{m,j} C_f \left(Y_{j,f} - Y_{j,s}\right) = 0 \tag{10.202}$$

D_l is the dispersion coefficient in m²/s, C_f is the total fluid concentration in mol/m³, Y_f and Y_S are mole fractions of the component in fluid and solid, respectively. The dispersion coefficient for laminar flow in a circular tube can be calculated from the Taylor–Aris equation:

$$D_l = D_{AB} + \frac{u_m^2 R^2}{48 D_{AB}} \tag{10.203}$$

where D_{AB} is the molecular diffusion coefficient of A in the mixture, u_m is average velocity and R is the tube radius. Equation 10.203 is valid provided that the following condition is met:

$$\frac{L}{R} > 0.16\frac{u_m R}{D_{AB}} \tag{10.204}$$

The boundary conditions for Equation 10.200 are the so-called Dankwerts condition. They are

$$\left(-D_l C_f \frac{dY_{j,f}}{dz}\right)_{0^+} = u_m\left[Y_{j,f0} - Y_{j,f}(0^+)\right] \quad \text{at } z = 0 \tag{10.205}$$

$$\frac{dY_{j,f}}{dz} = 0 \quad \text{at } z = L \tag{10.206}$$

In Equation 10.205, the term 0^+ refers to the position at $z = 0$ but inside the reactor. The mole fraction, $Y_{j,f0}$, is the value for species j in the feed. Unless there is significant reaction near the reactor entrance, a constant inlet mole fraction can be used as the boundary condition instead.

In the solid, at steady state the rate of mass transfer to the surface is equal to the rate of reaction. The solid phase mole balance is thus:

$$k_{m,j} C_f \left(Y_{j,f} - Y_{j,s}\right) = \eta\left(-R_j\right)_S \tag{10.207}$$

The effectiveness factor, η, is included to account for diffusion resistance in the porous washcoat. Its value must be calculated in a submodel, which will be described shortly.

As with any catalytic system, attention must be paid to the units of reaction rate. In Equation 10.207, the rate has units based on the surface area of the catalyst, that is, the interfacial contact area between the fluid and solid. This definition of the rate has been used for monolith reactor modeling in several studies reported in the literature. Although this basis offers some convenience, it is not necessarily the best reference to use, and a rate based on the catalyst volume is often preferred. If the volume of washcoat is V_W, and the interfacial area is denoted A_S, then the mole balance is written as

$$A_S k_{m,j} C_f \left(Y_{j,f} - Y_{j,S} \right) = \eta \left(-R_j \right)_V V_W \tag{10.208}$$

The ratio of the surface area to the catalyst volume is equal to the channel perimeter divided by the cross-sectional area of the washcoat. Obviously the form of this ratio depends on the geometry. For example, for a circular channel of diameter, D_H, with an annular washcoat layer of outside diameter D_{WC}, the mole balance becomes

$$\left(\frac{4 D_H}{D_{WC}^2 - D_H^2} \right) k_{m,j} C_f \left(Y_{j,f} - Y_{j,S} \right) = \eta \left(-R_j \right)_V \tag{10.209}$$

The energy balance in the gas phase includes terms for the bulk flow and conduction.

$$\frac{d}{dz} \left(k_I \frac{dT_f}{dz} \right) - u_m \left(\rho C_P \right)_f \frac{dT_f}{dz} + \frac{4}{D_H} h \left(T_S - T_f \right) = 0 \tag{10.210}$$

k_I is an effective axial thermal conductivity defined by analogy with the Taylor–Aris diffusion coefficient:

$$k_I = k + \frac{\left[u_m R \left(\rho C_P \right)_f \right]^2}{48 k} \tag{10.211}$$

The boundary conditions are also diffusive conditions at the inlet, but it is usually acceptable to impose a fixed inlet temperature and a zero flux outlet condition.

The two fluid phase conservation equations are usually written as steady state, as is the solid phase mole balance. Even for transient operation, the gas phase response times are so much lower than the solid that the fluid is invariably in a pseudo steady state, unless very small time steps are used. However, the solid phase energy balance may be either steady state or transient, depending on how the reactor is operated. The energy balance includes for axial conduction, accumulation (if transient), exchange of energy with the fluid, and the generation of thermal energy owing to reaction. The 1D solid energy balance uses average values over the wall thickness. The conduction term thus depends on the total cross-sectional area of the solid, A_C, and the reaction term depends on the cross-sectional area of the washcoat, A_W. The energy balance can be written as

$$A_C \frac{d}{dz} \left(k_w \frac{dT_S}{dz} \right) - Ph \left(T_S - T_f \right) + \left(-\Delta H_R \right) \rho \left(-R_j \right)_V A_W = A_C \left(\rho C_P \right)_S \frac{dT_S}{dt} \tag{10.212}$$

The area terms again depend on the geometry. For our circular channel of diameter, D_H, with an annular washcoat layer of outside diameter D_{WC}, and with a total diameter of D_S, the energy balance can be written

$$\frac{d}{dz}\left(k_w \frac{dT_S}{dz}\right) - Z_a h\left(T_S - T_f\right) + Z_b \left(-\Delta H_R\right)\rho\left(-R_j\right)_V = \left(\rho C_P\right)_S \frac{dT_S}{dt} \tag{10.213}$$

Where k_w is the effective thermal conductivity of the wall in W/(m · K) and the constants are

$$Z_a = \left(\frac{4D_H}{D_S^2 - D_H^2}\right) \quad \text{and} \quad Z_b = \left(\frac{D_{WC}^2 - D_H^2}{D_S^2 - D_H^2}\right) \tag{10.214}$$

If heat transfer from the ends is ignored, then zero flux boundary conditions can be imposed at each end. The effective thermal conductivity will be a combination of the effective thermal conductivities of the washcoat and substrate, taking into the relative thickness of each layer.

It is necessary to have correlations for the external heat and mass transfer coefficients. Because we have laminar flow in a duct, the values of Sherwood and Nusselt number are similar in magnitude to the classical values for laminar flow. We recall that for laminar fully developed flow the Nusselt and Sherwood numbers are constant, whilst in the developing flow region they depend on the inverse Graetz number. Because the fluid is a variable density gas, with a chemical reaction, the values are not exactly the same as for a fluid of constant properties, but are close. Typical generalized correlations for a fluid with a Prandtl number of 0.7 have the following form, where the value of the constant depends on the cell shape and the boundary conditions imposed on the channel walls.

$$Nu = C_1 \times \left[1 + C_2\sqrt{Gz}\exp\left(-\frac{50}{Gz}\right)\right] \tag{10.215}$$

The Graetz number is defined as

$$Gz = \frac{D_H}{z}\,\text{Re}\,\text{Pr} \tag{10.216}$$

Two different values of the Nusselt number exist for the two classical wall boundary conditions. For a constant wall temperature, we define Nu_T and for a constant wall flux we define Nu_H. The local Nu value in the system can be approximated by interpolating between Nu_T and Nu_H using the interpolation formula of Brauer and Fetting (1966):

$$\frac{Nu - Nu_H}{Nu_T - Nu_H} = \frac{Da\,Nu}{(Da + Nu)\,Nu_T} \tag{10.217}$$

The Damköhler number, Da, is defined for an arbitrary reaction of component A by

$$Da = \frac{\eta\left(-R_A\right)_S D_H}{4C_{A,S}D_{AB}} \tag{10.218}$$

Note that the rate is defined in terms of the washcoat interfacial area. This formula does not give the exact values, but is sufficient to provide a reasonable estimate in most cases. The value for the constants depends on the channel geometry.

Channel Shape	Nu_T, Sh_T		Nu_H, Sh_H	
	C_1	C_2	C_1	C_2
Circle	3.657	2.7	4.364	3.0
Square	2.977	3.6	3.095	4.8
Triangle	2.47	4.0	1.89	4.8
Hexagon	3.34	3.0	3.862	3.9

The Sherwood number was obtained by substituting the Schmidt number, Sc, for the Prandtl number in the equation for Graetz number, that is, Equation 39.

10.13 Reactors for Gas–Liquid–Solid Systems

There are two main types of reactor used for reacting mixtures of liquids and gases over solid catalysts. These are the slurry reactor and the trickle-bed reactor, both of which were introduced in Chapter 7. The slurry reactor resembles a stirred tank or a fluidized bed, whilst the trickle-bed reactor is a fixed-bed reactor with two phase flow. Illustrations of the two reactor types are given in Figures 7.8 and 7.7, respectively.

10.13.1 Slurry Reactors

The slurry reactor consists of a stirred tank of liquid through which a gas is passed by means of a sparger. The solid catalyst is suspended in the liquid in much the same way that the solid catalyst is suspended by the flowing gas in a fluidized bed. A key feature of the slurry reactor is that it permits a relatively uniform temperature in much the same way that a well-mixed stirred tank does. For very exothermic reactions, this provides good temperature control. Also, the high heat capacity of the liquid provides an excellent heat sink. The good mixing also means that the external diffusion resistance can be reduced significantly. Small particles of catalyst are normally used, both to achieve a good level of suspension without excessive stirrer speeds, and to eliminate internal diffusion resistance. This latter point can be significant, because the rate of diffusion in liquids is much smaller than in gases. One drawback of the use of a very fine catalyst is that it may be difficult to contain the catalyst in the system.

Modeling of slurry reactors is more complex than for gas phase fixed-bed reactors, and is not discussed in detail. The mass transfer resistances are generally significant, and additional complexities are introduced by the need to have simultaneous contact of liquid, solid, and gas. There are general descriptions of slurry reactor analysis in most advanced reaction engineering textbooks, but for the latest models the reader is advised to consult the research literature.

10.13.2 Trickle-Bed Reactors

The trickle-bed reactor is nothing more than a fixed-bed reactor through which a mixture of gas and liquid reactants flow. Because it is a packed bed, the catalyst sizes are larger than for the slurry reactor. The catalyst is thus easy to hold in the bed, although the internal diffusion resistances will naturally be more significant. Most trickle-bed reactors operate in co-current downward flow mode. Some units use co-current upward flow, a mode of operation that is referred to as flooded flow. Counter current flow of gas and liquid is usually not used.

The most common use for the trickle-bed reactor is in the hydrotreating of petroleum fractions. Hydrotreating is a process used to remove sulfur from the feed. In many cases, the sulfur contained in petroleum is chemically bound within larger molecules. Simply to extract all of the molecules containing sulfur atoms would waste a lot of product. Rather, the hydrocarbon molecules containing sulfur are cracked and the sulfur reacted with hydrogen to form hydrogen sulfide, which is subsequently extracted from the product. Numerous other chemical reactions occur at the same time.

The feed to a hydrotreating reactor may consist of any one of a number of petroleum fractions. When a lighter naptha is to be treated, it is usually partially vaporized prior to entering the reactor, thus the down flowing gas is a mixture of hydrogen and hydrocarbons. Heavier petroleum fractions will not be significantly vaporized prior to entering. The reactors are typically operated at high pressure, up to 70 bar. The catalyst is usually either a mixture of cobalt and molybdenum oxides, or nickel and molybdenum oxides. The catalytically active for is sulfide, and the oxide catalysts are pretreated *in situ* with hydrogen sulfide prior to use.

Analysis of trickle-bed reactors is also a complicated task. Not only are the mass transfer steps important, but the hydrodynamics of the multiphase flow are not completely understood. As with the slurry reactor, the detailed consideration of the trickle-bed reactor is outside of the scope of this text.

10.14 Summary

In this chapter, we have considered a variety of models for catalytic reactors, focusing on packed beds. These models range from the relatively simple to the very complex, and one should always aim to use the simplest possible model that will give reliable results. Also, as noted elsewhere in the text, there are many different types of catalytic reactors in use, and there analysis is usually not straightforward. When faced with such analysis challenges, the reader should consult some of the many advanced books available on this subject.

PROBLEMS

10.1   A first-order isothermal reaction occurs in a bed of 3 mm diameter spherical catalyst pellets. The first-order rate constant is 0.3 s^{-1}. The effective diffusivity in the catalyst pellets is 3×10^{-7} m^2/s. The reactor length is 2 m. It is desired to replace the 3 mm diameter particles with 6 mm diameter particles. Calculate the length of the reactor required to maintain the same conversion as before. Assume that the bulk densities of the two beds are the same.

10.2

    a. Consider an isothermal fixed-bed reactor with axial mixing superimposed on plug flow (axial dispersion model). Show that, for a given set of operating conditions, the effect of axial diffusion decreases with increasing length.

    b. For the following set of parameters, compute the axial concentration profile as a function of total length and compare it to the profile obtained for pure plug flow.

    Mean bulk velocity of 0.01 m/s

    Bed porosity of 0.4

    Axial Peclet number of 2 (based on the particle diameter)

    Pellet diameter of 4 mm

    Bulk density of the bed 1200 kg/m^3

    First-order reaction rate constant $k = 1 \times 10^{-5}$ m^3/kgcat $\cdot$ s

    c. Verify for this case whether or not a bed length of 50 pellet diameters is sufficient to eliminate the effects of axial diffusion.

10.3 Consider a catalytic fixed-bed reactor. The process stream consists of a small amount of an impurity, which is destroyed by catalytic oxidation. Because the concentration of impurity is small the reactor may be considered to be isothermal. The intrinsic reaction rate is very high and as a result the reaction is completely controlled by the rate of mass transfer to the external surface of the catalyst pellets. Therefore, the concentration at the external surface of the catalyst is essentially equal to zero and the internal diffusion is not important. A current reactor is operating with a fraction conversion of 0.6. A new reactor is to be built in which the following changes will be made.

    a. The catalyst diameter will be one-third the present diameter.

    b. The mass of the catalyst will be doubled, but the reactor diameter will be the same.

    c. The reactant flow rate will be increased by a factor of 5.

    Assume that the bed density remains the same and that in the new scenario the reaction is still mass transfer controlled. Calculate your best estimate of the conversion from the new reactor. State clearly all of your assumptions.

10.4 Consider a tubular reactor of circular cross-sectional area through which a fluid is flowing in plug flow. There is a catalyst coated on the wall and there is no homogeneous reaction. The reaction and rate are

$$A \rightarrow B \quad (-r_A) = k C_A \text{ mol/m}^2\text{s}$$

    Note that the rate is based on the surface area of the wall. Consider the two extremes of operating conditions. (a) The fluid velocity is very high and the rate constant is small and (b) the fluid velocity is low and the reaction rate is extremely fast (the rate constant is very large). Derive a single mole balance equation for each of cases (a) and (b). Clearly state the meaning of all of your symbols and show that the units are consistent.

10.5   Consider a tubular reactor with a nonporous catalyst coated on the wall. A liq-
        uid solution containing reactant A flows through the tube in laminar flow with
        an initial concentration, C_{A0}. The reaction

$$A + B$$

occurs homogeneously via a first-order reaction and heterogeneously via a
Langmuir–Hinshelwood mechanism:

$$A + S \rightleftharpoons A \cdot S$$

$$A \cdot S \rightleftharpoons B \cdot S$$

$$B \cdot S \rightleftharpoons B + S$$

where S represents a surface site and the surface reaction is rate controlling. The
reactor is heated electrically which gives a constant heating flux over the entire
reactor. Derive the relevant PDE, which describe the reactor (2-D model) and
give the necessary boundary conditions.

References

Bird, R.B., W.E. Stewart, and E.N. Lightfoot, 2002, *Transport Phenomena*, 2nd Ed., Wiley, New York.

Brauer, H.W. and F. Fetting, 1966, Stofftransport bei wandreaktion im einlaufgebiet eines strömung-
    srohres, *Chemical Engineering Technology*, **38**, 30–35.

Carberry, J.J., 1976, *Chemical and Catalytic Reaction Engineering*, McGraw-Hill, New York.

Churchill, S.W., 2007, The interaction of reactions and transport Part II: Combined reactors and
    exchangers, *International Journal of Chemical Reactor Engineering*, **5**, 1542–6580.

Darcy, H.P.G., 1856, *Les Fontaines Publiques de la Ville de Dijon*, Victor Dalmont, Paris.

De Wasch, A.P. and G.F. Froment, 1972, Heat transfer in packed beds, *Chemical Engineering Science*, **27**,
    567–576.

Dixon, A.G and D.L. Cresswell, 1979, Theoretical prediction of effective heat transfer parameters in
    packed beds, *AIChE Journal*, **25**, 663–676.

Ergun, S., 1952, Fluid flow through packed columns, *Chemical Engineering Progress*, **48**(3), 89–94.

Feyo de Azevedo, S., M.A. Romero-Ogawa, and A.P. Wardle, 1990, Modelling of tubular fixed-bed
    catalytic reactors: A brief review, *Transactions of the Institute of Chemical Engineers, Part A:
    Chemical Engineering Research and Design*, **68**, 483–502.

Forchheimer, P.H., 1901, Wasserbewegung dutch doden, *Zeitschrift des Vereines Deutscher Ingenieure*,
    45, 1782–1788.

Froment, G.B., K.B. Bischoff, and J. DE Wilde, 2011, *Chemical Reactor Analysis and Design*, 3rd. Ed.,
    Wiley, New York.

Hayes, R.E., A. Afacan, and B. Boulanger, 1995, An equation of motion for an incompressible
    Newtonian fluid in a packed bed, *Transport in Porous Media*, **18**, 185–198.

Hayes, R.E. and S.T. Kolaczkowski, 1997, *Introduction to Catalytic Combustion*, Gordon and Breach,
    Reading.

Incropera, F.P., D.P. DeWitt, T.L. Bergman, and A.S. Lavine, 2006, *Introduction to Heat Transfer*, 6th Ed.,
    John Wiley & Sons, New York.

Kunii, D. and O. Levenspiel, 1991, *Fluidization Engineering*, 2nd Ed., Butterworth-Heinemann,
    Newton.

Levenspiel, O., 1999, *Chemical Reaction Engineering*, 3rd Ed., John Wiley & Sons, New York.

MacDonald, I.F., M.S. El-Sayed, K. Mow, and F.A.L. Dullian, 1979, Flow through porous media—The Ergun equation re-visited, *I&EC Fundamentals*, **18**, 199–208.

Radcliffe, S.W. and R.G. Hickman, 1975, Diffusive catalytic combustors, *Journal of the Institute of Fuel*, Dec, 208–214.

Schlünder, E.U., 1978, Transport phenomena in packed bed reactors, *Chemical Reaction Engineering Reviews—Houston*, ACS, 110–161.

Wakao, N. and S. Kaguei, 1982, *Heat and Mass Transfer in Packed Beds*, Gordon and Breach, London.

Wilhelm, R.H., 1962, Progress towards the *a priori* design of chemical reactors, *Pure and Applied Chemistry*, **5**(3), 403–422.

Yagi, S. and N. Wakao, 1959, Heat and mass transfer from wall to fluid in packed beds, *AIChE Journal*, **5**, 79.

11

Experimental Methods in Catalysis

The development of any new process or catalyst begins with a research program, in which efforts are made to gain an understanding of how a catalyst will operate in a full-scale unit. Although most practicing engineers will not be involved in such studies, it is necessary to have an understanding of how they are conducted so that the results from these investigations can be interpreted correctly.

Experimental studies in catalysis can be divided into three broad types. The first type is kinetic investigations designed to gather sufficient data to develop a rate expression which can be used in reactor analysis and design. These types of studies are always performed when a new catalyst is developed for a process, or improvements to the performance of a catalyst are sought. In these experiments, measurements are performed on catalysts that may be in the form of a powder or pellets, or coated onto a support. Ideally, experimental conditions should cover the range of temperatures and pressures likely to be encountered in the reactor where the catalyst will be used—in practice, because of cost and other factors; this is not always the case.

The second type of investigations can be described as catalyst characterization studies. These studies can be further divided into those used to obtain general physical information, such as effective diffusivities, pore size and surface area analysis, and so on, and what are classified as surface science studies. The latter use very sophisticated techniques to determine information about the chemical state of the catalyst and its active constituents. In these experiments, measurements are made to understand the surface structure of a catalyst at the atomic level, and how the catalyst works at the molecular level.

The third type of study can be called catalyst performance or screening trial. These are experiments conducted to study the suitability of catalyst formulations and to compare the performance of different catalysts. Experimental conditions are usually not as rigorously controlled as in kinetic studies. The objective in such tests may be to determine catalyst longevity, to study the selectivity of different formulations, or simply to determine which is the best catalyst among a variety of contenders. Data from catalyst screening trials are usually not suitable for formulating rate expressions.

The design of an experimental program in catalysis usually requires much specialized knowledge, which is beyond the scope of this book. The purpose of this chapter is to provide an introductory overview of some of the important considerations that must be considered in any such program, and to give sufficient information to understand reports of them. Some references that give more details on experimental methods include Wjingaarden et al. (1998).

11.1 Kinetic Investigations

The analysis and design of a reactor requires that the reaction rate be expressed as a suitable rate expression. We have discussed rate functions at length in the previous chapters, and we have seen that rate equations can be based on the underlying reaction mechanism or be completely empirical. Regardless, as we have stated before, and repeat here for emphasis, essentially all rate expressions contain constants that must be determined experimentally. These rate constants are normally determined from experimental data obtained in the laboratory or, less commonly, pilot-scale equipment. In Chapter 5, we considered some of the methodology that is used for modeling homogeneous reactions, the collection of rate data, and its subsequent analysis. For catalytic systems, the general methodology for testing rate equations and determining the parameters thereof is similar. The challenge in catalytic systems is to collect data that can be transformed easily into reaction rates at known values of concentration or partial pressure.

In Chapters 7 through 9, we discussed at length the steps in the catalytic reaction, consisting of transport and reaction steps. What is usually desired from a kinetic investigation is the reaction rate determined in the absence of external and internal heat and mass transfer limitations. We have seen in Chapter 9 how these effects can alter the observed rate. In technical terms, we desire the *intrinsic* reaction rate, which can be described as the rate in the absence of transport limitations, as opposed to the *global* rate, which is the actual rate that may include such transport effects. We have seen in Chapter 10 how the intrinsic kinetics and the transport terms are combined in full-scale reactor analysis. The transport effects can often not be determined *a priori*, and therefore a set of experiments is usually performed to test for the presence of transport limitations.

The danger of extrapolating engineering models of any type is well known, and care should be taken if it is essential to do so. That said, it is prudent to perform the kinetic experiments over as wide a range of operating conditions as possible. The expected range of operation of the full-scale reactor should be encompassed if possible. Even if the model is based on the underlying mechanism, the mechanism can change with temperature, so care should be taken.

The initial objective of the experimental work is to generate a table of reaction rates and the corresponding reactant and product concentrations and temperature. These data are then used to fit kinetic models using the same techniques discussed in Chapter 5.

11.1.1 Laboratory Reactor Types

There are many factors to consider when selecting a laboratory-scale reactor for the collection of rate data. Some of these factors include ease of construction and operation, the ease with which transport effects are quantified and eliminated, the ease of data analysis, the accuracy of the measurement system, and many more. The choice is rarely simple, and all factors should be considered before making the final decision. In the following, we discuss some of the possibilities, and weigh the pros and cons of each. Usually, catalytic kinetic data are acquired from experiments conducted in a continuous flow reactor, operating at an assumed steady state, where either isothermal or adiabatic conditions. In an isothermal system, the temperatures of the fluid and solid are assumed to be uniform throughout the reactor. In an adiabatic system, it is assumed that there is no heat loss or gain at the reactor wall. For reactions with a large enthalpy change of reaction, isothermal operation may be difficult to achieve. However, because laboratory reactors are usually small, it is also difficult to achieve true adiabatic operation. For simplicity of data analysis, isothermal operation is usually preferred.

Although there are potentially many designs of reactors that can be used to obtain data for kinetic analysis, we will limit the discussion to two broad types, and discuss their various modes of operation. These types are the tubular reactor, which can be near plug flow in performance, and reactor types that behave like continuous stirred tank reactors. The use of either of these two mixing patterns allows for the use of the mole and energy balance equations developed in Chapters 2 and 3. We should point out that the transformation of conversion data into rates of reaction is predicated on the correctness of the underlying reactor model and its assumption. These assumptions of perfect mixing or plug flow can be tested by doing transient experiments without reaction to measure the residence time distribution curves, as discussed in Chapter 6. Comparison of observed and experimental curves will give an indication of the flow patterns within the reactor, and allow a judgment to be made as to the validity of the assumptions. There is a fairly extensive literature on the use of various laboratory reactor designs, and the challenges involved in designing a rigorous experimental program. The review of Doraiswamy and Tajbl (1974) is a good starting point.

11.1.2 Experimental Tubular Reactor

In Section 5.6.8, it was stated that the tubular reactor was seldom used for obtaining reaction rate data. While this statement is generally true for homogeneous systems, the tubular reactor is widely used for catalytic reactions.

Tubular reactors potentially offer many advantages over other reactor types. It is conceptually the simplest form of a reactor to build and to operate. Furthermore, because there is so much experience available in the form of literature results, their behavior is the best understood. In this type of reactor, a small quantity of the catalyst is positioned inside a tube through which the reactants are passed. The catalyst is usually held in place by metal gauze or quartz wool. Conversion across the reactor is determined by measuring inlet and outlet concentrations of the reactants and products. Usually, these reactors are operated in plug flow, which requires a uniform velocity across the reactor cross section with no significant channeling/by-passing of the gas between the catalyst and the inside wall of the reactor. This latter effect is not always easy to avoid. When a tube is packed with catalyst pellets or powders, the wall introduces some local ordering to the packing which increases the local porosity, causing some channeling. A general rule for minimizing the significance of channeling is to ensure that the reactor diameter is greater than 10 times the diameter of the catalyst pellets. To minimize the effects of axial dispersion, the reactor length should be greater than 50 particle diameters. It should be noted that, depending on the catalyst particle diameter, these criteria may lead to relatively large experimental reactors. For reactions with strong heat effects, it may be difficult to ensure isothermal operation.

As mentioned earlier, the reactor should be operated without transport limitations. To reduce internal transport limitations, as discussed in Chapter 9, a small catalyst particle is indicated. The external transport effects can be reduced by the use of a suitably high fluid flow rate. Experimental methods for determining the presence or absence of these limitations are discussed shortly.

An attempt should be made to design the reactor to operate in either adiabatic or isothermal mode, the analysis of data from isothermal operations being considerably easier.

11.1.2.1 Isothermal Conditions Maintained in Reactor

If isothermal conditions are to be maintained in the reactor, then it is necessary to remove excess heat from the reactor. This requires good heat transfer from the center of the bed to the wall which in turn implies that small-diameter reactors be used. A rough guideline is that the reactor diameter should be less than five particle diameters. It should be noted that this criterion conflicts with the one mentioned earlier to avoid channeling, and obviously a compromise must be made. In a packed bed, inert material may also be added to the bed to act as a heat sink and aid the distribution of the heat generated by the reaction. If this material is added, it should be done uniformly, to avoid complexities in the data analysis.

11.1.2.2 Adiabatic Operation

In adiabatic operation, the reactor is insulated so that (ideally) there is no heat loss from the reactor. The reactor temperature therefore changes with position. Due to the exponential dependence of the reaction rate on temperature, it can vary considerably over the reactor length. In addition, the high heats of reaction associated with combustion reactions together with heat transfer from the bed could make this mode of operation difficult to sustain. In any case, the mole and energy balance equations must both be solved simultaneously before reaction rate data may be deduced, which is considerably more complex than for the isothermal case, in which only a mole balance equation must be solved. We shall not discuss this mode of operation further.

There are two generally accepted methods of operating a tubular reactor for a kinetic investigation. Each method has its strengths and weaknesses, and the method of data reduction is different in each case. The two modes of operation are called differential or integral. A differential reactor is always operated in the isothermal mode, while an integral reactor can be either isothermal or adiabatic.

11.1.2.3 Differential Operation

In the differential mode of operation, the reactor is operated at low conversions, typically <5%, although conversions as high as 10% have been reported. The main advantage of this method of operation is the ease of extracting reaction rate data from the measured conversion. Recall the mole balance equation for species A in a PFR

$$-\frac{dF_A}{dW} = F_{A0}\frac{dX_A}{dW} = (-r_A) \tag{11.1}$$

We now make the key assumption of this method, and express the differential term by a simple difference formula, based on the inlet and outlet conversions. Assuming that the inlet conversion is zero, the approximate equation is

$$F_{A0}\frac{X_A}{W} \approx (-r_A) \tag{11.2}$$

It is seen that the rate is simply calculated from the conversion data. The simplicity of this calculation will be better appreciated after the discussion of integral reactors, which follows. When using this reaction rate in a regression analysis, it should be noted that the concentration of reactants to be used for this rate value is the average of the inlet and outlet concentrations.

One of the advantages of differential operation is that the small conversion helps to maintain an isothermal system. However, there are also some drawbacks to this mode of operation that must be pointed out. The first relates to its primary advantage, the low conversion. Because the difference between the inlet and outlet concentrations is low, it is necessary to have extremely accurate and precise measurements of the concentration, to avoid large errors in the resulting rate.

Another feature of the differential reactor is both an advantage and a disadvantage. Because the conversion is low, there are few products formed, so their influence on the rate is not so significant. This fact can be considered an advantage because it means that the rate in the absence of products can be determined, which can help in the data modeling. However, if the products do affect the rate, it is necessary to conduct experiments with reaction products present in the feed. Sometimes this requirement can pose experimental difficulties. We also note that each experiment gives one data point for the resulting regression analysis, potentially leading to a larger number of experiments than required for the integral method.

11.1.2.4 Integral Operation

When a tubular reactor is operated in integral mode, the conversions are allowed to be large; typically values between 20% and 80% are used. These high conversion levels offer the immediate advantage in the chemical analysis. The accuracy and precision required in the gas analysis are much less than those required for the differential method to give the same level of accuracy in the resulting rate value. A drawback is that the higher conversion levels may lead to larger problems in maintaining isothermal operation, if that is what is desired.

The analysis of the data from an integral reactor is more complicated than that used for differential operation. The techniques used for data reduction and rate modeling are analogous to those used for the data obtained from a constant volume batch reactor, which were discussed thoroughly in Section 5.6. Those techniques will be briefly summarized here.

Typically, the experiments performed in an investigation will involve the determination of conversion at different feed flow rates at a variety of operating conditions. That is, the inlet concentration and temperature would be fixed, and then the conversion will be measured as a function of the total feed flow rate. This step is repeated for a variety of inlet conditions, which would result in a set of curves like those illustrated in Figure 11.1. The interpretation of these curves can then be done in one of the two ways, using either a differential or an integral method of analysis.

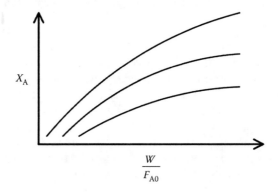

FIGURE 11.1
Conversion curves obtained in a tubular reactor under different conditions.

In the differential method of analysis, each of the curves is differentiated to determine the rate. The PFR equation can be written as

$$\frac{dX_A}{d(W/F_{A0})} = (-r_A) \tag{11.3}$$

It is evident that the slope of the curve at any point equals the reaction rate that corresponds to that level of conversion. The challenge is to find a reliable method of determining the slope. Methods include the finite difference formulae discussed in Section 5.6, or fitting a curve to the data and differentiating it. The pros and cons of these methods are discussed in Section 5.6 for the CVBR.

In the integral method of analysis, a rate equation is proposed and then substituted into the mole balance equation. The mole balance equation is then integrated, and optimization techniques are used to see if there exists a set of parameters that will give a match between experimental and predicted conversions. The integral method is more appropriate for simple rate functions, and most of the time, the differential method of analysis is used.

The integral method of reactor operation offers additional advantages. With high conversion levels, the products are present naturally in the flow stream, and thus the effect of products can be determined. The reaction rate in the absence of products can be calculated by extrapolating the curve to zero conversion. Furthermore, each curve can generate a multitude of data points for the regression analysis. Although a large number of experiments may need to be done to generate each curve, it is generally easier to simply vary the total flow rate with the other conditions held constant.

To summarize the use of the tubular reactor, it has been seen that there are a number of options both for operation and data analysis. The best choice for a given situation will depend on the nature of the reactions and the catalyst. There is not single best method that covers all eventualities.

11.1.2.5 Testing for Mass Transfer Limitations

We have stated that intrinsic reaction kinetics are usually desired, that is, internal and external mass transfer limitations should not be present. Although it is sometimes possible to estimate these effects computationally, it is usually necessary to perform a set of experiments.

The determination of the presence or absence of internal diffusion limitation is relatively straightforward. We know from Chapter 9 that the Thiele modulus is directly proportional to the catalyst size. It follows that if a set of experiments are performed in which the only variable is the catalyst size, the onset of diffusion limitation will be apparent. Figure 11.2 shows a typical result that might be obtained if conversion is plotted as a function of increasing pellet size. The conversion is constant as the diameter increases, until a critical value is reached where the conversion starts to decrease. At this point, internal diffusion resistance starts to become significant. These experiments would normally be repeated over the temperature range of the experiments. When the temperature increases, the diameter at which the onset of diffusion limitation occurs decreases, owing to the increase in the rate constant with temperature.

There are two sets of experiments that can be used to determine the presence or absence of external diffusion limitation. Both are predicated on the fact that the external resistance

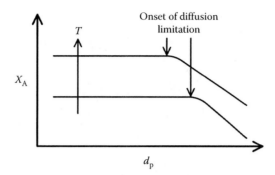

FIGURE 11.2
Testing for internal diffusion resistance with increasing particle diameter. If the fractional conversion is independent of the particle size then internal resistance is negligible.

decreases as the flow rate increases. The first way to test for these effects is to conduct a series of experiments with a constant ratio of catalyst mass to flow rate over a range of flow rate. Obviously, when the flow rate is changed, the catalyst mass must also be changed to keep the ratio constant. If the external resistance is insignificant, then the conversion is the same in all cases. Figure 11.3 shows what might be observed in practice. At low flow rate (and corresponding low catalyst mass), the conversion is seen to increase with flow rate, until it reaches a plateau. This point corresponds to the elimination of external resistance.

The preceding set of experiments requires many different catalyst charges to the reactor, of different mass. To avoid changing the catalyst so many times, an alternative set of experiments may be performed. The alternative is to vary the flow rate for a given catalyst mass, and to measure the conversion. This set of experiments is repeated for a different catalyst mass (usually twice or one half the previous value). A set of curves like those illustrated in Figure 11.4 is thus generated. If the curves for the two catalyst masses do not coincide, as shown in Figure 11.4a, then mass transfer limitation is present. If the curves do coincide, then diffusion limitation is not significant. The flow rate and catalyst mass must be adjusted until a set of conditions is found where external resistance is negligible over the range of temperatures and reactant concentrations desired.

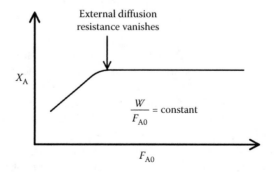

FIGURE 11.3
Testing for external diffusion resistance at constant W/F ratio with increasing flow rate. If the fractional conversion is independent of the flow rate, then external resistance is negligible.

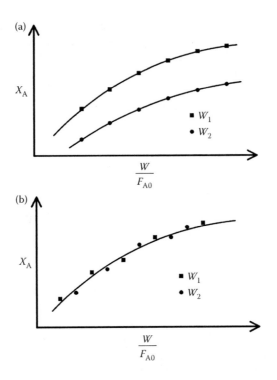

FIGURE 11.4
Testing for external diffusion resistance at variable W/F ratio with different catalyst mass. If the fractional conversion is the same for both masses of the catalyst, then external resistance is negligible. In (a), the conversion is not the same, so external diffusion resistance is important. In (b), both curves are coincident, so the external diffusion resistance is negligible.

11.1.3 CSTR-Like Reactors

In Chapter 5, we saw that calculating the reaction rate from fractional conversion in a CSTR was straightforward. From the mole balance for a CSTR, for species A, written in terms of the catalyst mass, W, we can write the reaction rate as

$$\left(-r_A\right) = \frac{WX_A}{F_{A0}} \tag{11.4}$$

Thus, if the inlet and outlet compositions and flow rates are known, the reaction rate easily follows. Any level of conversion can be used in the reactor, and products can be added to the feed as required. The perfectly mixed CSTR also has the advantage of having a uniform temperature. Because of the simplicity of data analysis, this type of reactor is preferred by many experimentalists. However, while it is relatively easy to envisage the design of a CSTR for homogeneous reactions, the presence of the catalyst makes the design of a catalytic CSTR more problematic.

The obvious challenge in a CSTR with a gas and solid catalyst is to achieve good gas mixing and to have uniform contact of the gas with the catalyst. In other words, all concentration gradients, including those across the catalyst boundary layer, must be effectively eliminated. Clearly, there must be an effective agitation, and the catalyst must be placed in the reactor in such a way to have uniform gas flow around each particle. Although there are many variations on this design theme, there are two approaches that are used. In the first, the catalyst is fixed and the fluid circulated through it. This type of reactor is often called a Berty reactor. In the second design, the catalyst is contained within a basket which forms the impeller; this basket is rotated rapidly through the fluid mixture. The latter design is typified by the spinning basket reactor developed at the University of Notre Dame (Carberry, 1976). Each of these types is discussed briefly in the following sections.

11.1.3.1 Berty Reactors

In the Berty reactor, the catalyst is held fixed within the reactor volume, usually by means of a wire basket. Alternatively, in some special applications, the catalyst may be coated

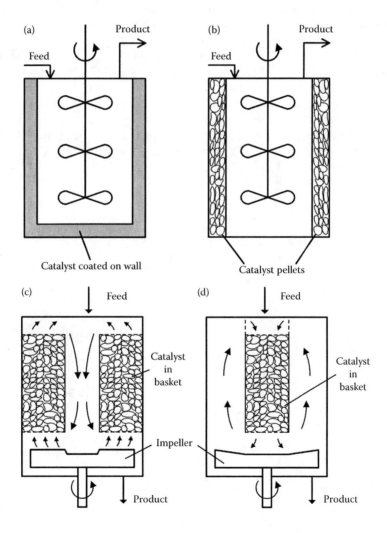

FIGURE 11.5
Some different configurations possible for Berty-type reactors. (a) Catalyst coated on the wall with centrally mounted impellers, (b) catalyst pellets in an annulus with central impeller, (c) catalyst in an annular basket with bottom mounted impeller, and (d) centrally mounted catalyst basket with bottom impeller. (Based on figures in Doraiswamy, L.K. and D.G. Tajbl, 1974, *Catalysis Reviews—Science and Engineering*, **10**, 177–219.)

onto the wall of the reactor. There are many variations on this reactor type, and four possible scenarios are illustrated in Figure 11.5, adapted from Doraiswamy and Tajbl (1974). Other examples are shown in that reference. Part (a) and (b) show a system with a centrally mounted impeller, while (c) and (d) show an arrangement with an impeller at the bottom. One of the requirements is to have a good penetration of the gas through the catalyst bed, without significant bypassing of gas. Usually, it is necessary to use relatively large catalyst particles, which may lead to internal transport resistance. One advantage of the fixed catalyst is that it may be possible to have a thermocouple inserted into the bed, which gives a more reliable indication of the catalyst temperature.

11.1.3.2 Carberry (Spinning Basket) Reactor

In a stirred or spinning basket reactor, the catalyst is contained inside a basket that is rotated inside a vessel—the basket behaves as a stirrer. Figure 11.6 shows a schematic of the design from the University of Notre Dame (Carberry, 1976). In this version, the catalyst is contained in the wire basket with four "paddles." Other designs of catalyst bed have been reported. Baffles and auxiliary mixing devises may be used as an aid to ensure uniform concentration and temperature. As with the Berty reactor, it is important to have good gas penetration within the bed, which again usually requires larger catalyst particles. Inside the stirred reactor, the composition and temperature of the reacting mixture is assumed to be uniform, and gas and catalyst temperatures are assumed to be the same. As the basket is rotating, it is difficult to measure the temperature of the catalyst bed.

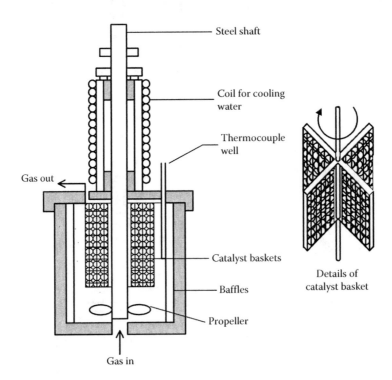

FIGURE 11.6
The Notre Dame spinning basket reactor. (Adapted from Carberry, J.J., 1976, *Chemical and Catalytic Reaction Engineering*, Mcgraw-Hill, New York.)

11.1.3.3 Transport Limitations in Catalytic CSTR

Both the Berty and spinning basket reactors can be operated in regimes where internal and external mass and heat transfer resistances may be significant. To test for internal resistances, one can vary the size of the catalyst pellets used, in the same manner that was discussed earlier for fixed beds. The drawback is that the general design limitations of these designs may not permit a sufficiently small catalyst size to be employed. It is relatively easy to test for the presence of external transport limitations. The usual practice is to observe the outlet fractional conversion as a function of the stirrer speed. Figure 11.7 shows how the conversion might vary as the speed increases: when the conversion becomes constant it is safe to assume that external resistance is negligible.

11.1.3.4 External Recycle Reactor

The CSTR systems described above have a certain complexity in both construction and operation, and some drawbacks with regard to catalyst size. Another type of reactor that combines some of the advantages of the tubular reactor with the data analysis simplicity of the CSTR is the recycle reactor. This system, which is illustrated in Figure 11.8, consists of a tubular reactor with a recycle loop. In the figure, points 1 and 2 represent positions at the inlet and the outlet of the reactor, respectively. A pump is used to recycle a fraction of the effluent from the reactor exit to its inlet. The recycle ratio is usually fairly large, which leads to a high flow rate in the bed. The high flow rate minimizes external transport resistances, and leads to a fairly small incremental conversion across the reactor. At the same

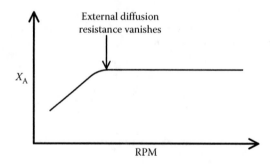

FIGURE 11.7
Testing for external diffusion resistance with increasing speed. If the fractional conversion is independent of speed, then external resistance is negligible.

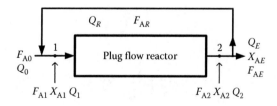

FIGURE 11.8
A recycle reactor. The reactor itself operates in plug flow, and a significant portion of the reactor effluent is recycled to the reactor inlet. At very high recycle rates, the reactor operates as a CSTR.

time, relatively high overall conversion can be achieved. This combination means that the reactor is essentially being run in differential mode with an integral conversion level. Since a fixed bed of catalyst is used, relatively small particles can be employed, which minimizes internal transport resistance. The drawbacks include the relatively large amount of piping used, which increases the dead volume. If homogeneous reactions are significant, this volume may be important. If the reaction is conducted at high temperature, then a large amount of cooling and subsequent reheating may be required to avoid damaging the recycle pump.

The mole balance around a recycle reactor is now developed. The process variables in Figure 11.8 are defined here for emphasis:

F_{A0} molar flow rate of reactant A in the fresh feed to the system
Q_0 volumetric flow rate of the fresh feed to the system
F_{A1} molar flow rate of reactant A to the reactor, a mix of recycle and fresh feed
Q_1 volumetric flow rate to the reactor, the sum of the fresh feed and the recycle
F_{A2} molar flow rate of reactant A from the reactor
Q_2 volumetric flow rate from the reactor
F_{AE} molar flow rate of reactant A from the system
Q_E volumetric flow rate from the system
F_{AR} molar flow rate of reactant A in the recycle stream
Q_E volumetric flow rate of the recycle stream
X_{A1} fractional conversion of reactant A in the feed to the reactor
X_{A2} fractional conversion of reactant A in the reactor effluent
X_{AE} fractional conversion of reactant A in the system effluent

The recycle stream has the same composition as the effluent from the reactor because the stream is simply split into two fractions. The inlet stream to the reactor (at position 1) will be composed of the fresh feed to the system and the contents of the recycle stream. Since the composition of the reactor effluent is the same as the system effluent, the values of X_{A2} and X_{AE} are the same. The recycle ratio, R, is defined as the ratio of the volumetric flow rates of the recycle stream and the system effluent, or

$$R \equiv \frac{\text{volume fluid returned to entrance}}{\text{volume leaving system}} = \frac{Q_R}{Q_E}$$

The mole balance for this system is now developed in some detail. We start with a mole balance on the reactor. The reactor is a PFR; therefore, the mole balance equation for the reactor written in terms of the molar flow rates and catalyst mass is in integral form

$$W = -\int_{F_{A1}}^{F_{A2}} \frac{dF_A}{(-r_A)} \tag{11.5}$$

Equation 11.5 contains molar flow rates at the reactor inlet and effluent. We want to develop a mole balance that depends only on the inlet and outlet conditions of the system. The total conversion over the system can be written as

$$X_{AE} = \frac{F_{A0} - F_{AE}}{F_{A0}} \tag{11.6}$$

Considering the molar flow rates around the reactor gives the following relationships:

$$F_{A1} = F_{A0} + F_{AR} \tag{11.7}$$

$$F_{A2} = F_{AE} + F_{AR} \tag{11.8}$$

The recycle ratio can be defined in terms of the molar flow rates, as follows:

$$R = \frac{F_{AR}}{F_{AE}} \tag{11.9}$$

Substituting Equation 11.9 into Equations 11.7 and 11.8 gives

$$F_{A1} = F_{A0} + RF_{AE} \tag{11.10}$$

$$F_{A2} = (R + 1)F_{AE} \tag{11.11}$$

Equations 11.10 and 11.11 can be substituted into the PFR mole balance equation (Equation 11.5) to calculate the outlet molar flow rate from the reactor. It is also instructive to write the mole balance equation in terms of the fractional conversion. When writing the mole balance in this form, it is important to be careful of the definition of fractional conversion that is used.

The differential form of the mole balance equation for a PFR is

$$-\frac{dF_A}{dW} = (-r_A) \tag{11.12}$$

The molar flow rate refers to the flow rate in the reactor. The flow rate of A in the effluent from the *system* is by rearrangement of Equation 11.6

$$F_{AE} = F_{A0}(1 - X_{AE}) \tag{11.13}$$

From Equations 11.11 and 11.13, and noting that $X_{A2} = X_{AE}$, the molar flow rate of A in the *reactor* effluent is given by

$$F_{A2} = F_{A0}(1 - X_{A2})(1 + R) \tag{11.14}$$

The general molar flow rate at any point in the reactor is related to the fractional conversion at the same point by

$$F_A = F_{A0}(1 - X_A)(1 + R) \tag{11.15}$$

Substituting Equation 11.15 into Equation 11.12, the mole balance, gives

$$(1 + R)F_{A0} \frac{dX_A}{dW} = (-r_A) \tag{11.16}$$

Equation 11.16 can be written in integral form as

$$\frac{W}{F_{A0}} = (1 + R) \int_{X_{A1}}^{X_{A2}} \frac{dX_A}{(-r_A)} \tag{11.17}$$

It is necessary to determine the limits on the integral. We have already noted that $X_{A2} = X_{AE}$, so it remains to find an expression for X_{A1}. Equation 11.15 is valid at any point in the reactor, including the inlet; therefore

$$F_{A1} = F_{A0}(1 - X_{A1})(1 + R) \tag{11.18}$$

Equation 11.10 is an alternative expression for F_{A1}. Substituting the definition of fractional conversion into Equation 11.10 gives the expression

$$F_{A1} = F_{A0} + R F_{A0}(1 - X_{AE}) \tag{11.19}$$

Equating Equations 11.18 and 11.19 and rearranging give an expression for X_{A1}:

$$X_{A1} = \left(\frac{R}{R + 1}\right) X_{AE} \tag{11.20}$$

The final mole balance equation for the recycle reactor can be written as

$$\frac{W}{F_{A0}} = (1 + R) \int_{\left(\frac{R}{R+1}\right)X_{AE}}^{X_{AE}} \frac{dX_A}{(-r_A)} \tag{11.21}$$

From Equation 11.21, the behavior of the recycle reactor may be deduced. If R is equal to zero, the reactor behaves as a normal single-pass PFR. As the value of R increases, the value of the lower limit (X_{A1}) will approach the value of the upper limit (X_{A2}). In the limit as R tends to infinity, the conversion in the reactor approaches zero, and the inlet reactor concentration will be the same as the outlet concentration. In other words, the system will behave as a CSTR. For practical purposes, when the recycle ratio has a value of about 20 or higher, the assumption of CSTR behavior is usually valid. The recycle reactor is usually operated at high recycle rates to simulate a CSTR. In such a case, the CSTR mole balance equation can be used instead of Equation 11.21 to analyze the results.

The recycle reactor is often used in kinetic studies of catalytic reactions. When used with a high recycle rate so that the reactor approaches CSTR behavior, the recycle reactor offers the advantage of using a fixed bed of small catalyst particles, which is easy to construct, but at the same time giving the computational advantages of CSTR analysis.

11.2 Measuring Physical Properties

There are a number of properties of catalysts that can be determined using different experimental techniques. These properties might be necessary for an understanding of the catalyst behavior or to calculate such quantities as the effectiveness factor. A few of the more common characterization methods are discussed in the following sections.

11.2.1 Calculation of Surface Area

Chemical and physical adsorption were both discussed in Chapter 8. In this section, we expand on the description of physical adsorption and describe how it forms the basis of one of the most popular and routine methods used to estimate the surface area of porous catalysts. The most common such method is based on the BET adsorption isotherm.

In Chapter 8, we presented examples of some adsorption isotherms. In fact, physical adsorption isotherms are typically classified into one of five possible types, which were first identified by Brunauer (1943). These are illustrated in Figure 11.9. This classification is commonly referred to as the BET classification. A notable feature of these isotherm types is that, with the exception of Type 1, these isotherms allow for an adsorption amount that exceeds one monolayer of adsorbed molecules.

The BET theory was based on similar assumptions as used in the development of the Langmuir isotherm, but extended by assuming that the localized adsorption for the

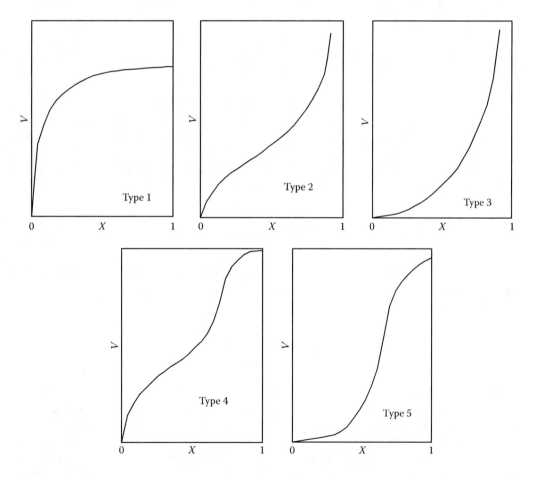

FIGURE 11.9
Examples of the five types of adsorption isotherms reported by Brunauer et al. (1940).

monolayer could be used for multiple adsorption layers. Each adsorbed species in the first layer is assumed to act as an adsorption site for the second layer, the adsorbed molecules in the second layer act as sites for the third layer, and so on. The key assumption was that the heat of adsorption of the second and subsequent layers was equal to the heat of lique-faction of the adsorbate. The heat of adsorption of the first layer was different. The original BET isotherm allowed for an infinite number of adsorbed layers. The theory was subsequently modified to allow for a limitation in the number of layers. This modification is reasonable because in small pores the maximum number of adsorbed layers is limited by the pore radius. Without discussing all of the underlying assumptions or giving the theoretical development, we state here that the following equation was proposed for multilayer physical adsorption:

$$v = \frac{v_m c x \left[1 - (n+1)x^n + nx^{n+1} \right]}{(1-x)\left[1 + (c-1)x - cx^{n+1} \right]} \tag{11.22}$$

In Equation 11.22, v is the volume of gas adsorbed and v_m is the volume of gas adsorbed in one monolayer. The constant c is related to the heats of adsorption of the first monolayer, H_1, and the heat of liquefaction, H_L:

$$c \approx \exp\left(\frac{H_1 - H_L}{R_g T} \right) \tag{11.23}$$

The variable x is the ratio of the pressure, P, of the adsorbate and its saturated vapor pressure, P_0, at the temperature of the system. This model can reproduce isotherm types 1, 2, and 3 by the correct use of v_m, n, and c. A more general form of the equation was later developed to explain isotherms 4 and 5.

For a monolayer coverage, that is, n equal to 1, Equation 11.22 reduces to the Langmuir adsorption isotherm:

$$\frac{P}{v} = \frac{P_0}{v_m c} + \frac{P}{v_m} \tag{11.24}$$

The most common form of Equation 11.22 is usually referred to as the BET equation, and describes type 2 isotherms. If we assume that n is unbounded, we obtain, after some rearrangement

$$\frac{1}{v(1-x)} = \frac{1}{v_m c} \frac{(1-x)}{x} + \frac{1}{v_m} \tag{11.25}$$

Equation 11.25 can be used to determine the surface area corresponding to a monolayer of adsorbed molecules. The experimental system to measure surface area is conceptually simple. A sample of the catalyst is placed in a sample holder of known volume, and any adsorbed species are removed by exposure to vacuum. Most surface area measurements are made using nitrogen gas at a temperature of 77 K (liquid nitrogen temperature). After the sample and holder are cooled to this temperature, a known amount of gas is admitted to the cell and the sample allowed to equilibrate. By measuring the gas pressure at equilibrium, the amount adsorbed at this pressure can be calculated. If we then plot the left-hand side of Equation 11.25 against $(1 - x)/x$, then a straight line should result, the intercept of which is the volume adsorbed at one monolayer coverage. If the area of an adsorbed molecule is known, then the area occupied may be calculated easily. The area occupied by a nitrogen molecule is 0.158 nm^2.

The range of pressure that is considered useful for surface area measurement corresponds to an x value between 0.05 and 0.35. There are many commercial surface area analyzers on the market today that make use of the BET principle. Usually, such systems are provided with a software that performs all the calculations and generates the surface area. As usual, one should appreciate the underlying equations used in this analysis, and not simply treat the equipment as a black box.

There are many weaknesses in the BET theory, and many of the assumptions made in its development are demonstrably false. However, the technique is widely used, and at the minimum provides a basis for comparison of different catalyst samples.

11.2.2 Pore Volume and Pore Size Distribution

The structure of porous catalysts was discussed in Chapter 7, and some illustrations of pore size distributions (PSDs) were also shown. A knowledge of PSD can be beneficial in understanding the performance of the catalyst, as well as assisting in computing parameters such as the effective diffusion coefficient and the effectiveness factor. There are two general methods of determining PSDs. One of these is based on adsorption/desorption isotherms, which in turn makes use of the phenomenon known as capillary condensation, and the other is based on measuring the force required to force a nonwetting liquid into the pores. A brief description of these two methods is given here. In practice, the application of these methods is now typically done using commercially available instruments; these instruments typically contain a software that performs all of the necessary calculations and generates the PSD. Therefore, the detailed calculations needed to compute the distributions are not given here.

11.2.2.1 Pore Size Distribution from Capillary Condensation Data

Condensation can occur in small capillaries below the vapor pressure of the gas, provided that the condensed phase wets the solid. The relationship that governs the pressure at which condensation occurs in a cylindrical capillary of radius r is called the Kelvin equation, after Lord Kelvin who developed it in 1871. The equation is

$$\ln\left(\frac{P}{P_0}\right) = -\frac{2\sigma V}{rR_g T}\cos(\phi) \tag{11.26}$$

where σ is the surface tension, ϕ is the contact angle, and V is the molar volume of the fluid. Taking a simpleminded approach, we observe that for any given pressure, P, the volume of liquid in the pores is known. The pore radius corresponding to that P is computed from the Kelvin equation. Once the total pore volume is known, then the distribution of radii can be computed. This procedure gives the essence of the method, although the reality is somewhat more complicated. The volume of liquid in the pores is actually a combination of multilayer adsorption on the walls of the pores not filled by Kelvin condensation, and those pores filled by Kelvin condensation. A correct calculation of PSD using capillary condensation data takes both of these effects into account. The details are beyond the scope of this text.

There are a number of approximations inherent in the use of capillary condensation data to calculate PSD. One of the biggest ones, and arguably the cause of most uncertainty in the data, is the assumption that the pores are uniform and cylindrical in shape. In most catalysts, the pores are not uniform in cross-section area along their length, and nor are they generally cylindrical. The result is that in some cases, the results generated may not be a good representation of reality. However, the method can be used for comparison purposes.

11.2.2.2 Pore Size Distribution from Mercury Porosimetry

If a nonwetting fluid surrounds an evacuated porous catalyst, the fluid will not enter the pores unless an external pressure is applied. The amount of force required depends on the surface tension of the fluid and its contact angle. The pressure required is given by

$$r = -\frac{2\sigma}{P}\cos(\phi) \tag{11.27}$$

The usual fluid that is used is mercury. The limitation of mercury porosimetry is that, for pores less than about 10 nm in diameter, the pressure required is too large for a practical instrument. Thus, only larger pores are measured; however, because large pores cannot be measured by Kelvin condensation, a combination of the two methods can be used. Again, the technique is conceptually simple. The volume of mercury forced into the pores at a given pressure is known, and the corresponding radius can be calculated from Equation 11.27.

11.2.2.3 Representations of Pore Size Distribution

PSDs are usually represented in one of the two ways, differential and cumulative. The differential method plots the ratio of the change in pore volume divided by the change in pore radius as a function of the pore radius. In effect, this shows a representation of the fraction of pore volume at a given pore size. The cumulative plot shows how the total pore volume increases as the pore radius increases. Both methods are useful. To represent the different methods, and also to illustrate the technique, we have taken an example distribution from the literature (Hayes et al., 2000) and offer some discussion on it. The samples used in this investigation were a monolith that was coated with a catalytic washcoat (as discussed in Chapter 10). The monolith substrate is porous, as is the catalytic washcoat. In this investigation, it was desired to have the pore size distribution of the washcoat alone, but it was not possible to measure it. The results in that paper make an interesting illustration of PSDs.

Pore size measurements were made on both the monolith substrate and the washcoated monolith. The samples used were small sections of washcoated monolith: the samples were not crushed. The pore volume distribution for the monolith substrate was obtained by mercury porosimetry and the cumulative pore volume distribution is shown in Figure 11.10. From these data, the PSD was plotted, as shown in Figure 11.11. Essentially, the cordierite substrate contains a group of pores with diameters between 1000 and 6000 nm, with an average pore diameter of 3500 nm. No micropore or mesopore struc-

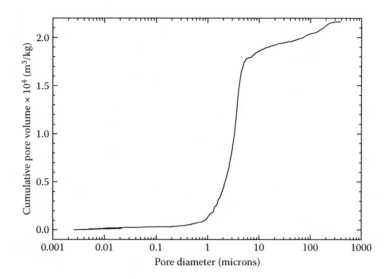

FIGURE 11.10
Cumulative pore volume for a monolith substrate.

ture is observed. A cumulative pore volume curve of washcoated monolith obtained using mercury porosimetry is given in Figure 11.12. This curve shows the combined influence of the support and catalyst/washcoat layers. It is clear from this figure that there are two distinct sets of pore sizes, one in the microrange and one in the macrorange. To determine the PSD in the washcoat alone, the pore volume of the support material can be subtracted from this curve. After deducting the effect of the substrate, and taking into account the relative mass fractions of substrate and washcoat, the PSD of the washcoat

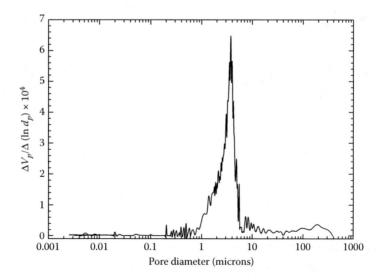

FIGURE 11.11
Pore size distribution for monolith substrate.

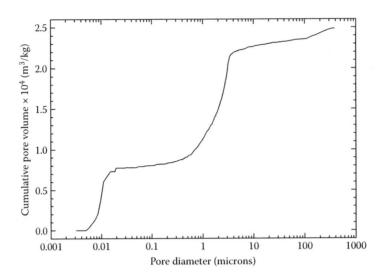

FIGURE 11.12
Cumulative pore volume for monolith with catalytic washcoat.

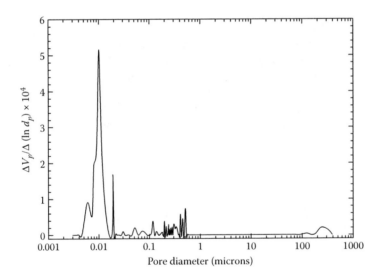

FIGURE 11.13
Pore size distribution for catalytic washcoat.

with the catalyst was calculated as shown in Figure 11.13. Here, it is seen that the actual catalyst has a single size distribution.

11.2.3 Effective Diffusivity

There have been a variety of experimental methods used to measure effective diffusion coefficients. The choice of method may depend on the type of catalyst of interest, for example, whether it is in the form of pellets or a washcoat of catalyst on the wall of a monolith channel. For catalyst pellets, a very classical method is the so-called Wicke–Kallenbach

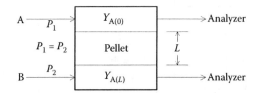

FIGURE 11.14
Schematic of a Wicke–Kallenback cell used for measuring effective diffusion coefficients.

cell, which is named after its creators. A diagram of a Wicke–Kallenbach cell is shown in Figure 11.14. The basic design is relatively simple. It consists of a cylindrical body that holds a catalyst pellet. On either side of the catalyst is a mixing chamber, which contains well-mixed gases. Then, to measure the effective diffusion coefficient of species A, a mixture of A in a reference gas (e.g., nitrogen) is fed to one chamber (marked A in Figure 11.14). The pure reference gas is fed to the bottom chamber (B in Figure 11.14). Both flow rates must be known accurately, and the pressure on either side of the catalyst should be the same, to avoid forced flow. Some A will diffuse through the catalyst and leave through the bottom chamber exit. By measuring the concentration, and knowing the flow rate, the diffusive flux can be calculated by a simple mole balance. Let Q be the flow rate from the bottom chamber and $Y_{A(L)}$ the mole fraction of A; the molar flow rate is given by

$$F_{A(L)} = QCY_{A(L)} \tag{11.28}$$

The diffusive flux, N_A, across the pellet is related to the cross-sectional area. Assuming a pellet of radius R, the flux is

$$N_A = \frac{F_{A(L)}}{\pi R^2} \tag{11.29}$$

If we assume a Fick's law diffusion, then we can write

$$N_A = -D_{\text{eff}} C \frac{dY_A}{dz} \tag{11.30}$$

If one-dimensional flow is assumed and if D_{eff} does not depend on concentration, then the differential can be written in terms of the concentration difference between the two chambers. Thus, substituting Equation 11.28 and after suitable rearrangement, we can write the expression for the effective diffusion coefficient as

$$D_{\text{eff}} = \left(\frac{Y_{A(L)}}{Y_{A0} - Y_{A(L)}} \right) \frac{QL}{\pi R^2} \tag{11.31}$$

Provided that the flow rates are known, D_{eff} can be calculated.

There are other techniques for measuring effective diffusion coefficient discussed in the literature (Hayes and Kolaczkowski, 1997). See, for example, Hayes et al. (2000).

11.3 Electron Microscopy

Electron microscopy is a very valuable tool for the detailed examination of surfaces. Magnifications can be up to about one million times, and thus many small details of the surface can be revealed. Electron microscopy is extensively used in catalyst characterization. One advantage of this method of characterization is that the catalyst surface is viewed directly. The methods discussed in the later sections of this chapter (the so-called surface science methods) are based on the interpretation of the response of the surface of various forms of radiation. There are two types of electron microscopy; transmission electron microscopy (TEM) and scanning electron microscopy (SEM). The following sections present a very brief (and somewhat simplistic) description of the two methods.

11.3.1 Transmission Electron Microscopy

In TEM, a very thin sample of the material to be studied is used. The sample is mounted on a suitable holder, and a beam of electrons is passed through it. The beam of electrons is focused using magnetic lenses, as is the captured stream of electrons leaving the sample. An image is formed on a fluorescent screen which can subsequently be recorded. The level of magnification is controlled using the magnetic lenses. The size of the various features on the catalyst surface can be measured directly from the resulting pictures, although below sizes of about 2 nm there is no direct correspondence between the image size and the actual size. Nonetheless, the method is very useful for the determination of morphological features of the catalyst surface.

When interpreting the electron micrographs, one has to be careful in generalizing the features found to the entire catalyst surface. As a result of the high magnifications used, only very small portions of the surface can be viewed. For example, if the magnification is 100,000 times, then a 20 cm by 25 cm micrograph represents a catalyst surface area of about 5×10^{-12} m^2. Because only a limited number of frames can be made, extrapolation and generalization is to be viewed with caution.

11.3.2 Scanning Electron Microscopy

SEM works on a different principle than TEM. Here, the sample is bombarded with a very thin beam of electrons, <2 nm in diameter. A number of different types of emissions occur from the surface. These emissions are collected using different detectors, and each gives information about the nature of the surface. The sample must be an electrical conductor. Because many catalyst support materials are nonconducting, they are coated with a very thin film of gold, of the order of 20 nm thickness. The maximum resolution of SEM is about 10 nm.

11.4 Surface Science Studies

There is a huge gamut of surface science techniques that can reveal very detailed information about catalysts at the molecular level. These studies are usually carried out under high-vacuum conditions and require very specialized (and very expensive) equipment. Some of these methods are performed using well-defined crystals as the sample. In this case, relating the observed phenomena to a real supported catalyst may not always be obvious. Many of these techniques are known by their acronyms, and we mention a few of them here. A good reference source that gives a more detailed description of many of these methods is Thomas and Thomas (1996).

Low-energy electron diffraction (LEED) is a technique that is used to probe the surface layers of single crystals. The results can be used to obtain important structural information about the surface, although the single crystals used are somewhat different from real catalysts, as noted earlier. An important class of methods falls under the general term "electron spectroscopy." These techniques vary depending on the nature of the incident radiation and the resulting emissions. X-ray photoelectron spectroscopy (XPS), also known as electron spectroscopy for chemical analysis (ESCA) uses x-rays as the incident radiation, and measures the kinetic energy of the resulting electrons that escape from the top few nm of the material. XPS can measure the elemental composition of materials, the chemical or electronic state of the surface elements, and other features. Another form of electron spectroscopy uses the Auger electrons, which are also emitted when a sample is bombarded with x-rays. These electrons can also give useful compositional information about the surface. This technique is called Auger electron spectroscopy (AES). Electron microscopes have also been built that use Auger electrons, and these are called scanning Auger microscopes (SAM). There are literally dozens more surface science techniques, and the reader should consult suitable references for further information about them.

11.5 Summary

This chapter has provided an overview of some of the experimental techniques used in catalysis research. With the information contained here, the reader should have the ability to evaluate experimental data and to have a reasonable understanding of the results of published papers. However, the development of a detailed experimental program in catalysis requires much specialized knowledge. There are many excellent books that discuss experimental catalysis in great detail, and such references should be consulted if one wishes to embark on an experimental program of their own.

PROBLEMS

11.1 Nitrogen is adsorbed at low temperature on a catalyst that showed a type 4 isotherm. The amount adsorbed at STP was 20 cm^3 at a relative pressure (P/P_0) of 0.285. Estimate the surface area of this catalyst.

11.2 Low-temperature nitrogen adsorption was carried out on a sample of alumina. The results in the following table were obtained. Estimate the surface area of the catalyst.

P/P_0	Adsorbed Nitrogen (cm³/g)	P/P_0	Adsorbed Nitrogen (cm³/g)
0.096	23.13	0.245	29.37
0.113	24.03	0.270	30.32
0.136	24.91	0.305	31.74
0.158	25.87	0.349	33.44
0.185	27.02	0.405	35.91
0.222	28.42	0.472	38.99

11.3 The effective diffusivity of a porous slab of catalyst was measured in a Wicke–Kallenbach cell. The slab thickness was 0.5 cm, the diameter was 2.5 cm, and the measured diffusion rate was 0.03 mol/h. The total pressure was 150 kPa and the temperature was 600 K. The molar mass of the diffusing species was 2. The mole fraction of this species on one side of the plate was 0.8 and on the other it was 0.3. The porosity of the catalyst was 0.75, the specific gravity was 2.0, and the total surface area of the pores was 100 m²/g. Calculate the tortuosity of the catalyst slab. Assume that the diffusion in the pores is only Knudsen diffusion.

References

Brunauer, S., 1943, *The Adsorption of Gases and Vapours*, University Press, Princeton.

Brunauer, S., L.S. Deming, W.E. Deming, and E. Teller, 1940, On a theory of the van der Waals adsorption of gases, *Journal of the American Chemical Society*, **62**, 1723–1732.

Carberry, J.J., 1976, *Chemical and Catalytic Reaction Engineering*, McGraw-Hill, New York.

Doraiswamy, L.K. and D.G. Tajbl, 1974, Laboratory catalytic reactors, *Catalysis Reviews—Science and Engineering*, **10**, 177–219.

Hayes, R.E. and S.T. Kolaczkowski, 1997, *Introduction to Catalytic Combustion*, Gordon and Breach, Reading.

Hayes, R.E., S.T. Kolaczkowski, P.K. Li, and S. Awdry, 2000, Evaluating the effective diffusivity of methane in the washcoat of a honeycomb monolith, *Applied Catalysis, Part B: Environmental*, **25**, 93–104.

Thomas, J.M. and W.J. Thomas, 1996, *Principles and Practice of Heterogeneous Catalysis*, VCH, Weinheim.

Wicke, E. and R. Kallenbach, 1941, *Kolloid Z*, 97: 135, as referenced in Smith, J.M., 1981, *Chemical Engineering Kinetics*, 3rd Ed., McGraw-Hill, New York.

Wjingaarden, R.J., A. Kronberg, and K.R. Westerterp, 1998, *Industrial Catalysis: Optimizing Catalysts and Processes*, Wiley-VCH, New York.

Appendix 1: Numerical Methods

Reaction engineering problems and engineering problems in general require the solution of various types of equations. Broadly speaking, these can be divided into algebraic equations and differential equations. In linear algebraic equations, the power on the independent variable is one, whereas in nonlinear equation, the power is other than one.

Differential equations relate the rate of change in some quantity (the dependent variable) with respect to another (independent variable). An *ordinary differential equation* (ODE) has a single independent variable. The steady-state PFR mole balances and the transient CSTR mole balance are both examples of ODE. ODE may be *initial value problems* (IVP) or *boundary value problems* (BVP). An IVP is usually a first-order ODE, and its solution requires only the value at the beginning of the solution domain in space or time. A BVP is usually a second- or higher-order equation that requires that values of the variables or their derivatives be known at the beginning and the end of the solution domain. The one-dimensional steady-state axial dispersion model is an example of an ODE that is also a BVP.

It was stated in Chapter 1 that many problems in chemical reaction engineering require numerical solutions, and that exact (analytical) solutions are restricted to simple linear and low-dimensional problems involving few variables. For most other situations, a numerical solution must be sought. In the following sections, some of the techniques used to solve various types of equations are given. This appendix is meant only to give an overview, and the reader should consult specialized textbooks for details.

A1.1 Nonlinear Algebraic Equations

Many engineering problems lead to nonlinear equations which may be written in the form

$$f(x) = 0 \qquad \text{(A1.1)}$$

The challenge is to find a value of x that satisfies the equality. The most widely used method for solving nonlinear equations is the Newton–Raphson method, which is described hereafter. Consider the Taylor series expansion of a function $f(x)$ about some initial point $x = x_0$:

$$f(x) = f(x_0) + (x - x_0)f'(x) + \frac{(x - x_0)^2}{2!} f''(x) + \frac{(x - x_0)^3}{3!} f'''(x) + \cdots \qquad \text{(A1.2)}$$

If $x - x_0$ is very small, we can set all of the quadratic and higher terms to zero, and solve the linear approximation of $f(x) = 0$ for x to obtain a better approximation as

$$x_1 = x_0 - \frac{f(x_0)}{f'(x_0)} \qquad \text{(A1.3)}$$

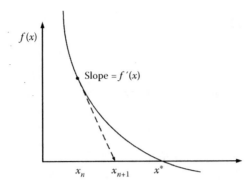

FIGURE A1.1
Newton–Raphson approximation.

Subsequent iterations are defined in a similar manner as

$$x_{n+1} = x_n - \frac{f(x_n)}{f'(x_n)} \tag{A1.4}$$

Geometrically, x_{n+1} can be interpreted as the value of x at which a line, passing through the point $(x_n, f(x_n))$ and tangent to the curve $f(x)$ at that point, crosses the y axis. Figure A1.1 provides a graphical interpretation of this situation.

The Newton–Raphson method can easily be extended to multidimensional systems. Consider the case where $n = 2$ (two-dimensional systems, 2D)

$$f_1(x_1, x_2) = 0 \tag{A1.5}$$
$$f_2(x_1, x_2) = 0$$

If we expand f_1 and f_2 around an approximate solution $(\tilde{x}_1, \tilde{x}_2)$, and retain the linear terms only, we obtain

$$f_1(x_1, x_2) \approx f_1(\tilde{x}_1, \tilde{x}_2) + \frac{\partial f_1}{\partial x_1}(x_1 - \tilde{x}_1) + \frac{\partial f_1}{\partial x_2}(x_2 - \tilde{x}_2)$$

$$f_2(x_1, x_2) \approx f_2(\tilde{x}_1, \tilde{x}_2) + \frac{\partial f_2}{\partial x_1}(x_1 - \tilde{x}_1) + \frac{\partial f_2}{\partial x_2}(x_2 - \tilde{x}_2) \tag{A1.6}$$

Set Equation A1.6 equal to zero and denote $\Delta x_1 = (x_1 - \tilde{x}_1)$ and $\Delta x_2 = (x_2 - \tilde{x}_2)$:

$$0 = f_1(\tilde{x}_1, \tilde{x}_2) + \frac{\partial f_1}{\partial x_1} \Delta x_1 + \frac{\partial f_1}{\partial x_2} \Delta x_2$$

$$0 = f_2(\tilde{x}_1, \tilde{x}_2) + \frac{\partial f_2}{\partial x_1} \Delta x_1 + \frac{\partial f_2}{\partial x_2} \Delta x_2 \tag{A1.7}$$

We can rewrite Equation A1.7 as

$$\frac{\partial f_1}{\partial x_1} \Delta x_1 + \frac{\partial f_1}{\partial x_2} \Delta x_2 = -f_1$$

$$\frac{\partial f_2}{\partial x_1} \Delta x_1 + \frac{\partial f_2}{\partial x_2} \Delta x_2 = -f_2$$

(A1.8)

Equation A1.8 can be written in matrix form as

$$\begin{bmatrix} \dfrac{\partial f_1}{\partial x_1} & \dfrac{\partial f_1}{\partial x_2} \\ \dfrac{\partial f_2}{\partial x_1} & \dfrac{\partial f_2}{\partial x_2} \end{bmatrix} \begin{Bmatrix} \Delta x_1 \\ \Delta x_2 \end{Bmatrix} = \begin{Bmatrix} -f_1 \\ -f_2 \end{Bmatrix}$$

(A1.9)

Equation A1.9 denotes a system of linear equations in the unknowns Δx_1 and Δx_2. By solving (A1.9), we can find Δx_1 and Δx_2. The matrix collection of partial derivatives is called the Jacobian, and is denoted $\mathbf{J}$

$$\mathbf{J}(x_1, x_2) = \begin{bmatrix} \dfrac{\partial f_1}{\partial x_1} & \dfrac{\partial f_1}{\partial x_2} \\ \dfrac{\partial f_2}{\partial x_1} & \dfrac{\partial f_2}{\partial x_2} \end{bmatrix}$$

If we denote the current iterate by $x^{(i)}$ and the next approximation by $x^{(i+1)}$, we can write

$$x_1^{(i+1)} = x_1^{(i)} + \Delta x_1^{(i)}$$

$$x_2^{(i+1)} = x_2^{(i)} + \Delta x_2^{(i)}$$

Writing Equation A1.9 in vector notation

$$\mathbf{J}\Delta x = -\mathbf{F}$$

(A1.10)

We can arrange this for different iterates, i, to obtain

$$\Delta x^{(i)} = -\left[J^{(i)} \right]^{-1} f^{(i)}$$

(A1.11)

which yields the following equation, noting that $\Delta x^{(i)} = x^{(i+1)} - x^{(i)}$:

$$x_{i+1} = x_i - \left[J(x_i) \right]^{-1} f(x)$$

(A1.12)

Equation A1.12 is the Newton–Raphson equation for multidimensional systems. We can solve such a system using the following algorithm:

- Start with an initial guess $\vec{x}^{(0)}$
- Determine $\vec{F}(\vec{x}^{(0)})$ and $\mathbf{J}(\vec{x}^{(0)})$
- Solve the linear system in the unknown $\Delta \vec{x}^{(0)}$: $\mathbf{J}(\vec{x}^{(0)}) \cdot \Delta \vec{x}^{(0)} = -\vec{F}(\vec{x}^{(0)})$
- The new estimate for $\vec{x}$ now becomes $\vec{x}^{(1)} = \vec{x}^{(0)} + \Delta \vec{x}^{(0)}$
- Repeat until convergence

A1.1.1 Convergence

To study how the Newton–Raphson scheme converges, we consider the Taylor series expansion expand $f(x)$ around the root $x = x^*$

$$f(x_i) + (x_{i+1} - x_i)f'(x_i) = 0 \tag{A1.13}$$

Suppose we are very close to the true root x^* (and so $f(x^*) = 0$), so that we can write a Taylor expansion up to the second derivative

$$f(x_i) = f(x^*) + (x_i - x^*)f'(x^*) + \frac{1}{2}(x_i - x^*)^2 f''(x_i) \tag{A1.14}$$

Subtracting Equation A1.13 from Equation A1.14 and setting $f(x^*) = 0$, we obtain

$$(x_{i+1} - x^*)f'(x_i) = \frac{1}{2}(x_i - x^*)^2 f''(x_i) \tag{A1.15}$$

If we define the error as $E_i = (x_i - x^*)$, Equation A1.15 becomes

$$E_{i+1} = \frac{(1/2)E_i^2 f''(x_i)}{f'(x_i)} \tag{A1.16}$$

This equation says that from one iteration to the next, the new error reduces with the square of the old error, whence quadratic convergence. The presence of the derivative term in the denominator shows that the scheme will not converge if $f'(x)$ vanishes in the neighborhood of the root.

A1.2 Linear Algebraic Equations

Linear equations also arise in the solution of engineering systems. One classic example is the balancing of chemical reactions. A chemical reaction can be expressed by a stoichiometric equation with the following form:

$$\sum_{i=1}^{NC} v_{i,j} A_i = 0 \quad j = 1, 2, \dots, NR \tag{A1.17}$$

where A_i stand for the chemical symbols and v_i stands for stoichiometric coefficients, which are negative for reactants and positive for the products, NR is the number of independent reactions in the system, and NC is the number of elements or atoms in the system. We can use the conservation of atoms (elements) in the system to determine the unknown stoichiometric coefficients. We delineate three matrices from Equation A1.17, namely, matrix comprising the formula, $[A]$, a column vector consisting of the stoichiometric coefficients, $\{v\}$, and a null vector, $\{0\}$, written as

$$[A]\{v\} = \{0\} \tag{A1.18}$$

The solution of Equation A1.18 for the unknown coefficients gives a balanced reaction system. There are a number of methods which may be used, some direct, while others are iterative in nature and provide only approximate solutions. Direct methods include the Cramer's rule, Gauss elimination, matrix inverse, or LU factorization. Gauss elimination is described as a representative method in the following section.

A1.2.1 Gauss Elimination Method

In the Gauss elimination method, the unknowns are eliminated progressively until the system is reduced into an upper triangular matrix. The solution may then be found by back-substitution. Let us consider solving a system of three equations given by A1.19 for illustration.

$$a_{11}x_1 + a_{12}x_2 + a_{13}x_3 = b_1$$
$$a_{21}x_1 + a_{22}x_2 + a_{23}x_3 = b_2 \tag{A1.19}$$
$$a_{31}x_1 + a_{32}x_2 + a_{33}x_3 = b_3$$

Step 1: Assuming that $a_{11} \neq 0$, we can eliminate x_1 from the second and the third equation by subtracting a_{21}/a_{11} times the first equation from the second equation and a_{31}/a_{11} times first equation from the third equation, respectively. This results in the system

$$a_{11}x_1 + a_{12}x_2 + a_{13}x_3 = b_1$$
$$a'_{22}x_2 + a'_{23}x_3 = b'_2 \tag{A1.20}$$
$$a'_{32}x_2 + a'_{33}x_3 = b'_3$$

Step 2: Assuming that $a'_{22} \neq 0$, we can eliminate x_2 from the third equation by subtracting a'_{31}/a'_{22} times the second equation from the third equation resulting in

$$a_{11}x_1 + a_{12}x_2 + a_{13}x_3 = b_1$$
$$a'_{22}x_2 + a'_{23}x_3 = b'_2 \tag{A1.21}$$
$$a''_{33}x_3 = b''_3$$

Step 3: we can now evaluate the unknowns, x_1, x_2, and x_3 from the reduced set of Equation A1.21 by performing back-substitution.

It should be noted that the method would not work if any of the pivot elements, a_{11}, a'_{22}, and a''_{33} becomes zero. This could be avoided by rewriting the equations in such a manner that nonzero pivots are used.

A1.2.2 Jacobi and Gauss Seidel Iterative Methods

The system of Equation A1.19 could also be solved using iterative methods, such as Jacobi or Gauss Seidel, where a simple rearrangement and an initial guess of the equations are used to improve the initial estimates repeatedly. For example, using the Jacobi method, the equations are rewritten as

$$x_1 = \frac{1}{a_{11}}(b_1 - a_{12}x_2 - a_{13}x_3)$$

$$x_2 = \frac{1}{a_{22}}(b_2 - a_{21}x_1 - a_{23}x_3) \tag{A1.22}$$

$$x_3 = \frac{1}{a_{33}}(b_3 - a_{31}x_1 - a_{32}x_2)$$

An initial guess of $x = [0\ 0\ 0]$ is usually sufficient to obtain an improved solution for the Jacobi method, whereas the Gauss Seidel method utilizes any previously obtained value in subsequent estimations.

A1.3 Ordinary Differential Equations

In this section, we consider some of the common techniques used to solve initial value problems. For simple cases, we may obtain analytical solutions, but in general we need to use numerical solutions. We first consider the solution of a single differential equation that is an IVP.

A1.3.1 Single Differential Equation

Numerical methods for ordinary differential equations approximate solutions to initial value problems of the form

$$\frac{dy}{dt} = f(t, y) \tag{A1.23}$$

with initial condition $y(0) = y_0$. Equation A1.23 is an ordinary differential equation of the first order because the highest order of the dependent variable (y) is unity. The right-hand side function is a known function of t and $y, f(t,y)$. This is known as an initial value problem because the value of y is known at the starting point, $t = 0$. The objective is to find the family of values which satisfies the differential equation for the given initial conditions.

There are many sophisticated and efficient computer algorithms developed for the solution of initial value problems of ODE, usually based on Runge–Kutta methods or linear multistep methods. Multistep methods are more efficient for problems with large time intervals and near constant step sizes. On the other hand, if step sizes need frequent and large adjustments, Runge–Kutta methods are preferred.

A1.3.2 Euler's Method

Probably the simplest method for the solution of an IVP is Euler's method. Euler's method has two variations; the explicit method and the implicit method. Consider a first-order ODE, $dy/dt = f(y, t)$, with an initial condition $y = y^{(0)}$ at $t = 0$. We wish to compute numerically how y develops in time t. The simplest way to do this is to divide time into small time steps h (such that $t = ih$ with $i = 0, 1, 2, \ldots$) and writing the dy/dt in terms of its forward derivative:

$$\left.\frac{dy}{dt}\right|_i \approx \frac{y^{(i+1)} - y^{(i)}}{h} \tag{A1.24}$$

Then the ODE can be written as

$$\frac{y^{(i+1)} - y^{(i)}}{h} = f\left(y^{(i)}, t^{(i)} = ih\right) \tag{A1.25}$$

This gives us a rule for updating y:

$$y^{(i+1)} = y^{(i)} + hf\left(y^{(i)}, t^{(i)}\right) \tag{A1.26}$$

Equation A1.26 is called the explicit method because all values on the right-hand side are explicitly known. Since $y^{(0)}$ is known, we can determine $y^{(1)}$, and, with this, progressively determine subsequent values. The main shortcoming of the explicit Euler method is that it is only first order and the method can become unstable when large time steps are used. To cure the instability problem, implicit methods have been devised. In this case, the backward derivative is used to approximate the derivative, rather than the forward derivative:

$$\frac{y^{(i+1)} - y^{(i)}}{h} = f\left(y^{(i+1)}, t^{(i+1)}\right) \tag{A1.27}$$

The unknown $y^{(i+1)}$ appears both in the left-hand side and in the right-hand side of the equation, and thus it is more complex to develop the solution. However, the implicit method has a larger stability range which means that larger time steps can be used, although the accuracy remains first order.

A1.3.3 Runge–Kutta Methods

A much more accurate family of methods for IVP solution are the Runge–Kutta methods. Consider a first-order ODE with y as a function of t with t in the interval $[t^{(i)}, t^{(i+1)}]$ and $t^{(i+1)} - t^{(i)} = h$. We can write

$$\frac{y^{(i+1)} - y^{(i)}}{h} = \frac{dy}{dt}(\xi) \tag{A1.28}$$

With ξ being somewhere in between $t^{(i)}$ and $t^{(i+1)}$. The main idea of Runge–Kutta methods is to find the value of ξ and therefore estimate the value of the function at that point, $y(\xi)$. Once we have proper estimates for ξ and $y(\xi)$, we can evaluate $f(y(\xi), \xi)$. Since $f(y(\xi), \xi) = dy/dt(\xi)$, we then have a rule to update y:

$$y^{(i+1)} = y^{(i)} + hf\left(y(\xi), \xi\right) \tag{A1.29}$$

A1.3.3.1 First-Order and Second Runge–Kutta Methods

For first-order Runge–Kutta methods, $\xi = t^{(i)}$ and $y(\xi) = y^{(i)}$. This method is therefore the same as the Euler forward method. A second-order version of Runge–Kutta method denotes $\xi = t^{(i)} + h/2$ and, therefore, $y(\xi) = y^{(i)} + (h/2)(dy/dt)|_i = y^{(i)} + (h/2) f(y^{(i)}, t^{(i)})$. This method is second-order accurate in the time step h. The update rule in this case becomes

$$y^{(i+1)} = y^{(i)} + hf\left(y^{(i)} + \frac{h}{2} f\left(y^{(i)}, t^{(i)}\right), t^{(i)} + \frac{h}{2}\right) \tag{A1.30}$$

A1.3.3.2 Classical Fourth-Order Runge–Kutta Method

The classical Runge–Kutta method is fourth-order accurate and reads as follows:

$$y^{(i+1)} = y^{(i)} + \frac{1}{6} h \{k_1 + 2k_2 + 2k_3 + k_4\} \tag{A1.31}$$

The terms $k_{1\ldots4}$ are

$$k_1 = f\left(y^{(i)}, t^{(i)}\right)$$

$$k_2 = f\left(y^{(i)} + \frac{h}{2} k_1, t^{(i)} + \frac{h}{2}\right) \tag{A1.32}$$

$$k_3 = f\left(y^{(i)} + \frac{h}{2} k_2, t^{(i)} + \frac{h}{2}\right)$$

$$k_4 = f\left(y^{(i)} + hk_3, t^{(i)} + h\right)$$

RK methods can be readily generalized to systems of first-order ODE, $d\vec{y}/dt = \vec{f}(\vec{y}, t)$. For example, the fourth-order method as discussed above can be written in vector form as

$$\vec{y}^{(i+1)} = \vec{y}^{(i)} + \frac{1}{6} h \{\vec{k}_1 + 2\vec{k}_2 + 2\vec{k}_3 + \vec{k}_4\} \tag{A1.33}$$

where the recurrence relationships are now given in vector form

$$\vec{k}_1 = \vec{f}\left(\vec{y}^{(i)}, t^{(i)}\right)$$

$$\vec{k}_2 = \vec{f}\left(\vec{y}^{(i)} + \frac{h}{2}\vec{k}_1, t^{(i)} + \frac{h}{2}\right)$$

$$\vec{k}_3 = \vec{f}\left(\vec{y}^{(i)} + \frac{h}{2}\vec{k}_2, t^{(i)} + \frac{h}{2}\right) \tag{A1.34}$$

$$\vec{k}_4 = f\left(\vec{y}^{(i)} + h\vec{k}_3, t^{(i)} + h\right)$$

A1.3.3.3 Order and Stability Function

The method is said to be of order p with $y(t_1) - y_1 = O(h^{p+1})$ whenever $y(0) = y_0$ and f is sufficiently smooth. The local error introduced by one time step integration is of the order $O(h^{p+1})$. The global error $y(t_n) - y_n$ is formed by n such local errors and will be of the order $O(h^p)$ if $y_0 = y(0)$ and the function f is sufficiently smooth. Thus, a method of order p is convergent of order p when applied to a fixed, smooth ODE problem.

The stability properties of ODE methods are, to a large extent, determined by the behavior of the methods on the scalar test equation

$$y'(t) = \lambda y(t) \tag{A1.35}$$

Let $z = h\lambda$. Application of RK to the test equation gives $y_{n+1} = R(z)y_n$, with a rational function R (the so-called stability function). $R(z)$ is a polynomial of degree $\leq n$. For implicit methods, it is a rational function with degree of both denominator and numerator $\leq n$. If the RK method has order p, then $R(z) = e^z + O(z^{p+1})$, $z \to 0$. This can be seen by considering the scalar test equation with $y_0 = y(0)$ and $|\lambda| = 1$, since we know that $y(t_1) - y_1 = e^h - R(h)$. The exact solution of the test equation satisfies $y(t_{n+1}) = e^{h\lambda} y(t_n)$, so the solution does not grow in the modulus $\mathrm{Re}\lambda \leq 0$. For an A-Stable method, the numerical approximations have the same property no matter how large the step size is chosen. The stability function for an explicit method with $p = n$ (possible for $n \leq 4$) equals

$$R(z) = 1 + z + \frac{1}{2}z^2 + \cdots + \frac{1}{n!}z^n \tag{A1.36}$$

For $n = 1,2,4$, respectively, this gives the stability function for the forward Euler method, the second-order method of example B.1 and the classical fourth-order RK method.

The stability function for the backward Euler method is

$$R(z) = \frac{1}{1-z} \tag{A1.37}$$

A1.3.3.4 Implicit RK Methods

With implicit RK methods, one has to solve at each step a system of nonlinear algebraic equations. Usually, this is done by a modified Newton iteration where the Jacobian is held fixed during iteration. Although implicit methods are more expensive than explicit ones, they are often used for parabolic problems and stiff chemistry problems because of their superior stability properties.

A1.3.4 Linear Multistep Methods

Linear multistep methods are also used for the solution of differential equations. While the single step method like Euler's use only one previous step to find the next solution point, multistep methods attempt to gain efficiency by keeping and using the information from previous steps rather than discarding it. Consequently, multistep methods refer to several previous points and derivative values. In the case of linear multistep methods, a linear combination of the previous points and derivative values is used.

Multistep method will use the previous steps to calculate the next value. The general formula for linear multistep method can be written as

$$\sum_{i=0}^{k} a_i y_{n-i} = h \sum_{i=0}^{k} b_i f_{n-i} \tag{A1.38}$$

where h denotes the step size, f_i denotes the function $f(t_i, y_i)$, and the coefficients, a_i and b_i determine the particular linear multistep method. The method is designed such that the coefficients will exactly interpolate $y(t)$ when it is an nth-order polynomial. If $b_i = 0$, the method is "explicit," whereas when $b_i \neq 0$, the method is implicit. Iterative methods, such as the Newton–Raphson are usually used to solve the implicit formula.

A number of linear multistep methods are commonly used: Adams–Bashforth methods, Adams–Moulton methods, and the backward differentiation formula (BDF). Here, we describe the BDF method for illustration purposes.

A1.3.4.1 Backward Differentiation Formula Methods

The BDF is a family of implicit multistep methods that are especially suited for the solution of stiff differential equations. A BDF can be used to solve the initial value problem

$$\frac{dy}{dt} = f(t, y), \quad y(t_0) = y_0 \tag{A1.39}$$

The family of BDFs consist of the methods arising from the case $b_i = 0$ for $i > 0$. The general formula for a BDF can be written as

$$\sum_{i=0}^{k} a_i y_{n-i} = h b_0 f_n \tag{A1.40}$$

BDF methods are implicit and, as such, require the solution of nonlinear equations at each step. Typically, a modified Newton's method is used to solve these nonlinear equations.

A1.3.4.2 Predictor–Corrector Methods: Milne Method

Predictor–corrector methods are carried out by extrapolating a polynomial fit to the derivative from the previous points to the new point (the predictor step), then using this to interpolate the derivative (the corrector step). Whereas in the previous methods described for the solution of the differential equation over the interval $[x_i, x_{i+1}]$ only the value of y at the beginning of the interval was required, in the predictor–corrector method, prior values are used to determine the value of y at x_{i+1}. A predictor formula is used to predict the value of y at x_{i+1}, and then a corrector formula is used to improve the value. One such method used to achieve this is the Milne's method described hereafter.

Consider again the differential equation given by Equation A1.39

$$\frac{dy}{dt} = f(t, y), \quad y(t_0) = y_0$$

To find the approximate value of y for $x = x_0 + n\Delta x$ by using the Milne's method, we proceed as follows:

Given the value $y_0 = y(x_0)$, we can calculate the following values $y_1 = y(x_0 + \Delta x)$, $y_2 = y(x_0 + 2\Delta x)$, $y_3 = y(x_0 + 3\Delta x)$ by using Taylor series method. We next compute the values $f_0 = f(x_0, y_0), f_1 = f(x_1, y_1), f_2 = f(x_2, y_2)$, and $f_3 = f(x_3, y_3)$. To calculate $y_4 = y(x_0 + 4\Delta x)$, we use the relation

$$y_4 = y_0 + \int_{x_0}^{x_0+4\Delta x} f(x, y) \, dx \tag{A1.41}$$

where the function $f(x, y)$ is obtained by using Newton's forward interpolation formula

$$f(x, y) = f_0 + n\Delta f_0 + \frac{n(n-1)}{2}\Delta^2 f_0 + \frac{n(n-1)(n-2)}{6}\Delta^3 f_0 + \cdots \tag{A1.42}$$

thus obtaining

$$y_4 = y_0 + \Delta x \int_{x_0}^{x_0+4\Delta x} \left(f_0 + n\Delta f_0 + \frac{n(n-1)}{2}\Delta^2 f_0 + \frac{n(n-1)(n-2)}{6}\Delta^3 f_0 + \cdots \right) dx \tag{A1.43}$$

Setting the expressions $x = x_0 + n\Delta x$ and $dx = \Delta x \, dn$, we obtain

$$y_4 = y_0 + \Delta x \int_0^4 \left(f_0 + n\Delta f_0 + \frac{n(n-1)}{2}\Delta^2 f_0 + \frac{n(n-1)(n-2)}{6}\Delta^3 f_0 + \cdots \right) dn \tag{A1.44}$$

To obtain

$$y_4 = y_0 + \Delta x \left(4f_0 + 8\Delta f_0 + \frac{20}{3}\Delta^2 f_0 + \frac{8}{3}\Delta^3 f_0 + \cdots \right) \tag{A1.45}$$

Neglecting the fourth- and higher-order differences and expressing Δf_0, $\Delta^2 f_0$, and $\Delta^3 f_0$ in terms of function values, we obtain the predictor function

$$y_4 = y_0 + \frac{4\Delta x}{3}(2f_1 - f_2 + 2f_3) \tag{A1.46}$$

Having found y_4, we can calculate the first approximation to $f_4 = f(x_0 + \Delta x, y_4)$. We can then obtain a better approximation of y_4 using the Simpson's one-third rule

$$y_4 = y_2 + \frac{\Delta x}{3}(f_2 + 4f_3 + f_4) \tag{A1.47}$$

which is the corrector formula. We therefore have an improved value of f_4, for which again the corrector may be applied to find an even better approximation of y_4. Once f_4 and y_4 have been found to satisfaction, we can then use the predictor to find $y_5 = y(x_0 + 5\Delta x)$ as

$$y_5 = y_1 + \frac{4\Delta x}{3}(2f_2 - f_3 + 2f_4) \tag{A1.48}$$

from which $f_5 = f(x_0 + 5\Delta x, y_5)$ can be calculated. Then, we can obtain a better approximation to the value of y_5 from the corrector as

$$y_5 = y_3 + \frac{\Delta x}{3}(f_3 + 4f_4 + f_5) \tag{A1.49}$$

which may be repeated until a satisfactory value is obtained, before proceeding to y_6 and so on.

In order to improve the accuracy of the Milne's method, starting values must be improved as well as smaller intervals.

A1.4 Numerical Integration

There are two main reasons for you to need to do numerical integration: analytical integration may be impossible or infeasible, or one may wish to integrate tabulated data rather than known functions. Here, we are concerned with determining the integral $I = \int_a^b f(x)\,dx$. In some cases, we only know $f(x)$ as a collection of points $f_i = f(x_i)$. In other cases, we know a closed expression for $f(x)$ but the integral cannot be evaluated analytically. In such cases, we also base our numerical estimate of the integral on a number of function evaluations $f_i = f(x_i)$ as shown in Figure A1.2 for the trapezoidal rule, which is discussed in Section A1.4.1.

A1.4.1 Trapezoidal Rule

In trapezoidal rule, the function approximated with a set of linear functions, each portion with width h contributes

$$\frac{f(x_{i+1}) + f(x_i)}{2} \times h \tag{A1.50}$$

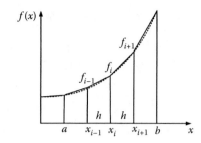

FIGURE A1.2
Illustration of the trapezoidal rule.

A1.4.2 Simpson's One-Third Rule

In this case, the function with a parabola passing through three points $f(x_{i-1})$, $f(x_i)$, $f(x_{i+1})$ with $h = x_{i+1} - x_i = x_i - x_{i-1}$. The portion of the integral related to the interval $[x_{i+1}, x_{i-1}]$ then becomes

$$\frac{h}{3}\left[f(x_{i-1}) + 4f(x_i) + f(x_{i+1})\right] \tag{A1.51}$$

A1.4.3 Simpson's Three-Eighth Rule

The Simpson's three-eighth rule uses four points $f(x_{i-1})$, $f(x_i)$, $f(x_{i+1})$, $f(x_{i+2})$ through which a third-order polynomial is fitted. The portion of the integral related to the interval $[x_{i-1}, x_{i+2}]$ then becomes

$$\frac{3h}{8}\left[f(x_{i-1}) + 3f(x_i) + 3f(x_{i+1}) + f(x_{i+2})\right] \tag{A1.52}$$

A1.4.4 Assessment of Accuracy of above Methods

As an example, we assess the error in the trapezoidal rule in the segment $[x_i, x_{i+1}]$ of width h. Let us approximate the function in this interval by a third-order expansion

$$f(x) \approx f(x_i) + (x - x_i)f'(x_i) + \frac{(x - x_i)^2}{2}f''(x_i) + \frac{(x - x_i)^3}{6}f'''(x_i)$$

Then

$$I_i = \int_{x_i}^{x_{i+1}} f(x)\,dx \approx hf(x_i) + \frac{h^2}{2}f'(x_i) + \frac{h^3}{6}f''(x_i) + \frac{h^4}{24}f'''(x_i)$$

The trapezoidal rule would give:

$$T_i = h\frac{f(x_i) + f(x_{i+1})}{2} = hf(x_i) + \frac{h^2}{2}f'(x_i) + \frac{h^3}{4}f''(x_i) + \frac{h^4}{12}f'''(x_i)$$

The differences between I_i and T_i start with the h^3 terms. In the leading (third) order, the error is $(h^3/12) f''(x_i)$. This is the error for a single segment. The total error in the integral over the interval $[a, b]$ then is the sum of the errors over all segments. Assume n segments, that is, $h = b - a/n$. The total error then is $E \approx (1/12)(b - a/n)^3 \sum_{i=1}^{n} f''(x_i)$. If we define the average value of the second derivative as $\overline{f''} = (1/n)\sum_{i=1}^{n} f''(x_i)$, then $E \approx (1/12)$ $(b - a/n)^2 (b - a)\overline{f''} = (1/12)h^2(b - a)\overline{f''}$. The trapezoidal rule apparently leads to second-order accuracy $(O(h^2))$.

A1.5 Numerical Differentiation

Spatial derivatives are often approximated using finite differences. The essence of the concept of finite differences is embodied in the standard definition of the derivative

$$f'(x) = \lim \frac{f(x + \Delta x) - f(x)}{\Delta x} \tag{A1.53}$$

Instead of passing to the limit as Δx approaches zero, the finite spacing to the next adjacent point, $x_{i+1} = x_i + \Delta x$ is used so that you get an approximation

$$f'(x) \cong \frac{f(x_{i+1}) - f(x_i)}{\Delta x} \tag{A1.54}$$

The difference formula can also be derived from Taylor's formula

$$f(x_{i+1}) = f(x_i) + \Delta x f'(x_i) + \frac{\Delta x^2}{2} f''(\xi_i); \quad x_i < \xi_i < x_{i+1} \tag{A1.55}$$

which is more useful because it provides an error estimate (assuming sufficient smoothness)

$$f'(x) = \frac{f(x_{i+1}) - f(x_i)}{\Delta x} - \frac{h}{2} f''(\xi_i) \tag{A1.56}$$

An important aspect of this formula is that ξ_i must lie between x_i and x_{i+1} so that the error is local to the interval enclosing the sampling points. It is generally true for finite difference formulas that the error is local to the stencil, or set of sample points. Typically, for convergence and other analysis, the error is expressed in asymptotic form as

$$f'(x) = \frac{f(x_{i+1}) - f(x_i)}{\Delta x} + O(\Delta x) \tag{A1.57}$$

This formula is most commonly referred to as the first-order forward difference. The backward difference would use x_{i-1}.

$$f'(x) = \frac{f(x_i) - f(x_{i-1})}{\Delta x} + O(\Delta x) \tag{A1.58}$$

Taylor's formula can easily be used to derive higher-order approximations. For example, subtracting

$$f(x_{i+1}) = f(x_i) + \Delta x f'(x_i) + \frac{\Delta x^2}{2} f''(x_i) + O(\Delta x^3) \tag{A1.59}$$

from

$$f(x_{i-1}) = f(x_i) - \Delta x f'(x_i) + \frac{\Delta x^2}{2} f''(x_i) + O(\Delta x^3) \tag{A1.60}$$

and solving for $f'(x)$ gives the second-order centered difference formula for the first derivative

$$f'(x) = \frac{f(x_{i+1}) - f(x_{i-1})}{2\Delta x} + O(\Delta x^2) \tag{A1.61}$$

If the Taylor's formulas shown are expanded one order further and added, and then combined with the formula just given, it is not difficult to derive a centered formula for the second derivative

$$f''(x) = \frac{f(x_{i+1}) - 2f(x_i) + f(x_{i-1})}{\Delta x^2} + O(\Delta x^2) \tag{A1.62}$$

Note that while having a uniform step size Δx between points makes it convenient to write out the formulas, it is certainly not a requirement. In general, formulas for any given derivative with asymptotic error of any chosen order can be derived from the Taylor's formulas as long as a sufficient number of sample points are used. However, this method becomes cumbersome and inefficient beyond the simple examples shown. An alternate formulation is based on polynomial interpolation: since the Taylor's formulas are exact (no error term) for polynomials of sufficiently low order, so are the finite difference formulas. It is not difficult to show that the finite difference formulas are equivalent to the derivatives of interpolating polynomials. For example, a simple way of deriving the formula just shown for the second derivative is to interpolate a quadratic and find its second derivative (which is essentially just the leading coefficient).

A1.6 Partial Differential Equations

Many reaction engineering problems fall into a class of equations called *partial differential equations* (PDE). PDE have two or more independent variables. A transient PFR is described by a PDE because it has both time and axial position as independent variables. Many PDE are also boundary value problems, and as such can be complex to solve. There are many numerical schemes used for the solution of PDE, including finite difference, finite volume, finite element, and orthogonal collocation. Some more esoteric techniques include boundary element and spectral methods. The choice of which method to use may depend on the characteristics of the problem and the available software.

One popular method for the solution of PDE is the method of lines, which can be applied to certain types of reaction engineering problems. The numerical method of lines is a technique for solving partial differential equations by discretizing in all but one dimension, and then integrating the semidiscrete problem as a system of ODE or DAE. A significant advantage of the method is that it allows the solution to take advantage of the sophisticated general-purpose methods and software that have been developed for numerically integrating ODE and DAE. For the PDE to which the method of lines is applicable, the method typically proves to be quite efficient.

It is necessary that the PDE problem be well posed as an initial value (Cauchy) problem in at least one dimension, since the ODE and DAE integrators used are initial value problem solvers. This rules out purely elliptic equations such as Laplace's equation, but leaves a large class of evolution equations that can be solved quite efficiently.

Consider the following problem:

$$\frac{\partial y}{\partial t}(x,t) = \lambda^2 \frac{\partial^2 y}{\partial x^2}(x,t) \tag{A1.63}$$

This is a candidate for the method of lines since you have the initial value $y(x,0) = 0$. Problem (A1.63) will be discretized with respect to the variable x using second-order finite differences, in particular using the following approximation:

$$\frac{\partial^2 y}{\partial x^2} \approx \frac{y(x + \Delta x, t) - 2y(x,t) - y(x - \Delta x, t)}{\Delta x^2} \tag{A1.64}$$

Thus, Equation A1.63 becomes

$$\frac{\partial y}{\partial t} = \frac{\lambda^2}{\Delta x^2} \big(y(x + \Delta x, t) - 2y(x,t) - y(x - \Delta x, t) \big) \tag{A1.65}$$

Once the boundary conditions have been incorporated, the equation can be solved as a system of first-order ODE in t. Even though finite difference discretizations are the most common, there is certainly no requirement that discretizations for the method of lines be done with finite differences; finite volume or even finite element discretizations can also be used.

Appendix 2: Thermodynamic Data

In this appendix, we give heat capacity and critical property data for selected species. Table A2.1 contains tabulated data for the enthalpy and free energies of formation at 298.15 K, and the constants in the heat capacity polynomial

$$C_P = a + bT + cT^2 + dT^3$$

for selected compounds. When computing heat capacity, temperature is in kelvin to give heat capacity in J/mol K. Based on data in Kyle (1999), Table A2.2 contains critical properties of selected compounds.

TABLE A2.1

Enthalpy of Formation, Free Energy of Formation and Heat Capacity Data for Selected Compounds

Substance	Formula	ΔH_f° (kJ/mol)	ΔG_f° (kJ/mol)	a	$b \times 10^2$	$c \times 10^5$	$d \times 10^9$	Temperature Range (K)
Nitrogen	N_2	0	0	28.85	−0.1569	0.8067	−2.868	273–1800
Oxygen	O_2	0	0	25.44	1.518	−0.7144	1.310	273–1800
Air				28.09	0.1965	0.4799	−1.965	273–1800
Hydrogen	H_2	0	0	29.06	−0.1913	0.3997	−0.8690	273–1800
Carbon monoxide	CO	−110.6	−137.4	28.11	0.1672	0.5363	−2.218	273–1800
Carbon dioxide	CO_2	−393.8	−394.6	22.22	5.9711	−3.495	7.457	273–1800
Water vapor	H_2O	−242.0	−228.7	32.19	0.1920	1.054	−3.589	273–1800
Ammonia	NH_3	−46.22	−16.6	27.524	2.5603	0.98911	−6.6801	273–1500
Nitric oxide	NO	90.44	86.75	29.29	−0.09380	0.9731	−4.180	273–1500
Nitrous oxide	N_2O	81.60	103.7	24.07	5.8537	−3.556	10.56	273–1500
Nitrogen dioxide	NO_2	33.9	51.87	22.91	5.706	−3.515	7.86	273–1500
Methane	CH_4	−74.90	−50.83	19.86	5.016	1.267	−10.99	273–1500
Ethane	C_2H_6	−84.72	−32.9	6.889	17.24	−6.395	7.273	273–1500
Ethene	C_2H_4	52.32	68.17	3.95	15.61	−8.331	17.64	273–1500
Propane	C_3H_8	−103.9	−23.5	−4.04	30.43	−15.70	31.68	273–1500
Propene	C_3H_6	20.4	62.76	3.15	23.79	−12.16	24.58	273–1500
n-Butane	C_4H_{10}	−126.2	−17.2	3.95	37.09	−18.31	34.94	273–1500
Methanol	CH_4O	−201.3	−162.6	19.02	9.137	−1.216	18.03	273–1000
Ethanol	C_2H_6O	−235.0	−168.4	19.9	20.93	−10.36	20.02	273–1500
Sulfur dioxide	SO_2			25.74	5.785	−3.805	8.598	273–1800
Sulfur trioxide	SO_3			16.38	14.56	−11.18	32.37	273–1300
Hydrogen chloride	HCl	−92307	n/a	30.28	0.7608	1.325	−4.330	273–1500

TABLE A2.2

Critical Properties of Selected Compounds

Compound	Formula	T_C (K)	P_C (atm)
Hydrogen	H_2	33.2	12.8
Nitrogen	N_2	126.2	33.5
Argon	Ar	150.8	48.1
Ammonia	NH_3	405.6	111.3
Ethane	C_2H_6	305.4	48.2
Ethene	C_2H_4	282.4	49.7
Methanol	CH_3OH	512.6	81.0
Carbon monoxide	CO	132.9	35.0

Reference

Kyle, B.G., 1999, *Chemical and Process Thermodynamics*, 3rd Ed., Prentice Hall, Englewood Cliffs.

Appendix 3: Useful Integrals

There are many common integrals that occur in reactor analysis. Some of these have analytical solutions, although the reader many have forgotten the techniques required to solve them. There are many good-quality books that give solutions to the common integrals, and we have reproduced here some of the more useful ones.

$$\int \frac{dx}{x} = \ln x$$

$$\int x^n dx = \frac{x^{n+1}}{n+1} \quad \text{for } n \neq -1$$

$$\int \frac{dx}{x(a+bx)^2} = \frac{1}{a(a+bx)} - \frac{1}{a^2} \ln \left[\frac{a+bx}{x} \right]$$

$$\int \frac{dx}{x(a+bx^n)} = \frac{1}{an} \ln \left[\frac{x^n}{a+bx^n} \right]$$

$$\int \frac{dx}{(1-x)(a-x)} = \frac{1}{a-1} \ln \left[\frac{a-x}{a(1-x)} \right]$$

$$\int \frac{dx}{x^2(a+bx)} = -\frac{1}{ax} + \frac{b}{a^2} \ln \left[\frac{a+bx}{x} \right]$$

$$\int \frac{dx}{(a+bx)} = \frac{1}{b} \ln(a+bx)$$

$$\int \frac{dx}{(a+bx)^2} = \frac{-1}{b(a+bx)}$$

$$\int \frac{dx}{(a+bx)^3} = \frac{-1}{2b(a+bx)^2}$$

$$\int \frac{dx}{X} = \frac{1}{\sqrt{-q}} \ln\left[\frac{2cx + b - \sqrt{-q}}{2cx + b + \sqrt{-q}}\right]$$

where $X = a + bx + cx^2$ and $q = 4ac - b^2$

$$\int \frac{x\,dx}{(a + bx)} = \frac{1}{b^2}[(a + bx - a\ln(a + bx)]$$

$$\int \frac{x\,dx}{(a + bx)^2} = \frac{1}{b^2}\left[\ln(a + bx) + \frac{a}{a + bx}\right]$$

$$\int e^{ax}dx = \frac{e^{ax}}{a}$$

$$\int x^n e^{ax}dx = \frac{x^n e^{ax}}{a} - \frac{n}{a}\int x^{n-1}e^{ax}dx$$

$$\int \frac{(1 + \varepsilon x)}{(1 - x)}dx = \left[(1 + \varepsilon)\ln\left(\frac{1}{1 - x}\right) - \varepsilon x\right]$$

$$\int \frac{(1 + \varepsilon x)}{(1 - x)^2}dx = \frac{(1 + \varepsilon)x}{1 - x} - \varepsilon\ln\left(\frac{1}{1 - x}\right)$$

$$\int \frac{(1 + \varepsilon x)^2}{(1 - x)^2}dx = 2\varepsilon(1 + \varepsilon)\ln(1 - x) + \varepsilon^2 x + \frac{(1 + \varepsilon)^2 x}{1 - x}$$

Appendix 4: Numerical Software

There are a number of commercial software tools, which can be used to solve mathematical problems both numerically and analytically. Here, we briefly describe three popular tools, namely POLYMATH, MATLAB®, and COMSOL Multiphysics. More details may be found from the online quoted sources.

A4.1 POLYMATH (http://www.polymath-software.com)

POLYMATH is a robust computational environment specifically designed for educational as well as professional use. It allows users to apply numerical analysis techniques in an interactive manner and results are presented graphically for easy understanding and for incorporation into papers and reports. It is mainly designed for linear and nonlinear algebraic equations, and for initial value problems.

Some of the current features include the ability to export POLYMATH problem to Excel™ with a single keystroke. Thus, problems can be solved completely in POLYMATH or exported to Excel for solution. A POLYMATH ODE Solver Add-In is included for solving ordinary differential equations in Excel. Automatic export to Excel includes all intrinsic functions and logical variables. Ordered equations can also be provided to assist with optional MATLAB solutions of POLYMATH problems.

The professional version problem-solving capabilities include

- Linear equations—up to 264 simultaneous equations
- Nonlinear equations—up to 300 simultaneous nonlinear and 300 additional explicit algebraic equations
- Differential equations—up to 300 simultaneous ordinary differential and 300 additional explicit algebraic equations
- Data analysis and regression—up to 200 variables with up to 1000 data points for each, with capabilities for linear, multiple linear, and nonlinear regressions with extensive statistics plus polynomial and spline fitting with interpolation and graphing capabilities

A4.2 MATLAB® (http://www.mathworks.com)

MATLAB is a programming environment designed for automated computation of numerous mathematical problems. Interactive analysis is carried out through a matrix-optimized environment, which is versatile enough to allow user development and implementation of complex algorithms with ease.

There are hundreds of mathematical, statistical, and engineering functions implemented, giving the ease and comfort of solving a number of problems. Some of the features include

- Linear algebra and matrix computation
- Statistical and Fourier analysis functions
- Differential equation solvers
- Trigonometric and other fundamental mathematical operations
- Graphical visualization of results
- Advanced visualization tool for surface and volume rendering

A4.3 COMSOL Multiphysics (http://www.comsol.com)

The COMSOL Multiphysics is an integrated engineering simulation software environment that facilitates all steps in the modeling process—from defining the geometry, meshing, specifying the physics, solving, and ultimately visualization of results. It uses the finite element method to solve partial differential equations in one, two, or three space dimensions.

A number of predefined physics interfaces enable quick model setup. The predefined physics interfaces for applications range from fluid flow to heat transfer and chemical reaction engineering. There are also numerous templates for many common problem types. Finally, one can also choose different physics and define the interdependencies or specify partial differential equations (PDEs) and link them with other equations and physics.

Index